ENERGY: HYDROCARBON FUELS AND CHEMICAL RESOURCES

ENERGY: HYDROCARBON FUELS AND CHEMICAL RESOURCES

Don K. Rider

Formerly–Head, Organic Materials Research and Development Department
Bell Laboratories
Murray Hill, NJ 07974

A WILEY-INTERSCIENCE PUBLICATION

JOHN WILEY & SONS
New York • Chichester • Brisbane • Toronto

Published by John Wiley & Sons, Inc.
Copyright © 1981 by Bell Laboratories, Inc.

Library of Congress Cataloging in Publication Data:

Rider, Don K., 1918–
  Energy hydrocarbon fuels and chemical resources.

  Includes index.
  1. Power resources. 2. Energy policy. 3. Feedstock.
I. Title.
TJ163.2.R52   333.8′2   81-196
ISBN 0-471-05915-3   AACR2

Printed in the United States of America

10 9 8 7 6 5 4 3 2 1

PREFACE

As an organic materials engineer (a chemist by training), I could not help but be aware of our dependency on petroleum for polymeric materials. During the oil embargo, when some plastics could only be obtained by "wheeling and dealing" and paying premium spot prices, it became apparent that the Bell System, consumer of plastics on a vast scale for use in the world's largest communications system, was running into serious spot shortages of key materials.

While others of my associates focused on developing technology for recycling plastic telephone housings and handsets, and plastic-insulated wire and cable, I was encouraged to examine the supply situation for these plastics and associated polymeric materials. This proved to be a stimulating undertaking which led to an internal report for the edification of my peers concerned with the materials aspects of the telephone plant.

When it was suggested later that I update the initial report and investigate the long-range supply prospects, it required little prompting to propel me into a more in-depth study of what the future might hold for the polymer industry and petrochemicals in general. As background, this led to a series of internal reports covering the primary raw materials for the petrochemical industry—crude oil and natural gas. Then came a report on a secondary raw material—coal, and another on converting coal to synfuels and other specific organic derivatives. When yet another report emerged on tar sands and shale oil, I found myself under some gentle prodding to go public, with a book encompassing the full treatment for a general audience having only a passing concern with the energy and feedstock worries of the telephone community.

I succumbed to the prodding some months before requesting early retirement, and carried the task into my first winter away from the ravages of the Middle Atlantic snows and cold. The remoteness of my new environment from familiar library facilities and helpful associates resulted in some slackening of my literary output, but it is probably just as well, as some of the dust must yet settle on the rapidly changing energy scene before anyone can hope to have any idea of where we are truly headed. One thing is certain. These are parlous times for all but the most perceptive of prophets. It would indeed be presumptuous on my part to try to tell the reader just how much oil and gas will be available, and for how long, and at what price. Even our Middle East suppliers have no inkling of the answers. It would also be purely fortuitous if my poor guesses as to the future impact of alternative hydrocarbons were to prove correct. There are just too many imponderables totally in the hands of the politicians and their bureaucratic handmaidens. The technological problems are largely solvable; the people problems are another matter.

This book, then, is an attempt to review policies, performance data and trends, new technological developments, and future options, in order to help chart the road ahead. With luck, this may assist readers past some of the potholes and detours along the way to planning their energy and feedstock futures.

It has been said that "no man is an island." This certainly has been true in this undertaking. The effort would never have been brought to a satisfactory conclusion without the patient understanding of my wife Lois; the catalyst of my former supervisor, Dr. David W. McCall (Chemical Director at Bell Laboratories); the perseverance of at least five young ladies who provided much of the secretarial support—Mesdames Claudia P. Jursik, Patricia A. Munsell, Sandra S. Malenky, Shirley Welch, and Debbie A. Flanagan; and helpful comments from numerous associates, particularly C. V. Lundberg, and T. D. Schlabach, for his critical review of the manuscript. I also wish to acknowledge the full support of the Bell Laboratories organization in this enterprise. For all this, I am deeply appreciative.

DON K. RIDER

Ocean City, New Jersey
Naples, Florida
March 1981

CONTENTS

ENERGY: HYDROCARBON FUELS AND CHEMICAL RESOURCES

1 AN OVERVIEW

Men in the economic sense exist solely by virtue of being able to draw on the energy of nature.

SIR FREDERICK SODDY

The subject of energy, in all its ramifications, has generated an enormous amount of heat from politicians, environmentalists, regulatory bodies, economists, industry, physical scientists, the press, and ill-informed citizens. It seems unlikely that any of these groups can be totally right or totally wrong in their analysis of the subject. Determining the bounds of the gray area between the two is perhaps the most challenging part of this literary undertaking, particularly in light of the many emotional issues involved. An effort will be made to expose the facts and speculations, properly identified as such, of sources believed to be reliable, and to isolate and describe novel energy proposals and critiques of them offered by experts.

Only past and future policy decisions by the administration in power can ultimately determine the situation with which we must all abide in the future. As noted by Rochlin,[1] "the real energy crisis comprises the set of decisions needed during the next several decades as the remainder of our hundred-million year old inheritance of fossil fuels is drawn down towards depletion."

The years remaining in the lifetimes of most readers of this book will see brought to light the road humankind must follow with full commitment, since the cost of second thoughts could be astronomical in dollars and catastrophic in lost time. These are decisive times, and the writer will attempt to identify, for the individual who has perhaps had neither the time nor the inclination to delve into the literature of the energy field, some of the statistics, options, and expert opinions that undoubtedly will play a role in determining our future policies.

One might well ask why any segment of the Bell System should be concerned with energy when communications is not noted as a particularly energy-intensive industry. It must be remembered that the Bell System is a major user of petrochemicals and other energy-intensive materials. These include plastics and metals particularly, but glass also may well become an

1

important factor, at such time as lightwave communications begin to play a large role in our technology. Projections of the availability of petrochemical feedstocks and fuel could become important in decisions regarding technological options open to our physical designers and systems engineers. Furthermore, through sheer size, the Bell System consumes vast amounts of fuel to provide space conditioning, process heat and mechanical drive, as well as propulsion for its large fleet of motor vehicles.

It would perhaps appear to some readers that since we are a fluid-fuel-oriented industry, a mere summary of the petroleum and natural gas prospects would be adequate for our purposes. This writer sees it differently. The continuing complete dominance of a single energy source, such as oil and associated natural gas, seems remote in the foreseeable future. Our energy producers are inevitably faced with the prospects of juggling many energy sources to meet our coming needs. This can be viewed as both a strength and a weakness in our society—it may deprive us of economies of scale but, by its very diversity, it should help free us in the future from becoming helpless pawns of any given political or business consortium. The particular energy mix(es) made available to us in the future by virtue of governmental strategies and energy-industry decisions can be expected to have considerable bearing on the availability of resources and on the cost of providing the Bell System and other industries with suitable energy and materials options.

It takes time to develop alternative energy sources, conversion technologies, and production facilities, and additional time to develop, test, and approve petrochemical or other energy-intensive products resulting from these new technologies. "Clearly our most serious shortage is time, and we have frittered this away by not appreciating and facing the [energy] problem early enough," according to Costa.[2] Energy and industry are inextricably entwined for all time, and it behooves us to recognize and accept this fact without delay, so that we may be prepared to develop materials options and fuel strategies expeditiously as needs dictate.

Quite apart from technological considerations, we shall see that capital and resource demands of various energy strategies may well provide serious competition for the capital and resource needs of the communications and other industries. All indications are that departures from our present dependency on fossil fuels necessarily will extract enormous amounts of capital from the marketplace that might otherwise be available for needed growth of the telephone system or other sectors of our economy. Detailed discussion of this subject will be left to the economists but, where available, recent data on the capital costs of various energy options will be noted.

For the sake of uniformity and clarity throughout this book, it should be noted that the historical connotation of Q (a large unit of energy) has been *quintillion* or 10^{18} Btu. More recently it has come to mean *quadrillion* (also *quads*) or 10^{15} Btu,[3] and on rare occasions 10^{15} Btu has been abbreviated q to avoid this confusion. Since virtually all current literature, including most government reports, reference *quads*, Q (10^{15} Btu), this book will employ the same convention. It will also use M to signify a *million* or 10^6 units and k for *kilo* or 10^3 units, instead of the confusing convention of M for 10^3 and MM for 10^6, used in many energy publications, particularly in reference to gas. But to avoid possible confusion between the two conventions, million will generally be written million or 10^6. In some papers, Mt signifies metric tons or tonnes (10^3 kg), while t stands for U.S. (short) tons (2000 lb). Again, to avoid confusion, we shall use t for U.S. (short) tons and te for tonnes (or metric tons) in combined units, although we generally spell these out when they stand alone.

We have become accustomed to speaking loosely of energy *consumption*; however, the laws of thermodynamics tell us that energy cannot be consumed. In a strict sense, energy ultimately is dissipated to the environment as waste heat, or is, at best, otherwise degraded to such a low level that, employing available technology, it has no further practical value. Because of the widespread acceptance of the term *consumption* with respect to energy usage, we shall continue this practice, recognizing, nonetheless, the term's inaccuracies.

RESERVES VS RESOURCES

Before embarking on a discussion of supply and demand and lifetimes or life indexes, it is essential that we make some attempt to clarify the matter of reserves vs. resources. In a crude sense, *reserves* have been referred to as "birds in hand" and *resources* as "birds in the bush."

Reserves are perhaps best defined as "identified deposits known to be recoverable with current technology under present economic conditions."[4]* The obvious next question is: "What does *identified* imply?" It means basically that the location, quality, and quantity are supported by engineering data.

The U.S. Geological Survey (USGS) and the Bureau of Mines (BuMines or BOM) go further and categorize reserves as measured, indicated, or inferred, depending on the degree of certainty of existing evidence and the current economics of recovery.

Measured reserves are identified resources from which a commodity can be extracted economically using current technology, and whose location, quality, and quantity are known. (Also called *proved* or *apparent* reserves.)

Indicated reserves are based partly on specific measurements and partly on projections (for a reasonable distance) of geological evidence. (Also known as *probable* reserves.)

Inferred reserves are based on geological knowledge without supporting engineering evidence, and are considered to be reserves extractable at a future

*See the Glossary, Appendix B.

date as technology and economics improve and engineering data become available. (Also known as *possible* reserves.)

Measured and indicated reserves taken together occasionally are called *demonstrated reserves*, while indicated and inferred reserves together are sometimes known as *potential reserves*.

Where clearly designated by the original source, this book will attempt to identify reserves as proved (measured) or potential. The former should be known with an accuracy of ±20 to 25 percent, while an estimated margin of error of 50 percent is generally assigned to potential reserves.

Resources combine reserves with materials identified but not currently extractable because of economic or technological constraints, and materials not yet discovered but reasonably expected to exist in favorable geologic formations.

Identified resources are specific deposits whose existence and location are known. If evaluated as to extent and grade, they are classed as reserves; otherwise, they are *conditional resources* that may become reserves given technological advances or a different economic climate.

Undiscovered resources are of two categories:

Hypothetical resources are undiscovered deposits reasonably expected to be found in identified regions.

Speculative resources include conventional deposits yet to be recognized.

A perusal of current energy literature will show that there is little agreement among authors regarding precise reserves and resources terminology. Table 1.1 summarizes the generally accepted descriptions.

Regardless of how we categorize our resources, one thing is clear: "The moment it takes more work to find, recover, process, transport, and apply a natural material than we can get out of that material or in exchange for it, it ceases to be a resource; how much of the material remains in the ground, in the sea, or on the moon, is irrelevant."[5]

It might seem that borings and test wells would provide the nation with accurate data on fossil fuel resources, but it must be remembered that it is the oil, gas, and mining producers and their stockholders who foot the bill for these exploratory ventures, and the price comes high. Furthermore, governmental bodies (local, state, or federal) may tax these companies on proved reserves of their holdings. It is understandable then that in the business world there would be a reluctance to make public full details of these explorations. Zimmerman notes: "A corporation knows reasonably well what its ore reserves are ... National reserves ... are largely a matter of conjecture."[6]

On the world scene, many countries consider their uranium reserves to be classified information; hence these disclosures must be considered suspect. In any reporting of reserves—uranium, fossil fuels, or other mineral ores—

Table 1.1 Reserves and Resources Summarized

Category	Description
Potential	
Proved[a]	Measured (with small error)
Probable[a]	Geological and judgmental extrapolation
Possible[a]	In producing regions, but not reasonable to assume as probable
Speculative	
Geologic[a]	Little information, but reasonable to expect existence
Economic	
Discovered and undiscovered	Potentially reclassifiable with improved economics, but too remote, too lean, or too difficult to extract

[a] Presently extractable economically.

there is often a question as to how the data are reported: for example, whether the figures allow for inherent losses in the recovery operation, which can vary considerably with the resource, the ore, the process, the location, the depth, and so forth. One source suggests that crustal abundance to a depth of 3 mi (4.8 km) below the earth's surface be adopted as an arbitrary but reasonable limit for assessing the resource base of our minerals.[7] No such convention, however, has found general or official acceptance.

In any case, classifications change as technology develops, market prices of the commodity change, economics of recovery improve, transport facilities expand, undeveloped areas open to exploration, and availability of capital improves. Resources become new reserves, and potential reserves become proved reserves. The reverse route also applies, in part. Development of substitute materials, introduction of environmental restrictions, falling prices, and so forth, may reduce reserves. Hence reserves are not concrete physical quantities depending solely on discoveries and extractions, and herein lies one of the main difficulties in assessing the lifetimes of our resources.

As if this were not enough, lifetimes are necessarily dependent on the rate at which we consume these materials, and this is something that cannot be estimated with any certainty. The OPEC oil embargo had an almost immediate depressing effect on the consumption of gasoline for transportation. This produced a trend toward the demand for smaller, more efficient automobiles. Shortly after the crisis, however, automakers abandoned assembly lines for compact cars for lack of buyer demand and increased production of the full-size models, as consumers seemingly forgot the long gasoline lines of the fall and winter of 1973 and 1974. As a result, gasoline consumption was soon on the rise again, despite the mandated improvements in gas mileage for new cars.

The recession following the oil embargo and accompanying the rapid increase in fuel prices had a drastic impact on the consumption of fuel and other resources. The return to the historic energy consumption growth rates generally has been slow to develop, in the light of uncertainties concerning the effect on the economy of the Organization of Petroleum Exporting Countries (OPEC) petroleum prices, the change in administration, the limited availability of venture capital, and the general business climate. All of these factors are highly speculative, yet economists must consider each in order to arrive at rational estimates of future consumption, hence lifetimes, of fuel resources.

If all energy forms, including wood and farm waste, were to be considered in terms of coal equivalent, *global* consumption would show a trend of an annual growth rate (AGR) of 2 percent over the past 120 years. On the basis of Btu's, the AGR of our total *domestic* energy consumption would show a trend of 3 percent over the past century. But if one adjusts the domestic AGR of about a century ago, for the then higher U.S. population growth rate, the energy consumption growth rate then becomes 2 percent, in keeping with the worldwide trend. A century ago, the energy base was virtually 100 percent nonfossil. Today, that figure is about 15 percent, so much of the growth rate of fossil fuels can be accounted for by their replacement of nonfossil fuels.[8]

Despite barriers to precision, there is near unanimity among the experts that worldwide production of oil will peak toward the end of this century, even as our domestic oil production has already peaked. This does not mean that there will be a final, abrupt halt to petroleum production because of totally depleted fields. The conclusion of the President's Materials Commission in 1952 was: "The reserves of fuels with which to meet our energy demands are relative. It is not a question of emptying the bin. We shall never do that. It is a question of how far down it is worth reaching in terms of economic cost. Such a question has no exact answer."[9]

Perhaps the most widely accepted values for lifetimes of our energy resources are now those attributed to Hubbert. These will be discussed in the specific chapters dealing with the particular resource. Set forth in a more personal context, the earth's nonrenewable fossil energy resources can be stated in terms of human life spans: petroleum and natural gas, 0.3 to 0.6 life spans, and coal and lignite, 3.0 to 6.0 life spans.

POLICIES AND STRATEGIES

The domestic demand for energy has been increasing* at a faster rate than the domestic supply (Table 1.2). This can lead to one of two consequences: enforced reduction of consumption or increased importation of foreign sup-

*With a brief deceleration brought on by the recession following the OPEC oil embargo (fourth quarter of 1973 through first half of 1974).

Table 1.2 Sources of Domestic Energy, 1950–1976

	Coal		Petroleum Products from:		Natural Gas		Electricity		
	Bitumi-nous/ Lignite	Anthra-cite	Crude	Other[a]	Dry	Natural Gas Liquids	Hydro-electric[a]	Nuclear	Total
1950	10.36	1.01	12.30	0.29	6.07	0.72	1.07	—	31.82
1955	9.65	0.60	15.96	0.26	9.11	1.12	1.25	—	37.95
1960	8.67	0.45	17.17	1.12	12.56	1.35	1.60	0+	42.92
1965	10.47	0.33	17.63	2.26	15.93	1.78	2.05	0.04	50.49
1970	11.76	0.21	23.11	3.86	21.82	2.45	2.55	0.23	65.99
1971	11.28	0.19	23.68	4.29	22.13	2.46	2.86	0.40	67.29
1972	11.78	0.15	24.82	5.21	22.47	2.69	3.03	0.58	70.73
1973	12.91	0.14	26.38	5.56	22.51	2.65	3.16	0.89	74.10
1974	12.60	0.14	25.80	4.96	21.67	2.57	3.42	1.21	72.37
1975	12.68	0.13	26.37	3.97	19.95	2.49	3.28	1.84	70.71
1976	13.62	0.13	28.48	4.25	20.23	2.54	3.16	2.04	74.75

10^{15} Btu (Quads)

Source: Reference 10.
[a]Includes imports.

plies, with all that implies for the stability of our economy. In the 1960s, the United States managed to maintain a positive trade balance of up to about $7 billion annually, with total exports in 1965 being $26.7 billion. Since 1970, we have had a trade deficit in all but two years, with petroleum imports alone rising from $3 billion in 1970[11] to about $90 billion annually up to mid-1980, and projected to increase to $94 to $95 billion annually following the July 1, 1980, OPEC price increase.[12] The alarming feature is that the deficit contribution by petroleum imports is on the rise and exceeding by far the prospects of other sectors of our trade (e.g., agricultural products and chemicals and allied products) to compensate for the increase. Our balance of trade is helped additionally by large exports of arms and technology to the OPEC nations, but Iran is no longer importing U.S. arms.

Our present energy crisis is basically a matter of too much consumption and too little domestic supply, coupled with continued, growing demand. To compound the problem, President Carter's National Energy Plan (NEP-1)—proposed April 20, 1977—which emphasized conservation, underwent radical surgery in Congress for about a year and a half before it was finally approved (as NEP-2) minus a conservation measure. In May 1980, House and Senate conferees finally agreed on a $20 billion, four-year energy package to seek increased fluid fuel production through federally subsidized synthetic fuel plants. A later phase of the program would make an additional $68 billion available, to meet President Carter's objective of 2.5 million bpd of synfuel from coal, oil shale, and tar sands. This could decrease our consumption of imported oil by about a third by 1990.[13]

Three recent studies (Harvard Business School,[14] Ford Foundation,[15] and Resources for the Future[16]) have supported conservation as a keystone of U.S. energy policy, an action that is strongly urged by this writer also.

We have a precious, but limited, heritage of nonrenewable fossil fuels, the supply of which is declining at a rate of about 6 percent per year, while demand in 1977 was increasing at a rate of about 5 percent per year. Crude oil and natural gas constitute 75 percent of our domestic energy consumption, but only 7 percent of our domestic energy reserves.[17] Yet we have no effective conservation plan in the offing. The federal energy policy, as of mid-year 1980, seemed aimed at increased production via conversion of coal to fluid fuels, but it ignored the most basic need—conservation—to enable us to regain control of our destiny. In the words of one expert: "Unless energy consumption can be reduced significantly and voluntarily, Americans may discover that they have consumed their political liberties."[18] This statement, made in 1975, is even more true today, after five additional years of over-consumption and trade deficits. Increasing oil production 2 million bpd, by bringing the Prudhoe Bay, Alaska, field on stream, could not even compensate for our oil trade deficit in 1974. Even the USGS informed legislators of the need for reducing consumption *in addition to* development of alternative sources of energy: "Energy conservation and vigorous attempts to develop other sources of energy must be important objectives. . . ."[19]

Development of these alternative energy forms is essential and has long been hampered by artificially low prices of domestic crude oil. Decontrol of domestic oil prices, to allow them to rise to international levels, should offer some modest stimulation of U.S. production but, more important, should encourage conservation measures and permit coal conversion, shale oil development, solar energy, and biomass to compete on the marketplace with oil purchases.[20]

Conservation is a low-cost "source" of new energy. Development of additional energy sources, whether coal, oil, or gas, can be a costly operation, and no business manager can afford to undertake such a venture unless he or she can be certain of an adequate return on investment. Most oil or gas well developments (outside the Mideast) would call for $4.5 billion to increase our annual output by 1 Q. Capital cost of coal would be somewhat less than $2 billion for 1 Q, but synfuels from coal would probably be $10 billion. The capital costs to conserve the same amount of energy by improving efficiency may run as low as $1.5 billion, but more typically, something less than $4 billion, and, rarely, as high as $11 billion. In any case, according to Schlesinger, then of the Department of Energy (DOE), conservation would enable us to save the equivalent of 1 bbl of oil at a cost of about $1.50,[21] and it is a near-term option.

An essential to increasing the output and efficiency of our energy supply process is streamlining government procedures. The proposed pipeline from Long Beach, California, to Midland, Texas, is a case in point. This pipeline is essential to the efficient use of Alaskan oil, yet the project has been in the works since 1974. It would ensure delivery, at reasonable cost, of Prudhoe Bay oil to midwestern refineries, where it is needed, but it has been plagued constantly by red tape. Finally, in mid-1979, the pipeline was being proposed as a major feature of NEP-3. The pipeline developer (Sohio) found that they had to obtain approval of nearly 700 permit applications to construct the pipeline—a monumental, costly, time-consuming, and largely unnecessary task. One Senate staffer reported: "California may now be institutionally incapable of permitting anything,"[22] and this charge extends to the federal level as well.

Nicandros has stated that our energy outlook for 1990 will be shaped by political or other human decisions more than by resource limitations or market forces.[23] He further hazarded a prediction that the share of oil and gas in our energy future would decline from 74 percent in 1977 to 62 percent in 1990, with coal [19 percent (1977) to 22 percent (1990)] and nuclear energy (3 percent vs. 12 percent) taking up the slack. Oil imports were expected to rise to 57 percent of our oil supply or 27 percent of our total energy budget. His predictions were made, of course, prior to the developments of 1979 and 1980 in the Mideast and at Three Mile Island and other nuclear sites.

At about the same time, another source was predicting that nearly half our 6.5 billion bbl oil demand in the year 2000 would have to be met by imported oil; natural gas consumption (15 tcf) would be down to three-fourths that of 1977, and coal production would have to be increased 250 percent. If 450,000 MW of nuclear power capacity cannot be met, oil production will have to be increased by over three-fourths of that amount, or coal production will have to be 450 percent greater than in 1977.

The Center for Strategic International Studies (Georgetown University) reported in 1980 that our 1979 energy budget had been 80.8 Q, approximately 46 percent (37.1 Q) of which came from oil (45 percent or 16.7 Q imported). Natural gas and coal accounted for 19.8 Q (24.5 percent) and 17.8 Q (22.0 percent), respectively. The remaining 6.1 Q (7.6 percent) came from hydroelectric, nuclear, and geothermal sources.

An additional consideration, which is usually ignored, is the fact that generation of such large amounts of electric power will require huge volumes of cooling water. One source states that as much cooling water as the total freshwater runoff in the lower 48 states would be needed to generate 450,000 MW.[24]

These energy usage figures assume a 2.1 percent energy growth rate from 1973 to 2000 compared with twice that rate in the previous decade. In any case, the prospects are frightening, because the doubling time for energy demand would be only about 33 years. If the old growth rate prevailed, the demand in 2000 would be up about 70 percent from the amount predicted.[25]

The hazards of predicting energy consumption, growth rates, and future contributions of various forms of energy, are exemplified by a 1976 government publication dealing with various energy scenarios: "If world oil prices decline to $8 per barrel, ..."[26] Over 300 pages of carefully researched text and tables were negated by a retrospectively implausible assumption, since prices for crude oil, by June 1980, had become: Mexican Isthmus light, $33.50 per bbl; Mexican

Mayan (heavier crude), $28 per bbl; Saudi Arabian, $28 per bbl; and other OPEC sources (average), slightly over $30 per bbl[27] and rising an additional $2 per bbl on July 1, 1980.

SUPPLY AND DEMAND

Table 1.3 shows 1976 Energy Research and Development Administration (ERDA) estimations of potentially recoverable domestic energy resources. The quantities are large, but accessibility, extractability, environmental impacts, economics, and other factors severely limit the recovery of most of the fuels. By 1965, we had extracted only about 17 percent of our domestic oil and gas resources,[29] yet they were so depleted, from a practical standpoint, that within about five years extraction rates were exceeding discoveries. Economics and recovery rates are the primary limiting factors.

In the past, we have depended on very large fields for about 85 percent of our oil and gas supplies. With the land mass of the lower 48 states having been peppered with over 2 million holes over the past 125 years, the prospects of discovering any new giant fields are highly unlikely. Our major prospect for additional domestic oil supplies lies primarily in enhanced recovery technology, which could possibly raise the recovery rates from approximately 32 percent to about 50 or 60 percent. Over a period of time, this could make available about 167 billion bbl of oil not otherwise economically recoverable, but at a much higher price.[30]

The Department of Energy expects enhanced oil recovery to result in about 2.7 Q additional oil in 1985 and 5.9 Q in 1990. Unconventional gas recovery (from western tight sands and Devonian shales) is expected to yield 1.8 and

Table 1.3 Potentially Recoverable
Domestic Energy Resources, 1976
(Estimated by ERDA)

Resource	Quads
Natural gas	1,030
Petroleum	1,100
Geothermal	3,434
Oil shale	5,800
Coal	13,300
Uranium	130,000
Fusion	3×10^9
Solar	Infinite[a]

Source: Reference 28.
[a]43,000 Q per yr.

6.2 Q, respectively, in the same years. Wood is expected to provide 1.8 and 2.4 Q. Other new sources of energy are expected to have nominal impacts.[31]

Demand for high-grade fossil fuels could be reduced considerably if we were to learn to tailor our technology to the needs of the application at hand. Approximately half of our energy budget is applied to some form of heating. About a third of this is for space heating, where low-grade (low-temperature) energy would be adequate and would result in dissipation of less waste heat.[29]

Our other major energy needs are for propulsion, industrial process heat, and electrical power generation. Nearly half the industrial demand is for direct process heat and for process steam, while almost 20 percent is for raw materials (about two-thirds coke and liquefied petroleum gas). Thirty-five percent of the industrial energy is from natural gas, 18 percent is from coal, and 16 percent is petroleum-based.[32]

Because of the historically low cost of our energy in the past, there has been no incentive to insulate structures or to design energy-efficient machinery. The result is tremendous energy waste: 41 percent in the home, 48 percent in commerce, 88 to 89 percent in transportation, 68 percent in utilities, and 49 percent in the industrial sector (the latter, a loss that has been considerably reduced since these data were compiled).[33]

SOCIOECONOMIC FACTORS

One consideration we must bear in mind is that we exist in a finite environment where we have limited resources to exploit and finite repositories for our wastes. Anything we do to upset the balance of our system—environmentally, economically, politically, militarily, and so forth—will ultimately affect even the most remote inhabitants of this earth. Possibly a billion of its people live under conditions so wretched as to be virtually beyond the belief of most readers of this book. Our waste of the earth's resources can only widen the breach between the have-nots and ourselves. Unfortunately, continued high population growth rates in the less-developed countries can only increase the gap while making it more improbable that we can ever even out the differences.[25] *Our* wanton use of resources and *their* high propagation rate are two of the major factors in the instability of the world today.

REFERENCES

1   G. I. Rochlin, *Phys. Today* **28**(7), 45–47 (July 1975).

2   L. D. Costa, *Mech. Eng.* **97**(8), 18–25 (Aug. 1975).

3   G. C. Szego, *Amer. Sci.* **64**(5), 480, 482 (Sept.–Oct. 1976).

4   *Energy Facts II*, Sci. Policy Res. Div., Libr. Congr., Serial H, U.S. GPO, Washington, DC, Aug. 1975, p. 495.

5   E. Cook, in *Resource Conservation, Resource Recovery, and Solid Waste Disposal,* report of U.S. Senate Public Works Comm., U.S. GPO, Washington, DC, 1972.

6 E. W. Zimmerman, *World Resources and Industries,* Harper, New York, 1951, p. 443; reported in G. J. S. Govett and M. H. Govett, Eds., *World Mineral Supplies: Assessment and Perspective,* Elsevier, New York, 1976, p. 29.

7 D. B. Brooks, in W. A. Vogeley, Ed., *The Economics of the Mineral Industries,* Amer. Inst. Min. Eng., New York, 1975; reported in G. J. S. Govett and M. H. Govett, Eds., *World Mineral Supplies: Assessment and Perspective,* Elsevier, New York, 1976, p.21.

8 C. Marchetti, *Phys. Technol.* **8**(4), 157–162 (July 1977).

9 President's Mater. Comm. (1952); reported in D. S. Halacy, *The Coming Age of Solar Energy,* Avon, New York, 1973, p. 13.

10 Calculated from common units (DOE data); reported in *Coal Facts: 1978–1979,* Natl. Coal Assoc., Washington, DC, p. 58.

11 DOC data.

12 *Phila. Inquirer* **302**(165), 10-A (June 13, 1980).

13 *Time* **115**(22), 59 (June 2, 1980).

14 R. Stobaugh and D. Yergin, Eds., *Energy Future,* Random House, New York, 1979.

15 H. H. Landsberg, Ed., *Energy: The Next Twenty Years,* Ballinger, Cambridge, MA, 1979.

16 S. H. Schurr et al., *Energy in America's Future,* Johns Hopkins Univ. Press, Baltimore, MD, 1979.

17 C. H. Rich, Jr., *Projects to Expand Energy Sources in the Western States—An Update of Information Circular 8719,* DOI, IC 8772, Washington, DC, Aug. 1977, p. 1.

18 E. S. Cheney, *Chemtech* **5**(6), 370–374 (June 1975).

19 V. E. McKelvey, Dir., USGS, before House Ways and Means Comm., Mar. 10, 1975.

20 R. Stobaugh and D. Yergin, "Conclusion: Toward a Balanced Energy Program," pp. 216–233, in reference 14.

21 *Bus. Week* (No. 2480), 66–72, 77, 80 (Apr. 25, 1977).

22 F. G. Garibaldi, *Energy User News* **3**(7), 22, 23 (Feb. 13, 1978).

23 C. S. Nicandros; reported in *Plast. World* **35**(4), 17 (Apr. 1977).

24 V. Dalal, *Energy Convers.* **13**(3), 85–94 (July 1973).

25 J. W. Simpson, *Public Util. Fortn.* **99**(12), 27–31 (June 9, 1977).

26 *1976 National Energy Outlook,* FEA, Washington, DC, Mar. 4, 1976, p. xxiv.

27 *Phila. Inquirer* (May 17, 1980), p. 6-B (Mexico); K. Goff, *ibid.* (May 15, 1980), p. 10-C (Saudi Arabia and OPEC).

28 ERDA (1976); reported by H. P. Harrenstein and W. Clark, in T. N. Veziroğlu, Ed., *Energy Conservation,* Forum Proc., Clean Energy Res. Inst., Univ. Miami, Coral Gables, FL, Dec. 1–3, 1975, pp. 529–552.

29 M. Altman et al., *Energy Convers.* **12**(2), 53–64 (June 1972).

30 J. Schanz; reported in *Civ. Eng.—ASCE* **47**(7), 67 (July 1977).

31 J. Gourald; reported in *Chem. Eng. News* **56**(47), 8 (Nov. 20, 1978).

32 *Energy User News* **2**(31), 5 (Aug. 8, 1977).

33 *Dun's Rev.* **109**(6), 54–57 (June 1977).

2 CRUDE OIL

The phenomenal growth of this country ... is due largely to the overlavish use of our resources, and the migrations of our pioneer people, in their depletion of the natural wealth of the country, have been not unlike the flight of a swarm of locusts over a fruitful land.

PRESIDENT THEODORE ROOSEVELT (1908)

If crude oil had become the subject of a serious radio or television talk show shortly after the 1973 oil embargo, the "buzz words" would most certainly have been *OPEC, divestiture,* and *decontrol.* There is no question but that the Organization of Petroleum Exporting Countries (OPEC), founded in 1960, has had a most profound effect on our economy—employment, balance of payments, energy supply, inflation rate, and so forth—and probably on the entire future course of American life.

In the eyes of many of our citizens and politicians, however, our salvation lay in price controls and vertical divestiture of the so-called *Seven Sisters** and possibly other giants of the oil industry. Debate over the pros and cons of such moves will be noted later in the chapter, as they cannot be totally ignored in a discussion of the energy situation. More than one "expert" felt that without the consent of the large multinational oil companies, the OPEC "cartel"† could not exist, since OPEC lacked the broad scope of economic muscle—in particular, technology, and marketing and distribution expertise and infrastructure—needed to sustain an increase in oil prices without support of the multinationals.[1]‡ In the eyes of others, there was (and still may be) a suspicion that

---

*Exxon, Mobil, Standard of California (Socal), Texaco, Gulf, Royal Dutch Shell, and British Petroleum.

†"Cartel" has been a misnomer for OPEC, since in its first 15 years or so, it lacked a prime requisite of such an organization—the ability to assign production quotas and prices, and to physically punish those members who failed to abide by them. Unfortunately for the consuming nations, OPEC is gaining strength and savvy, and to all appearances, is now fully in the driver's seat.

‡The *Seven Sisters* traditionally marketed 90 percent of OPEC output, 30 to 40 percent of it being distributed through the smaller oil companies.

OPEC is merely a front for the "real" petroleum cartel—the multinational oil companies. Perhaps a truer picture of their past relationship is that credited to Rand,[2] who has said, "OPEC and the seven majors resemble two muskets which cannot stay standing unless they lean against each other." Today, there is little evidence to support suspicion of collusion. OPEC has seemingly come of age and is standing tall—fully in command of the world's energy situation.

The only "weapons" devised to date to combat OPEC are the International Energy Agency (IEA)* and a budding organization of 20 Latin American countries preparing to take their case before the UN General Assembly.[3] The IEA was formed as an offshoot of the 24-nation, Paris-based Organization for Economic Cooperation and Development (OECD),† following the Arab oil embargo. Members, comprising the major oil-importing nations, first agreed to[4]:

1   Redistribute all available oil in case of a shortfall of 7 percent or more.
2   Establish tough energy conservation measures.
3   Stimulate adoption of other natural energy resources through incentives (e.g., guaranteed minimum prices).
4   Accelerate research into new energy sources (e.g., solar, fusion, hydrogen).

It remains to be seen how effective IEA will be. To date, its main thrust seems to have been excoriating the United States for its energy appetite and, recently, its $5 per bbl domestic subsidy on petroleum distillates.[5] The Latin Americans are gravely concerned with the effect of OPEC oil prices on their economies, charging OPEC with bankrupting the world.[3]

HISTORICAL OVERVIEW

The OPEC nations are not without their own problems, as can be seen from a capsule history of the Mideast.[6] From A.D. 600 to 1500, the Arab states were dominant international powers—ranging in influence from Spain to India. Their science, technology, and culture far surpassed those of Europe. The Crusades (from about 1096 to 1270) represented efforts both to install Christianity as the dominant religion in Jerusalem and to break the economic stranglehold the Arabs had on Europe. In the latter effort, nothing was gained immediately, but gradually the Arabs stagnated and lost their technical superiority. When Vasco DaGama sailed around the Cape of Good Hope in 1497–1498, the Arab trade monopoly with India was broken, and so the tables were turned (for at least 500 years). The Mideast found itself without maritime

---

* Australia, Austria, Belgium, Canada, Denmark, Great Britain, Greece, Spain, Ireland, Italy, Japan, Luxembourg, New Zealand, Norway, Spain, Sweden, Switzerland, The Netherlands, Turkey, the United States, and West Germany.
† France is not a member.

power and without resources, and the age of sail gradually gave way to the industrial revolution, which required both.

By the time of World War I, the Arab states were centuries behind the Western world and falling under the control of the Ottoman Empire and the British and French. A change finally came about as oil and nationalism increased the affluence and political awareness in this region following World War II.

Today, the Arabs find themselves blessed with an embarrassment of riches and with no apparent long-range plans to use it to best advantage. Almost without exception, the Arab states seem bent on sticking with an oil economy and importing everything else (except for a few petrochemicals), but with the "end" of oil only 50 years or so down the road, this is hardly an attractive policy for the long haul. Any Western success in the search for significant new energy sources could bring an even earlier demise of their oil economy by reducing the demand for their sole resource.

One way out of this situation would be for them to develop a higher-value-added petrochemical industry, refineries, or fertilizer production. Yet another option would be the direct reduction of iron by natural gas.* These are all capital-intensive undertakings that are now attracting the attention of most of the oil-producing states. The difficulty is that only a modest portion of the population of these states (less than 2 percent in the case of Saudi Arabia) could find employment in these ventures because of their lack of needed skills. The bulk of the high-paying jobs would have to be held by foreigners.

Another possible strategy would be to develop other specific parts of their economy, such as agriculture, but this would not be so attractive economically, and lack of technical skills could ultimately lead to destruction of the fragile soil. New labor-intensive industries could be developed, but other poor countries, such as China, Taiwan, or South Korea, could provide cheaper and more highly qualified labor. All in all, the modern world calls for skilled labor, an element almost totally lacking in the Mideast.[6] So the Arab nations find themselves at something of an impasse—having money beyond their wildest dreams,† but no currently viable means of employing it to the long-range betterment of their people.

Even education to develop the technical skills produces its share of pitfalls, as the scramble to the top by highly educated Arabs threatens to leave an overabundance of unemployed managers needing jobs and an excess of unskilled, uneducated people at the base of the economy. There is also the fatal danger of the people saying: "Why work, when everything you want you already have?"[6]

---

*A relatively new, low-capital process, which eliminates the need for limestone and metallurgical coke.

†The average per capita income of the United Arab Emirates is said to be about $16,000 per year—more than double that if only citizens are included.[7] The U.S. median income per household was $13,570 in 1977, and $15,060 in 1978, according to Bureau of the Census.[8]

THE MULTINATIONALS

Perhaps one of the most alarming aspects of our energy situation is the control the multinational "big oil" companies have over the reserves data on which our national security could depend.[9] This is one of the things that has made them targets for attempts at divestiture—either vertical or horizontal. Nationalization of the petroleum industry is not necessarily the answer, since conduct of the nationalized British Petroleum seems little different from Exxon, Texaco, and others,[1] and there is little or no evidence to suggest that a bureaucracy is capable of operating the petroleum or any other industry efficiently.

Without attempting to place divestiture on trial in these pages, it may be of interest to touch on the major points at issue. It is noted by those favoring vertical divestiture that the competitive strength of the independents lies in the areas of refining and marketing, where they have 22 and 26 percent of the business, respectively. They then charge that the price of crude production is manipulated by the majors, and profits from the rigged price are used to subsidize downstream activities that compete with the independents (i.e., refining and marketing).* It is charged further that dealers and jobbers are treated unfairly by the big oil companies (despite the fact that about 90 percent of major-brand gasoline is sold through franchised dealers).

In rebuttal, the majors reply that the industry is, in reality, highly competitive and has become increasingly so in the area of refining and marketing over the past decade. The company with the largest sales in 1978 (Exxon), had a little over one-fourth (27.4 percent) of the total sales of the top 10 oil companies. Sales of the top four oil companies amounted to two-thirds of the total for the top 10.[11] The majors claim that the market concentration (Table 2.1) has been less for the big four oil companies than for the average of all manufacturing.†

Table 2.1 Market Concentrations of Major U.S. Oil Companies, 1970

| | Share of Domestic, % | | | |
Companies	Reserves	Oil Production	Refining Capacity	Gasoline Market
Top 4	37.2	31.1	32.9	30.7
Top 8	63.9	50.5	58.1	55.0
Top 15	84.9	66.0	78.0	73.9

Source: Reference 12.

*In 1976, major-brand marketing cost was 12 to 16¢ per gal vs. 4 to 8¢ for private brands.[10]
†The Industrial Reorganization Act designates a monopoly as a group of four or fewer companies accounting for 50 percent or more of an industry's business.

Furthermore, their return on investment (ROI) in the early 1970s was only slightly higher than the average of all industry—13.4 vs. 13 percent. They cap their defense with the claim that vertical integration benefits all; there is a need for "bigness" in the industry to provide and manage the imposing capital requirements of our energy gluttony. It is claimed that "the scale and timing of energy problems are such that they appear to exceed the response capabilities of most private companies."[13] Despite the failure to sustain an indictable case against "big oil" on vertical divestiture,[14] there is still considerable political pressure for horizontal divestiture—aimed at preventing oil interests from expanding excessively into coal, nuclear, solar, or other energy ventures outside the petroleum field.

Independent economists seem to take a dim view of the arguments both pro and con either vertical or horizontal divestiture. They see a negligible effect of divestiture on the domestic energy situation or on the industry's relationships with OPEC. Their primary concern is that a legislative battle will make a political football of energy and divert attention from the real issues. The ensuing transition period will then create uncertainty, which will have an adverse impact on our economy, perhaps even prompting oil producers to divert capital to nonenergy uses, when it is needed to increase oil and gas production.[15] They feel, however, that if energy development projects become economically viable, even divestiture will not inhibit the ultimate capital flow required to bring them to fruition.[16]

The solution to this dilemma lies in the hands of the administration in power in Washington. One economist says: "On energy, there is little division among economists. The division is between economists and politicians."[17] Time alone will tell how well the administration in power manages to separate economic facts from the political necessities, and gets on to guiding the nation's energy destiny.

GEOLOGY AND GEOGRAPHY

Prior to embarking on a more detailed discussion of our petroleum situation, it might be worthwhile to dwell briefly on the geological and geographical aspects of "black gold."

To form petroleum, a process that took place in the Ordovician period (between 600 and 400 million years ago plus or minus 100 million years),[18] sufficient organic material had to be present and it had to be preserved geologically. It also had to be buried deeply enough in mud or clay to encounter the heat and pressure needed to generate petroleum, and then it had to be expelled from the site of its generation. Permeable carrier beds received the petroleum and allowed it to pass through to reservoirs or traps. These traps were such that the oil could accumulate; that means they had to be provided by nature with an impervious seal or roof rock. The largest traps are generally associated with anticlinal (domal) arches located in stable continental interiors. Geologic stability is essential for preservation.[19]

Although we like to think of the 1859 discovery of the Oil Creek field near Titusville, Pennsylvania, as the start of the petroleum industry, it is sobering to realize that supply shortages, prices, and oil contract complaints were noted in cuneiform records dating back to at least 1875 B.C. Supplies at that time were obtained from surface seeps and shallow, hand-dug wells. The first recorded discoveries of natural gas were in China, made while drilling wells with bamboo, in about 600 B.C., and the first deliberate explorations for gas produced the product in Szechwan Province as far back as 211 B.C.[19]

Up to 1976, slightly over 22,000 oil and gas fields had been discovered in the world, and approximately 73 percent of them were in the United States and Canada. It is predicted that ultimately 70,000 to 80,000 fields will be discovered and only 30 percent will be in the United States and Canada.[19] Thus it is apparent that the time is overdue for our seeking other sources of fluid fuels.

Fields are classified in the industry by size of recoverable reserves. This, in turn, is a function of the geologic history of the field. Normally, a *field* consists of several *pools* or *reservoirs* interconnected through carrier beds of permeable rocks in a common *trap*. Of the 94 *giant* oil fields* discovered in North America prior to 1974, 85 percent were in the United States, but these fields have already been severely depleted.[20] The world's 491 giant fields (as of 1976) are estimated to contain nearly 80 percent of the world's known petroleum reserves. Of the 22 or so *supergiant* fields† throughout the world, one (probable) exists in the United States, and that is in Alaska. There are 10 in the Middle East and eight in the U.S.S.R.[19] The total reserves in the 135 western hemisphere giant fields are matched by the reserves in only two Mideast supergiants (Chawar and Burgan).[19] This picture undoubtedly will be altered somewhat when two recently discovered giants in Mexico are fully evaluated.

In 1974, the U.S.S.R. supplanted the United States as the world's largest oil producer, with 16.4 percent of the world's output vs. 15.6 percent for the United States and 14.6 percent for Saudi Arabia.[21] At the beginning of 1979, the OPEC nations were estimated to have a total crude oil production capacity of 36.6 million bpd, but for the month of January they operated at only about 80 percent of capacity, compared with about 89 percent for the previous month. This loss in operating rate was due to the upheaval in Iran, where production fell from 6.1 million bpd to 0.7. Even an increase by Saudi Arabia to 10.1 million bpd (94.4 percent operating rate) could not offset the loss of Iranian production.[22]

More than 60 percent of the giant fields of the world exist in a crescent-shaped region encompassing North Africa, the Mideast, and west-central U.S.S.R. Furthermore, this region contains almost 68 percent of the world's proved gas and oil reserves.[19] All in all, the distribution of petroleum in the world gives us no cause for complacency, particularly in light of our growing dependence on fluid fossil fuels (see Table 1.2). It is reported that every man,

---

*Contains a minimum of 500 million bbl of recoverable oil. A giant gas field contains at least 3 tcf of recoverable natural gas.[19]

†Containing at least 10 billion bbl of recoverable oil or 60 tcf of natural gas.[19]

woman, and child in the United States uses, on the average, more than 3 gal of oil and almost 250 ft^3 of natural gas per day.[23]

ECONOMIC IMPACT OF OPEC

The world as a whole has suffered in many ways from the Arab oil embargo and the 14-fold increase in the price of crude from 1970 to mid-1980. The real sufferers, however, have been the 100 or so Fourth World or less-developed countries (LDCs) in which half of the world's population resides. For example, in 1972, Jamaica began to see the beginnings of prosperity from tourism and her rich bauxite deposits. At the time, her energy bill amounted to a manageable $49 million per year. Today, her dream has vanished—the annual inflation rate climbed to 15 percent by 1976, unemployment stood at 20 percent, her foreign reserves had shrunk to the equivalent of only two weeks' imports, in the first half of 1976—largely as a result of a 1975 energy bill of $215 million.[24] In India, the situation was fully as dismal. Her energy costs soared from $268 million in 1972 to $1.48 billion in 1974, with her petroleum-based fertilizer (and consequently food) increasing proportionately in cost, just as the "green revolution" was beginning to give her a degree of self-sufficiency in food.[24] Among the real sufferers are the Latin American countries, with their high population growth rate, inflation, and unemployment.

OPEC sources make much of the fact that, on a percentage-of-gross national product basis, they provide more foreign aid to the LDCs than the United States does. What they fail to mention is that the bulk of their aid goes to the relatively richer Arab or Moslem nations on whom they can count for continuing moral, political, and military opposition to Israel.[24] In 1977, the OPEC oil ministers pledged a total foreign aid commitment of $1 billion to LDCs, an increase of 25 percent over 1976,[25] against a non-OPEC world total oil bill in excess of $100 billion, and a projected 1977 OPEC windfall of $10 billion.

The wealth of the OPEC nations is almost beyond comprehension. Put in terms easy to visualize, albeit hard to believe, if all the oil income of the OPEC nations (on a 1974 basis) had been used to purchase major assets of the world, it would have required only 79 days' income to purchase Exxon Corporation, 143 days for IBM, 9.2 years for all stocks listed on the NYSE, or 15.5 years for all the personal wealth in Great Britain![26] A recent CIA survey indicated that from 1973 to 1976, the Arab nations invested a total of $34 billion in U.S. holdings, mainly in the nation's largest banks, possibly in an effort to influence support for any future action in the volatile Mideast situation.*

The full power of OPEC is not completely reflected in these figures. Schlesinger (Department of Energy) warned that if the United States persisted in re-

---

*From 1973 to 1976, OPEC members accumulated a surplus of $150 billion and added about $45 billion in 1977. Twenty-four First World, industrialized nations (including the United States, Japan, and most of Western Europe) ran about a $25 billion combined oil deficit in 1977, and the 100 Fourth World nations were not far behind with about a $22 billion deficit.[27]

taining price controls on domestic oil, OPEC would retaliate by still more price hikes.[28] In other words, the United States, the world's most powerful nation, is no longer in control of its own destiny. Garvin (president of Exxon Corporation) painted an even more gloomy picture by estimating that U.S. gasoline prices would be at $2 per gal in 1990, and now others are saying that Saudi Arabian light crude could go as high as $45 to $50 per bbl in 1990. The only benefit to be accrued from these prices is the fact that our synfuels based on coal, peat, oil shale, biomass, and waste could then become economically attractive and return to us the opportunity to regain some small degree of self-sufficiency. We are still faced with the prospect, however, that these alternative fluid fuel sources will still be able to account for only a small fraction of our energy needs,[29] but the market power generated by such options could decelerate further OPEC price increases.

OPEC oil (particularly that from the Persian Gulf) has marked cost advantages over crude oils from other sources, in that the capital and wellhead technical costs are appreciably less (Table 2.2).*

From 1972 to 1976, the per barrel price of our imported oil increased by a factor of about five, but our total national oil bill for imported products increased by a factor of nearly 7.5, because the volume of our oil imports rose 55 percent in the same period.[31] Oil price projections made in 1977 for 1980 and beyond are already out of date, since the estimate for 1980 of $15.90 per bbl[32] was exceeded in 1979. The 1990 projection was about $37.50 per bbl for both OPEC and domestic crude, equating to about $2.50 per gal for gasoline. The

Table 2.2   Capital and Technical Crude Oil Costs

	U.S. Dollars	
Oil Source	Capital Cost[a]	Wellhead Technical Cost[b]
Persian Gulf	100–300	0.10–0.20
Nigeria	600–800	0.40–0.60
Venezuela	700–1000	0.40–0.60
North Sea	2500–4000	0.90–2.00
Large deep-sea reservoirs	Over 3000 (?)	2.00–?
New U.S. reservoirs[c]	3000–4000	2.00–2.50

Source: Reference 30.

[a]Per bbl capacity.

[b]Cost per bbl exclusive of transportation, government revenues, and producer's profits.

[c]Not too remote.

*Because of the rate of inflation, these data should be considered relative rather than absolute.

high domestic price for crude assumes depletion of "old" oil and only a modest increase (20 percent) in coal production, and apparently also reflects the phasing out of price controls in 1981.

There is one aspect of oil acquisition that defies all rules. This is the spot market operating out of the Caribbean, Singapore, Italy, and particularly Rotterdam. These are all large refining centers. On the spot market, crude oil and petroleum products are invariably available if the buyer (merchandiser) can pay the price demanded.[33] Spot market prices in the first quarter of 1979 were reported to be as high as $26 per bbl vs. an official OPEC price of $13.34,[34] and in the fourth quarter of 1979 they were typically $37 to $38 per bbl vs. $18 for the official base price of Saudi benchmark crude. In that quarter, spot prices went as high as $50 per bbl for some oil put on the market by the National Iranian Oil Company (NIOC).[35] In the past, spot prices have usually been below OPEC prices.[36] Prior to the Iraqi-Iranian conflict, NIOC was probably the largest single source of supply for the spot market,[34] but Italy, where 30 independent refineries are located, usually supplies the bulk of the products traded.[36]

This market caters to traders needing a product quickly, regardless of price. The product may be on the high seas when traded by telephone or telex to meet a market demand. Typically, it becomes available when the original buyer, who purchased it months before, finds that it no longer has a need for the particular product.[36]

Rotterdam is a natural focus for this sort of operation, with its location at the convergence of the Rhine River, the English Channel, and the North Sea. It is the world's most active seaport and largest refinery center, and the river basin is the site of Europe's largest concentration of chemical companies.[36]

OIL RECOVERY

Some people tend to assume that once an oil trap is discovered, it can be pumped down to the last barrel. Actually, much more oil remains in the ground than can be pumped out during a reservoir's producing life. Table 2.3 shows the story of recoverability of U.S. reserves through 1975. Much of the oil remains bound in microscopic pores between sand grains, by viscosity and capillary forces. The normal recovery factor for most U.S. oil reservoirs in the 1930s was only 15 percent;[38] in 1977, it was about 20 percent for primary recovery and about 10 percent additional for secondary recovery, thus bringing total recovery to approximately 30 percent. Another source[39] shows the estimated apparent average recovery factor for U.S. oil fields discovered in 1960 to be 28.6 percent, for the period 1966–1974, but the trend was steadily upward, from 26.9 to 29.9 percent.

It might also be thought that drilling new wells into remote reaches of the reservoir would yield increasing amounts of crude. Actually, only one well per reservoir is needed to achieve essentially maximum ultimate recovery, since the

Table 2.3 Domestic Oil Recovery Through 1975

	Billion bbl
Total discoveries[a]	418
Total production through 1975	109 (26.1%)
Additional recoverable reserves[b]	28 (6.7%)
Total recoverable (est.)	137 (32.8%)
Balance unrecoverable	281[c]

Source: Reference 37. Reprinted with permission of the copyright owner, The American Chemical Society.
[a]Excluding Alaskan North Slope.
[b]Extractable by conventional methods.
[c]*Chem. Eng. News* rounded this figure to "nearly 300," possibly assuming additions resulting from revisions of reserves as production proceeds.

oil, under in-place conditions, is generally highly mobile, having a viscosity less than that of water under normal conditions.[39] Ultimate recovery, however, may be somewhat responsive to extraction rate. Under optimum production controls, slightly higher yields may be achieved at higher rates of drawdown; hence many wells will normally be drilled into a single producing trap. When the influx of water from conterminous aquifers or from deliberate injection of water (waterflooding), takes place during drawdown of the crude, isolated recovery factors as high as 50 percent may be achieved. The same can be said for the use of gas injection techniques. Recovery factors of 40 percent are assumed, perhaps optimistically, for North Sea fields.[39]

To permit a credible estimate of recoverable reserves, it is necessary to consider many geological factors: dimensions of the zone; pore volume of the rocks; permeability of the pore space by oil, gas, and water; physical characteristics of the oil (and gas); depth to the trap ceiling; and temperature and pressure of the reservoir.[40] From these data, the petroleum geologist can determine the type of "drive" to be expected and then estimate the recovery factor. Various drive mechanisms in primary recovery of oil from U.S. fields result in wide-ranging recovery factors (from 5 to 80 percent). *Solution gas drive* recovery factors average about 20 percent, but may be as low as 5 percent. *Water drive* averages 50 percent, but may go as high as 80 percent in isolated cases. *Gas cap expansion* results in an average recovery factor of about 32 percent.[40] (Recovery factors for natural gas are much higher than for oil, falling in the range 60 to 90 percent, because gas is able to flow more readily to the well bore.)

One effect of the decontrol of domestic oil prices would be to provide an incentive for costly *enhanced oil recovery* (EOR) techniques. After natural solution gas drive or gas cap expansion had yielded 20 percent or more recovery, waterflooding (secondary recovery) is generally employed to raise the recovery to about 30 percent. Yields can be pushed even higher by three different EOR techniques: thermal (steam injection or in situ combustion to reduce viscosity), carbon dioxide injection, or chemical flooding to reduce surface tension.

Chemical flooding entails the use of special polymers (polyacrylamides, polysaccharides, xanthan gum),[41] surfactants (*micellar flooding* with isopropyl or higher—C_4 to C_6—alcohols and petroleum-based sulfonates), or alkali. (Terminology will differ with individuals; polymer and alkali flooding might be considered by some as augmented waterflooding or secondary methods.)[37]

The normal procedure for micellar or surfactant flooding is to inject a "slug" of a surfactant mixture of alcohol and sulfonate to reduce interfacial tension and modify surfactant adsorption. This is followed with a larger slug of water-dispersed polymer. The combination improves mobility and sweep efficiency. The goal is to match the rheological characteristics of the injected material with the in-place petroleum.

The cost effectiveness of EOR is highly dependent on the pricing of crude, in part because most of the chemicals used are petroleum-based. About 17 percent of the enhanced yield would have to be fed back into the loop to provide the petrochemical surfactants, alcohols, and polymers necessary for tertiary recovery.[42] Current indications are that the chemical slug must be tailored to the specific reservoir, thus adding to the economic uncertainties.[37]

In 1977, the National Petroleum Council (NPC) estimated that surfactant flooding could not be cost effective until oil exceeded what came to be the second quarter 1979 official OPEC price of $14.55 per bbl. Table 2.4 shows the effect of oil pricing on tertiary recovery. The volume and cost of chemicals required is considerable, as can be seen in Table 2.5. It has been estimated that as much as 500,000 tons of surfactant would be required for tertiary recovery in Brent, a single North Sea giant field.[39]

Stripper wells, that is, wells that produce no more than 10 bpd of crude and petroleum condensates, including natural gas liquids (NGL), are the source of

Table 2.4 Tertiary Recovery as Function of Crude Oil Price

Crude Oil Price, $/bbl	Tertiary Recovery Method(s)	Ultimate Tertiary Recovery through Year 2000, Billion bbl	
		Estimated Range	Best Estimate
5	Thermal	1–11	2
10	Thermal; carbon dioxide	n.a.	7
15	Surfactant flooding	n.a.	2.1
20	Surfactant flooding	n.a.	6.4
25	Surfactant flooding	2–13	8.3

Source: Reference 37. Reprinted with permission of the copyright owner, The American Chemical Society.

Table 2.5   Chemical Cost of Micellar Flooding

	Crude Oil Price	
	$15/bbl	$25/bbl
Oil recovery, 1976–2000, 10^9 bbl	2.1	8.3
Chemical requirements, lb/bbl		
Surfactant[a,b]	17.5	17.5
Alcohol[c,d]	5.25	5.25
Polymer[e]	1.0	1.0
Total chemicals, 10^9 lb		
Surfactant[a]	36.8	145.3
Alcohol[c]	11.0	43.6
Polymer[e]	2.1	8.3
Total	49.9	197.2
Average price per lb, dollars		
Surfactant[a]	0.43	0.59
Alcohol[c]	0.20	0.27
Polymer[e]	2.10	2.52
Chemical market, 10^9 dollars		
Surfactant[a]	15.8	85.7
Alcohol[c]	2.2	11.8
Polymer[e]	4.4	20.9
Total	22.4	118.4

Source: Reference 37. Reprinted with permission of the copyright owner, The American Chemical Society.

[a]100% active.
[b]GURC estimates 10 +5, −3 lb/bbl.
[c]Assumes oil-based isopropyl alcohol, but higher alcohols are probable.
[d]GURC estimates 3 lb/bbl; Union Carbide estimates 2 to 11 lb/bbl.
[e]Assumes equal volumes of polyacrylamides and polysaccharides.

significant overall yields only because of their large numbers (nearly 370,000 in 1975). The reserves of oil for production by stripper wells is about 4.8 billion bbl, but the recovery factor is low. The average production per stripper well (2.93 bpd) is so low as to seem uneconomic, but the stripper product has been on the free market (i.e., without price controls); hence the continually increasing price of crude is conducive to continuing production of the strippers below the level where such wells normally would have been abandoned in past years.[43]

DOMESTIC SUPPLY

Although many people tend to place the blame for the present oil situation at the feet of "big oil," the Arab oil barons, or the federal bureaucracy, the real culprits are consumers themselves, who are using oil faster than new domestic discoveries can be made. The result is dwindling reserves and a rising consumption rate*—disregarding the effect of the Arab oil embargo and sporadic fluctuations in the business climate (Table 2.6 and Fig. 2.1).†

The excess of production over discoveries (Table 2.6) can only lead to a drawdown of domestic reserves and an increase of imports. Figure 2.2 shows our growing dependence on imported oil, particularly from OPEC sources. In 1979, our largest sources of imported oil were Saudi Arabia (18 percent), Nigeria (14 percent), and Libya (10 percent). The Saudi product is a sour, heavy crude, whereas Nigeria and Libya produce light, low-sulfur crudes. The quality difference is reflected in the official March 1979 prices: Saudi Arabia, $14.55 per bbl; Nigeria, $20.98; and Libya, $21.09.[49] Recent figures show that U.S. oil imports climbed above 10 million bpd for the first time in history during the week ending February 25, 1977. We imported 3.13 million bpd of refined products and a record 6.95 million bpd of crude, for a total of 10.08 million bpd, an

Table 2.6 Proved Recoverable Reserves of Liquid Hydrocarbons,[a] United States (Estimated)

	Million bbl			
Year	New Discoveries[b]	Production	Proved Reserves Year's End (Est.)	Net Change Over Previous Year
1950	3,329	2,171	29,536	1,158
1955	3,385	2,740	35,451	646
1960	3,090	2,903	38,429	188
1965	3,880	3,242	39,376	639
1966	3,858	3,453	39,781	405
1967	3,892	3,682	39,991	210
1968	3,140	3,826	39,305	(686)[c]
1969	2,401	3,931	37,775	(1,530)
1970	12,996[d]	4,067	46,704	8,920
1971	2,666	4,002	45,367	(1,337)
1972	1,796	4,037	43,126	(2,241)
1973	2,555	3,926	41,754	(1,371)

Source: Reference 44.

[a]Crude oil and NGL. Numbers may not be exact due to rounding.
[b]Including revisions and extensions of known fields.
[c]( ) signifies decline.
[d]Reflects assessment of 1968 Prudhoe Bay strike.

---

* Average annual growth rate of +4.1 percent from 1965 to 1974,[46] and expected to remain positive at least through 1990.[47]
† The trends are unmistakable. The differences in absolute numbers between Table 2.6 and the Fig. 2.1 may reflect inclusion of NGL in the higher figures shown in the table.

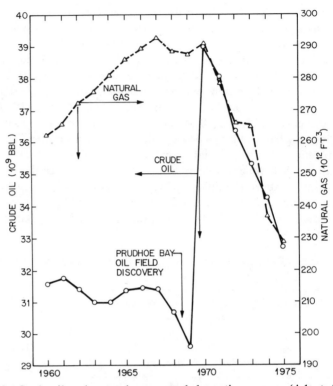

Figure 2.1 Crude oil and natural gas proved domestic reserves. (Adapted from reference 45; reprinted with permission of the copyright owner, The American Chemical Society.)

increase of 45 percent over the same week in 1976. In early 1977, crude oil imports alone amounted to almost as much as the total petroleum product imports of a year earlier.[50]

Crude oil imports dropped 25 percent in June 1980, according to the Department of Energy, with gasoline consumption for the first five months of 1980 being about 8.1 percent less than for the same period of 1979. These decreases are thought to be due to the recession, rising fuel prices, insulation of homes, and increasing fuel efficiency of the nation's car fleet. Such indicators suggest that oil imports may decrease from 8.5 million bpd in 1979 to 7 million bpd in 1980 (average) while the growth of GNP is maintained (+2.3 percent in 1979).

A perusal of exploration activities in 1976 shows that a great deal of effort has been going into discovering new domestic sources. Nearly 40,000 wells were drilled, with over 17,000 (43 percent) oil strikes and over 9000 (23 percent) gas strikes resulting. About one-third were "dusters" or dry holes.[51] Data on "new reserves" uncovered by this exploration were not disclosed.

Data on 1973 reserves show that just three states held three-fourths of our

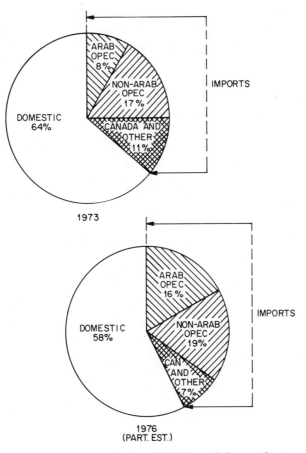

Figure 2.2 U.S. dependence on foreign oil. (Adapted from reference 48; material furnished by the Economic Analysis Section of the American Telephone and Telegraph Company.)

domestic reserves: Texas, 33.3 percent; Alaska, 28.6 percent; and Louisiana, 13.0 percent. Combined with the other major petroleum states (California, 9.9 percent; Oklahoma, 3.6 percent; and Wyoming, 2.6 percent), they held 91 percent of our reserves. The total domestic reserves were 35.3 billion bbl, and Texas reserves alone were 11.75 billion bbl.[52]*

Worldwide proven reserves data show that Saudi Arabia has over twice as much oil (165.7 billion bbl) as the second-place nation (U.S.S.R., 71.0 billion bbl). The United States is seventh on the international list with 28.5 billion bbl, and China is ninth with 20.0 billion bbl.† The other seven of the top 10 coun-

---

* Another source shows Texas reserves of 14.5 billion bbl in 1973 and 8.5 billion bbl in 1978.[53]
† This must be considered a questionable figure, because such data are not readily available from China.

tries are OPEC nations. The top three countries (Saudi Arabia, U.S.S.R., and Kuwait) control nearly 50 percent of the 618 billion bbl of global oil reserves.[54] It must be remembered that such data are very imprecise estimates, hence will tend to vary widely from source to source.

DECONTROL

It seems in order at this point to try to put into proper perspective the interactions of OPEC and U.S. policies. Up until the early 1960s, the United States controlled approximately 70 to 80 percent of the world's petroleum and mineral resources, but this picture has changed. About 1970, the State Department tried to warn Congress and other leaders that we would be forced to import half of our oil supply by 1980 unless we took immediate action to forestall this situation. It also advised that two-thirds of these imports would have to come from the unstable Mideast. Today, we are beginning to see the truth of these predictions, because the necessary actions were not taken.[55]

Following the 1973 oil embargo, and at a time (December 1975) when it should have been obvious that we needed to conserve energy and to convince the public that a serious energy crisis was at hand, Congress passed legislation that continued the maximum limit on the price of domestic crude oil* through May 31, 1979, at the discretion of the president. Since this action held the price down, it only served to show the nation and the world that the United States saw no need to conserve or to provide incentives for increasing domestic production.[57] Relief from OPEC's pricing policies can be expected only when we dedicate our policy to significant conservation and increased domestic production.[58] Of course, the long-range hazard of increased domestic production is the accelerated drawdown of our own limited reserves.

The price control extension put President Carter in a box from which there was no escape without antagonizing some element of our society. The alternatives were: (1) immediate decontrol, (2) graduated decontrol with excess profits tax, (3) controls with rationing, (4) controls with allocation, or (5) controls with improved regulation. Maintaining controls would aggravate the oil shortage by encouraging consumption and merely mire us deeper in an untenable regulatory jungle. But of more compelling concern, it would provoke OPEC further and stimulate support within the organization for yet another round of price hikes.

Immediate decontrol would provide the world with proof that the United States was dedicated to conservation and increased domestic production—the major steps toward solving our energy problems. It would also have a very prompt and drastic impact on our economy and lead to large windfall profits for the petroleum industry. Graduated decontrol with an excess profits tax would

* "Old" oil (oil discovered prior to 1973) was pegged at $5.82 per bbl, and "new" oil (discovered after 1973) was controlled at $12.98.[29] The average price for domestic crude was $9.98 vs. the official OPEC price of $14.54 per bbl.[56]

result in higher prices at the gas pump and have an adverse effect on inflation and the economy in general. Long-term effects, however, should produce some conservation and stimulate oil production and development of synfuels, but this would not be apparent for possibly as long as several years.

Controls with rationing would serve as the most effective message for the world, and could save the most energy, but it would provide no incentive for increased production or synfuel development. The costs of the bureaucracy necessary to administer the policy would be inflationary, but less so than decontrol. Controls with allocation is little different from what we have had for many years. It provides limited capability for dealing with a shortage, but affords no cure for the shortage. We are familiar with this option and have the federal bureaucracy to handle it, but it conveys a message to OPEC which that organization does not want to hear—namely, that the United States is unwilling to alter its gluttonous appetite for energy. Also, it is a relatively inflexible system incapable of dealing with a severe shortfall situation. Controls with improved regulation is a rather improbable and unpopular scenario which would allow "big brother" to intrude even more deeply into private affairs.[57]

The Congressional Budget Office (CBO) estimated the effect on domestic production of several crude oil price control options. Continuation of price controls begun in 1973 would lead to production of 6.75 million bpd (in 1990). A House-approved tax plan would result in 7.08 million bpd. A Senate Finance Committee decontrol-plus-tax plan would yield 7.63 million bpd. Without a windfall profits tax, 7.92 million bpd domestic oil could be produced in 1990.[59]

The president adopted graduated decontrol with an excess profits tax as the most palatable of the various alternatives. Any windfall profits could be dealt with in at least three different ways: (1) the funds could remain with the oil companies, presumably for exploration and/or new facilities; (2) they could be paid out to stockholders as taxable dividends; or (3) they could be paid directly to the government as a penalty tax.[60] A combination of options 1 and 3 appeared to the Carter Administration to be the most practical. It would allow the oil companies to retain only about 30 percent of the windfall, while contributing the remaining 70 percent to the government for funding various energy programs.*

There can be little doubt that price controls had been ineffectively administered, failed to provide appropriate feedback relative to supply and demand, and offered no incentive for increased production or development of alternative fuels. Time alone will tell what comes of graduated decontrol, which began in June 1979. The full impact will be unknown until at least October 1981, when controls are scheduled to be phased out completely.

More than one source has suggested that stimulus of the development of alternative sources could probably be achieved without the use of federal money, by the simple expedient of requiring that 10 percent of all gasoline sold in the United States be of nonpetroleum origin. This would "get the alternative

---

* The actual windfall profits tax bill passed by Congress channeled just 15 percent of the revenues into new energy development.

energy industry off the ground, unendangered by politics or paper work," and would cost taxpayers nothing[61] (except at the gas pump).

TRANSPORT

In 1977, about 95 percent of our oil imports were shipped in foreign-registered bottoms, largely Panamanian or Liberian,[62] to avoid the high cost of American taxes, labor, and registry regulations. The Jones Act, however, requires that all oil shipped between U.S. ports be moved in U.S.-registered ships. Shipment of Alaskan oil alone was expected to require 35 tankers of average 100,000-dwt capacity, far in excess of the capabilities of our domestic maritime fleet at that time.

Congress wanted a law requiring that 30 percent of our oil from foreign ports be shipped in U.S. bottoms, but the bill was vetoed. Meanwhile, the three-year construction lead time for large tankers was frittered away pending decisions affecting tanker design.

Probably the biggest question concerns the pros and cons of double bottoms. The U.S. Coast Guard initially favored this type of construction, but on further study, sided with Japan and France in vigorously opposing it. Environmentalists favor double bottoms because they feel minor collisions and groundings will not open up the oil tanks to create disastrous spills. Most maritime experts, however, claim that double-bottom construction is such that a rupture of the outer hull could cause a number of tanks to open up; also, such accidents would cause a double-hulled tanker to ride deeper in the water and probably cause it to ground fast and ultimately break up completely in high seas.[62] But the major argument against double hulls is that a small leak between the outer compartment and the oil tanks would create a floating bomb because of the accumulation of explosive gases. This would require installation of gas inerting systems.

Other proposals would call for various safety regulations largely involving sophisticated navigational aids, operational details, and redundant safety features. One problem is how best to exert effective control on tankers entering U.S. navigable waters without violating existing international maritime treaties and recommendations of the UN-sponsored Inter-governmental Maritime Consultative Organization (IMCO). The route taken has been to require an effective inspection program and to grant the Coast Guard the authority to keep unsafe vessels out of U.S. waters.

Use of supertankers for delivery of crude to U.S. ports is not now possible, because no U.S. ship channel is deep enough to accommodate these carriers, which require well over 50 ft of water for a fully loaded 250,000-dwt ship.[63] Transfer of oil from supertankers to smaller carriers (called "lightering") is carried out in the Gulf of Mexico* or at Caribbean ports, to effect maximum savings in transoceanic transport. There is a fringe benefit to lightering. The U.S.

---

*About 1 to 1.5 million bpd in 1978.[63]

Customs Bureau collects a tonnage tax on all vessels arriving in the United States from foreign ports. Since a Very Large Crude Carrier (VLCC) is not considered a foreign port, the lightered crude enters untaxed. The tax is modest—2¢ per net ton on ships arriving from North American ports and 6¢ per net ton on ships from other foreign ports. In any given year, a single VLCC would probably pay less than $250,000 in Customs taxes, if subject to the tax.[64]

The ability to off-load directly in U.S. ports could save up to 37¢ per bbl on import costs. A man-made 1.4-million-bpd deepwater port, the Louisiana Offshore Oil Port (LOOP), is expected to begin operation in 1981. It will connect with a pipeline system with access to 25 percent of all U.S. refinery capacity.[63] Another offshore port, the Texas Deepwater Port, was granted a construction license from the Department of Transportation in mid-1979.

OFFSHORE OIL

Exploration for offshore oil began in 1897, but the first producing well out of sight of land (10 mi off Los Angeles) did not become a reality until 50 years later.[65] Over this period of time, and continuing to the present, offshore exploration has grown rapidly worldwide, with the rest of the world owing a large debt to the United States both for the technology and the hardware employed.

It is obvious that despite the considerable technical difficulties of offshore drilling, it is done solely with the expectation of discovering economically recoverable oil and/or gas. Since 1958, on the average, every 97 exploratory wells off the coast of Louisiana have uncovered one field containing 100 million bbl of oil or 600 billion scf of gas. Onshore, the nationwide average for this period was 5100 exploratory wells for the same yield, and the total onshore discoveries have been only one-tenth those of offshore fields.[66]

In 1971, the average domestic onshore oil well was 4787 ft deep, while offshore the average depth was 10,189 ft.[67] Natural gas tends to be in deeper traps than oil does, but it generally exceeds oil in quantity (on an energy-equivalent basis) in offshore reservoirs.[68] Because of the adverse marine environment, the capital cost of offshore rigs comes high: for example, approximately $300 to $500 million each for the North Sea rigs (about 1976). In the decade 1969–1979, oil/gas drilling costs have risen by a factor of about 2 to 3, with the greatest increase (a factor of about 3.2) being in OCS fields. A 10,000-ft offshore well in 1979 might have cost $1,750,000 vs. $200,000 for a 5000-ft onshore Texas well.[69]

It is estimated that more than half the offshore oil and gas lies on the outer continental shelf (OCS), while one-third occurs on continental slopes and rises and in small ocean basins. Less than 2 percent is thought to be on deep ocean floors.[70] In view of the relatively low occurrence of petroliferous deposits in the extensive deep-ocean areas, only a very small fraction of the large water area of the earth constitutes what could be considered a "potentially favorable area" for exploration, and these are largely in the continental margins (Table 2.7). Nonetheless, the total area of the continental margins is sufficient to be the

Table 2.7   Worldwide Recoverable Oceanic Oil Resources (Estimated)

	Total Area	Potentially Favorable Area	Total Recoverable Oil,[a] Billion bbl	
			Minimum	Maximum
	Million mi^2			
Water area	139.4	7.9 (6%)	500 (34%)	720 (32%)
Land area	57.5	18.0 (31%)	990 (66%)	1500 (68%)
Earth area[b]	196.9	25.9	1490	2220
OCS				
0–200 ft[c]	5.9	3.3	130	200
0–666	9.5	5.2	290	430
0–1000	11.4	6.3	440	660
0–2000	14.4	7.9	500	720
>2000	125.0	0	0	0

Source: Reference 65.

[a]Discovered and undiscovered.
[b]Water plus land area.
[c]Depth.

source of an estimated third of our recoverable oil resources. Table 2.8 indicates our estimated domestic offshore oil and gas resources. In 1976, about 17 percent of our crude production came from the OCS, although only about 2 percent of the shelf area had been explored.[72]

The United States has four offshore regions for exploration. Of these, only three have produced to date: Alaska, Pacific coast, and Gulf of Mexico. The cumulative numbers of wells drilled in these regions through 1974 are shown in Table 2.9, and the drilling activity and the discoveries during 1974 are shown in Table 2.10.

To date, the prospects for exploitable discoveries in the Baltimore Canyon, along the Atlantic coast, are not too encouraging. Five major oil companies already have drilled 13 dry holes and withdrawn from the scene. Even with the announcement of an oil strike by Tenneco Oil Exploration and Production Company, on June 5, 1979, the others have not yet reversed their decisions.

Tenneco reported a find of high-quality, sweet crude, in a 3-ft vertical zone, 8318 ft below the seabed, about 105 mi east of Atlantic City, New Jersey. The flow was 630 bpd. On land, a flow of 250 bpd would be considered a good find (the average onshore well now produces only 19 bpd), but even 630 bpd may not be enough, in light of the $1.5 million-per-mile cost for the 105-mi pipeline and the additional cost of a huge production platform to operate in that location.[75]

The expected production of oil (estimated in 1976) from all four coastal regions is indicated in Table 2.11, but the Atlantic coast estimates are already out of line, since *no* production can be expected in 1980. The Gulf of Mexico is expected to be the most productive of the lower 48 regions, but only a little over

Table 2.8 U.S. OCS Petroleum Resources, End 1974[a]

	OCS Cumulative Production Billion bbl (tcf)	Percent of Onshore	Additional Proved OCS Resources Billion bbl (tcf)	Percent of Onshore	Undiscovered Recoverable OCS Resources Billion bbl (tcf)	Percent of Onshore
Crude oil	6.090	6	3.220	10.3	10–49	27–60
	34.8 Q		18.4 Q		57–280 Q	
Natural gas	(33.976)	7.6	(35.956)	17.8	(42–181)	16–36
	34.0 Q		36.0 Q		42–181 Q	
Total quads[b]	68.0 Q		54.4 Q		99–461 Q	

Source: Reference 71.

[a]200-m depth or less.

[b]Conversions: quad = 0.175 billion bbl of crude oil = 1 tcf of natural gas.

Table 2.9 U.S. OCS Wells Drilled[a]

Province	Total Drilled
Alaska	305
Pacific coast	3,324
Gulf of Mexico	15,500
Total	19,126

Source: Reference 73.

[a]Cumulative, end of 1974.

Table 2.10 U.S. OCS Drilling Activity, 1974

Province	Number of Wells		
	Total	Producers	Dry Holes
Alaska	11	11 (100%)	0
Pacific coast	71	54 (76%)	17
Gulf of Mexico	857	379 (44%)	478
Total	939	444 (47%)	495

Source: Reference 74.

half as rich as the Alaskan coastal region. A "major" new crude oil and natural gas find was made in mid-1978 in the Alaskan North Slope area east of Prudhoe Bay. This is just outside the Arctic National Wildlife Range. The state scheduled a lease sale in this area, but its laws prohibit disclosure of the exploration data for two years from the date of the strike.[77] It should be noted that by 1990, production is expected to be very much on the decline in all four offshore regions.

The important Atlantic OCS province is felt to be a geologically safe region for exploration and production, because it offers no high subsurface pressures that could cause natural oil seeps or blowouts. Furthermore, it could serve a highly industrialized area that is now forced to import two-thirds of its light refined products from the Gulf coast, and 90 percent of its crude and heavy fuel oils from foreign sources.[78]

The Atlantic coast is expected to be the least productive of the four U.S. offshore regions*—perhaps for the very same geologic reasons that make it relatively safe. Nonetheless, from the standpoint of economics, any amount of production that can be achieved is preferable to buying foreign oil. Table 2.12 shows the extent and depths of three different areas to be explored along the East coast. The overall offshore exploration projections for 1975 to 2000 are presented in Table 2.13.

ENVIRONMENTAL EFFECTS OF OIL SPILLS

In the course of day-to-day living, oil finds its way into the marine environment by various routes:

1 Admixture with urban wastes.
2 Association with industrial wastes.
3 Intentional discharge from ships cleaning their storage tanks.
4 Accidental marine spills.
5 Seeps from natural sources.
6 Stream and storm sewer runoff.

Some guess that the annual spill volume is on the order of 1 billion gallons, but this is a highly speculative figure which should suggest only that the volume is significant.[80]

Accidental marine spills (item 4 above) have been receiving a lot of attention from environmentalists in the wake of a number of such spills during the winter 1976-1977. These events arise both from OCS oil production and from transport.

---

*Maximum probable recoverable oil is 4 billion bbl. The probability of finding over 2 billion bbl is only 75 percent, and there is expected to be a 5 percent probability of finding none.[79]

Table 2.11 Estimates of Domestic Offshore Crude Oil

| | Production Forecasts (Annual)[a] | | | | | |
| | 1980 | | 1985 | | 1990 | |
Province	10^6 bbl	10^{12} Btu	10^6 bbl	10^{12} Btu	10^6 bbl	10^{12} Btu
Atlantic coast	80	464	145	841	94	545
Gulf of Mexico	91	528	197	1143	170	986
Pacific coast	60	348	165	957	141	818
Alaskan coast	8	46	465	2697	396	2297
Total	239	1386	972	5638	801	4646

Source: Reference 76. Reprinted with permission of the copyright owner, The American Chemical Society.

[a]Assumes wellhead price of $12 per bbl. Higher prices would effectively increase reserves and potential production.

Table 2.12 OCS Mobile Rig Exploration Forecast for Atlantic Coast

Area	Total Acres, 10^6	Water Depths, ft	Acres to Be Tested,[a] 10^6 (est.)
Georges Bank basin	2.816	250–660	0.929
Baltimore Canyon Trough	3.520	200–660	1.173
Blake Plateau Trough	3.000	600–3000	1.000

Source: Reference 65.

[a]GURC consensus.

Our OCS oil represents about 17 percent of our total oil production, or 10 percent of our total domestic consumption, yet yields about half of our oil spills and one-fifteenth the volume of our total spills.[80] Natural questions are, then: (1) whether it is worthwhile to expose the environment to this abuse for the return gained, and (2) whether the "environmental costs" would be reduced by importing additional foreign oil in place of producing domestic OCS product. Events of early 1979 make the answer patently clear. We must learn to eliminate the environmental problems insofar as possible, and then adapt to those problems we cannot solve.

Offshore oil has long been anathema to environmentalists, who fear its effect on the marine flora and fauna and on the aesthetics of the environment. Since 1953, over 18,000 OCS wells have been drilled in U.S. waters with only 11

Table 2.13 U.S. OCS Exploration, 1975–2000

Province	Prime Areas, 10^6 Acres[a]	Number of Wells Required[b] Drilling from:[c]	
		Mobile Rig	Platform
Alaska	117.264	43,820	87,640
Atlantic coast	7.529	6,190	2,250
Gulf of Mexico	13.861	7,250	8,290
Pacific coast	2.708	2,230	810
Total	141.364	59,490	98,990

Source: Reference 65.

[a]GURC consensus.

[b]To explore prime areas

[c]Assumes 60% seafloor completions on Atlantic coast, 40% on Pacific coast, 20% in Gulf of Mexico, and average of one-third Alaskan wells drilled from mobile rigs, and balance from fixed structures.

major oil spills (exceeding 5000 bbl or about 210,000 gal each). None occurred between 1972 and 1976. The infamous Santa Barbara spill in 1969 released less than 15 percent of the estimated total spills from tanker operations in the same year, but it did so close to shore, in a populated area of small geographical size. It was due primarily to disregard of local geological faulting, and a shallow reservoir under abnormally high pressure, which made it difficult to control at shallow depth.[78] Spills of this type have virtually no chance of occurring on the East coast because of geologic differences, and technological advances since 1969 have greatly reduced the risk of such spills occurring in any OCS fields.

Nevertheless, there is considerable concern about the effect oil spills would have in the Georges Bank basin, off the New England coast, one of the world's richest fishing grounds. It is estimated that there is a 5 percent probability of discovering up to 2.4 billion bbl of crude and 12.5 tcf of natural gas there.[81]

Spills in open seas tend to favor rather rapid evaporation of hydrocarbons with molecular weights under 300, but this may be retarded by very low water temperatures.[82] The residue undergoes oxidative processes, enhanced by exposure to solar radiation, to form more soluble, surfactantlike products. In stormy seas, some dissipation may also occur over large areas via sea spray. The remaining hydrocarbons take on a puddinglike consistency but ultimately break up under wave action to form tar balls having a specific gravity close to 1. By various means, their specific gravity may increase sufficiently in time for them to sink to the ocean floor. Tar balls are estimated to have a lifetime of about one year.

Many marine animals and microorganisms have a great capacity to metabolize or detoxify hydrocarbons—the straight-chain types more readily than the naphthenic and aromatic types. There is no indication at this time that petroleum hydrocarbons concentrate in food chains.[82] Marine birds, particularly diving birds, and shellfish are the most visible forms of marine life known to be vulnerable to oil spills. Natural oil seeps from Coal Oil Point in the Santa Barbara (California) Channel have been studied for sublethal effects of petroleum on marine organisms. The study disclosed "no change in total biomass or in biomass of major groups" as a result of petroleum hydrocarbons present in the sediments. Sea urchins and mussels did contain petroleum hydrocarbons, but the muscle tissue of lobsters and abalone did not.[83]

A seven-year study of a No. 2 fuel oil spill in Buzzards Bay (West Falmouth, Massachusetts) concludes that in some environments, bottom-dwelling organisms may display some short-term effects of hydrocarbon contaminants, and that oil residues in marshland sediments can become incorporated in marine-organism tissues and have a serious effect on survival rates over many years.[84] Major production areas of eelgrass and clams off the coast of North Carolina have yet to recover fully from the apparent effects of some lipid-soluble trace metals present in oil spilled from the large number of tankers sunk by German U-boats in World War II,[85] but this is a rather singular spill situation.

The damage to marine life by the *Torrey Canyon* spill off the southwest coast of England in March 1967 was due, in large measure, to toxic aromatic solvents used in dispersants employed to break up the oil slick. The oil that escaped the dispersants caused relatively little ecological damage.[86] According to Abelson, "talk of an ecological catastrophe thus far is only talk. It has no factual basis."[82] Actually, the environmental impact of oil spills is not yet clear. Recent studies suggest persistence of physiological and community disruption for at least 10 years.[87]

Biosynthetic production in nature is estimated to range from 3 to 10 million tpy. In 1971, a similar amount of hydrocarbons in the marine environment was contributed by petroleum sources, of which 2.1 million tons resulted from losses in transport and a nearly equal amount came from urban and stream runoff.[82]

Coast Guard and U.S. Geological Survey data on oil spills indicate that older OCS wells (>15 years) experience more spills than do newer wells, probably due to corrosion with age. About three-fourths of the spillage is attributed to the gathering net or subsea pipeline system. Fires, hurricanes, and subsea blowouts account for most large spills.[80] Because of the distance from shore, some of the newer OCS leases will replace pipelines with shuttle tankers loading at single-buoy moorings (SBMs).[88]

Using present technology, the spillage associated with importing the equivalent of all offshore production should be less by a factor of about 4, than that resulting from producing the oil offshore and transporting it to the refinery (for the number of occurrences exceeding 1000 gal).[80] From purely ecological considerations, increasing imports appears to be a lesser evil than producing an equivalent amount of offshore oil, but this totally ignores the economic aspects and the necessity of maintaining a balance of payments.

PRUDHOE BAY

Looking to the future, much has been written about our only supergiant field*—Prudhoe Bay—on Alaska's North Slope. Two of the more interesting and detailed accounts of this field are given in references 89 and 90.

The Prudhoe Bay field, covering only 200 mi^2 (500 km^2) of the North Slope's 100,000 mi^2 (260,000 km^2), is estimated to have 9.6 billion bbl of crude reserves plus 26 tcf (700 billion m^3) of associated natural gas. Elsewhere in Alaska, there are thought to be other undiscovered, recoverable accumulations ranging between 12 and 76 billion bbl of oil and 29 and 440 tcf of gas.[89]

The Prudhoe Bay oil exists at 24 drill sites, having an initial total of 154 wells. The oil, which lies about 10,000 ft (3000 m) below the tundra, is recoverable under its own pressure at a temperature of about 180°F (80°C). By the time it reaches the TAPS pipeline, its temperature has dropped to 80 to 145°F (25 to 65°C). While being pumped 800 mi (1290 km) through the 48-in. pipeline at up to 1180 psi, the friction generated will maintain the temperature at about 135°F (57°C) for up to 21 days at an air temperature of −60 to +90°F (−51 to +32°C).[90] Although the journey takes only 4½ days, allowance had to be made for emergency shutdowns, and the oil had to be protected with insulation from congealing. At the same time, the pipeline design and layout had to be such that the permafrost underlying the tundra would not be thawed out; caribou would be free to continue their migrations; three mountain ranges, three major rivers, and about 300 lesser streams would be traversed; and automatic block valves would halt the flow automatically in an operating emergency, with loss of no more than 15,000 bbl. The entire pipeline and terminal system must also be capable of withstanding earthquake shocks up to 8.5 on the Richter scale. The pipeline has seismic sensors located at 11 critical points along its length, with connections to a master computer center at the terminal.[91] These features have made construction of the pipeline fantastically complex and costly. It is estimated that 20 percent of the cost of the entire pipeline system went into protection of the environment.

The southern terminal of the pipeline, in Valdez, one of the few ice-free ports in Alaska, will cover 1000 acres (400 hectares) and have (initially) 18 holding tanks of 510,000-bbl capacity, from which the oil can be pumped to tankers up to 250,000-dwt capacity.

With all the time, money, labor, and materials poured into these facilities, it is hard to conceive that when production began in late 1977, there was no effective way established to transport the oil from the West coast to areas of the United States where it was most needed. Geographically, the logical tanker terminal in the lower 48 would be somewhere along the Pacific coast, at a point from which the oil could be pumped through pipelines to the oil-starved Midwest and industrial East.

Sohio (which, along with British Petroleum, controls 49 percent of the oil)

---

*If it does prove to be a supergiant, it will probably be only marginally so.

preferred to build a 200-mi segment of pipeline* from Long Beach to Blythe, California, to gain access to its existing pipeline system serving the Midwest, by way of Midland, Texas. It should be noted that Sohio has no marketing outlets on the West coast. Arco (21 percent interest) has sufficient West coast refinery capacity to process 200,000 bpd of the sour Alaskan crude, but it would have liked to process some of its 230,000 bpd in its Bellingham, Washington, refinery and pipeline the products from Port Angeles, Washington, to Clearbrook, Minnesota, through a new 1500-mi pipeline (which could not be completed until the 1980s). Exxon (20 percent) considered buying into either or both Sohio's or Arco's refining and distribution systems since it had dragged its feet on converting its own California refinery to process sour Alaskan crude. It also looked into a possible northern pipeline route from Kitimat, British Columbia, to Edmonton, Alberta, which would require a new 470-mi section, but would feed into other Canadian pipelines going south, east, and west. Another four oil companies planned to market their small shares on the West coast.[92] As of the beginning of 1980, the proposal for the Northern Tier pipeline appeared to be nearing approval, despite its 1500-mi length and its estimated cost of $1.6 billion.[93]

The dilemma is that a large excess of Alaskan oil arrived on the West Coast by the end of 1977, with no terminal facilities to handle it, inadequate refinery capacity to process it, and insufficient bottoms or pipelines to transport it to where it was needed. To make matters worse, California and Washington environmentalists wanted to block it out of their ports, the Jones Act prevented the use of foreign tankers, Congress would not allow its sale to foreign countries, even on a one-to-one swap basis, the Panama Canal was too small for supertankers, and Cape Horn was too costly a route to the Gulf ports, even if we had domestic tankers to move the oil.[94] A December 1976 tanker explosion in Los Angeles harbor did nothing to alleviate the anguished pleas of the environmentalists.

If such was the muddled situation at that time, one might wonder what we can expect in later years, when some of the additional undiscovered Alaskan petroleum resources become much-needed recoverable reserves. All in all, there seems to be little relief in sight from our energy gluttony, from our declining reserves, or from the constant counterproductive battles between environmentalists and producers or politicians and economists.

Coupled with the environmental, extraction, and political problems inherent in using Alaskan oil was the initial high cost of getting it to market. This was alleviated, in the case of Alaskan oil, by exempting it from entitlement payments, which otherwise would have had to be paid to refiners of high-priced OPEC oil to compensate them for the high cost of their raw materials. Under a new rule, however, made retroactive to May 1, 1980, the Department of Energy has required that refiners of Alaskan oil now pay entitlements of $12 per bbl to refiners of the more expensive OPEC oil. This reflects the improving transport costs of Alaskan oil.

---

*This would require two to three years to build.

OVERTHRUST BELT

The last energy frontier in the United States appears to be the Overthrust Belt of the Rocky Mountains. This is a petroleum formation averaging a little over 70 miles wide and meandering from the Arizona–New Mexico border with Mexico, 2200 mi north to the Idaho border with Canada. In the search for a new oil or gas bonanza, led by Amoco Production Company and Chevron, the Rocky Mountains had been peppered with 500 dusters, some 17,000 ft deep and costing up to $8 or $9 million each, before the first strike was made in 1974. Now there are 16 significant new fields within 25 miles of Evanston, Wyoming, with more in the offing. Two new large natural gas processing plants are under construction, and three pipelines, including the 800-mi *Trailblazer,* are under construction or under review by the federal government, with $1 billion capital at stake.[94a]

Proven reserves to date amount to 1 billion bbl of a rich, waxy crude and 9.7 tcf of natural gas. Estimates of potential reserves range from 14 billion bbl of oil and 52 tcf of gas, by the Rocky Mountain Oil and Gas Association (RMOGA), to 7.5 billion bbl of oil and 30 tcf of gas (USGS). At best, this will not make us energy independent. The sobering aspect is that such a "bonanza" will be barely sufficient to maintain the *decline* of our fluid fossil reserves at a tolerable rate. Still larger finds would be necessary every three years to turn us around with respect to our energy needs. Furthermore, the Overthrust Belt passes through or borders on some of the nation's most treasured scenic areas—the Grand Tetons, Jackson Hole, the Bob Marshall Wilderness in Montana, and the Palisades region of Idaho. The battle lines have been drawn between the environmentalists and the energy industry. When the dust has settled, Denver could well supplant Houston as the nation's energy capital.[94a]

SHORTAGE OF PETROLEUM PRODUCTS

A recent special, 4000-word investigative news report analyzed the reasons for the gasoline shortage in the summer of 1979 and the projected shortage of distillate fuels—heating oils and diesel fuel—for the following season. It is significant to note that at no point did the study suggest "overconsumption" or "waste" as a cause for the shortages.[95] This seems to be symptomatic of the thinking of most people. It is difficult for people to point an accusing finger at themselves. It is much easier to assign the cause of troubles to more remote, impersonal sources—big business, big government, or a foreign power.

Basically, it was noted in the report that the major causes of the shortages were questionable policies set forth by the federal government and ill-advised actions taken by the oil industry in 1978. These flaws were aggravated by the failure of Congress to establish an effective energy policy in six years of debate. The Iranian revolution was a minor irritant that need not have had a significant impact, had the other elements been in order.

Simply put, in 1978 the world had a glut of oil, the portent of which was lost on leaders in both industry and government. Inventories and imports* of gasoline, distillates, and crude oil, and refinery operating rates were all reduced in the face of a generally rising demand.† Large transfers of crude were also made to the Strategic Petroleum Reserve. The result was that, from a glut, supplies of petroleum products suddenly plummeted to record low levels (a 29-day supply of gasoline and a 69-day supply of distillate fuels).

A basic tenet of good business is to raise prices when the customer is in short supply and is left with no choice but to pay. This is precisely what OPEC did, with the result that in the latter part of the first half of 1979, prices on crude (including surcharges) rose to as high as about $21 per bbl‡, while the spot market could demand close to $40 per bbl.

In mid-1979, the Department of Energy (DOE) responded by allowing a $5 per bbl subsidy on distillate fuels, bringing on international competition for higher-priced oil on the spot market. This did not result in increased supplies—only in higher prices and angry friends in Western Europe, who found themselves forced to meet the higher prices. To aggravate the situation still further, DOE had, for years, been keeping refinery profits on unleaded gasoline so low that the oil industry was discouraged from adding new refinery capacity.§ Furthermore, DOE allowed more profit on other petroleum products, so refineries inclined to reduce the output of gasoline. The government then vacillated in issuing orders to industry to produce more or less gasoline or distillate fuels, at the expense of the other product line.

Basically, these factors have led to wildly escalating prices and shortages of crude oil and petroleum products, additional inflation, the poorest balance of payments in our history,¶ and a worsening of relations with friendly nations in

---

*In a dramatic reversal in early June 1979, imports of crude oil increased to 6.6 million bpd, up from 6.048 million bpd in 1978, but still below the 6.694 million bpd for the 1977 average.[96]

†Demand for petroleum products declined in the first 4 months of 1979, to 19.366 million bpd from 19.470 million bpd for the corresponding period in 1978.

‡It is impossible in a publication of this sort to provide "current" fuel prices, particularly in the case of crude oil. At the end of 1979, the average OPEC oil price was $25.50 per bbl. By mid-May 1980, the average was nearly $31. In the June 9-10, 1980, meeting of the OPEC oil ministers, the price was raised again, to a "theoretical" base price of $32 per bbl and a ceiling of $37. (The $5 per bbl differential accommodated the varying qualities of the crudes.) This increase amounted to about a $2 per bbl advance for approximately half the then current 28 million bbl per day output of OPEC. This translated to about a 3 percent crude oil price increase for the non-Communist world as a whole, and to an increase of 2 to 3¢ per gallon in the U.S. prices of heating oil and gasoline. As in the past, OPEC "accord" was not unanimous, with Saudi Arabia refusing and the United Arab Emirates postponing the new increases, and Iran and Algeria refusing to reduce their respective prices from $35 and $38.21 per bbl.[97]

§Domestic refinery capacity in the first half of 1979 was 17.3 million bpd and the operating rate was 84.5 percent (June 4-11, 1979), as opposed to a potential level of 90+ percent.[96]

¶The major deficit is expected to stem from the $60 billion petroleum import bill in 1979, up from an earlier Treasury Department estimate of $52 billion,[96] well in excess of the 1978 estimate of $42 billion.

the Western world. Unfortunately, the only likely way to reduce demand now is through the recession now being experienced or through conservation.

In crude oil, gas (the methane-to-butane fraction) is the lowest-molecular-weight distillate (boiling range -259 to $+31°F$). Gasoline is the next major fraction (5 to 12 carbon atoms, boiling range 31 to 400°F), followed by jet fuel (10 to 16 carbon atoms, 356 to 525°F), gas oil (15 to 22 carbon atoms, 500 to 700°F), and lubricating oil (19 to 35 carbon atoms, 640 to 875°F). The residuum has 36 to 90 carbon atoms.[98] Over recent years, misguided regulation of our refinery operations has led to reduced availability of even residual fuel oils for power generation.[99]

In the refining of a typical 91 RON, unleaded gasoline, a liquid volume (LV) charge of 100 units of crude plus 6.4 fuel-oil-equivalent liquid volumes of C_1 to C_4 generally will yield about 49 units of gasoline, plus 30 units of distillates, 13 units of No. 4 low-sulfur fuel oil, 7 units of jet fuel, and 6.9 units of by-products. Some of these will be used as process fuel and the balance will be sold. Only about 0.5 liquid volume unit is lost in the process. Figure 2.3 shows the process flow of a typical refinery.[100]

In the second quarter of 1979, assuming a cost for Saudi light crude of 35¢ per gal ($14.70 per bbl), the retail prices of gasoline at the pump and heating oil to the home were 77.7 and 62¢ per gal, respectively.* Refining provided the largest element of cost after the raw material: 20.4¢ (including wholesaling and marketing) for gasoline and 14¢ for heating oil. Taxes of 13.2¢ per gal applied

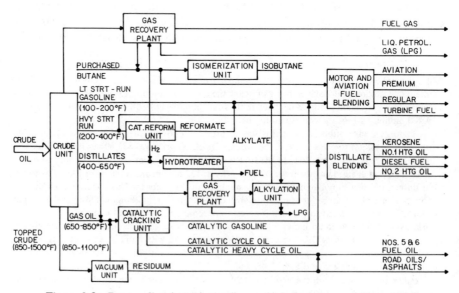

Figure 2.3   Process flow in typical refinery. (Adapted from reference 100.)

*National averages, April 1979.

to the gasoline only, and retailing costs of 9.7¢ were added. Wholesaling and retailing of the heating oil amounted to 14¢ per gal, but no tax was involved.[54] Crude oil cost therefore represented 45.3 and 56.5 percent of the costs of gasoline and heating oil, respectively. The detailed figures have doubtlessly changed considerably, with subsequent crude oil price increases and federal regulation changes, but the current relative figures are probably similar.

Despite our heavy consumption of refined petroleum products, the United States is in no way the dominant force in the field of refining, as can be seen in Table 2.14. Our domestic capacity (1973) for various petroleum processes was widely varied. At the beginning of 1973, we had a charge capacity of 13.46 million bpcd for distilling crude oil, 4.50 million bpcd for catalytic cracking, and 3.23 million bpcd for catalytic reforming. Our capacities for coking, alkylation, hydrocracking, and thermal cracking ranged down from about 1 to less than ½ million bpcd.[102]

The growth rate in domestic refining capacity has been negligible in recent years compared with eastern Canada, Italy, the Caribbean, and the Mideast, despite our growing demand for refinery products. Reasons given are all economic and political—the high cost of capital, uncertainties in product pricing, and marginal refining profits. The two- to three-year lead time required from contract to startup to increase refining capacity affords little hope for offsetting our need to import much of our desulfurized No. 6 fuel oil from the Bahamas, the Caribbean (Netherlands Antilles, Trinidad, Venezuela), and Italy.[99] Environmental constraints now require desulfurized residual fuel oil in major use areas, but most domestic refineries are not designed to process sour crudes. Nearly all the residual fuel oil used on the East coast comes from foreign sources.[99] Alaskan crude will be of little help for this market, because it is a sour crude and, furthermore, transport difficulties make it unlikely that it will find its way to the East coast.

Table 2.14 Worldwide Refining Capacity, January 1, 1973

Region	Crude Oil Charge, 10^6 bpcd
Western Europe	17.42 (29.9%)
United States	13.46 (23.1%)
Western Hemisphere (less U.S.)	8.07 (13.9%)
Asia and Far East	7.88 (13.5%)
Soviet Union and Eastern Europe	7.79 (13.4%)
Middle East	2.70 (4.6%)
Africa	0.86 (1.5%)
World total	58.18 (100%)

Source: Reference 101.

SUMMARY

Overall, there is little or no cause for optimism relative to our crude oil and petroleum products situation. There is nearly universal agreement that supply will become critical by the end of the century. The United States and the rest of the free world have become pawns of OPEC—with no say in pricing or supply policies. Our energy gluttony has continued to grow, virtually unabated, following a temporary reversal after the Arab embargo. A second reversal in the first half of 1980 may or may not be the start of a new trend toward significant, lasting energy conservation. Pressure by environmentalists is cutting into our exploration for new oil sources; at the same time, it is squeezing us from the other end by hindering the availability of alternate energy—coal and nuclear.

There have been many charges leveled at irresponsible "big oil" for creating a gasoline shortage by withholding oil from the market.[103] Such charges stem from official sources as well as beliefs of the man in the street. The DOE, in the words of deputy secretary O'Leary, testifying before a House subcommitte, has claimed: "If there is a massive rip-off, it is at the hands of those who spend 25 cents to produce a barrel of oil and sell it at $16 to $18 and occasionally up to $30 a barrel"—in other words, OPEC.[104] People are inclined to see the billions of dollars in profits of "big oil" and not take into consideration what sort of return is appropriate for an industry whose total investment in our economy is approaching half a *trillion* dollars.

One must admit that the Arab world has some justifications on its side. Because of our greed for their oil, we are expecting them to part with their only marketable heritage at a rate that will leave them with nothing but memories within possibly 50 years. Their response is to set the price high enough to discourage waste and delay the inevitability of the loss of their one resource, while ensuring as large a return as possible to bank against the future. But when their high prices bring about a reduction in our demand, they raise their prices again to ensure a continuing high return for themselves.

So just who or what is to blame for the fix in which we find ourselves? Basically, the blame must rest squarely on whoever, at anytime in the past 40 years or so, has done anything to encourage our wasteful use of energy. This includes, in part: the oil companies which made low-cost oil so freely available to us, the special-interest politicians, who insisted on trying to revoke the economic law of supply and demand by controlling oil prices, the automakers who convinced us we needed to encase ourselves in two tons of steel to go to the corner market, the people who insisted on the big car, the electric toothbrush, electric can opener, air conditioner, and so forth. In short, we have no one to blame but ourselves and our greed for enjoying such opulent luxuries.

Of course, there have been other errors as well. We have alienated the major supplier of our foreign oil, as well as our main competitors for the limited supplies. We coddled the Shah of Iran, convinced the Saudis against their better judgment that they should do likewise, and when they did so, we embarrassed them by withdrawing our support. We then ignored their pleas to send a naval task force into the Persian Gulf area as a display to the Soviets and Iran of our determination to look after our oil interests. Finally, we ignored the warnings of the Saudis that our support of the Camp David accord would alienate their Arab neighbors and force the Saudis to moderate their pro-U.S. stance.[105] Regardless of one's personal views on any one of these specific issues, it must be acknowledged that each one undoubtedly had an adverse effect on the outcome of the March 26, 1979, meeting of OPEC in Geneva, in which Saudi Arabian oil minister Sheik Ahmad Zaki Yamani apparently supported the OPEC oil surcharges.[106]

In the latter part of May 1979, the Administration announced a $5 per bbl subsidy on distillate fuels (diesel fuels, heating oil, etc.). This move, aimed at making distillates more available from Caribbean refineries, placed us in a position of bidding up limited supplies in competition with OECD nations and virtually ensuring that OPEC, in the next session, would increase the price of their oil so that this added bonanza would end up in their own bank accounts.[5]

We, as a nation, seem to have developed a remarkable track record for compromising ourselves in some very sensitive areas. These are highly controversial topics which possibly have no place in such a book as this. They are noted here in passing, simply to suggest the many-faceted aspects of our energy dilemma.

REFERENCES

1 J. M. Blair, *The Control of Oil*, Pantheon, New York, 1977; book review by R. E. Müller, *N.Y. Times* 126(43,485), Sect. 7: 1, 22, 23, (Feb. 13, 1977).

2 C. T. Rand, *Environment* 18(3), 12–17, 25, 26 (Apr. 1976).

3 I. A. Levi, *Phila. Inquirer* 300(119), 11-E (Apr. 29, 1979).

4 *Bus. Week* (No. 2436), 72J (June 14, 1976).

5 *Time* 113(25), 52, 53 (June 18, 1979).

6 R. N. Farmer, *Bus. Horiz.* 20(1), 74–80 (Feb. 1977).

7 L. T. Lounsbury and J. A. Morris, Jr., *Courier-News* (Bridgewater, NJ) (Feb. 21, 1976), p. C-4.

8 *Miami-Herald* 70(121), 28-A (Mar. 30, 1980).

9 Alleged by R. Helms (CIA) in reference 1.

10 F. C. Allvine, *Bus. Horiz.* 19(4), 41–51 (Aug. 1976).

11 *Fortune* 99(9), 270–289 (May 7, 1979).

12 *Investigation of the Petroleum Industry*, U.S. Senate Comm. Gov. Oper., U.S. GPO, Washington, DC, July 12, 1973.

13 H. Kelly, *Chemtech* 7(1), 32–35 (Jan. 1977).

14 M. Beiser, *Energy User News* 3(15) (Apr. 10, 1978).

15 *Time* 109(8), 58–60 (Feb. 21, 1977).

16 *Bus. Week* (No. 2445), 93, 96, 98 (Aug. 16, 1976).

17 A. Greenspan; reported in reference 15.

18 J. H. Marshall, *Chemtech* 6(9), 550–553 (Sept. 1976).

19 A. A. Meyerhoff, *Amer. Sci.* 64(5), 536–541 (Sept.-Oct. 1976).

20 D. A. Holmgren el al. (1974); updated and reported in reference 19.

21   API (1976); reported in reference 19.

22   DOE; reported in *Bus. Week* (No. 2577), 28, 29 (Mar. 19, 1979).

23   API; reported in *Public Util. Fortn.* **98**(1), 28 (Nov. 18, 1976).

24   C. T. Rowan, *Reader's Dig.* **109**(653), 152–156 (Sept. 1976).

25   UPI, in *Star-Ledger* (Newark, NJ) **64**(9), 30 (Mar. 2, 1977).

26   WASH-1400 Rep., Natl. Tech. Info. Serv., Springfield, VA, Nov. 1975; reported in S. S. Penner and J. P. Howe, "Nuclear Energy and Trade Deficits," in S. S. Penner, Ed., *Energy*, Vol. 3, *Nuclear Energy Policies*, Addison-Wesley, Reading, MA, 1976, p. 558.

27   W. J. Eaton, *Phila. Inquirer* **296**(135), 5-C (May 15, 1977).

28   *Bulletin* (Phila.) **133**(43), 1, 7 (May 24, 1979).

29   M. W. Karmin and D. Hess, *Phila. Inquirer* **300**(161), 1-A, 12-A (June 10, 1979).

30   A. B. Looius, "Energy Resources," presented at U.N. Symp. on Popul., Resour., Environ., Stockholm, 1973.

31   *Time* **109**(14), 58–67 (Apr. 4, 1977).

32   Sherman H. Clark, Assoc.; reported in *Courier-News* (Bridgewater, NJ) **93**(279), C-3 (Apr. 25, 1977).

33   A. Siegert, *Phila. Inquirer* **300**(161), 13-A (June 10, 1979).

34   *Bus. Week* (No. 2577), 28, 29 (Mar. 19, 1979).

35   *Time* **114**(21), 85 (Nov. 19, 1979).

36   M. Seeger, *Miami Herald* **69**(122), 7-B (Apr. 1, 1979).

37   E. V. Anderson, *Chem. Eng. News* **55**(4), 12, 13 (Jan. 24, 1977).

38   J. J. Simpson; reported in *Technol. Rev.* **79**(3), 25 (Jan. 1977).

39   C. Wall and R. Dawe, *New Sci.* **73**(1039), 407, 408 (Feb. 17, 1977).

40   H. F. Keplinger, *Public Util. Fortn.* **99**(3), 18–23 (Feb. 3, 1977).

41   *Bus. Week* (No. 2356), 32D (Nov. 1, 1976).

42   *Ibid.* (No. 2436), 31, 32 (June 14, 1976).

43   C. Bowlin; reported in *Public Util. Fortn.* **98**(12), 65 (Dec. 2, 1976).

44   AGA, API, and Can. Petrol. Assoc. (1973); reported in *Energy Facts II*, Sci. Policy Res. Div., Libr. Congr., Serial H, U.S. GPO, Washington, DC, 1975, p. 297.

45   API and AGA; reported in *Chem. Eng. News* **54**(16), 14 (Apr. 12, 1976).

46   Counc. Econ. Advisers (1975); reported by D. E. Sander, "The Price of Energy," in J. M. Hollander and M. K. Simmons, Eds., *Annual Review of Energy*, Vol. 1, Annual Reviews, Palo Alto, CA, 1976, p. 392.

47   W. T. Slick, Jr., *Energy User News* **3**(15), 22, 23 (Apr. 10, 1978).

48   *Bus. Cond.* (AT&T) (Dec. 1976), pp. 2, 3.

49   *Time* **113**(24), 70, 71 (June 11, 1978).

50   FEA; reported by UPI, in *Star-Ledger* (Newark, NJ) **64**(10), 16 (Mar. 3, 1977).

51   IGT; reported by UPI, in *ibid.* **63**(365), I-10 (Feb. 20, 1977).

52   BuMines; reported by N. J. W. Thrower, Ed., *Man's Domain: A Thematic Atlas of the World*, McGraw-Hill, New York, 1975, p. 42.

53   J. F. Bookout, "Energy," in *Texas Business: Texas in the 1980's*, 1979, pp. 17, 18.

54   *Time* **113**(19), 70–79 (May 7, 1979).

55   J. C. Harsch, *Phila. Inquirer* **300**(162), 11-A (June 11, 1979).

56   *Phila. Inquirer* **300**(126), 8-M (May 6, 1979).

57   N. Kempster, *Miami Herald* **69**(122), 6-A (Apr. 1, 1979).

58   *Time* **113**(1), 57, 58 (Jan. 1, 1979).

59 *Ibid.* **114**(22), 84 (Nov. 26, 1979).

60 C. A. Nickerson, *Bulletin* (Phila.) **133**(19), 15 (Apr. 30, 1979).

61 P. Caldwell, *Time* **113**(25), 53 (June 18, 1979).

62 S. Scheibla, *Barron's* **57**(9), 9, 10, 12, 23 (Feb. 28, 1977).

63 E. Crawford, *Energy User News* **3**(19), 14 (May 8,1978).

64 D. L. Barlett and J. B. Steele, *Phila. Inquirer* **301**(36), 1-A, 22-A (Aug. 5, 1979).

65 E. A. Lohse, *Inventory of U.S. Companies' Offshore Petroleum and Related Activity,* Rep. No. GURC-146, AD-A029930, DOT, Washington, DC, Mar. 1976.

66 *EOS, Trans. Amer. Geophys. Union* **58**(3), 128 (Mar. 1977).

67 E. Edelson, *Popul. Sci.* **202**(4), 82, 83, 163, 164 (Apr. 1973).

68 *Dun's Rev.* **108**(2), 65, 67 (Aug. 1976).

69 Chevron U.S.A. Inc. data.

70 Natl. Petrol. Counc. (1974); reported by F. F. H. Wang and V. E. McKelvey, "Marine Mineral Resources," in G. J. S. Govett and M. H. Govett, Eds., *World Mineral Supplies: Assessment and Perspective,* Elsevier, New York, 1976, p. 237.

71 USGS Circ. 725 (1975); reported in reference 65.

72 K. O. Emery, *Technol. Rev.* **78**(4), 31–38 (Feb. 1976).

73 API (1975); reported in reference 65.

74 AAPG (1975); reported in reference 65.

75 *Phila. Inquirer* **300**(158), 3-B (June 7, 1979).

76 Arthur D. Little, Inc., reported in *Chem. Eng. News* **54**(52), 7, 8 (Dec. 20, 1976).

77 *Phila. Inquirer* **299**(30), 8-A (July 30, 1978).

78 W. B. Travers and P. R. Luney, *Science* **194**(4267), 791–796 (Nov. 19, 1976).

79 USGS; reported in *Bus. Week* (No. 2446), 24, 25 (Aug. 23, 1976).

80 R. J. Stewart, *Technol. Rev.* **78**(4), 46–59 (Feb. 1976).

81 *Technol. Rev.* **79**(4), 19 (Feb. 1977).

82 P. H. Abelson, *Science* **195**(4274), 137 (Jan. 14, 1977).

83 D. Straughan; reported in *Chem. Eng. News* **55**(11), 20 (Mar. 14, 1977).

84 C. T. Krebs and K. A. Burns, *Science* **197**(4302), 484–487 (July 29, 1977).

85 L. G. Williams, *ibid.* **195**(4279), 636 (Feb. 18, 1977).

86 R. A. Cochran, "Oil Spill," in D. N. Lapedes, Ed., *McGraw-Hill Encyclopedia of Energy,* McGraw-Hill, New York, 1976, pp. 718–722.

87 V. H. Vandermeulen et al., *Science* **202**(4363), 7 (Oct. 6, 1978).

88 *Mech. Eng.* **99**(3), 69 (Mar. 1977).

89 N. Reuth, in *ibid.* **97**(11), 21–41 (Nov. 1975).

90 B. Hodson, *Natl. Geogr.* **150**(5), 684–717 (Nov. 1976).

91 Reference 66, p. 127.

92 B. Dietrich, *Courier-News* (Bridgewater, NJ) **93**(127), B-1 (Oct. 29, 1976).

93 C. Bukro, *Phila. Inquirer* **301**(169), 2-D (Dec. 16, 1979).

94 *Ibid.* **93**(126), C-1 (Oct. 18, 1976).

94a L. Eichel, *Phila. Inquirer* **303**(55), 1-A, 14-A (Aug. 24, 1980); *ibid* **303**(56), 1-A, 10-A (Aug. 25, 1980).

95 D. L. Barlett and J. B. Steele, *Phila. Inquirer* **300**(168), 1-A, 12-A, 13-A, (June 17, 1979).

96 *Phila. Inquirer* **300**(168), 6-E (June 17, 1979).

97 *Ibid.* **302**(163), 1-A (June 11, 1980).

98 J. J. McKetta and H. L. Hoffman, "Petroleum Processing," pp. 552–558, in reference 86.

99 S. S. Miller, *Environ. Sci. Technol.* **7**(6), 494–496 (June 1973).

100 E. K. Grigsby, E. W. Mills, and D. C. Collins (NPC, 1973); reported in K. P. Anderson and J. C. DeHaven, *Long-Run Marginal Costs of Energy,* Natl. Sci. Found., PB-252504, Feb. 1975, p. 80.

101 *UN Statistical Yearbook, 1973* (1974), pp. 273–275.

102 BuMines (1973); reported in reference 97.

103 A. Dougherty (FTC); reported in *Phila. Inquirer* **300**(164), 11-A (June 13, 1978).

104 J. E. O'Leary; reported in *ibid*.

105 J. Anderson, *Bulletin* (Phila.) (May 7, 1979), p. 44.

106 *Miami Herald* **69**(115), 28-A (Mar. 25, 1979).

3 NATURAL GAS

Natural gas is our most efficient fossil fuel, used directly by the consumer without processing or transformation, and it delivers more primary energy than any other fuel.

G. H. LAWRENCE

Natural gas is estimated to have amounted to less than 3 percent of the total initial fossil energy resources of the world. Still, it has assumed a dominant role in our domestic energy picture, and is increasing in importance throughout the rest of the world, despite the fact that its future seems limited.[1]

Natural gas was formed under essentially the same geologic conditions that produced petroleum—anaerobic decomposition of organic matter under heat and pressure, assisted by bacteria.[1] The bulk of all natural gas (about 65 percent) was formed in Mesozoic strata. About one-fourth is found in Paleozoic era reservoirs, and the balance (about 10 percent) was formed in the Tertiary epoch (or early Cenozoic era).[2] Nearly 90 percent of the world's proved and prospective reserves are in giant fields, the majority of which were discovered before 1940.[3]

Natural gas comprises basically 75 to 99 percent methane, CH_4, the lowest-molecular-weight hydrocarbon. The methane content of producing fields generally falls in the higher end of the range, although there is said to be no such thing as a "typical" natural gas. A natural gas containing 85 percent methane might include ethane (up to 10 percent), propane (up to 3 percent), and any combination of other hydrocarbons, hydrogen sulfide, carbon monoxide or dioxide, hydrogen, nitrogen, helium, and water to make up the balance.[1] In an isolated case, helium has run as high as 7.5 percent in a particular U.S. field, but this is far higher than normal.[2] Domestic natural gas generally contains no olefinic hydrocarbons, carbon monoxide, or hydrogen.[4] Methane and ethane are the primary combustibles of natural gas, and carbon dioxide and nitrogen are the major inert components.

Natural gas is of basically two types.

Associated or *casinghead gas* is dissolved in crude oil in the reservoir under pressure or forms a pressurized gas cap over the oil pool, and ultimately is separated on extraction of oil from the well.

Nonassociated or *dry gas* is found in wells containing no crude oil.[1] The associated gas provides pressure for solution gas drive or gas cap expansion in the primary recovery of crude oil. (See Chapter 2 for more on crude oil recovery.)

The ratio of associated to nonassociated gas varies from field to field. In California, 60 percent of the gas is associated with oil. In western Texas, the figure is 40 percent. The domestic average is only about 23 percent. About half the known U.S. associated-gas reserves lie along the Gulf coast of Texas and Louisiana.

Natural gas may also be classified as: *dry and lean* (largely methane), *wet* (high in other hydrocarbons),* *sour* (high in hydrogen sulfide), *sweet* (low sulfide), or *residue* (following removal of higher hydrocarbons).[2]

The popularity of natural gas stems from the facts that: (1) on an energy-equivalent basis, it has been price-controlled below its chief competitor, oil;† (2) it is the cleanest-burning hydrocarbon fuel (the major air pollutants being carbon dioxide and oxides of nitrogen); (3) it is the most convenient fuel for stationary use; (4) it is easy to transport on demand (over contiguous land areas);‡ (5) it has a high flame temperature; (6) it has a high heat content (about 1000 to 1100 Btu per scf for pipeline quality gas);§ and (7) it is an efficient fuel, since it is used in its natural state, without the need for being generated from other fuels.[4]

THE GREAT DRAWDOWN

Initially, natural gas was looked upon as a useless by-product of crude oil production. It was extracted generally in relatively remote regions, far removed from population centers and industries, and there were no means for transporting it to potential marketplaces, so it was simply vented to the atmosphere or flared off by burning. Today, the Mideast oil fields are experiencing the same situation (as shown in Table 3.1). Its inherent value was recognized by some, but its exploitation was not possible until advances in metallurgy resulted in development of the technology for welding thin-walled steel pipe that could contain natural gas under pressure.[7] The first pipeline was laid in 1930.[1] After this hurdle was passed, production began to grow rapidly in a race to keep up with demand. Production finally peaked in 1973. Prior to that time, production growth was at a rate of nearly 5 percent per year (Table 3.2) a doubling time of approximately 15 years.

Most of our domestic natural gas reserves were discovered prior to World War II. As the war drew to a close, natural gas furnished only 12 percent of our energy requirements. Estimates of domestic natural gas resources by various

*Dry natural gas has less than 0.1 gal natural gasoline condensate per 1000 ft³; wet natural gas has more than 0.1 gal.[4]
†Based on a 1954 decision of the Warren Supreme Court.
‡More will be said later about intercontinental transport.
§One thousand cubic feet will heat the average house for about three days.[5]

Table 3.1 Natural Gas Flared in Persian Gulf, 1973

Region	Billion m³
Iran	40
Saudi Arabia	25
Kuwait	10
Neutral zones	1
Other	1
Total	77[a]

Source: Reference 6.
[a]Potential feedstock for 80 million tons of methanol.

Table 3.2 U.S. Production of Natural Gas, 1945–1976[a]

	Billion ft³
1945	4,042
1950	6,282
1955	9,405
1960	12,771
1965	16,040
1970	21,921
1971	22,493
1972	22,532
1973	22,648
1974	21,601
1975	20,109
1976	19,952

Source: Reference 8.
[a]Wet natural gas; 1094 Btu per scf.

agencies vary widely depending on assumptions made regarding price, recovery factor, drilling depth, water depth, and other factors. Best estimates of U.S. natural gas resources, as of December 31, 1977, are given in Table 3.3. Globally, ultimately recoverable reserves of natural gas have been estimated by the National Academy of Sciences (NAS) at about 7700 tcf. Cumulative production through 1975 was about 843 tcf. The future discovery potential is expected to be 4580 tcf.[10] There are no worldwide figures on the amount of natural gas used for repressuring or wasted by venting or flaring.

The 1954 Supreme Court decision authorizing the Federal Power Commission (FPC) to control interstate gas prices at the lowest intrastate price made gas

Table 3.3 U.S. Dry Natural Gas Resources,
December 31, 1977

	Proved and Currently Recoverable	Estimated Remaining Recoverable
Tcf	209[a]	760[b]–1170[c]
Quads	210	780–1190

Source: Reference 9. Reprinted by permission of
the Institute of Gas Technology.
[a]AGA (Apr. 10, 1978).
[b]Rounded estimate of NAS, USGS, Exxon, Mobil,
and others (1974–1976).
[c]Potential Gas Comm., Colorado School of Mines
Found. (1976, 1977).

so attractive that it even undercut No. 6 heavy residual fuel oil, and forced the
oil companies to market the residual oil at a loss. This pricing policy led to excessive consumption and waste and brought on the undercutting of more abundant but less desirable fuels. In 1975, domestic natural gas cost about 38¢ per
million Btu, compared with about 63¢ for coal and 90.5¢ for crude oil, for the
same heating value.[11] In the fourth quarter of 1976, low-sulfur fuel oil cost
about 1.7 times as much as interstate natural gas, and diesel fuel and propane
cost 2.5 and 4 times as much, respectively, on an equal-energy basis.[12]

It is estimated that natural gas at $2 per 1000 scf is equivalent (on an energy
basis) to fuel oil at $12 per bbl (28.6¢ per gal). Since the delivered price (to the
United States) of OPEC crude was about $13.25 prior to the 1977 price rise, it is
obvious that natural gas was well underpriced.[13] At 1979 year-end prices (about
$2.60 per 1000 scf for interstate natural gas and about $25.50 per bbl for crude
oil), the difference was further magnified.

The first quarter 1977 prices of Algerian liquefied natural gas (LNG) were
$1.25 per 1000 scf for preembargo contracts and $3.30 for postembargo contracts. The latter price compared with the then current price of $1.44 for interstate natural gas. Prices of Algerian natural gas are expected to continue to
escalate along with OPEC oil prices.[14] Prices of our "future" gas were expected
to be nearly 1.5 times as high as interstate natural gas obtained by enhanced
recovery, up to nearly twice as much for Alaskan gas, and two to three times as
much for gas from coal, shale, biomass, and so forth.[5]

In 1978, domestic consumption of natural gas had dropped to about onefourth of our national energy budget compared with a high of one-third at about
the time of the production peak in 1973. Table 3.4 shows domestic consumption
in 1972 by sector of the economy. Manufacturing used nearly 42 percent of our
gas budget and obtained nearly 60 percent of its fuel needs from gas. The pattern was quite different in the Northeast, where imported oil historically has

Table 3.4 U.S. Natural Gas Consumption by Sector, 1972

Sector	Percent
Manufacturing	41.9
(Natural gas as % of total fuel used in manufacturing)	(59)
Residential	25.4
Commercial and other	12.5
Electric utilities	20.2
Total	100.0

Source: Reference 15. Material furnished by the Economic Analysis Section of the American Telephone and Telegraph Company.

played the major role in manufacturing. In the West-South Central region, electric utilities and industry have been large consumers and would normally be targets for sharp curtailments except for the fact that most of their supplies are intrastate.[15]

In the early 1970s, residential use of gas took priority over industrial applications, with the result that industry had to begin scrambling for supplies. At this point, substitute natural gas (SNG) plants for cracking naphtha, propane, and crude oil to gas began coming onto the scene.

The commercial history of natural gas through the last 20 or so years has a rather unbusinesslike aura about it. Large amounts of fluid hydrocarbons were extracted in one region. The liquid crude oil was saved, but the associated gas was flared* or vented to the atmosphere. The liquid was then transported to another region, where some of it was converted to gas with a resulting large loss of energy in the conversion process. The gas found a ready market because of its numerous attractive characteristics. But if it was so valuable, why should it be flared off at its natural source, be generated with an energy loss from oil extracted from the same wells, and be sold at prices regulated below the crudest form of oil-derived fuel when delivered to its final destination?

By 1963, natural gas had outstripped petroleum to become the principal source of our domestic energy. Production increased until 1973, when it peaked out and began to decline (Table 3.2). Since 1967, only a third to a half as much new gas has been added to our reserves as the amount we have been consuming. The year 1968 has been the only one in the past quarter century that has yielded enough new discoveries† to cover our record production of gas in 1973 of 22.6 tcf.[16]

In 1946, the domestic reserves-to-production ratio was 32.‡ By 1963, it had

---

* In Saudi Arabia alone, about 5 billion scf per day (or 5 trillion Btu, worth $1 million per day), 90 percent in associated gas, equal to about one-fourth of the U.S. gas production rate in 1970.[1]

†Although discovered in 1968, additions to reserves from Prudhoe Bay are generally recorded for 1970, probably reflecting the time required to assess them.

‡Reported ratios vary with estimates of reserves employed.

fallen to 20,[17] and by 1973, it was 10:1. After rising to 10.6:1 in 1974, it has been on a steady decline: 9:1 in 1975 and 8.5:1 (estimated) in 1976.[12]

In the United States, 90 percent of our natural gas production comes from six states: Texas, Louisiana, New Mexico, Kansas, Oklahoma, and California,[1] with over 70 percent coming from Texas and Louisiana alone.[2] Alaska is expected to become another major exporting region, with the discovery of recoverable associated gas reserves of 26 tcf plus 5 tcf of nonassociated gas in the Prudhoe Bay field alone.[18]

Between 1966 and 1972, the average number of successful exploratory wells in southern Louisiana and the major producing area of Texas, our two most productive states, fell 42 and 54 percent, respectively. Even sharply increased drilling activity in 1973–1974 failed to help, with 1974 production being 5.7 percent below 1973. Production in 1975 continued the decline (7.5 percent off the 1974 figures). The ratio of significant finds to exploratory wells has declined about 50 percent from the prewar period, and the size of the average find has also decreased.[16]

It should be mentioned that recoverable gas reserves have a higher degree of uncertainty than petroleum reserves, in part because small price fluctuations have a major effect on supply economics, hence affect recoverability. It is perhaps worthy of note that per capita U.S. reserves (life index of 10.4 years[17]) are now at a very low level, particularly in light of our high consumption rate relative to the rest of the world. It should be noted that one of the reasons for large differences in reserves reported by the American Gas Association (AGA) and federal agencies is that the former requires that data be based on test production or conclusive geological formation tests, whereas the U.S. Geological Survey (USGS) does not.[17]

The U.S.S.R. has the world's largest proven recoverable natural gas reserves, with about 918 tcf. Iran is second with 330 tcf, followed by the United States with 220 tcf and Algeria with 126 tcf. A significant factor in these statistics is the fact that the United States has by far the lowest reserve-to-population ratio, yet we have the highest per capita consumption.[19]

Annual worldwide gas production figures for 1973 are shown in Table 3.5. Gross production means all gas extracted at the wellhead. Marketed production is that gas sold to pipeline companies, petrochemical and LNG producers, and so forth. Some is "lost" in the processing (and sale) of liquefied petroleum gas (LPG) or natural gas liquids (NGL).* Some is fed back to oil wells for repressuring in secondary oil recovery. Small amounts must be vented or flared because economical recovery and production are not always possible in isolated situations. Typically, up to 5 percent of the gross national production will be vented or flared, and 4 to 5 percent will be used for well repressuring, leaving 90 to 95

---

*NGL data are normally neglected, although proved recoverable reserves of NGL amounted to 6.3 billion bbl and remaining recoverable reserves were estimated at about 248 billion bbl as of December 31, 1975. These estimates were based on an average recovery rate of 30 bbl of NGL per million scf (168.4 m^3 per million m^3) of natural gas.[21]

Table 3.5 World Natural Gas Production, 1973

Region	tcf	
	Gross	Marketed
United States	24.1	22.6 (93.7%)
Sino-Soviet bloc	10.8	9.6 (88.9%)
Other western hemisphere	7.1	4.8 (67.6%)
Western Europe	5.2	5.1 (98.0%)
Mideast	5.0	1.2 (24.0%)
Africa	1.7	0.3 (17.6%)
Far East and Oceania	1.1	0.8 (72.7%)
Total	55.0	44.4 (82.7%)

Source: Reference 20.

percent to be marketed.[17,21] With our extensive gathering, transmission, and distribution pipeline networks, relatively little production is lost.

A compilation by various sources* of domestic production projections to 1985 shows a probable reduction to about 16 tcf output in that year.[22] There is little reason to expect any improvement in our gas supply, based on conventional sources. The major reason for this is that new additions to reserves have not been keeping pace with new production since 1967, except for the Prudhoe Bay discovery, assessed in 1970. Even this find is modest when one considers that it could supply only about 6 percent of the nation's projected needs after 1980, and would be depleted in 20 years—assuming a suitable pipeline were completed and functioning at design levels.

Primary gas recovery typically ranges from about 60 to 90 percent of the resource in place.[23] Most of this is recovered without lifting, because nonassociated gas flows naturally throughout the producing life of the well, and this gas represents about two-thirds of domestic production.[24] Abandonment pressures in depleted fields fluctuate widely. Secondary recovery, involving acidizing, hydraulic fracturing, or the use of explosives, is expected to account ultimately for 300 to 600 tcf additional yield from existing gas wells.[25]

Some private entrepreneurs have undertaken a different sort of gas recovery ploy. Over the past 10 or 20 years, many wells were shut down as production costs began to exceed the then low price of gas. This new breed of producer, emboldened by price rises up to $2.60 per 1000 scf for intrastate gas, has undertaken to lease abandoned wells, drill out the concrete plugs, install new compressors, and pump the gas to market, at a fraction of the cost required to drill new wells.

*Including FPC, Institute of Gas Technology and Congressional Budget Office.

One such operator soon owned the logs and records of some 35,000 wells in an eight-county area of south Texas. In mid-1976, he was producing 3 million scf of gas per day. As an example of the economics of such an operation, one well was abandoned in 1965 while producing 200,000 scf per day. The selling price (at that time) of 15¢ per 1000 scf allowed the well to net only $30 per day, which was not sufficient to pay for the operating costs of the compressor. The well was refurbished for $21,000 and, with intrastate gas selling for $2, it yielded $400 per day. Recoverable reserves of this well, at the higher 1976 price, would possibly have been worth $700,000. Drilling the same well from scratch would have cost about $150,000 on the 1976 market.[26]

Domestic natural gas resources have been drawn down to about 62 percent of the original recoverable resource in place vs. 91 percent for the world as a whole.* World natural gas resources are thought to be adequate for another 60 years, at current rates of consumption, but production is expected to peak in the mid-1980s unless a major new source is brought on stream soon.[22] Alaskan gas could ultimately provide 3 tcf per year to the lower 48, but the pipeline cannot be completed before the late-1980s.[27]

One wildcatter claims that only about 2 percent of our acreage rated "favorable" or "potential" for the discovery of gas or oil have been drilled, and only 1 percent of the wells drilled in 1976, for example, went below 15,000 ft. About 90 percent of the wells drilled in 1975 were located in mature, shallow fields containing only 30 percent of our undiscovered resources.[28] The deep levels are where high temperatures and pressures are favorable to the natural "cracking" of other fossil hydrocarbons to methane.[29] These lower levels, and the "favorable" areas, are the regions that must be explored for new, near-term supplies.

NEW AND UNCONVENTIONAL SOURCES

One of the more promising sources for natural gas is in very deep onshore reservoirs. One of these, discovered by Socal (Chevron) in August 1977, is located northwest of Baton Rouge at a depth of 21,346 ft. This formation was assessed at 140 million scf per day. It is part of the Tuscaloosa Trend, extending 200 miles along an east-west line just below the Mississippi border. To date, one-plus producing wells have been developed for every three wildcats drilled. In this total tract, Shell has 500,000 acres, Chevron and Gulf have 300,000 acres, and Standard of Indiana (Amoco) has 200,000 acres.[30] This tract is in the center of a vast, existing pipeline network. It is unique because it lies at such a great depth (16,000 to 20,000+ ft vs. less than 5000 ft for most U.S. oil and gas wells). There are two other deep zones: the Permian Basin and the Deep Andarko Basin.† The Colorado School of Mines estimates that the three zones are the sites of huge gas reservoirs that could contain 65 tcf, or one-third of our

---

*Compared with 54 percent and 81 percent, respectively, for U.S. and global crude oil resources.
†The Andarko Basin is in Oklahoma and Texas just southeast of the Panhandle.

proven domestic reserves, and could yield one-fourth of our energy prior to the mid to late-1980s when Alaskan gas and coal gasification could come on stream. These deep wells tend to be large producers (5 to 6 billion scf per year—equivalent to about 1 million bbl of crude oil on an energy basis). No new technology or transmission systems would be required for their development. These producing areas are claimed to have resulted from the Natural Gas Policy Act of 1978, which allowed higher prices on new gas and on gas produced at greater risk or greater expense. This gas was scheduled to be deregulated as of November 9, 1979. Higher-priced gas will be rolled into the price of "old" gas, up to 130 percent of the cost of No. 2 fuel oil, on an energy-equivalent basis. Deep Tuscaloosa wells cost $5 to $6 million to drill, and Andarko wells cost about $3.5 million vs. $250,000 for a normal, shallow well. The main reason for the high cost of these deep wells is that they take six months to drill. About 200 deep wells have been drilled in the Permian Basin and a score in the Andarko Basin.[30]

Natural gas reserves also exist in at least four different types of formations that have barely been probed at this date: coal seams, Devonian shale, tight sands, and geopressured zones.[31] Estimates of the potential resources in these formations are: coal beds, about 300 to 800 tcf; Devonian shale, about 500 to 600 tcf; tight sands, 600 tcf; and geopressured gas, 2940 to 49,977 tcf.[32]

All coal seams contain methane—from 33 to 641 scf per ton (1 to 20 m^3 per tonne) of coal, but probably averaging about 224 scf (7 m^3). Known U.S. coal deposits should therefore contain at least 272 tcf (8.5 trillion m^3) of methane.[31] One reference claims that the total is 800 tcf (22.7 trillion m^3), with 267 tcf (7.6 trillion m^3) being recoverable.[32] Domestic coal mines annually vent 72 billion scf (2.25 billion m^3)[31] to 91 billion scf (2.58 billion m^3)[33] to the atmosphere, a small amount in terms of our annual consumption of natural gas, but enough to make a significant contribution to the local energy budget. Properly managed, possibly 1 million scf per day could be recovered from 200 active bituminous coal mines.[34] This is the same gas that causes many mine explosions—one of the major hazards of deep mining—and results in about half the accidental deaths of coal miners.[35]

Methane can be released from coal seams by: (1) drilling a gas well intersecting the seam, (2) fracturing the coal seam to permit free flow of the gas, or (3) drilling a large shaft to the seam and then drilling lateral channels in the seam. The first option is inefficient because of the poor permeability of the coal seam to gas. The second option creates a potential hazard in subsequent mining, because it may weaken the mine shaft ceiling. The best alternative is to employ the third option several years in advance of opening the mine, and collect and market the gas to pay for the digging of the shaft. Extraction of the methane should also increase mining safety and save money by reducing ventilation demands* and reducing downtime.

---

*Mine safety practices call for keeping the mine atmosphere at a minimum 99 (air):(methane) ratio.[33]

Devonian or brown shale is an almost coal-like shale, formed 350 million years ago, that occurs in 13 eastern and midwestern states. The deposits are spread over 163,000 mi² of the Appalachian Basin, in formations ranging from a few feet to over 8000 ft in thickness.[36] The gas content ranges between 20 and 30 scf per ton (0.63 and 0.95 m³ per tonne).[31] The Office of Technology Assessment (OTA) estimates that 1 tcf per year of this gas could be produced over the next 20 years,[36] but about 69,000 widely scattered wells would be required because of the low permeability of the shale.[37] The USGS estimates are more optimistic; they claim that the shale contains a total of about 454 tcf (14 trillion m³). The flow rate of the gas is slower than conventional gas wells, but flow can be increased 50 percent by fracturing the shale formation.[31]

The so-called tight sands in the San Juan Basin straddling the New Mexico–Colorado border[28] are actually clay, chalk, or sandstone beds interspersed with shale and existing below conventional gas fields. Their low permeability calls for massive hydraulic fracture of the strata.* Estimates by USGS are for 600 tcf (17 trillion m³) in the Fort Union and Mesaverde reservoirs of the Rocky Mountains alone, and an equal amount in other locations.

Potentially the largest, but also the most speculative unconventional source of natural gas is the geopressured zone encompassing all of Louisiana and much of the adjacent Gulf coast states and northeastern Mexico. These deposits lie at a depth of 8000 to 26,000 ft.[38] These aquifers contain concentrations up to 45 scf (1.28 m³) of natural gas per barrel of hot [over 300°F (150°C)] salt water, at pressures up to 11,000 psi.[31] A USGS study estimates a total content of 24,000 tcf (or quads) of gas under Texas and Louisiana,[39] but other estimates range between 3000 tcf (85 trillion m³)[40] and 49,400 tcf (1400 trillion m³).[41] Offshore Gulf coast sites probably contain equal amounts.

Exploration and development of this resource will probably be slow and en·cumbered by a reluctance on the part of the energy industry to risk the technical, economic, and environmental gamble, even with the huge potential payoff. If ways could be developed to tap these aquifers, the gas, the heat, and the pressure could all be used to generate power or provide other forms of energy. Salt water disposal and land subsidence pose serious environmental questions. Because of subsidence, it is felt that it may be possible to withdraw only 5 percent of the water.[41] The question, then, is whether the pressure would be reduced enough to exsolve the remaining gas and allow it to accumulate in artificial gas caps where it could be tapped, or whether it would remain dispersed in the hot water and remain unavailable.[31] In the fourth quarter of 1979, the Department of Energy approved an $8.5-million, 24-month project for Dow Chemical Company to test the feasibility of recovering geopressured gas. Whatever the outcome of this study these aquifers are technically challenging, potential sources of significant energy.

At least one scientist has defied tradition and suggested that the earth ac-

---

*Large amounts of water and sand are pumped in under high pressure to fracture the formation. The sand particles prop open the fractures when pressure is released.[31]

tually contains vast reserves of methane, of nonbiogenic origin, which would last us over 1 million years at our present rate of consumption.[42] His presentation has been reviewed by Ratterson.[43] It is so speculative that it will not be dealt with further in these pages, except to say that it contains some intriguing and possibly plausible explanations for observations that cannot be dismissed too lightly.

WITHHOLDING, CURTAILMENTS, AND SUCH

As any follower of the news knows by this time, the large producers have been accused, in the past few years, of withholding gas wells from production in order to apply pressure for higher prices. This writer is not privy to unpublished data of producers or their accusers, hence cannot assess the charges and countercharges. Instead, an attempt will be made here merely to summarize and air the published claims.

A preliminary survey of five gas fields by the Department of the Interior (DOI) in 1977, disclosed that of the 244* reservoirs of gas represented, only 19 (or less than 8 percent) had been brought into production since 1974. This was far below the level promised by producers. Furthermore, the average production rate for the producing wells was only 58 percent of that deemed to be the "maximum efficient rate." In addition to these observations, several producing wells had been shut down for no obvious reason. These facts were felt to be ample justification for a full-blown investigation by the federal government.[5]

During the first quarter of 1977, 10 major energy companies were sharply criticized by DOI for not producing as much gas as they could from federal leases. The companies accused were: Tenneco, Gulf, Mobil, Amoco, Union, Texaco, Cities Service, Getty, Atlantic-Richfield, and Continental. The four specific gas fields, involving possibly a total of 0.4 tcf, are off the coast of Louisiana.[44] Most of the gas was in shallow, easily tapped, "behind-the-pipe" reservoirs† that could not be produced profitably without first exhausting the reservoir below. Development would require about six months.[45]

There are laws that spell out the ground rules for operating under federal leases. Lease holders are obligated to begin production within five years of being granted the lease unless an extension has been obtained.[46] Producers are required to show "due diligence" in getting federal lease gas to the consumer.[47]‡

There are ways of "withholding" gas that are legal, if not strictly ethical.[48] The simplest is for the producer to delay final exploration. The first exploratory step is to drill a wildcat well to determine what lies beneath the surface. This well is then plugged up for as long as several years. As many as 62 Gulf coast offshore fields were in such a state and had exceeded the federal five-year limit without being renewed. Producers can plead delays in completing their analysis

---

*Also reported elsewhere as 188 or 202.

†Pockets above reservoirs in production.

‡One-fourth of all U.S. natural gas comes from such leases.

or in obtaining necessary equipment. They can also shut down wells for an indeterminate period of time pending "repairs." A single platform will support as many as 25 wells, and when one is shut down, all others on the same (and perhaps nearby) rig taken out of service, in the interest of safety.

In the few years prior to the passage of the Natural Gas Policy Act of 1978, producers have shown a preference for selling to the higher-priced (about $2 per 1000 scf) intrastate market rather than to the interstate market (maximum of $1.44 per 1000 scf). In 1967, interstate pipelines received two-thirds of all new gas. In 1975, the figure was about one-eighth. Producers do have the right under law to withhold fields from production and to withhold gas from interstate pipelines until they deem the price to be economically sound, but only as long as it underlies private land.[47]

The FPC found a total of 7 tcf in nonproducing reservoirs in offshore and onshore federal leases, but gasmen claimed that the wells must be tapped at a "judicious rate" to ensure optimum long-range recovery.[47] On record, they opposed a "drain-America-now" policy, and pointed out that fields were geared to produce over a 20-year contract life.[45]

A somewhat different type of charge against producers was that some had failed to meet contractual delivery commitments to interstate pipeline companies, thus shorting local gas companies and forcing them to buy higher-priced gas from other sources, the cost of which then had to be passed on to the consumers.[47] This practice allegedly cost New Jersey consumers an additional $17.5 million in 1974–1975.[45] This form of withholding generally was not reported by the pipeline companies for fear they would be blacklisted for all time by producers.[47] However, one pipeline company, Transco Companies, Inc., supplier of 69 percent of the gas to New Jersey, New York, and Pennsylvania, claimed to have been shut out of buying onshore gas in the first quarter of 1977, because the producers were getting up to $2.15 for the gas intrastate. Other producers "sat on their gas"* to avoid "vintaging" it prematurely and thus committing it permanently to sale at lower prices.[49]

In light of the shortage of gas and resulting increasing curtailments, it became necessary (early in 1977) for the FPC to establish consumer priorities. These categories, in order of importance, were: residential and commercial space heating, chemical and fertilizer feedstocks, industrial plant maintenance, utility use for electricity generation, and industrial boiler fuel.[17] The FPC placed "priority 1" needs at 5 tcf per year. The "priority 2" consumption at that time was in excess of 1 tcf per year.[50] Electric power generation was placed low on the list because: (1) gas is used inefficiently in large boilers, (2) other fuels can be substituted more readily and with less economic dislocations in large power plants than in smaller units and, (3) it is more practical to employ pollution abatement equipment for dirty fuels in large installations.[51]

From 1973 through the first quarter of 1977, the trend of curtailments (on an energy basis) increased to about 25 percent of our total consumption for the

---

*Withheld it from production.

time period. The regions most seriously affected in the severe winter of 1976–1977 were New England and the East-North Central states. The apparent reduction of curtailments on a percentage basis was probably due to progress made in substituting other fuels (e.g., propane).[52]

Despite expectations of severe effects on chemical and allied industries due to natural gas curtailments, the impact was surprisingly small. Only about 1 percent of the work force was affected, and then for only 2 to 10 days, on the average, as management scrambled for alternative fuels. The major exceptions were the ammonia producers, who, as an example, lost about 25 percent of their production early in February 1977. Textile finishing operations were also affected, since suitable alternative fuels were not available for their operations.[53] Over one-fourth of the total employment in the Ohio and West Virginia glass industry was affected for varying periods of time, but the dislocations were not as severe as expected.[54]

Substitution of propane or oil saved the day in many specific situations. Plans have since been made by many members of the glass industry to save gas in the future by: switching to lehrs with inside belt returns; experimenting with coal-firing and dual-firing (oil and gas) equipment; installing capital-intensive, but energy-saving float processing; and electric boosting on glass tanks, to supplement gas.[54]

The industrial use of gas by major industrial sectors is shown in Table 3.6.

STORAGE

One of the more serious outgrowths of the heavy drawdown of gas supplies during the winter of 1976–1977 was the problem of refilling the storage system prior to the arrival of the next heavy drawdown period. The drawdown on the interstate pipelines was estimated by the FPC to be 20 percent greater than in 1976, and it was expected to amount to 1.5 tcf by season's end.[56] The utilities were in at least as critical a state. This meant that full deliveries to industry

Table 3.6 Industrial Use of Natural Gas

Sector	Percent of Total
Electric, gas, and sanitary services	28
Chemicals	17
Primary metals production	12
Petroleum refining and related industries	11
Stone, clay, glass, and concrete industries	10
Other	22
Total[a]	100

Source: Reference 55.
[a]About 7 tcf (1975) by Bureau of Mines estimate.

could not be restored until such time as some assurance was felt that storage could be replenished within the few months remaining before the next season started. The alternative was an even more severe economic and personal dislocation the next year, even if the winter weather should moderate.[57]

Storage reservoirs comprise: low-pressure, aboveground tanks; buried batteries of high-pressure pipe and bottles; water-sand aquifers; depleted or partially depleted oil and gas reservoirs; and LNG storage containers.[57] Also of significance as a storage container is the domestic trunk pipeline network itself. This system is 225,000 mi (363,000 km) long. One mile of 3-ft diameter pipe will contain 1 million scf of natural gas at 400 psi.[1] At the end of 1974, 3.969 tcf of natural gas was stored in the United States. The total storage capacity was 6.36 tcf, mainly in depleted oil and gas traps.[1] Despite the fact that the following winters (until 1976–1977) were relatively mild, a net drawdown on stored gas still occurred.

EXPLORATION

Exploratory and development drilling* increased steadily from 1945 to 1956, but since that time the trend has been generally downward. The suspicion is that the ceilings imposed on gas prices by the FPC in 1954 had no immediate effect because of prior commitments to drill and bring wells into production. Since 1957, however, FPC field pricing apparently had an inhibitory effect on the drilling of wells,[17] at least until the flurry of activity in 1976, which may have been stimulated by the improvement in pricing in that year. Even then, despite the increased drilling activity, marketed production and sales of interstate gas fell, in comparison with the first quarter of 1975.[58] Footage of offshore drilling decreased sharply, while the total number of wells drilled rose.

In 1973, exploratory drilling had its second highest year. There were 900 natural gas discoveries, but 43.8 percent of them (394) were onshore, in Texas, where intrastate prices were attractively high. In 1974, 447 new gas fields were discovered, for an increase of 7.5 percent over 1973. But these discoveries were largely in old producing areas of Texas, Louisiana, and Oklahoma, where new discoveries are notably small in yield.

Exploratory drilling at the low FPC mid-1976 price ceilings of $0.52 to $0.53 per 1000 scf was carried out in established, low-cost regions. The prices ceilings were not conducive to exploratory drilling in the more-capital-intensive regions offshore, or in Alaska, the Rocky Mountains, or western Texas.[17]

Statistics show that only about 1 of 10 exploratory wells is successful; hence one "producer" must carry the load for the cost of drilling the nine dry holes.[59] Only one-third of the successful offshore wells in the Gulf of Mexico were in production in 1977.† In one offshore area of the Gulf, a total of $1.5 billion had

---

*Exploratory drilling refers to drilling in previously unexplored areas, while development drilling means that done to extend the limits of known reservoirs.

†Offshore production goes into interstate commerce, where pricing is not yet as attractive as intrastate sales.

been spent on drilling by a number of firms, and the only "discovery" to date was salt water.[48] Some Gulf coast drilling facilities were operating 110 mi from shore in 1800 ft of water, and drilling through 23,000 ft of mud, shale, and rock. In 1976, the Gulf provided 14 percent of U.S. gas production; projections were for 30 percent in 1985.[48]

To support such costly undertakings having such low rates of success, drillers must be able to drill large numbers of wells to spread their risks. (There are over 10,000 oil and gas producers in the United States.) Utilities are often limited partners or nonoperating joint venturers in gas exploration. As such, they may contract to pay a "carried interest" of 100 percent of the cost of exploratory drilling. When a well begins producing, the utility normally holds 75 percent of a working interest in the well. The operator or producer holds the remaining 25 percent. (This is referred to as "quarter carry."[59]) The exact details of such contracts will vary depending on the risk involved, available capital, and so forth.

Interstate gas transmission companies are also involved in exploration and development. Some producers acquire their venture capital through advance payment agreements (in effect, interest-free loans) by pipeline companies, in exchange for the right to buy the gas discovered. One such transmission company, Transco, supplies New York, New Jersey, and Pennsylvania through an 1800-mi pipeline from Texas and Louisiana.* During the winter of 1976–1977, this pipeline operated at only two-thirds capacity because of the gas shortage.[60]

Exploration is becoming increasingly expensive and speculative. Exxon erected a 90-story high, $74 million drilling platform (dubbed *Hondo*, Spanish for "deep") for operation in 850 ft (260 m) of water off Santa Barbara, California, in the Hondo field. This platform was designed for drilling 28 directional wells. The total investment, when production started, was expected to be about $500 million.[61] This rig was dwarfed by Shell's *Cognac* rig, which towers 1265 ft from the Gulf floor, 100 mi from the southeast coast of Louisiana (off the mouth of the Mississippi River) in 1020 ft of water. Fifty-six wells can be drilled from this rig. (It might be noted that the Empire State Building is "only" 1250 ft high.[62])

Drilling deep holes in weak rock formations, rates may average as high as 5 ft per hour, but conventional rotary bits have an operating life of only 10 to 20 hr and it takes an equal length of time to withdraw the bit and replace it. Drilling technology has advanced very slowly, and it is now one of the weak links in our oil and gas supply chain, but considerable research is now under way to develop novel drilling methods.[63]

As current oil and gas fields are depleted, the energy industry must look to new fields—offshore, more remote and deeper onshore sites, geopressured zones, Alaska, and so forth—where drilling costs are higher. Onshore drilling in *proved* reserves (1978) cost $20 to $50 per foot under normal conditions. Our *potential* oil and gas reserves generally lie in deep formations which lead to drilling costs of $125 to $172 per foot. A shortage of drilling capacity could result in

*This pipeline has a capacity of 3 billion scf per day.

the 1980s from the higher cost of difficult drilling conditions and a possible shortage of materials and of skilled labor.[64]

A report of the Department of Energy[65] indicates that beginning January 1, 1976, the cumulative footage drilled must reach 930 million feet by 1980 if we are to meet projections for our oil and gas production needs. By 1995, our footage needs will be 9.6 billion feet. At the current growth rate of the drilling industry, it would be able to drill only 720 million cumulative feet by 1980 and 5.6 billion feet by 1995; hence we might expect a shortfall of up to 40 percent of our needs. Technological increases in drilling speeds would free up more rigs for use elsewhere and reduce capital requirements.

One of the prime areas for exploratory drilling is along the Atlantic coast, in the Baltimore Canyon—a huge geological OCS zone extending from Long Island, New York, to North Carolina—where the results of exploratory drilling have been "encouraging." Late in November 1979, Texaco made its third gas strike in the same 15,600-ft well, about 100 mi east of Atlantic City, New Jersey. One flowed at 18.9 million ft^3 per day (sufficient to heat about 50,000 homes). Tenneco had earlier announced a strike which also yielded some oil. A preliminary USGS survey discovered gas at the edge of the continental shelf. Fourteen other tracts have yielded only dry holes. Exploration to date suggests the possibility that there may be several separate reservoirs of gas. It is estimated that pipelines for bringing the gas to shore cannot be put into service until six or seven years after gas is discovered in commerical quantities. The magnitude of such a discovery will have to be at least 1 tcf (proven reserves) with a flow of 200 million ft^3 per day* to warrant expenditure of $400 million on the required pipeline.[66]

There is no really sharp line of demarcation between "gas wells" and "oil wells." Gas-to-oil ratios in so-called oil wells may range from a few cubic feet to many thousand cubic feet of gas per barrel of oil. A "gas-condensate well" will yield large volumes of gas with droplets of light, volatile hydrocarbon liquids entrained in the gas stream. This entrained fluid is condensed and sold as NGL. Most gas wells contain recoverable quantities of propane and butane, which become LPG, and more stable, volatile liquids, which become "natural gasoline," used for blending with refinery gasoline.[24]

Corrosion, in various forms, occurs in oil and gas wells, and adds to the maintenance costs[24]:

1 *Sweet corrosion* (in the absence of oxygen and hydrogen sulfide) is caused by carbon dioxide and fatty acids in sweet oil and condensate products.

2 *Sour corrosion* occurs in the production of oil or gas containing even trace amounts of hydrogen sulfide, and may be accelerated by oxygen, carbon dioxide, or organic acids.

3 *Oxygen corrosion* takes place wherever there is exposure to air, most com-

---

*At 1979 prices, these reserves would yield $2.9 billion per year (at retail).

monly in offshore wells, shallow producing wells, and brine-handling and injection equipment.

4 *Electrochemical corrosion* occurs in soil.

In addition to corrosion, paraffin deposits tend to clog piping and foul pump ("sucker") rods and must be removed with solvents, heated oil, or steam, or by scraping.[24]

ALASKAN GAS

As noted earlier, there is an estimated recoverable reserve of 26 tcf of associated gas (and possibly 5 tcf of nonassociated gas) waiting to be produced on Alaska's North Slope. Unfortunately, it is also awaiting a means to get to market in the midwestern and northeastern states. This reserve amounts to only 10 percent of the total "lower-48" reserves, but even so, its probable value is assessed at about $40 billion.[67]

Proposals for three different transmission systems were filed with the FPC by their sponsors and, following a 19-month study, one proposal was endorsed by the FPC Administrative Law Judge, in February 1977. This was the proposal of the Alaskan Arctic Gas Pipeline Company. The Commission then had to recommend it to President Carter and he, in turn, had until December 1, 1977, to approve it (without its being rejected by Congress). It also had to be approved by the Canadian National Energy Board (NEB) and by Prime Minister Pierre Trudeau.

In his endorsement, the FPC judge cited the Arctic Gas pipeline as being the quickest and most effective means of getting the gas to midwestern, eastern, and Pacific coast markets. It was a well-thought-out plan based on well-established technology. It offered the lowest cost to the consumer and the shortest lead time, and accomplished its goals with minimal environmental impact.[68]

On the debit side of the ledger, it was not an all-U.S. route. It depended on picking up Canadian gas from the Mackenzie Delta by means of a 195-mi link, to make it economically viable. Gas discoveries in this region at the time were only sufficient to provide half the flow of Canadian gas originally anticipated.[67] Canada's major new gas finds have been in the high Arctic Islands northwest of Hudson Bay, and will eventually call for a pipeline parallel to the Arctic Gas system and several hundred miles to the east of it.

Also, despite the fact that its overall environmental impact was relatively minor, the Arctic Gas pipeline would cross the 9-million-acre Arctic National Wildlife Range of Canada. It received a great deal of opposition from environmentalists, because of three noisy compressor stations and the regular low-level overflights that would be required for security and maintenance.[69] Furthermore, it required accord of the Indians and Eskimos of the Northwest Territories, on the issue of land rights. Canadian government approval also proved to be a major barrier to the Arctic Gas plan, despite the inclusion of Canadian firms in the consortium.[67]

It would have been a 4175-mi, 48-in.-diameter pipe constructed at an estimated cost of about $8.5 billion by a consortium of 16 U.S. and Canadian firms. It would have had a capacity of 4.5 billion scf per day and would have carried gas to the Pacific coast, the Midwest, and as far east as Pittsburgh, where it would have interconnected with the existing network. Even with a not unlikely 20 percent cost overrun, it would still have been economically competitive. It would have permitted the lowest delivery costs of the three systems. Probably 5 years would have been required for its construction, once all approvals were granted.[70]

A competing El Paso Alaska Company proposal was rejected by the judge, but it was still considered a viable alternative. The strength of this system lay in its strong support by local environmentalists, who saw its avoidance of any wildlife range and its use of an 809-mi overland corridor common to the TAPS oil pipeline as compelling factors in its favor.[69] It also enjoyed the approval of Alaskans, labor, and western governors, who looked with favor on the fact that it was an all-U.S. route, free of any possible future pressures from Canada.

Its major shortcoming probably was that it would depend on the handling and transport of liquefied natural gas (LNG) by about 1000-ft-long cryogenic tankers from Gravina Point, Alaska, 42 mi southeast of Valdez, through potentially treacherous waters, to the Pacific coast of the lower 48 states. The Jones Act requires shipment in U.S. tankers, of which there were none available at the time.[71] Furthermore, the proposal provided for delivery only to the gas-rich Pacific coast.* Although it had general approval by Alaskan environmentalists, it was strongly opposed by California environmentalists, who looked on the tankers as being floating bombs posing a danger from explosion and fire, as well as an environmental threat to air and water.[73] (More will be said on LNG transport later in this chapter.) The proposal included construction costs of six cryogenic tankers, but the project probably would have required 11 at a cost of about $2 billion.[69]

The judge cited the total package as being only marginally economically competitive, and lacking in flexibility and credibility. It would have required a liquefaction plant at Gravina Point and a regasification plant in southern California (possibly Point Conception, Oxnard, or Long Beach). The $3.5 billion cost estimate (including pipeline, tankers, and liquefaction and regasification plants) appeared unrealistic. Actual capital cost was more likely to be about $8 billion.[74] The cost of delivery of the product, as pipeline quality gas, would be the highest of the three proposed systems. Another factor that had to be taken into consideration was the reputation of the company behind the proposal.[75]

The third proposal, made by Alcan Pipeline Company [a group comprising Northwest Energy Company, Foothills Pipe Lines Ltd. (a subsidiary of AGTL†),

---

*A shortage is expected to develop there in 1980-1981.[72]

†Alberta Gas Trunk Line Company.

and Westcoast Transmission Company], was looked on by Canadian energy planners as being the least intrusive for the Indians and Eskimos, and the least disruptive of the Canadian economy and politics.[67] It would parallel TAPS to Fairbanks and then branch off along the Alcan Highway through Canada.

It was cited by the FPC Administrative Law Judge as being an undersized* and inefficient system that would require 1200 miles more pipeline than the Arctic Gas system, because of the circuitous route taken to stay in U.S.-controlled territory as much as possible. The judge also found the lead time (three years) and cost ($4.7 billion) not to be credible. It was expected that the cost would rise to as high as $10 billion when inflation and an appropriate U.S. distribution system were added in.[76] The delivery cost of gas would have been intermediate between the other two proposals.[68]

More recent developments resulted in abandonment of the Arctic Gas proposal, largely because of environmental and sociological pressures from Canada, which led to threats to delay construction for 10 years to permit studying the impact of the project. The Alcan proposal received the necessary approvals, although the El Paso approach was still considered a viable alternative. Alcan is planned for transport of both Alaskan and Canadian gas, while El Paso offered strictly an all-U.S. operation.[77] Alcan Pipeline Company will be responsible for the 731-mi Alaskan segment. Foothills Pipe Lines Ltd. will build the 2029-mi Canadian segments. The pipeline will then feed into a 911-mi Canada-to-San Francisco link (Pacific Gas Transmission and Pacific Gas and Electric Company) and a 1117-mi Canada-to-Chicago link (Northern Border Pipeline Company).[78]

The Alcan proposal has escalated to an estimated $14 billion for the 4800-mi project.[78] As now planned, the southern portion would be built first and be used to transmit gas from Alberta fields to the United States until the Alaskan gas begins to flow.[79]†

The El Paso plan would undoubtedly have resulted in higher transportation costs (estimated at $1.83 per 1000 scf on the West Coast, $2.27 in the Midwest, and $2.48 on the East Coast), compared with Alcan ($1.36 West Coast and $1.67 Midwest and East Coast).[81]‡

Still up in the air is the siting question for a West Coast LNG terminal and regasification plant. It would be essential if the El Paso plan were to be revived, but it could also handle LNG imports from Indonesia. The California Public Utilities Commission did consider a remote site at Point Conception. Such a terminal would be expected to have a capacity of 900 million scf of LNG annually—about 55 percent from Indonesia and the balance from Alaska.[82] Deliveries would be made at the initial rate of 146 million scf per day, ultimately peaking at 500 million. Double-hulled tankers, with all void areas and insulation space purged with inert gas, and transfers carried out without venting to

---

*A modified proposal increased the capacity nearly 50 percent and overcame this objection.

†Startup target date for gas transmission was initially 1983, but U.S. regulatory officials have already delayed start of construction to late 1984.[80]

‡Estimated in 1977.

the atmosphere, should prevent formation of flammable mixtures.[83] The major drawback to the El Paso proposal is its inefficiency. Liquefaction of natural gas is only about 70 percent energy-efficient.[84] At the other end--during regasification—no means has been developed for making use of the refrigerating capabilities of the incoming LNG, a process that could reduce overall operation costs.

LIQUEFIED NATURAL GAS

As noted earlier, natural gas is an essential factor in our present energy (and feedstock) picture, yet our domestic reserves are being depleted faster than new discoveries are being made. The obvious answer offers an option of importing more or consuming less. Unfortunately, natural gas is the fossil fuel most difficult to transport between noncontiguous land areas, and imports from our only contiguous foreign sources are declining.* Futhermore, there is the problem of balance of payments if imports are increased.

Liquefaction

It is not practical to ship natural gas as a vapor, but it can be reduced in volume by a factor of 600 to make it a liquid, practical to transport by surface vessels. Since methane has a low critical temperature ($-100°F$), it cannot be liquefied at ambient temperature by merely increasing pressure as is done with LPG (propane and butane). It must be cooled to about $-260°F$ ($-162°C$) to liquefy it at slightly above atmospheric pressure. It must then be maintained at these conditions during storage and transport, and then be regasified before distribution through the conventional gas pipeline network.[86]

A very modest fraction of our gas requirements is met by imports of LNG. In 1976, less than 0.1 percent was imported as LNG. About 10 billion scf (regasified volume) or 10.8 trillion Btu were shipped to Massachusetts alone from Algeria. This amounts to about 12 shiploads.[86] The FPC approved imports of 400 billion scf by 1980 and had further approval requests for 3 tcf. The Energy Research Council recommended, however, that import limits from any one nation be set at 0.8 to 1.0 tcf per year, and total annual imports be limited to 2 tcf.[87]

It is estimated that it will be 1982 or possibly 1987 before appreciable quantities are imported.[14] During the winter of 1976-1977, despite our critical gas supply situation, which put more than 850,000 people out of work for various periods, we were net exporters of LNG, shipping about five times as much LNG from Alaska to Japan as we were importing from Algeria. Furthermore, the

---

*Imports from Canada are declining, and imports from Mexico have been nonexistent. Importation of Mexican gas was finally approved on December 29, 1979, but the price will be high (initially $3.625 per million Btu) and the amount will be limited to 300 million ft^3 per day. Furthermore, a portion of the pipeline is yet to be designed and built.[85]

price of the exports ($1.95 per 1000 scf) was less than the pipeline cost of emergency domestic supplies of natural gas from the Gulf of Mexico ($2.15),[71] and far less than postembargo import contracts of OPEC LNG (possibly $3.30).[14]

An agreement was reached between Phillips Petroleum Company and Columbia Gas Company to ship LNG from Phillips' Kenai, Alaska, liquefaction plant through the Panama Canal to Columbia's Everett, Massachusetts, regasification plant. Each specially constructed tanker would hold about 1.17 million ft^3 or 0.7 billion scf (regasified volume) of gas and require 20 days per round trip. Each shipload would amount to only about 10 percent of what Columbia could distribute *daily* through its pipeline system.[71] This reflects a fundamental difficulty in the distribution of LNG.

Hazards

The major problems in handling LNG involve suitable materials and structures to contain it during transport, and storage and safety measures to employ in case of accidental release. In 1977, there were about 35 cryogenic tankers for LNG in service, having an average liquid capacity of about 1.6 million ft^3 (46,000 m^3) or nearly 1 billion scf regasified vapor. Forty-one more were under construction or in the design stage.[72] These had a mean capacity of 4.4 million ft^3 (124,000 m^3), or 2.64 billion scf regasified. Plans were being drawn up for 5.8 million ft^3 (165,000 m^3), or 3.48 billion scf (vapor) tankers.[86]

Many reports have tagged LNG tankers as floating bombs and, in fact, when these ships* approach Boston Harbor once a month, bound for the LNG terminal in Everett, they are carefully inspected and then escorted into the harbor by Coast Guard vessels which keep all marine traffic at a distance of 2 mi ahead and 1 mi astern. Meanwhile, police cordon off all roads leading to the terminal area during the transfer operation.[88] Actually, the hazards of handling LNG are perhaps best exemplified by the fact that after 95 ship-years of such operations, the insurance rates on LNG tankers are currently lower than for other ships of comparable size.[83] Nonetheless, as long as Murphy's law prevails, there is probably little sense in tempting fate by building terminal facilities in crowded harbors in heavily populated areas. Accidents—from both natural and manmade causes—do occur in all phases of endeavor.

The tankers are very special ships. Each carries up to five welded, spherical, high-nickel steel or special aluminum alloy tanks. The tanks are double-walled and 1 m thick overall, with the intervening space filled with perlite insulation. The main deck is only 65 to 82 ft (20 to 25 m) above the water line, and the ship will have a draft of about 33 ft (10 m). (The density of the cargo is only about one-half that of seawater.)[75] The ships measure 900 to 1000 ft (275 to 305 m) in length and cost up to $180 million each.[76]

---

*Usually, a particular French ship of 50,000 m^3 capacity—enough to meet the needs of a city of 30,000 for a year.

Tankers are only one of the links in the transport and distribution of LNG. Huge liquefaction plants (about 370 billion scf vapor per year) must be constructed at a cost of about $850 million each.[76]* Completion of liquefaction plants lagged, due to inflation and construction delays, which added to financing problems. Meanwhile, cryogenic tankers laid idle around the world, waiting for cargoes.[76]

Liquefied natural gas has a few unique characteristics that are worth mentioning[75]:

1 *Flameless vapor explosions* can occur when LNG (containing sufficient ethane, propane, and nitrogen) comes into contact with water. The thin layer of gas in contact with water will superheat sufficiently for homogeneous nucleation of vapor to occur very rapidly. The chance of severe damage occurring is very small, however, since the pressure developed is only about 100 psi.

2 A phenomenon known as *rollover* can occur in storage if mixtures of LNG having different compositions, hence different densities, are brought together in such a manner that higher-density, warmer LNG is trapped below a layer of lower-density, colder LNG. The denser bottom layer as it picks up heat from the bottom and walls of the tank will not have sufficient buoyancy to enable it to penetrate the upper layer. Ultimately, instability occurs and rapid mixing takes place with a sudden build-up of pressure. Rollover generally can be avoided by making additions at the top of partially filled storage tanks.

3 Spilled LNG will spread rapidly over the much warmer land or water and vaporize quickly. Because of its high density, this vapor will stay close to the surface initially. As it mixes with the atmosphere, the air cools and moisture condenses into a visible cloud. As the cloud dissipates downwind, it warms and rises. The maximum distance downwind at which the vapor cloud is flammable will be determined by size and rate of the spill, the various factors that contribute to dispersion, and the energy input rate to the cloud. Attempts have been made to model the dissipation over land and water.[89]

Fire is a cause for concern. The liquid will burn much like gasoline in the presence of oxygen (air), but if not ignited before it manages to vaporize, the vapor cloud could drift downwind and become ignited. The flame will then burn back along the path of the vapor to its source. As long as the cloud is not confined, it is unlikely to detonate with sufficient force to cause serious blast damage, but large volumes of ignited vapor can cause severe fire damage (as noted below). Stored LNG is dangerous only if vaporized, mixed with air, and ignited.

Much has been written about fears of great catastrophes if LNG were to be

---

*Six are planned for Algeria alone.

shipped into heavily populated areas. These fears are based largely on two events: (1) A firestorm in Cleveland on October 20, 1944, led to 131 deaths and gutted 29 acres of buildings. In this incident, LNG had escaped from two storage tanks containing 2 million gal of the liquid, a fraction of the 33 to 42 million gal carried in modern cryogenic tankers.[90] (2) In 1973, another storage tank, this one on Staten Island, New York, collapsed and killed 40 workers engaged in repairing it.

The first accident resulted from failure to provide adequate dikes to confine the escaped liquefied gas around the storage tanks. The second involved what would now be considered unacceptable tank design. The Staten Island tank comprised an outer concrete wall insulated inside with polyurethane foam lined with an aluminized polyester plastic film. After two years of operation, the polyester film developed a leak. The LNG was pumped out and nitrogen gas was blown in to exclude air and to sweep out the gas residue. Workers then entered the tank and somehow fire broke out. The resulting increased pressure lifted the tank roof, in accordance with its design intent, but then it collapsed on the workers, resulting in the tragedy. Nonflammable insulation is now used in LNG tank structures.

From 1965 to 1974, about 150 million ft^3 per day of LNG was shipped to England from Algeria, and additional LNG was shipped to other Western European countries, Japan, and the United States. Over 100 LNG storage facilities have been built throughout the world.[91] No major LNG storage or tanker accident has occurred in this time, but it must be acknowledged that only 14 tankers were in service in 1974, and the new breed of tankers will be much larger.[69] Nonetheless, in over 2000 transoceanic voyages of LNG tankers, the safety record has been exemplary.[92]

More reasoned heads have estimated that the hazard of bringing LNG into Staten Island, with its high population density, would be about the same as handling gasoline, bottled gas, or natural gas. This would be approximately equivalent to the risk of being struck by lightning or one-tenth the hazard of a fire fatality in the home.[86] Transport of natural gas in vapor form is not without its problems. Trunk and feeder pipeline leaks do occur, although the record is improving with advances in technology. Table 3.7 indicates major causes of natural gas pipeline failures.

On balance, the overall hazards caused by handling LNG are probably less than those claimed by the news media and environmentalists, but possibly greater than those involved in the distribution of domestic natural gas, which lacks the problems of liquefaction, confinement at cryogenic temperatures, and regasification.

ALTERNATIVES TO NATURAL GAS

Alternatives to natural gas are necessary for two reasons: existing supplies are often remote and difficult to transport, and there is a growing shortage of acces-

Table 3.7   Gas Pipeline Failures

Cause	Percent 1950–1956	Percent 1971
Tractor gouges	30.0	57.0
External corrosion	13.4	7.4
Longitudinal seam failures	12.0	—
Other external causes	11.6	9.0
Acetylene girth welds	6.6	3.0
Other causes	23.4	23.6
Total	100.0	100.0

Source: Reference 93.

sible natural gas. That gas which is available in remote locations can be converted to more readily transportable products, and gaseous fuels can be obtained from other sources—at a cost.

Ammonia

A certain amount of ammonia (together with other by-products) is extracted in the process of cleaning up natural gas for feeding into the pipeline network. (See Table 3.8 for a breakdown of by-products in a typical 250 million cfd pipeline gas processing facility.) Ammonia, our third largest volume chemical,* about three-fourths of which is consumed in the manufacture of fertilizers, is

Table 3.8   Typical Quantities of Major By-Products from 250-Million-cfd Pipeline Gas Process

By-Product	Quantity
Sulfur (primarily as H_2S)	300–450 long tons
Ammonia	100–150 tons
Hydrogen cyanide	0–2 tons
Phenols	10–70 tons
Benzene	50–300 tons
Oils and tars	trace–400 tons

Source: Reference 94. Reproduced, with permission, from the *Annual Review of Energy*, Vol. 1. Copyright 1976 by Annual Reviews, Inc.

*Behind sulfuric acid and lime.[95]

synthesized commercially by steam reforming methane (natural gas) and catalytically reacting with nitrogen from the air.[96]

$$CH_4 + H_2O \rightarrow CO + 3H_2 \quad \text{(reforming)} \tag{1}$$

$$CO + H_2O \rightarrow CO_2 + H_2 \quad \text{(water-gas shift)} \tag{2}$$

$$N_2 + 3H_2 \overset{\text{cat.}}{\rightleftharpoons} 2NH_3 \quad \text{(catalytic combination)} \tag{3}$$

The resulting ammonia has no direct application as a commercial fuel, but the process could gainfully consume natural gas now flared off in the Mideast and convert it to a form that can be transported easily and safely, thus taking some of the pressures off dwindling domestic natural gas supplies which are currently needed for energy and feedstocks.

Methanol

Methanol is another potential product of natural gas which can be used directly as a fuel or can be methanated to re-form methane. Making methanol from natural gas is almost as costly as liquefaction. It comprises reforming methane and catalytically converting it to methanol[97]:

$$CH_4 + H_2O \rightarrow CO + H_2 \quad \text{(reforming)} \tag{4}$$

$$CO + 2H_2 \overset{\text{cat.}}{\rightleftharpoons} CH_3OH \quad \text{(catalytic conversion)} \tag{5}$$

Transport is less expensive because methanol can be shipped in normal liquid-container vessels without refrigeration. Methanation back to methane is a different matter. It entails considerable loss of efficiency:

$$CH_3OH + H_2 \rightarrow CH_4 + H_2O \quad \text{(methanation)} \tag{6}$$

About 60 percent of our current output of methanol ends up in polymers,[98] but it can be produced commercially as above, or from biomass, in sufficient quantity to use as a liquid fuel or fuel supplement. As in the case of ammonia from natural gas, methanol would not provide a direct replacement for natural gas, but it could use Mideast gas not now readily available to us, transport it to our shores in a safe form, and provide some relief from our consumption of gas for petrochemical production, as well as stretching gasoline supplies. Some thought has been given to the possibility of employing floating methanol plants to aid the development of small, remote natural gas fields, and to permit the plant to be moved as fields are abandoned.[99] More is said about methanol in Chapter 8. Plans in 1976 for the recovery and use of Saudi Arabia gas for petrochemicals had a projected on-stream date of 1979.[100]

Propane

Propane (bottled gas), from natural gas and refined crude oil, is already a functioning alternative to natural gas, particularly in areas beyond the reach of the existing pipeline network. It was also employed by the glass and other industries as a substitute for natural gas during the winter of 1976–1977. This increase in consumption added to the already critical supply situation brought on by the limited propane distribution system, which makes supply tenuous at best, even in more normal times. One additional problem in substituting propane for natural gas is that it must be stored under pressure to keep it liquefied, and few end users are equipped for this kind of storage.[101]

Propane is a seasonal fuel, with two-thirds of its consumption occurring in the first and fourth quarters. There has been a steady decline in domestic production since 1972, with little hope for a reversal in the trend. The only hope is for increased imports, but here the problems are a shortage of available tankers, the reduction in fossil fuel exports by Canada, and the balance-of-payments specter. The supply situation, price, and distribution system being what they are, propane cannot be considered a general substitute for natural gas.

Hydrogen

Hydrogen, based on the electrolysis of water, has been touted as a possible fuel replacement for natural gas. It could be distributed through the existing pipeline network at the rate of 10 tcf per year, but storage could pose a problem, as it remains to be seen whether present natural gas reservoirs (capacity 5.5 tcf) are capable of retaining the more evanescent hydrogen.

The major stumbling block, however, would be the capital requirements. Assuming that output for a hydrogen economy had the energy equivalency of natural gas consumption (22 + Q or tcf), about 70 tcf of hydrogen would be required per year. At the present state of the art, this would call for over 1 million MW of electric power, versus our present total domestic electricity generating capacity for all consumers amounting to about 360,000 MW.[102] Such a commitment would hardly be worthy of consideration.

NATURAL GAS POLICY

The Natural Gas Policy Act of 1978, passed by Congress in October 1978* after months of bitter wrangling, at last cleared the way for slow, partial deregulation† and for importation of gas. Deregulation has apparently already stimulated some renewed drilling activity in high-risk but potentially high-yield areas. The Department of Energy has estimated that by 1985, the impact of

---

*Effective December 1, 1979.
†DOE estimates that nearly 40 percent of our gas supply will still be under controls after 1987.[103]

deregulation on natural gas supplies will amount to about an additional 2 tcf annually (exclusive of Alaskan gas). This would be the energy equivalent of approximately 1.4 million bpd of crude oil.[103] The brusque, insensitive manner in which the DOE attempted to deal with the Mexican government petroleum company, Petroleos Mexicanos (Pemex), in December 1978 threatened to alienate our southern neighbors irrevocably. Fortunately, these abortive negotiations merely led to a year's delay and a 40 percent price escalation in producing an agreement allowing importation of Mexican gas, as noted earlier in this chapter.[104]

The new act appears to be tailored to make industrial users bear the brunt of the price increases, thus allowing the private sector (i.e., the voters) some relief from this burden. Lawyers and independent consultants are the only ones likely to be completely satisfied with the new gas policy act and its 66-page statute plus 364 pages of regulations. Compliance is going to be very difficult at best, because of ambiguous definitions and extensive paperwork demands. This latter aspect is pointed up by FERC's request for 300 additional personnel (at a cost of $11 million) to relieve startup problems.[103]

SUMMARY

A few things seem certain out of all this welter of facts and figures, charges and countercharges. Most of the easy, shallow gas has been found. Now we have to go deeper and farther afield in more remote areas. The costs of drilling rise dramatically with depth, and the cost of getting new gas to market also goes up as the distance increases from the producing wells to existing population centers, highways, and the pipeline network. Pre-1976 prices on interstate gas did not permit the more costly exploration and development of the new gas with the expectation of normal profits, and profit is the legally recognized keystone of the capitalistic system. Even at $2.60 per 1000 scf, gas is priced on an energy basis at the level of $15.60 per bbl of oil, which is nearly half the third quarter 1980 price of $32 per bbl of OPEC crude. There can be only one reasonable answer: peg the price of gas (on an energy basis) at the level of its competition—oil—over which we, as a nation, exercise no control. Even then, it may be necessary to consider an additional premium for gas, to account for the fact that it is the cleanest-burning, most convenient fuel available.

Oil at $32 per bbl (the third quarter 1980 price) would be equivalent to new gas at $6.24 per 1000 scf on an energy basis.* Adding a 15¢ premium would raise new gas to $6.39, over 250 percent of the unregulated price for interstate gas. To ease the burden to the consumer, the new gas could be blended with old, regulated ("glowing") gas in interstate pipelines, to sell at an intermediate price. Preferential pricing favoring the largest users of natural gas has been eliminated by the Natural Gas Policy Act of 1978. The aim must be to provide an inducement for industry and the individual consumer to use our more abundant fuels in preference to natural gas.

---

*At that time, interstate natural gas was actually selling at about $2.60 per 1000 scf.

REFERENCES

1   A. L. Simon, *Energy Resources,* Pergamon, New York, 1975, pp. 67–73.

2   M. T. Halbouty, "Natural Gas," in D. N. Lapedes, Ed., *McGraw-Hill Encyclopedia of Energy,* McGraw-Hill, New York, 1976, pp. 472, 473.

3   D. A. Holmgren et al., Panel Disc. No. 1(4), 1975, 9th World Petrol. Congr., Tokyo; reported in A. A. Meyerhoff, *Amer. Sci.* **64**(5), 536–541 (Sept.-Oct. 1976).

4   J. Huebler, "Fuel Gas," pp. 299–301, in reference 2.

5   *Sci. News* **111**(9), 135 (Feb. 26, 1977).

6   *On the Trail of New Fuels—Alternative Fuels for Motor Vehicles,* Bundesminist. Forsch. Technol., Bonn, 1974.

7   D. E. Carr, *Energy and the Earth Machine,* Norton, New York, 1976, p. 6.

8   *Coal Facts: 1978–1979,* Natl. Coal Assoc., Washington, DC, 1976, p. 56.

9   J. D. Parent, "U.S. Fossil Fuel Resources," in *Energy Topics,* Inst. Gas Technol., Chicago, May 22, 1978.

10 J. D. Parent and H. R. Linden, *A Survey of United States and Total World Production, Proved Reserves and Remaining Recoverable Resources of Fossil Fuels and Uranium as of December 31, 1975,* Inst. Gas Technol., Chicago, 1977, p. 19.

11 R. B. Kalisch, *Public Util. Fortn.* **98**(7), 40–43 (Sept. 23, 1976).

12 *Bus. Week* (No. 2451), 66–72 (Sept. 27, 1976).

13 *Chem. Eng. News* **54**(51), 12, 14 (Dec. 13, 1976).

14 *Time* **109**(11), 46 (Mar. 24, 1977).

15 *Bus. Cond.* (AT&T), Jan. 1975, p. 4.

16 E. F. Renshaw, *Public Util. Fortn.* **99**(9), 37 (Apr. 28, 1977).

17 A. J. Dahl, *ibid.* **98**(9), 24–32 (Oct. 21, 1976).

18 BuMines; reported by N. J. W. Thrower, Ed., *Man's Domain: A Thematic Atlas of the World,* McGraw-Hill, New York, 1975, p. 42.

19 *Oil Gas J.*; reported in reference 14.

20 H. Enzer et al. (DOI, 1975); reported in *Energy Facts II,* Sci. Policy Res. Div., Libr. Congr., Serial H, U.S. GPO, Washington, DC, 1975, p. 189.

21 *Reserves of Crude Oil, Natural Gas Liquids, and Natural Gas in United States and Canada and United States Productive Capacity as of December 31, 1975,* Vol. 30, AGA, API, and Can. Petrol. Assoc., May 1976.

22 G. H. Lawrence, *Energy User News* **3**(22), 22 (May 29, 1978).

23 H. F. Keplinger, *Public Util. Fortn.* **99**(3), 18–23 (Feb. 3, 1977).

24 R. L. Chenault, "Oil and Gas Field Exploitation," pp. 509–518, in reference 2.

25 FEA; reported in *Chem. Eng. News* **52**(50), 27 (Dec. 16, 1974).

26 *Bus. Week* (No. 2443), 42N–42Q (Aug. 2, 1976).

27 *Miami Herald* (Mar. 4, 1979), p. 28-A.

28 AGA; reported by E. Faltermayer, *Fortune* **96**(2), 156–170 (Aug. 1977).

29 L. F. Pitts; reported in reference 28.

30 J. Flanigan, *Forbes* **123**(9), 45 (Apr. 30, 1979).

31 T. H. Maugh II, *Science* **191**(4227), 549, 550 (Feb. 13, 1976).

32 AGA; reported by C. Schroeder, *Energy User News* **2**(39), 18 (Oct. 3, 1977).

33 *Public Util. Fortn.* **99**(11), 58, 59 (May 26, 1977).

34 *Civ. Eng.—ASCE* **47**(1), 43–47 (Jan. 1977).

35 Reference 7, p. 62.

36 J. P. Smith, *Courier-News* (Bridgewater, NJ) **94**(118), B-12 (Dec. 6, 1977).

37 *Public Util. Fortn.* **100**(3), 28 (Aug. 4, 1977).

38 W. M. Brown, *Fortune* **94**(4), 219, 222 (Oct. 1976).

39 B. Crider, *Evening Bull.* (Phila.) **131**(67), 42 (June 17, 1977).

40 B. R. Hise; reported in reference 31.

41 P. H. Jones; reported in reference 31.

42 T. Gold, "Terrestrial Sources of Carbon and Earthquake Outgassing," presented at Imperial College, London, week of June 22, 1978.

43 D. Patterson, *New Sci.* **78**(1109), 896-898 (June 29, 1978).

44 T. O'Toole, *Courier-News* (Bridgewater, NJ) **93**(222), C-1 (Feb. 17, 1977).

45 T. Raum, *ibid.* **93**(227), C-1 (Feb. 23, 1977).

46 *Star-Ledger* (Newark, NJ) **63**(363), 1, 18 (Feb. 18, 1977).

47 *Time* **109**(8), 60-65 (Feb. 21, 1977).

48 G. Taber, *ibid.* **109**(11), 47 (Mar. 14, 1977).

49 D. Wald, *Star-Ledger* (Newark, NJ) **64**(10), 5 (Mar. 3, 1977).

50 J. Cloherty and B. Owens, *Courier-News* (Bridgewater, NJ) **93**(207), A-1 (Jan. 31, 1977).

51 *Public Util. Fortn.* **98**(9), 49 (Oct. 21, 1976).

52 FPC; reported in *Chemtech* **6**(8), 479 (Aug. 1976).

53 *Chem. Eng. News* **55**(9), 7, 8 (Feb. 28, 1977).

54 *Glass Ind.* **58**(3), 21, 22, 27 (Mar. 1977).

55 *Chem. Eng. News* **55**(3), 16, 18 (Jan. 17, 1977).

56 *Public Util. Fortn.* **99**(5), 46 (Mar. 3, 1977).

57 P. G. Burnett, "Oil and Gas Storage," pp. 518-520, in reference 2.

58 FPC (1976); reported in *Public Util. Fortn.* **98**(9), 23 (Oct. 21, 1976).

59 J. W. Amos, *ibid.* **96**(6), 15-18 (Sept. 9, 1976).

60 *Bus. Week* (No. 2418), 68 (Feb. 9, 1976).

61 *Civ. Eng.—ASCE* **47**(3), 51-54 (Mar. 1977).

62 *Ibid.,* pp. 80-82.

63 J. B. Cheatham, "Oil and Gas Well Drilling," pp. 524-525, in reference 2.

64 *Public Util. Fortn.* **102**(8), 6 (Oct. 12, 1978).

65 *A Blueprint for a Comprehensive Federal Drilling Technology Development Program,* DOE, Washington, DC, 1978.

66 R. L. Waters, *Phila. Inquirer* **301**(169), 1-E, 5-E (Dec. 16, 1979).

67 M. Panitch, *Science* **195**(4284), 1308-1311 (Mar. 25, 1977).

68 *Public Util. Fortn.* **99**(4), 32-33 (Feb. 17, 1977).

69 L. J. Carter, *Science* **190**(4212), 362-364 (Oct. 24, 1975).

70 *Time* **109**(7), 10, 11 (Feb. 14, 1977).

71 R. J. Wagman and S. Engelmayer, *Star-Ledger* (Newark, NJ) **63**(365), Sect. 1, 37 (Feb. 20, 1977).

72 T. Bush, *Bus. Week* (No. 2468), 28B-28D (Jan. 31, 1977).

73 *Ibid.,* pp. 23B, 23D.

74 *Chemtech* **6**(6), 348 (June 1976).

75 Reference 7, p. 12.

76 *Bus. Week* (No. 2532), 36 (May 1, 1976).

77 P. McCarthy, *Energy User News* **2**(33), 1, 12 (Aug. 22, 1977).

78 *Bus. Week* (No. 2511), 102-108 (Nov. 28, 1977).

79   *Ibid.* (No. 2553), 155, 158 (Sept. 25, 1978).

80   *Miami Herald* (Mar. 4, 1979), p. 28-A.

81   C. Frey, *Energy User News* **2**(33), 6 (Aug. 8, 1977).

82   M. Beiser, *ibid.* **2**(34), 12 (Aug. 29, 1977).

83   E. J. Seniura, *Cryogenics* **17**(3), 131-134 (Mar. 1977).

84   E. Cook, *Man, Energy, Society,* Freeman, San Francisco, 1976, pp. 124, 125.

85   *Phila. Inquirer* **301**(183), 2-E (Dec. 30, 1979).

86   E. Drake and R. C. Reid, *Sci. Amer.* **236**(4), 22-29 (Apr. 1977).

87   *Bus. Week* (No. 2451), 72B, 72F, 72J (Sept. 27, 1976).

88   D. Q. Haney, *Courier-News* (Bridgewater, NJ) **94**(380), A-6 (Aug. 23, 1977).

89   D. N. Gideon and A. A. Putnam, *Cryogenics* **17**(1), 9-14 (Jan. 1977).

90   A. Cockburn and J. Ridgeway, *Parade* (Feb. 20, 1977), p. 23.

91   A. W. Francis, "Liquefied Natural Gas (LNG)," pp. 412, 413, in reference 2.

92   H. A. Proctor, *Bus. Week* (No. 2453), 8 (Oct. 11, 1976).

93   K. P. Anderson and J. C. DeHaven, *Long-Run Marginal Costs of Energy,* PB-252504, Natl. Sci. Found., Washington, DC, Feb. 1975, p. 96.

94   FPC; reported by H. R. Linden et al., "Production of High-Btu Gas from Coal;" in J. M. Hollander and M. K. Simmons, Eds., *Annual Review of Energy,* Vol. 1, Annual Reviews, Palo Alto, CA, 1976, p. 85.

95   BuCensus, BuMines, ITC, and C&EN; reported by E. V. Anderson, *Chem. Eng. News* **55**(18), 31-41 (May 2, 1977).

96   *Chem. Eng. News* **55**(5), 9 (Jan. 31, 1977).

97   Reference 7, pp. 171, 172.

98   P. G. Robinson, "Methyl Alcohol," in D. M. Considine, Ed., *Chemical and Process Technology Encyclopedia,* McGraw-Hill, New York, 1974, pp. 733-735.

99   *Chem. Eng. News* **55**(17), 6 (Apr. 25, 1977).

100 W. MacQuade, *Fortune* **94**(3), 112-115, 186, 188, 190 (Sept. 1976).

101 *Bus. Week* (No. 2471), 34 (Feb. 21, 1977).

102 Reference 8, p. 64.

103 A. Stuart, *Fortune* **99**(3), 86-89 (Feb. 12, 1979).

104 *Time* **114**(15), 50-59 (Oct. 8, 1979).

4 COAL

There are only two things wrong with coal today. We can't mine it and we can't burn it.

S. D. FREEMAN

Complex mixtures of carbon compounds are continuously being formed, transformed, and decomposed in our biosphere. Through a unique ability to acquire and use the energy of the sun, plants in the sea and on the land transform carbon dioxide (and water) present in their environment to a variety of organisms, each having a precise structure. In its simplest form, the process used, *photosynthesis,* is

$$CO_2 + H_2O + \text{light energy} \rightarrow [CH_2O] + \tfrac{1}{2}O_2 \tag{1}$$

In a reverse companion reaction, carbon dioxide is released by a process known as *respiration:*

$$[CH_2O] + O_2 \rightarrow CO_2 + H_2O + \text{energy} \tag{2}$$

The nominal current concentration of carbon dioxide in the atmosphere (as derived from six sources[1-6]) is approximately 320 ppm,* but at sunset, the concentration near ground level may rise to 400 ppm.[2]

The net rate of fixation (net productivity) of carbon dioxide will vary widely with different types of organisms. It is estimated that land plants fix 20 to 30 billion tonnes of carbon annually.† The amount of carbon (as CO_2) assimilated annually by phytoplankton in the oceans (40 billion tonnes) is in the same order of magnitude as the carbon dioxide consumed by land vegetation; however, the marine environment largely retains its components of the carbon cycle.[2]

The world's forest lands are the earth's main carbon dioxide consumers and also the main repository of biologically fixed carbon, except for fossil fuels, which essentially have been removed from the carbon cycle.‡ The earth's at-

---

*This is not a precise figure.
†Individual estimates range from 10 to 100 billion tonnes.[2]
‡Except for that relatively small portion returned to the cycle by human beings through fossil fuel combustion.[2]

mosphere contains about 700 billion tonnes of carbon as carbon dioxide, with the forests containing 400 to 500 billion tonnes or approximately two-thirds as much. Assuming the average tree is 30 years old, this would mean that about 15 billion tonnes of carbon (as CO_2) is transformed into wood each year.[2]

The carbon dioxide content of our deep oceans (as bicarbonate) has remained remarkably stable through the ages,* and has thereby exerted a stabilizing effect on the carbon dioxide content of the atmosphere (at a level of about 290 ppm),[3] until we started burning fossil fuels during the industrial revolution. A more detailed discussion of the environmental impact of CO_2 will be presented later in this chapter under the heading "Social Costs of Coal Combustion—Carbon Dioxide."

FORMATION AND COMPOSITION

The carbon cycle has been functioning on earth for at least 2 billion years. Abundant marine life began about 600 million years ago, followed by abundant life on land beginning some 300 million years later.[7] A small percentage of the dead organic matter from plants and animals was preserved under anaerobic conditions at elevated temperatures and high pressures, to form the fossil fuels. Living matter was converted to proteins (carbohydrates) or lipoids (hydrocarbons). The proteins were depolymerized microbiologically to amino acids, polycondensed to humic acids and the like, and then accumulated to form the kerogen deposits or beds of peat and successive stages of coal. The kerogen was probably thermally degraded and cracked to petroleum or natural gas and to residual kerogen. In a parallel path, the lipoids were mineralized microbiologically to crude oil and cracked to natural gas.[8]

Coal was formed largely from forest growth (trees, ferns, mosses, etc.) deposited in the tidelands of ancient seas 250 to 400 million years ago (during the Paleozoic era). It is a complex polymeric organic product resulting from geologic decomposition of this organic matter. It consists of aromatic clusters of fused rings held together by assorted hydrocarbon and heteroatom (O, N, S) linkages. If there were such a thing as "typical" coal, it probably would have the composition $CH_{0.8}O_{0.1}$.[7]

Individual coal seams† are found deposited on fine sedimentary material (underclays) formed from subsidence of coastal areas as swamps were being created. These underclays are largely the source of the mineral ash found in coal. Each coal field will normally have many individual coal seams measuring from a few inches to possibly a few hundred feet in thickness. Most commercial mining will be carried out on seams $2\frac{1}{2}$ to 8 ft thick. The minimum minable seam thicknesses generally are considered to be 18 in. for strip mining and 30 in. for underground mining, but the precise limit will also depend on depth below the surface, inclination, and other factors. In general, each foot-thick seam represents 5 to 8 ft. of rotted ferns, wood, and other forest plants.

*The deep oceans turn over only every several centuries.[3]
†Also called *beds* (or less appropriately, *veins*).

Coal and coal-like products vary widely in properties and composition, depending on age, source, geological conditions of formation, and so forth.[9] The progression in coal formation is:

1 *Peat* a brown, porous, high-moisture-content coal precursor that still retains visible plant remains.[10]
2 *Lignitic coals* a low-grade, brownish coal, intermediate between peat and bituminous coal.
3 *Subbituminous* and *bituminous coals* high-carbon, low- to high-volatile, "soft" coals, intermediate between lignitic coals and anthracite.
4 *Anthracite* hard, clean-burning, low-volatile, lustrous black coal generally formed in deepest deposits, under the most severe geologic conditions.

During carbonification of the organic matter into coal, the geologic pressure occurring in the process affects mainly hardness, physical strength, optical properties, and porosity, while temperature affects chemical composition (oxygen and hydrogen content), volatiles, fixed carbon, and heat contents.[9] Coals are classified by *rank* (U.S.) or *type* (international), in accordance with their stage of carbonification. United States designations of "types" have no international counterparts. The usual U.S. classification is by rank—anthracite, bituminous, subbituminous, and lignitic coals—and by groups within these ranks—metaanthracite, anthracite, and semianthracite, and low- through high-volatility bituminous—and so forth.[11]

Rank is usually established on the basis of chemical composition and heat content (Table 4.1). Basically, the rank of coal is higher in proportion to fixed carbon content, and inversely related to moisture and volatile hydrocarbon contents. *Grades* designate beneficiation treatment employed and mesh size. More will be said later about coal processing and about elemental analysis as it affects environmental impact.

Table 4.1 Typical Characteristics of Major Coal Ranks

| Ranks | Gross Heating Value[a] | | Fixed Carbon Content, wt % | Moisture Content, wt % |
	kcal/kg	Btu/lb		
Anthracite	Up to 8,900	Up to 16,020	86–98	1–3
Bituminous	6,100–8,300	10,980–14,940	50–86	3–12
Subbituminous	4,400–6,700	7,920–12,060	40–60	20–30
Lignite	3,100–4,400	5,580–7,920	<40	Up to 40

Source: Reference 12.
[a]Includes value of condensation of steam formed during combustion.

RESERVES AND RESOURCES

Over 95 percent of the world's coal resources are located in the northern hemisphere, above 30° latitude. About 90 percent is to be found within the boundaries of the U.S.S.R., United States, and China. It has been estimated that of the total coal resources of the world, about 5 percent is recoverable and only about 0.5 percent is economically exploitable under present economic conditions and using current technology.[13]

Although the scale, continuity, and immobility of coal deposits should, and undoubtedly do, make coal a more easily quantifiable resource than crude oil or natural gas, there is still a wide disparity in estimates even of "proved" recoverable domestic reserves. Estimation of recoverable domestic reserves traditionally has fallen in the province of the Bureau of Mines' [BuMines; Department of Interior (DOI)] responsibilities, just as the U.S. Geological Survey (USGS; also DOI) has been responsible for determining our domestic resources. Together, these two organizations have developed the only authoritative, official reckoning of our coal deposits. Yet another branch of the government, the FPC (Federal Power Commission) of the DOC (Department of Commerce),* felt that BuMines gave inadequate consideration to the economics of recovery, hence produced unrealistically high estimates.[14] In the view of DOC, the coal is there, but not so much of it is economically recoverable under present conditions. As in the case of oil and gas reserves/resources estimates, authors of most sources do not offer sufficient details to permit an independent assessment of their data, nor do they always provide an adequate description of the category reported. The coal data apparently held in highest esteem by the "experts" are those of Averitt (USGS), but they are now over 10 years old and have been further revised, in some literature sources, by Hubbert, also formerly of the USGS.†

Department of Interior (BuMines) and Department of Commerce (FPC) data are rather grossly inconsistent in that the former (Table 4.2) show the United States possessing about 32 percent of the world's proved (measured, demonstrated) recoverable reserves, whereas the latter (Table 4.3) indicate that we have about 68 percent of the world total. If FPC claims are to be believed, the Bureau of Mines reserves data for the United States are high by nearly 20 percent. Data for the rest of the world are greatly at odds. The Department of Interior overstates the FPC reserves figures by nearly a factor of 5. Perhaps the best data for purposes of this publication are those screened by the National Coal Association (NCA) (Tables 4.4 and 4.5). They show the United States with about 25 percent of the world's proved coal reserves.

Some government data have been reported by others in the literature as "estimated reserves," at levels about 800 percent of the FPC figures. On close inspection, however, these estimates appear to be "identified resources" data. This is the sort of situation that often arises in trying to "label" properly data presented in various references. (It should be noted that the sources of both

*Now FERC (Federal Energy Regulatory Commission) of DOE (Department of Energy).
†More recently a consultant for Shell Oil Company.

Table 4.2 World Recoverable Coal Reserves, 1973

Region	Billion Tons	Percent
Sino-Soviet bloc	307	46
United States	217	32
Western Europe	72	11
Far East/Oceania	46	7
Africa	17	3
Other western hemisphere	9	1
Total	668	100

Source: Reference 15.

Table 4.3 Recoverable Coal Reserves

| Region | Remaining | | |
	Billion Tons	Quads	Percent
United States	184	4,054	68
Rest of world	97	1,940	32
Total	281	5,994	100

Source: Reference 16.

Table 4.4 Remaining Recoverable U.S. Coal Reserves

	Billion Tons	Quads
Demonstrated reserve base[a]	436.7	
Known recoverable reserves[b]	218.4	4,980[c]
Ultimately recoverable reserves (est.)	1,037–1,789[d]	23,644–40,789[c]

Source: Reference 17.

[a]Known or indicated deposits economically recoverable by present technology.
[b]Fifty percent of demonstrated reserve base.
[c]Heat value 22.8 million Btu per ton.
[d]Lower value from USGS Bull. 1412 (adjusted for 1974-1975 production); excludes thin beds. Higher value includes thin beds as recoverable in future.

Tables 4.2 and 4.3 are quoted on facing pages of a publication prepared for the guidance of our congressional leaders in establishing policy, and with no adequate qualification of the differences.)[15,16] Table 4.6 shows proved U.S. reserves by rank (FPC). Of our "underground" coal,* 68 percent is located east of the

*The source[19] calls the data "recoverable reserves," but comparisons with data from other sources suggest that they may actually be demonstrated reserve base data.

Table 4.5 World's Coal Reserves (Estimated)[a]

Region	Billions Net Tons		
	Recoverable Reserves	Total Reserves	Total Resources
U.S.S.R.	150.6 (23.1%)	301.2 (19.2%)	6,298.2 (53.1%)
United States[b]	200.4 (30.7%)	400.8 (25.6%)	3,223.7 (27.2%)
China	88.2 (13.5%)	330.7 (21.1%)	1,102.3 (9.3%)
West Germany	43.6 (6.7%)	109.7 (7.0%)	315.4 (2.7%)
Australia[c]	26.8 (4.1%)	81.9 (5.2%)	218.9 (1.8%)
United Kingdom	4.3 (0.7%)	109.0 (7.0%)	179.5 (1.5%)
Canada	6.1 (0.9%)	10.0 (0.6%)	119.9 (1.0%)
Rest of Europe	66.9 (10.3%)	91.0 (5.8%)	108.0 (0.9%)
India	12.8 (2.0%)	25.5 (1.6%)	91.5 (0.2%)
Poland	25.0 (3.8%)	42.9 (2.7%)	66.8 (0.6%)
South Africa	11.7 (1.8%)	26.7 (1.7%)	48.9 (0.4%)
Latin America	3.1 (0.5%)	10.1 (0.7%)	36.3 (0.3%)
Rest of Asia	6.6 (1.0%)	19.1 (1.2%)	27.6 (0.2%)
Rest of Africa	5.6 (0.9%)	6.7 (0.4%)	16.0 (0.1%)
Rest of Oceania	0.2 (—)	0.4 (—)	1.2 (—)
World total	651.7 (100.0%)	1,565.6 (100.0%)	11,854.1 (100.0%)

Source: Reference 18.

[a]Excludes peat.

[b]Differs from BuMines and USGS because of the use of different seam and overburden thickness criteria.

[c]Does not include Queensland.

Table 4.6 U.S. Coal Reserves

Rank	Proved—Recoverable[a]		
	Billion Tons	Percent	Quads
Anthracite[b]	1	6	(234)
Bituminous/subbituminous	155	85	3436
Lignite	28	9	384
Total	184	100	(4054)

Source: Reference 14.

[a]Allows 50% recovery factor. Takes into account economics of recovery, unlike BuMines data.

[b]Numbers, as reported in the source, do not yield proper heating value for anthracite (about 26.7×10^6 Btu/ton). Energy value appears to be about one order too high, but the effect on the overall picture is minor.

Mississippi River, whereas only 32 percent of our "surface" coal is in the East. Only about 25 percent of our total coal reserves are low sulfur (i.e., < 1 percent), and approximately 70 percent of this is in our western fields.[19] One of the difficulties in assessing coal data is that criteria contributing to recovery factor are not always noted.*

To confuse the issue further, some sources refer to the *reserve base,* which means simply the quantity of in-place coal calculated on the basis of specified depth and thickness criteria, with no regard for recovery factor. It yields a quantity intermediate between "resources" (the total quantity in the ground) and "reserves" (the quantity recoverable). The potentially minable reserves for the various grades of coal, by region and state, are shown in Table 4.7 (surface minable) and Table 4.8 (deep minable).[20] Surface-minable reserves are generally about 32 percent of the reserve base, and the deep-minable ratio is only about 25 percent.

Available coal resources are in somewhat better agreement, perhaps merely because there are fewer data sources to draw on. Table 4.9 shows FPC estimates of proved recoverable coal resources of the United States and the rest of the world.

Tables 4.10 and 4.11 report U.S. coal resources data in reference to coal rank and sulfur content, respectively. The bulk of our coal is low in sulfur, but most of it is not too accessible to major markets, nor is its extraction acceptable to environmentalists.

Conventionally, resources/reserves are reported as tonnages in place. "Standardized" resources/reserves are also sometimes referred to. These are tonnages normalized for equivalent heat contents (or possibly sulfur contents), which constitutes a more meaningful statistic, since it is the energy value, not the raw mass that concerns us. The normalized data show only about 25 percent of the conventional resources and 10 percent of the conventional reserves.

Throughout this discussion of reserves and resources, it has been all too apparent that there are many considerations that must be factored in to provide reliable data. It is unfortunate that so few references provide enough background details to make even an offhand assessment of their validity possible. The range of the figures raises the suspicion that not all the source data are based on the same premises.

Among the important factors to be considered are: the limiting depth and thickness of the coal seam(s); geographic location; topography; water supply; accessibility of transport; sulfur, ash, and heat contents; cost of competing fuels; minability and recovery factors; environmental impact; and reclamation costs. Without careful consideration of all these items, and others, no reserves estimate can be reliable, yet no compilation of reserves provides any indication of the weighting applied in reference to these factors.

---

*In a given underground mine, worked by the room-and-pillar technology, the recovery factor is about 50 percent. In practice, however, substantial areas lying beneath cities and between mines, small isolated pockets, and severely faulted zones, all of which preclude mining operations, lead to an overall recovery of about 25 percent (20 to 35 percent) of the in-place reserve base.

Table 4.7 Potentially Surface Minable U.S. Coal Reserves, January 1, 1976

| | Reserve Base, Million Tons | | | |
Region/State	Anthracite	Bituminous	Sub-bituminous	Lignite
Appalachia				
Alabama		284		1,083
Kentucky (eastern)		4,468		
Ohio		6,140		
Pennsylvania	143	1,392		
West Virginia		5,149		
Other (Maryland, North Carolina, Tennessee, West Virginia)		1,361		
Total	143	18,794	0	1,083
Midwest				
Illinois		14,841		
Indiana		1,774		
Kansas		998		
Kentucky (western)		3,900		
Missouri		3,596		
Texas				3,182
Other		534		26
Total	0	25,643	0	3,208
West				
Alaska		80	641	14
Montana			33,843	15,767
New Mexico		601	1,847	
North Dakota				10,145
Wyoming			23,725	
Other (Arizona, Colorado, Oregon, South Dakota, Utah, Washington)		1,270	634	3,400
Total		1,951	60,690	29,326
Grand total	143	46,388	60,690	33,617

Source: Reference 20.

One problem in estimating domestic reserves is that the largest owners of our in-place coal are oil companies,* utilities, and railroads, each of whom generally considers its data proprietary. Of the 10 largest owners of U.S. coal reserves in 1977, four were oil companies, three were coal companies, two railroads, and one a steel company. Together, they owned 71+ billion tons of

*Fifteen of the 50 largest coal producers in 1977 were oil companies.[22] They accounted for 20 percent of domestic coal output.[23]

Table 4.8 Potentially Deep Minable U.S. Coal Reserves, January 1, 1976

Region/State	Reserve Base, Million Tons		
	Anthracite	Bituminous	Subbituminous
Appalachia			
Alabama		1,724	
Kentucky (eastern)		9,072	
Ohio		13,090	
Pennsylvania	6,967	22,336	
Virginia	138	3,277	
West Virginia		33,457	
Other (Georgia, Maryland, North Carolina, Tennessee)		1,572	
Total	7,105	84,528	0
Midwest			
Illinois		53,128	
Indiana		8,940	
Iowa		1,737	
Kentucky (western)		8,510	
Missouri		1,418	
Other (Arizona, Michigan, Oklahoma)	89	1,318	
Total	89	75,051	0
West			
Alaska		617	4,806
Colorado	26	8,468	
Montana		1,385	69,574
New Mexico	2	1,259	889
Utah		6,284	1
Wyoming		4,002	27,645
Other (Oregon, Washington)		258	849
Total	28	22,273	107,736
Grand total	7,222	181,852	107,736

Source: Reference 20.

recoverable coal reserves.[24] Coal is widely distributed through about 30 states. The major coal sources (including lignitic deposits) are shown in Table 4.12.

Much has been said in the popular press about the hundreds of years of coal reserves that exist in the United States, but little has been said about the effect of consumption growth rate on the lifetime of these reserves. Bartlett has calculated the life expectancy of our coal reserves at various annual consumption growth rates and at two assumed levels of reserves—430 and 1640 billion tons. At a consumption growth rate of about 6.7 percent (the average for the 50 post-Civil War years), the lower level of reserves would last about 60 years,

Table 4.9   Coal Resources

Region	Proved Recoverable		
	Billion Tons	Quads	Percent
United States	1,577	31,124	16
Rest of world	7,923	158,460	84
Total	9,500	189,584	100

Source: Reference 16.

Table 4.10   U.S. Coal Resources

Rank	Billion Tons[a]
	(FPC, 1975)
Anthracite	15 (1.0%)
Bituminous	725 (46.0%)
Subbituminous	389 (24.7%)
Lignite	448 (28.4%)
Total	1,577 (100.1%)

Source: Reference 21 (FPC).
[a] Reduced by 50% recovery factor.

Table 4.11   U.S. Coal Resources by Sulfur Content

Sulfur Content	Remaining—Identified[a]		
	Billion Tons	Percent	Quads
Low sulfur (0-1%)	1,022	57	17,854
Medium sulfur (1.1-3%)	240	18	5,451
High sulfur (>3%)	315	25	7,819
Total	1,577	100	31,124

Source: Reference 21.
[a] Allowing 50% recovery factor. Takes into account economics of recovery.

while the higher level of reserves would provide a lifetime of about 80 years. At the recent (mid-1970s) annual worldwide consumption growth rate, the low- and high-level reserves would have life expectancies of 105 and 150 years, respectively. With coal slated for synfuel production, as well as increased direct combustion for electric power generation, growth rate could even exceed the higher-consumption-growth scenario.[26] It has taken nature millions of years to

Table 4.12 Remaining Identified Coal Resources by States[a]

	Billion Tons[b]
North Dakota	350.6[c]
Montana	291.6[d]
Illinois	146.0
Wyoming	135.9
Alaska	130.1
Colorado	128.9
West Virginia	100.2
Pennsylvania	82.8
Kentucky	64.3
New Mexico	61.4
Ohio	41.2
Indiana	32.9
Missouri	31.2
Utah	23.4
Arizona	21.2
Kansas	18.7
Texas	16.3
Alabama	15.3
Virginia	9.6
Oklahoma	7.1
Total	1,708.7

Source: Reference 25.

[a] Twenty largest only (98.7% of total U.S.).
[b] Resources in ground; about half should be recoverable. Includes bituminous coal and anthracite beds at least 14 in. thick, and subbituminous coal and lignite at least 2.5 ft thick.
[c] All lignite.
[d] Nearly 40% lignite.

produce the earth's vast coal deposits. It now appears that it will take mankind only centuries or, at most, millennia to deplete these reserves.

MINING AND TRANSPORT

There are three major reasons not to use coal as our primary source of energy. They are economics, environmental impact, and logistics, not the least of the three being the latter item. In 1976, the United States had a total of 6161 coal and lignite mines.[27] The bituminous mines employed an average of 202,280 workers that year to produce 678.7 million net tons. The average productivity per worker-day was 14.46 tpd for surface and underground mines taken together.[28]*

To achieve total independence from foreign oil, we would need possibly 3 billion tpy production by 1985. This means 28,000 mines and 1 million mine workers would be required.[29] To replace our oil imports with synfuels from coal by the year 2000, we should need to open 56,000 new coal mines (2400 new mines per year), and 2 million new miners would be required. This would mean training about 100,000 new miners per year—almost like calling up an army for combat.[30] Furthermore, approximately 2700 unit trains of 100 hopper cars each may be required to transport the coal.[31]† Thus it becomes easy to understand the concern with the logistic and economic aspects of a coal economy in these times of high energy consumption, dwindling oil and natural gas supplies, and growing opposition to nuclear power.

These are only a few of the problems dealing with coal. Deposits in the Northern Great Plains (Montana, Wyoming, North Dakota, and South Dakota) are thick—typically 50 to 75 ft—and are quite shallow (30 to 40 ft),[32] but rainfall is so skimpy (< 10 to 20 in. per year)[33] as to make process and land-reclamation water inadequate. Rainfall in the northern part of the Mountain province (Idaho, Utah, and Colorado) is in the same range, but in the southern part of the Mountain province (Arizona and New Mexico) is even lower (< 10 in.). The Interior (Iowa, Kansas, Missouri, Illinois, Indiana, Oklahoma, Arkansas, and western Kentucky) and Eastern (Ohio, Pennsylvania, West Virginia, eastern Kentucky, eastern Tennessee, and Alabama) provinces have adequate rainfall (20 to 60 in.), but seams are much thinner (< 10 ft) and deeper than in most of the western region.

To better understand the problems, it would be well to review mining technology, productivity, preparation or beneficiation, and transport of coal.

Anthracite

Anthracite is only a minor part of our energy picture, but its mining involves some unique problems, so it will be dealt with briefly and separately. The use of continuous-mining machines is out of the question because most anthracite seams are wavy and/or steeply pitched, in addition to which the coal is so hard as to yield only to hand picks, pneumatic drills, and blasting. Furthermore, most anthracite seams are so deep that the high cost of pumping out water often makes recovery uneconomical. Hourly labor costs in the anthracite mines are considerably lower than in soft-coal mines,‡ but the delivered price of anthracite is relatively high, because of the inherent difficulties of producing it and the resulting low productivity.§ As a result, production fell to 6.3 million

---

*About 9.10 tpd per worker in deep mines, and 26.40 tpd in surface mines.

†See the later discussion on transport.

‡Anthracite miners belong to the United Mine Workers, but they bargain separately from their counterparts in the bituminous coal mines and, lacking the numbers clout, they fail to do as well.

§About half that of the bituminous coal mines.[34]

tons in 1973, and most of that was recovered from strip mines and old spoil banks.[35] The Bureau of Mines estimates that in Pennsylvania alone, there are 800 anthracite spoil banks containing a total of up to 5 quads of recoverable energy. Technical proposals are being sought by the Department of Energy to recover this energy.[36] One possibility is to use fresh anthracite blended with anthracite culm as a boiler fuel.[37]

Two major mining techniques are employed for producing coal. They are indicated in Table 4.13, together with variations. (See the Glossary, Appendix B, for definitions.)

Underground Mining

Deep or *underground mining* seems like a throwback to Dickensian times, with all its worst connotations. The underground extraction of 300 million tons of domestic coal in 1973 cost 107 lives (six times the number lost in surface mining).* About three times as many workers are required in deep mining to produce the same tonnage extracted by safer surface techniques. Spoil banks containing billions of tons of gob and coal residues loom over coal mining towns and threaten the inhabitants with landslides, floods, and fires. As air and water deteriorate the coal and pyritic shale pillars left as support in the mines, the pillars weaken and allow the mine roofs to collapse, causing massive surface subsidence. An unpublished report of the Bureau of Mines showed that some 6 million acres (an area exceeding that of the state of Massachusetts) has been undermined by deep mining of coal in the United States. About one-third[38] of

Table 4.13   Coal-Mining Technologies

Underground mining (deep, deep-pit, pit)
   Room-and-pillar (pillar-and-block)
   Longwall (advance)
   Retreat
   Shortwall
   Down-dip
   Up-dip

Surface mining (strip, open-cast, open-pit, open-cut)
   Area
   Contour
   Auger
   Dredging
   Haul-back
   Multiseam

Hydraulic mining

---

*About 1000 lives have been lost in American coal mines since the turn of the century.[38]

this area has subsided* and another one-eighth (0.75 million acres) is expected to collapse by the year 2000.[35]

Why, then, does underground coal production still remain an important factor (Table 4.14), providing over 40 percent of our production? Extensive deposits of low-sulfur (less than 1 percent) coal exist in "Appalachia," relatively close to eastern industrial markets, but the preponderance of this resource is too deep for surface mining. Vast deep deposits of medium- to high-sulfur bituminous coal also exist in the midwestern states,[40] but their exploitation now lags due to air pollution restrictions on combustion.

Underground mining of coal is the most hazardous major industrial occupation, but it is also one of the highest paid.† Since the 1950s the miner has been assisted on the level seams (in large mines) by continuous-mining machines costing up to $300,000.[41] Even so, productivity in these mines has been on the decline since peaking in 1969, the year the Coal Mine Health and Safety Act was enacted. Table 4.15 shows productivity in U.S. coal mines. The trend is not attributable solely to the restrictions of this act, since many new, inexperienced miners were taken on in the late 1960s. Nonetheless, downtime for the "continuous" mining machines, mandated by the act in order to monitor methane,

Table 4.14 U.S. Bituminous Coal Production, 1951–1976

	Million Net Tons			
Year	Under-ground	Auger	Surface	Total
1951	416.0	—	117.6	533.7
1955	343.5	6.1	115.1	464.6
1960	284.9	8.0	122.6	415.5
1965	332.7	14.2	165.2	512.1
1970	338.8	20.0	244.1	602.9
1971	275.9	17.3	259.0	552.2
1972	304.1	15.6	275.7	595.4
1973	299.4	15.7	276.6	591.7
1974	277.3	*a*	326.1	603.4
1975	292.8	*a*	355.6	648.4
1976*b*	294.8	*a*	383.9	678.7

Source: Reference 39.

*a*Included with surface mining.

*b*Estimate.

*Some subsidence has occurred under cities (e.g., Scranton and Wilkes-Barre, Pennsylvania), causing buildings and streets to collapse.

†With benefits, the employer's outlay for one worker-day in a medium-size mine was about $110 in 1972,[41] of which $60 were wages.[42] Steel workers' wages were the only ones in other industries comparable to those of the coal miners.[43]

Table 4.15 Productivity of U.S. Coal Mines

	Tons/Worker-Day			
	Bituminous Mines			Anthracite
Year	Underground	Surface	Average	Mines
1940	4.86	15.63	5.19	3.02
1945	5.04	15.46	5.78	2.79
1950	5.75	15.66	6.77	2.83
1955	8.28	21.12	9.84	3.96
1960	10.64	22.93	12.83	5.60
1965	14.00	31.98	17.52	6.55
1970	13.76	35.96	18.84	7.10
1971	12.03	35.69	18.02	6.30
1972	11.91	35.95	17.74	6.88
1973	11.66	36.67	17.58	7.15
1974	11.31	33.16	17.58	7.87
1975	9.54	26.69	14.74	7.45
1976	9.10	26.40	14.46	7.19

Source: Reference 34.

coat newly exposed surfaces to eliminate dust, and insert anchor bolts in the roof, means that the machine is out of service at least 80 percent of the time.[35] Haulage of the coal to the surface is a major bottleneck in taking full advantage of the machine. A major producer has been investigating the possibility of crushing the coal behind the continuous miner and conveying it hydraulically through a 10-inch pipe. This could reduce underground labor requirements by 25 percent and yet enable conveyors to keep up with the continuous miner.

The ultimate improvement in U.S. mining technology would be more widespread adoption of *longwall mining,* which is widely used in Great Britain. The longwall face may extend several hundred feet, with a special machine shearing or planing off the coal in a continuous motion as it advances. Hydraulic roof supports or jacks are automatically emplaced as the machine progresses. They are then advanced remotely as mining proceeds, thus allowing the roof to collapse gently behind the shearing machine. This technique recovers over 80 percent of the coal in place (Table 4.16), eliminates roof bolting and its attendant hazards and labor commitment,* and allows the overlying land to subside quickly and uniformly. It does, however, call for a capital investment of about $2 million per mine, so it cannot be employed in all mines. It offers a side benefit, in that personnel can be trained in less time to use the longwall machine than to use pick-and-shovel techniques.[47] The machine is in use only about half the time overall, because of the extended time required to

---

*Fifteen to 20 percent of underground labor.

Table 4.16   Coal-Mining Recovery Factors

Technology	Recovery Factor, %[a]
Undergound mining	
Longwall, U.S.	>80%
Longwall, U.K.	87–92
Room-and-pillar	50
Surface mining	
Area	80–90
Auger	40–60

Sources: References 35 (longwall, U.S.), 44 (longwall, U.K.), 45 (room-and-pillar and area), and 46 (auger).

[a]For a given mine. Seams lying in isolated pockets, in heavily faulted zones, under cities, etc., reduce overall average. See the text, "Reserves and Resources."

set up a new panel. Very large volumes of air are necessary to ventilate a longwall mine adequately, because of the amount of dust and methane released.

An American variant of this method is *shortwall mining,* which uses the less expensive continuous miner, employs a narrower panel, and sets fewer jacks. The need for this modified technique is dictated by the fact that overlying strata in many American geological formations do not collapse predictably. It is estimated that about one-third of our future deep mines will probably be adaptable to longwall mining and its variants. This is particularly true of the thicker seams that cannot be mined efficiently by the *room-and-pillar method.* *

Retreat mining, another variant of longwall ("advance") mining, is now attracting considerable attention in Great Britain. Normal longwall shearing is done as the machine advances along the panel in a direction away from the shaft. In retreat mining, service tunnels are driven ahead, and the machine cuts the coal face on its return toward the shaft.[48] These techniques, together, are spoken of as *advanced technology mining* (ATM), in that they employ coal cutting machines, automatic roof support systems, and conveyors. The retreat mining technique is reported to result in doubling the output per worker.

In underground mining, there are upper and lower limits on seam thickness for economical recovery of coal. The usual minimum is about 2½ ft and the maximum is about 16 ft. Over 16 ft it becomes too difficult to crib up the roof. Inclination of the seam can also affect economics of mining. Level seams permit the use of continuous miners. Seam slopes over 40° generally make deep mining impractical. Seams in the western United States, which may measure over 100 ft thick,† defy mining by present technology. The only alternative will probably be

---

*Generally, the thicker the seam, the lower the recovery factor by underground mining, because larger supporting pillars must be left in place.

†Up to 223-ft thick lignitic or subbituminous beds have been encountered in Wyoming.[9]

to open the seam up by drilling a network of holes, making a "Swiss cheese" of the bed, and then using controlled combustion to produce a low-energy gaseous fuel for minemouth power generation.[45]

Underground mining is usually employed where the coal seam is too deep (over 225 or 250 ft) for stripping, where the incline is too steep to permit economical removal and retention of overburden, or where the ground surface is too valuable to disturb or impossible to reclaim. Deep mining production costs are generally three to four times those of surface mining (including land restoration costs), and production per mine (and per worker) is considerably lower. Most of the smaller mines are family-owned (two or more miners digging a few tons per day), while many of the larger mines (up to 1000 miners and daily production up to 10,000 tons) are "captives,"* owned by steel or utility companies. These mines have the best safety records in the industry.[50]

Up-dip and *down-dip mining* refer to specialized underground mining techniques. Both involve mining of sloping seams outcropping on opposite sides of a mountain. In the past, it was standard practice to dig into the lower end of the seam and work up the slope ("up-dip" mining). This was to take advantage of gravity in draining the mine of water seepage and in bringing the coal to the minemouth, but it also means a long-term problem with *acid mine drainage* (AMD).

The current emphasis is on down-dip mining, that is, digging into the high end of the seam, to avoid AMD. This ensures that the water will be trapped in the mine for all time, as long as suitable care is taken to prevent excessive hydrostatic pressure from building up. This can be accomplished by leaving an 800-ft-long plug of coal in place for every 5 mi of seam.[51]

Surface Mining

Surface mining technology is applied where relatively level, extensive seams at least 18 in. thick lie near the earth's surface. Surface mining cannot be conducted where more than about 225 ft of overburden must be removed. In coal fields of the northern Great Plains, overburden typically measures 30 to 40 ft, overlying a 50- to 75-ft coal seam—ideal conditions for area strip mining. The largest surface coal mine produces 300,000 tpd,[52] and the record for deep mine production is 12,700 tpd.

In western coal fields, surface mining comprises six major steps[52]:

1 Topsoil and other unconsolidated materials are removed by scrapers and placed on spoil piles.
2 The consolidated overburden (shale, sandstone, limestone, slate, etc.) is drilled and blasted.
3 The loose overburden is removed by means of huge power shovels or draglines and added to the spoil banks.

---

*In recent years, captive mines have been producing about 15 percent of our output.[49]

4 The berm (exposed coal seam) is drilled and blasted.

5 The coal is loaded into trucks and hauled away for cleaning or for transfer to long-distance transport.

6 The spoil banks are then regraded by bulldozers.

Economics and logistics are dominant factors when it comes to exploiting the large deposits of low-sulfur subbituminous coal in the western states. In 1973, productivity was so high per worker (Table 4.15) that the coal was extracted at about $3 per ton, but the mines were so remote from the marketplace that transportation costs raised the delivered price in Chicago to $12 per ton. Since the subbituminous coal had only 75 percent of the energy content of midwestern or eastern coal, the latter could be delivered at $16 per ton and still remain competitive on an energy basis.[35]

In eastern and midwestern fields, which tend to be smaller and covered with thick overburden, a 10-ft-thick coal seam would be considered exceptional. At one mine in Oklahoma, 95 ft of overburden had to be removed to expose an 18-in. coal bed. The midwestern and eastern surface mining operations are basically the same as for the large western mines, but the mechanized equipment is smaller and more mobile. Western strip mines have remained quite stable at about 50 units, with 35 of these (70 percent) accounting for 90 percent of production. Three-fourths of the output comes from mines no more than about 15 years old.[52]

The latest development in strip mining is called *multiseam mining.* It consists of stripping deep into one section of land in successive stages, to maximize extraction of the coal. It results in better utilization of the land, more complete extraction of the coal, and lower energy costs, even though the technology is more complex. The multiple seams are separated by only a thin layer of weak rock called interburden. The difficulty lies in using huge draglines at such depths, without becoming "coal-bound," that is, being unable to coordinate removal of interburden with digging and transport of coal to the surface.[53]

The eastern and midwestern strip mines are necessarily less efficient than the great western mines, but even so, their productivity on a worker-shift basis greatly exceeds the most advanced deep mines. Over the past 40 years, the number of Appalachian strip mines has increased from approximately 30 to about 2100.

The bottleneck in deep mines is in conveying coal from the stope to the shaft. In surface mining, it is removing the overburden, and the limiting factor here is the capacity of the machinery, which is already so monstrous that the ground will barely support it.[32] In some of the major fields, the reserves are so large that mechanical monsters for removing the overburden will be operated for decades in one area.[54]

In hilly terrain, where the coal seam forms an outcrop underlying a hill, *contour mining* may be used,[55] with the overburden being dumped down the slope and spoil from subsequent furrows being dumped into trenches left by removal of the coal. A variation of contour mining is *haul-back mining,* where a peripheral strip of overburden on a mountainside is removed sequentially and hauled back into the void left by the advancing coal mining operation.[51]

Where the uphill slope is too great to permit removal of the overburden, or the geological formation is such that underground mine roofs cannot be sustained, *auger mining* will be used. Huge drills form closely spaced holes up to 8 ft in diameter (but more commonly 5 ft), driven to a depth of up to 300 ft into the coal seam. The augers extract up to 40 to 60 percent of the coal in the process.[32,56]

Dredging and *hydraulic mining* are other techniques used to a limited extent throughout the world, but the methods discussed above account for the great bulk of coal production.

Preparation or Beneficiation

Initially, all coal was mined by pick and shovel, resulting in coal clean enough to be sold without further preparation, in the absence of environmental regulations. About half of the coal extracted from mines today is sufficiently free of foreign material to be loaded for delivery directly to the consumer. This product is known as *run-of-mine* (ROM) coal. The other half is treated by some means to prepare it for the market.[51] This treatment may be a simple washing or a higher-level beneficiation. The beneficiation is generally applied at the mine-mouth to clean up the coal or to upgrade its properties for economic reasons. Historically, this was done to reduce ash, but its primary purpose now is reduction of sulfur to permit conformance to environmental regulations.[57]

The procedure must be kept as simple as possible because of cost considerations. Preparation, in its simplest form, can consist of using compressed air to remove dust,[58] or using flotation techniques to float the coal away from the heavier mineral contaminants. In this case, sizes under $^3/_8$ in. must be dried afterward to reduce moisture content to an acceptable 7 to 8 percent or less.[45] At the very least, tests will be conducted to determine the wettability and float-sink characteristics ("washability") of the coal before a decision is reached on preparation.

Washing and other preparation techniques reduce the tonnage of coal that reaches the market.* Part of this loss occurs because continuous miners cannot discriminate between coal, slate, or roof and bottom materials. Slight variations in seam thickness or slope can cause increases in the amount of dirt mined. Typically, about 15 to 20 percent of raw coal may be incorporated in the refuse during cleanup. This refuse (rocks, minerals, coal) has a heating value of 3000 to 5000 Btu per lb, but there is no technology available for making use of it.[59] It is easy to understand why so many slag piles are burning out of control throughout various mining regions.

It is estimated that at least one-third of the coal refuse from Appalachian mining operations could be recovered by crushing and fine cleaning.[60] This has

*Currently, *net production* figures reflect the smaller amount of cleaned coal rather than the total mined.

not yet proven cost-effective in the United States, but it is done routinely in Europe. Fine cleaning can be accomplished by jig or froth flotation. Jig flotation is a sophisticated, mechanized version of the "panning" technique carried out by gold prospectors over a century ago. It should remove pyritic fines (0.0003-in. diameter) from $<\frac{1}{4}$-in. mesh samples. Froth flotation requires a finer grind ($\frac{1}{60}$ in.). Chemicals are added and injected air bubbles carry the coal powder up into the froth, where it is removed by scraping. The cost should be less than stack gas scrubbing.

Coals differ widely in the gains to be realized by beneficiation. Since preparation affects the cost of the product, care must be employed to establish an appropriate cost-effective level. The major factors affecting preparation costs are[57]: plant location, capacity, and production, coal properties (before and after), coal-handling facilities, raw coal cost, and waste characteristics.

Level A beneficiation (A for "absence") is not a common practice. Level B (for "breaking") controls the maximum size of the coal and removes some trash. This method is generally used for ROM coal. Level C (for "coarse" beneficiation) calls for washing $+\frac{3}{8}$-in. coal and shipping zero-to-$\frac{3}{8}$-in. coal dry. Level D ("deliberate" beneficiation) washes all $+28$-mesh coal and either discards or dewaters and ships the fines with clean coal. Level E ("elaborate" beneficiation) washes all fractions and thermally dries the zero-to-$\frac{1}{4}$ in. fraction. Level F ("full" beneficiation)* reduces size further and produces two or more product streams, generally a clean coal stream† for steam generation and a middlings stream for noncombustion industrial use.[57]

In general, coal yield may decrease with higher-level cleaning to as low as 60 percent, but recovery of heating value stays at 80 percent or higher. Ash removal is fair to good at Level C and above, but sulfur removal is generally somewhat less efficient. The heating value of the clean coal may be up to about 25 percent higher than the raw coal. Full cleaning may come close to tripling the cost of clean coal on a tonnage basis.[57] In general, the benefits (with regard to properties and economics) will offset the modest added cost of beneficiation.[58]

Chemical comminution of coal could be in the offing. A contract was authorized to design and operate a pilot plant that would fracture the coal along natural fault lines by treating it with ammonia and methanol. This could eliminate mechanical crushing preliminary to sulfur and ash removal.[61]

Transport

The Carter Administration's goal is to increase coal production two-thirds by 1985. This will mean attaining an annual production level of about 1.1 billion tons at that time.‡ A major problem is going to be transporting this quantity of

---

*Also called deep or full cleaning, multicoal cleaning strategy, or multiproduct or multistage cleaning.

†Over 25 percent of the total product.

‡Net 1977 coal production reached 688.6 million tons, a 1.5 percent increase over 1976's 678.7 million tons.[62]

coal from mine to consumer. A summary of the recent coal transport situation is shown in Table 4.17. Nearly two-thirds of our output was shipped by rail, but rail transport has been severely hampered by deteriorating roadbed conditions that force "slow rules" in many cases (particularly in the Northeast)—increasing delivery and turnaround times on trains. Hopper cars are in short supply, and locomotives are often antiquated and underpowered. "Institutional factors"—featherbedding work rules and regulatory practices—contribute heavily to the deteriorating rail transport situation. To compound the problem further, the mines lack storage facilities in case of a backup of deliveries due to weather, strikes, or technical difficulties.

Water transport is the next largest mode of long-distance coal transport (about 11 percent), but it is hampered (with regard to both capacity and cost) by the need for physical transfer of the coal between hopper car or truck and barge. Waterways are also of limited capacity, particularly during a freeze-up such as occurred in the winter of 1976–1977 and during drought periods such as were in effect through much of the same years in certain parts of the country.* Other transport methods are largely short-haul and more costly.

Unit trains, generally 100-hopper car units pulled by dedicated locomotives and serving a single round-trip route between mine and large consumer (e.g., a power generating plant), provide the most efficient possible rail transport for coal, despite the fact that the trains return empty. Nationwide, such trains are expected to move coal an average of 430 mi per trip by 1985.[65]

Typical of what can be done in modern coal handling and transport, is a totally integrated rail–ship transport and transfer system recently built to fulfill a contract between Detroit Edison (Michigan) and Decker Coal Company

Table 4.17   Domestic Coal Transport, 1975

Mode	Million Net Tons	Percent of Production
Rail	418.1	64.5
Water	69.1	10.7
Motor vehicle	79.4	12.2
Used in minemouth generating plants	73.5	11.3
Other uses at mine[a]	8.3	1.3
Total	648.4	100.0

Source: Reference 63.

[a]Used by mine employees, locomotive tenders, for mine power and heat, etc.

*On the Great Lakes, shippers figure that every 1-in. decrease in water level means that about 100 tons less cargo can be carried. As a result of drought conditions starting in 1976, the Great Lakes was down to 3 in. below "datum," a condition that starts with the watershed feeding the lakes.[64]

(Montana) for delivery of 200 million tons of low-sulfur western coal over a 26-year period.[66]*

An all-rail land route between the Montana mine and the Detroit consumer would have been a logical course, but no single railroad had the necessary continental track rights. This meant that an all-rail route would call for switching tracks, locomotives, crew, and procedures to pass through already congested railyards for the 1500-mi journey. The decision was reached to haul the coal 800 mi by rail to Superior, Wisconsin (on Lake Superior), by 110-car, 11,000-ton unit train, then transship by 44,000- and 68,000-dwt† coal carriers through the Soo Locks and Lake Huron to the power plants.

To ensure reliable supplies through the cold winters when the lakes are frozen, a high-capacity transshipment terminal had to be built in Superior, covering a 200-acre waterfront site. The arriving unit trains travel around a loop through an 800-ft thaw shed to skin-thaw the coal when necessary.‡ The cars are emptied by a 3500-tph rotary dumper without uncoupling, and the coal is carried by conveyor belt to a "live" stockpile (capacity 120,000 tons). Over the winter, the incoming coal is accumulated within the loop-track area in a 50-ft high pile holding up to 7 million tons. From there it is fed by multiple plow feeders through an 8-ft diameter tunnel to an 11,000-tph traveling gantry shiploader, which operates from a 1215-ft dock without need to move the ship. Of course, not all contracts can justify such an elaborate dedicated scheme for delivering coal to a single customer.§

There is total disagreement between the Association of American Railroads (AAR) and the Slurry Transport Association (STA) over the preferred mode and the cost of coal transport. Environmentalists have also become involved over the "social costs" of transport.

There is perhaps an even wider disparity between estimates of the number of new hopper cars and locomotives needed to haul the coal output than there is between the estimates of coal resources. The estimates for transport range from 9700 new hopper cars per year[69,70] to as many as 20,800.[31,71,72] The estimates for new locomotives run from 110 to 485. The revised AAR estimate,[70] which is lowest in hopper cars and highest in locomotives, places the capital cost at $4.3 billion per year over the 8-year period to 1985.

There were 394,000 hopper cars (each of 100-ton capacity) in service in 1970. Four years later, there were 40,000 fewer (354,000), but only 334,000 of them were serviceable. The National Academy of Engineering (NAE) has estimated that to meet future needs, 150,000 new hopper cars will be required.[72] In 1973, only 2600 new cars were put into service; future requirements seem to

---

*Four million tons in 1976, increasing to 20 million tons per year.

†The largest coal carriers ever used on the Great Lakes.

‡There were news reports during the cold winter of 1976–1977 concerning mined coal supplies which were so solidly frozen that they could not be unloaded from hopper cars or could not be transferred from stockpiles to boilers. A "Frozen Coal Cracker" (single-roll crusher) has since been developed to break coal as it is unloaded from hopper cars.[67]

§It has been reported that Detroit Edison ultimately will pay Decker Coal Company $1 billion for the coal, but the transport bill will be $2 billion![68]

point to nearly 20,000 additions per year. One thing is clear; large quantities of new rolling stock will be necessary to keep coal moving by rail on schedule, and the cost will be high.

The alternative, in the minds of some, is slurry pipelining. This method is characterized by its supporters as being inherently less expensive and more environmentally acceptable than rail transport, and it provides the security of a backup transport system.

It is generally conceded that our rail system is rapidly decaying, and the railroads claim that they need the coal traffic to sustain the nationwide rail system.* As a result, they are fighting hard to keep pipelines from making inroads. At this time, they appear to be in the driver's seat because they have eminent domain over their rights-of-way, and thus can effectively block the slurry pipelines from breaking through these barriers. The final decision will be in the hands of the legislative and judicial branches of the government. They also have a strong, well-entrenched lobby, and now they even find themselves with rather unusual bedfellows—the environmentalists—the same people who have chastized the coal-carrying railroads in the past for polluting the environment with coal dust.

This time the environmentalists have a sturdier club to wield: it requires about 1 yd^3 of water to "flush" a yard of pulverized coal to market,[74] and the coal that would benefit most from this low-cost transport—the low-energy, low-sulfur western coal—comes from an area with limited water supplies.

A proposed 1000-mi pipeline operating between a Powder River Basin mine in Montana and power stations in Arkansas would carry 23 million tons of coal per year—the equivalent of 2500 unit trains per year†—but would use 15,000 acre-ft of water to do so. Environmentalists fear that the drawdown on water supplies would seriously affect the Madison Formation, an underground aquifer beneath Wyoming and South Dakota,[69] but many current pipeline plans call for the use of brackish waters unfit for any other uses.[75]

To rebuff the railroads, involved pipeline advocates‡ have proposed to construct a 200-mi, $200 million water main from a reservoir on the Missouri River in North Dakota to the Wyoming border. It would pick up 30,000 acre-ft per year, leave 10,000 acre-ft along the way for South Dakota farmers, and deliver the remaining 20,000 acre-ft to the Madison Formation. Their chief plea is that a new technology should not be stifled, especially when a new pipeline can deliver the coal at a lower cost than a new railroad.§ Even though the cost dif-

---

*Coal comprised about 25 percent of the freight handled by Class I railcarriers in 1973, and nearly that percentage of the total freight carried on the inland waterways.[73]

†Another source describes the pipeline as measuring 1378 mi and designed to carry 25 million tpy. An operational date of 1983 is projected, but two (of 67) Nebraska court cases remain to be settled at this writing.[75]

‡Energy Transportation Systems, Inc. (ETSI).

§The capital cost of the ETSI pipeline was estimated at $0.75 billion in 1974, but was not revised upward.[74] The capital cost of a brand new railroad serving the same route would be approximately twice that figure.[76] Based on other pipeline estimates, it is only fair to assume that actual current costs of either a pipeline or a new railroad would be higher than indicated.

ferential appears to be acknowledged by the railroads, an independent University of Illinois study funded by NSF[76] claimed that the railroads offer numerous advantages over coal slurry pipelines[77]:

1   Even elaborate refurbishing of an old railroad would cost only about half as much as installing a new pipeline.
2   Railroads are more flexible because they can shunt traffic between other lines and to other destinations.
3   They contribute more to employment.
4   A railroad system reduces environmental impact, particularly with regard to water.
5   Railroads use less scarce resources.

The pipeliners depreciate the environmental impact and emphasize the ease of installation, the high safety factor, the weatherproof characteristics, the hands-off operation, and the relative freedom of a pipeline system from the effects of strikes and the effects of inflation.[78]

One proposed coal pipeline system (San Marco pipeline) would pump 15 million tons of coal per year through a 900- to 1100-mi pipe* from southern Colorado to Houston, Texas. In this case, the backers (Houston Natural Gas Corp.) have announced their intention to use brackish or saline water unfit for agriculture or for human consumption. The water, removed from the slurry by centrifuging or filtering, would then be used to cool the power plant. Mention has been made, in another case, of laying dual pipelines—one to flush coal to market and the other to return the water to the mine.

One operating coal slurry pipeline now exists in the United States and has been in service as an economically viable facility for seven years. This is the Black Mesa Pipeline, carrying 4.8 million tpy of coal 278 mi across the rugged northern Arizona terrain. It operates between Peabody's Black Mesa Mine near Kayenta, Arizona, and the 1500-MW Mohave Power Station of Southern California Edison near the southern tip of Nevada.[75,79] The system cost $39 million and took three years to build. It is basically an 18-in.-diameter pipeline, but it drops down to 12 in. in the last 12-mi stretch as it loses 3000 ft of elevation. The 48 percent solids slurry of $^3/_{16}$-in. wet-ground coal is pumped at a rate of 4000 gpm, using water from 3000-ft deep wells near Kayenta. The slurry completes its journey in 67 hr and is then pumped into 7.8-million-gal holding tanks at its destination, where it is kept in suspension by giant agitators. There 75 percent of the water is centrifuged out,† and the resulting wet cake is dried, further pulverized, and burned. It is worth noting that only about 1 ton of water is required to slurry 1 ton of coal, but 7 to 8 tons is required per ton of coal in the power plant operation.[75]

---

*Two sources differ over length of the projected pipeline.
†About 1400 gpm of water is recovered and used as part of the 15,000 gpm makeup water for boiler cooling. It is claimed to be free of pollutants.

In 1975, the pipeline was on slurry 75 percent of the time, on water for flushing 7 percent of the time, and shut down for the balance of the time, as deliveries that year (4 million tpy) were only about 80 percent of capacity (5 million tpy). Management claims actual pipeline availability of 99.5 percent.

The battle lines are drawn between railroad and slurry advocates over the issues of water resources, eminent domain, and the need or lack of need for a backup coal delivery system. The outcome is in the hands of the courts and the policymakers, with the main issues being rights-of-way and water supply.[75]

A different approach to easing the problem of water supplies and transport is under investigation at the University of Texas. The researchers there suggest the use of a *methacoal* suspension (a coal–methanol slurry), which can be burned without drying or other processing and without the need for water to transport it.[80]

Still another alternative, suggested more as a means of stretching oil rather than conserving water, is the use of coal-in-oil slurries for boiler feed. These 40 percent coal slurries can be burned in existing oil-fired boilers. At the same time, it is estimated that 1 million bbl per day of oil could be saved.[81] The major technological concern is slurry stability, because it is not feasible for each user to have its own blending equipment. This problem may be overcome by use of surface-active additives presently under test. The goal is a slurry that can be shipped and stored like No. 6 fuel oil. The big hurdles are economic. Existing oil-fired installations would probably have to be provided with electrostatic precipitators. Furthermore, the lower energy content of the coal slurry means that, on a thermal output basis, the cost will be higher than coal: for example, in one series of tests, $2.44 per million Btu for slurry vs. $1.00 for coal and $2.50 for fuel oil. (The latter figure has probably escalated at a faster rate than the slurry as a result of recent OPEC oil price increases.) The use of additives will raise the slurry costs even more. Combustion efficiency* is sensitive to the particular coal rank, its maceral structure, and the flame pattern.[82]

A variation of this fuel alternative is the use of a coal/water/oil emulsion. Approximately 50 percent pulverized coal is suspended in an ultrasonically generated emulsion of 1 part water and 4 parts oil. Over a month's stability has been demonstrated by some samples without the use of additives. It has also been found that sulfur dioxide emissions can be reduced 50 percent by the addition to the slurry of pulverized limestone.[83]

A factor to be considered is that, in the national interest, we must have access to the vast reserves of low-sulfur western coal. Unfortunately, its economically competitive geographical range is limited to the western and midwestern states,† in part due to the distance and in part because of its low heat content. Any economic factor that offers the opportunity for making western coal more competitive with eastern coal, while extending its geographical range, should not be passed over lightly.

---

*Ranging from 50 percent to greater than 90 percent.
†Only about as far east as the western Ohio border.

PRODUCTION AND CONSUMPTION

Coal reached the zenith of its importance during the years between the Civil War and World War I. Its production was growing worldwide at a reasonably steady rate of about 4.4 percent per year, doubling about every 16 years.[84] Following World War I, oil came into its own, and coal production tapered off at an annual growth rate of less than 1 percent, with a projected doubling time of nearly a century. From the end of World War II until the mid-1960s, coal again took off, but at a more modest rate (about 3.6 percent). Today we are in a state of flux, with the rate of change being erratic due to the still uncertain future of coal. The environmentalists and the energy interests have taken opposing stands, but the Administration is straddling the issue—espousing the need for more and more coal but maintaining that the environment must be preserved. The economic and technical possibilities of achieving both goals are still in doubt.

In 1973, the Sino-Soviet bloc was producing a little over 50 percent of the world's coal, while U.S. output was a little less than 25 percent.[85] Tonnage figures, however, do not tell the whole story, because nearly one-fourth of the U.S.S.R.'s, almost all of East Germany's, and nearly three-fourths of the Czechoslovakian output is poor-quality lignite or brown coal.[84] Here, the energy content of the coal produced would be a more meaningful statistic. Domestic lignite production is only about 1 percent of the higher ranks on a tonnage basis, but on an energy basis it is less than $1/2$ percent. Table 4.18 presents 1977 bituminous coal production data by states. Output is dominated by the eastern and midwestern states, with Wyoming and Montana just becoming major producers in the last few years.

Table 4.18   U.S. Production of Bituminous Coal by States, 1977

State	Million Net Tons	Percent of Total
Kentucky	142.9	20.8
West Virginia	95.4	13.9
Pennsylvania	83.2	12.1
Illinois	53.9	7.8
Ohio	46.2	6.7
Wyoming	44.5	6.5
Virginia	37.8	5.5
Montana	29.3	4.3
Indiana	28.0	4.1
Alabama	21.2	3.1
Others	106.0	15.4
Total	688.6	100[a]

Source: Reference 62.

[a]Not exact, due to rounding errors.

In light of the need for increased coal production and the concomitant need for capital, a number of economists have gone through the exercise of trying to estimate the year when coal production should peak. Eliminating those estimates with the least tenable assumptions, the most reasonable concurrence of opinions would appear to be about 2020,[86] but so much hinges on unknown political and economic factors that it remains anyone's guess.

Production and consumption figures for the United States are shown in Table 4.19, together with import, export, bunkers,* and other data for the years 1969–1972. No significant trends are evident over this brief period. Anthracite is only a minor factor in our solid fuels picture, accounting for less than 1 percent of our coal and lignite consumption.[88]

Electric power generation is far and away the largest use sector of our coal output, consuming about two-thirds of our production (Table 4.20). Coking coal for steelmaking and other industrial uses accounts for about 20 percent of our coal consumption. Retail sales and exports are small. Coal consumption for power generation is expected to increase rapidly over the next quarter century. Regional differences in the use of coal for electric power generation are marked (Table 4.21). The West coast, the oil and gas states, and the New England states all depend heavily on nuclear power and/or fluid fossil fuels for this purpose.

The petroleum refining industry is the only one of the major energy-consuming industries that uses no coal. The primary-metals industry accounts for about half of our industrial coal consumption, with chemicals and allied products consuming about 12 percent. The overall coal consumption by manufacturing industries was about 5.6 Q in 1969.[91]

Table 4.19 U.S. Coal/Lignite Consumption and Production, 1969–1972[a]

	Million Tons						
						Consumption	
Year	Production	Imports	Exports	Bunkers	Additions to Stocks	Total	lb/ Capita[b]
1969	567.7	0.266	58.4	0.06	(5.6)	515.1	5084
1970	608.7	0.164	74.0	0.07	15.9	518.9	5066
1971	556.6	0.214	58.7	0.04	0.1	498.0	4810
1972	592.2	0.214	57.9	0.03	10.3	524.2	5020

Source: Reference 87.

[a]Converted from tonnes given in reference.

[b]Converted from kilograms given in reference.

*"Bunkers" refer to fuel for outgoing ships. They are not considered to be exports.

Table 4.20 Domestic Bituminous Coal Consumption and Exports, 1977

	Million Tons	Percent of Total
Electric power utilities	474.8	70.5
Coking coal	77.4	11.5
Retail deliveries	7.0	1.0
Steel and rolling mills	3.2	0.5
Other manufacturing, bunker, and mining	57.2	8.5
Subtotal—consumption	619.6	92.0
Exports	53.7	8.0
Total	673.3	100.0

Source: Reference 89.

Table 4.21 Coal Share of Steam Electric Utility Fossil Fuel Market, 1976[a]

Region (State)	Percent
East North Central (Wisconsin, Illinois, Indiana, Michigan, Ohio)	94
East South Central (Kentucky, Tennessee, Mississippi, Alabama)	93
West North Central (North Dakota, South Dakota, Nebraska, Kansas, Minnesota, Iowa, Missouri)	82
Mountain (Idaho, Montana, Colorado, Nevada, Utah, Wyoming, Arizona, New Mexico)	79
South Atlantic (West Virginia, Maryland, Delaware, Virginia, North Carolina, South Carolina, Georgia, Florida)	67
Middle Atlantic (New York, Pennsylvania, New Jersey)	60
Pacific (Washington, Oregon, California)	7
West South Central (Texas, Oklahoma, Arkansas, Louisiana)	7
New England (Connecticut, Rhode Island, Vermont, New Hampshire, Massachusetts, Maine)	4

Source: Reference 90.

[a] On energy basis.

A postembargo projection of U.S. energy consumption (Table 4.22) shows coal contributing just over one-fourth of our energy budget in 1981, with natural gas and oil accounting for about 20 percent and 15 percent, respectively, assuming oil imports are limited to 8 million bpd. The resulting coal production capacity requirements are very large (Table 4.23) and call for annual additions to capacity of 118 to 135 million tpy to 1981. In light of the long lead times

Table 4.22 U.S. Energy Consumption, 1981 (Estimated)

Source	Quads
Domestic oil	14.9
Natural gas	19.4
Nonfossil electricity[a]	8.5
Miscellaneous[b]	1.0
Imported oil[c]	16.8
Coal[d]	25.4
Total	86.0

Source: Reference 92.

[a] Nuclear, geothermal, hydroelectric.

[b] Wood, shale oil, solar, etc.

[c] Assuming oil imports are limited to 8 million bpd.

[d] Assuming other energy imports are negligible. Using a factor of 40 million tons per quad, coal consumption would be 1016 million tpy.

Table 4.23 U.S. Coal Capacity Requirements[a]

Year	Million Tons	
	Total[b]	Additions
1977	600	n.a.
1978	720	120
1979	855	135
1980	981	126
1981	1098[c]	118

Source: Reference 92.

[a] Assuming oil imports are limited to 8 million bpd and other energy imports are negligible.

[b] Data not precise, due to rounding.

[c] This writer's 1981 estimated total was 1016 million tpy (see Table 4.22, footnote *d*), using standardized energy-to-tonnage conversion factor. Numbers quoted here are Decker's.

for opening new coal mines (five to seven years), these goals appear impossible to attain. To compound the problem, indications are that the Administration may have been overly optimistic concerning the effectiveness of oil and gas conservation measures, so even more coal may be required, on a short-haul basis, to make up the difference.[93]

ACID MINE DRAINAGE*

Most coal-mining operations initiate a severe environmental problem known as acid mine drainage (AMD). This occurs when pyritic (FeS_2) minerals associated with the coal are exposed to air and water, by mining procedures. With time, the following (simplified) reaction occurs:

$$2FeS_2 + 2H_2O + 7O_2 \rightarrow 2H_2SO_4 + 2FeSO_4 \tag{3}$$

The acid leaches into streams and causes plant and fish kills that destroy the watershed.

About 75 percent of all AMD originates in deep mines. The seepage may continue for as long as 100 years. Deep mines extending below the water table, and properly operated down-dip mines, will flood totally and thus exclude the oxygen needed to form the acid. If properly conducted, strip mining allows burial of acid-forming materials under layers of benign soil and rock.[35]

All pyritic materials associated with coal do not necessarily result in AMD, as some pyrites are stable in water.† The major offending pyrite appears to be the reactive, fine-grained framboidal pyrite.‡ It reverts to ferrous iron and sulfate ions and forms hydrogen ions in the process. Abetted by three different iron-oxidizing bacteria, the ferrous ions then oxidize to ferric iron, which combines with hydroxyl ions from the water to precipitate "yellow boy" or ferric hydroxide.[94]

In the western coal fields there generally is no AMD problem, because there is seldom any pyritic shale in these formations and the coal is a low-sulfur grade. The problem there is more likely to be one of alkaline drainage, because of the arid climate.[95] Strip mining of western lands, although less damaging than deep mining from the standpoint of AMD causes soils high in salt concentration to be stirred up to the point where the leachate drains into wells and the watershed, where it sickens cattle, stunts plant growth, and kills marine life.[96]

In 1966, in Appalachia alone, 4 million tons (about 1 billion gal) of acid was discharged into 5700 miles of streams, with disastrous ecological results. All told, about 11,000 mi of U.S. streams have been contaminated to varying

*This topic is perhaps more appropriate for discussion in the later section on social costs, but it is a factor of such major concern that it will be dealt with separately.

†The most common pyrite, FeS_2 or iron disulfide, is reactive.

‡A form containing microscopic aggregates of pyrite grains, often in spheroidal clusters.

degrees by AMD. The cost of cleaning up this past damage is estimated at $6 billion.[68]

Lime [$Ca(OH)_2$] or limestone ($CaCO_3$) is normally used to deal with the problem. The effluent is treated, aerated, and allowed to settle[51]:

$$Fe^{2+} + \frac{1}{2}O_2 + 2H^+ \rightarrow Fe^{3+} + H_2O \tag{4}$$

$$Fe^{3+} + 3H_2O \rightarrow Fe(OH_3) + 3H^+ \tag{5}$$

These reactions generate 3 moles of hydrogen ion for each ferric ion precipitated; hence enough lime must be used to neutralize the original acid, as well as to neutralize that produced by the oxidation, hydrolysis, and precipitation of Fe^{3+}. The colloidal sludge formed with lime treatment is difficult to settle and dispose of, while the reactivity and efficiency of treatment with limestone decreases as pH rises.[97]

To effect more efficient treatment, various other procedures are being developed. One is a microbial oxidation process that is estimated to cost 4¢ per 1000 gal of drainage.[94] Researchers at Penn State achieved reductions in total iron and in ferrous iron of 92 and 99.8 percent (vs. 20 and 37 percent for lime treatment), respectively, by addition of hydrogen peroxide at the treatment facility.[98]

A team of researchers from Wright State University also developed a promising high-efficiency method of removing iron. Since AMD results from exposure of pyritic minerals to air and water, the removal of iron is an essential step. In their new phosphate-lime process, the iron is precipitated as phosphate rather than hydroxide:

$$Fe^{3+} + PO_4^{3-} \rightarrow FePO_4 \tag{6}$$

without the generation of hydrogen ion. On the contrary, 1 mole of H^+ is consumed for every mole of precipitate, thus lowering the acidity of the drainage; hence 44 percent less lime is required. Materials for the phosphate-lime treatment* should cost about the same as the lime treatment (1.1¢ per 1000 gal), but sludge disposal costs (1.8 to 2.0¢ per 1000 gal for the lime method) should be substantially less, because the iron phosphate forms a dense, flocculant precipitate that may be used as a soil conditioner.[97]

Yet another AMD treatment procedure is shown schematically in Fig. 4.1. A broad spectrum of treatments is available.[68] Before-the-fact methods include: diverting water away from the pyritic residues, sealing the mines (thus excluding the water and oxygen necessary to form the acid), restoration of the surface contours, and revegetation. Posttreatment of the drainage waters is another approach. This option includes neutralization, distillation, reverse osmosis, ion exchange, freezing, and electrodialysis.[68]

---

*Ordinary phosphate rock (fluoroapatite) is used as a trickle bed.

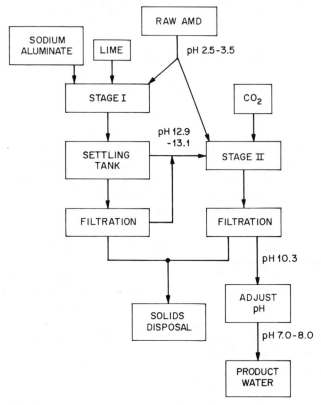

Figure 4.1 Alumina treatment for AMD (acid mine drainage). (Adapted from reference 94.)

SOCIAL COSTS OF COAL EXTRACTION

Psychologists have noted that the debasing features of nineteenth-century boomtown life are being visited upon occupants of today's western coal-mining towns. Kohrs has typified these features as the "4-D's"—drunkenness, depression, delinquency, and divorce. The original residents are not prepared for the influx of large numbers of newcomers accustomed to different life-styles and standards of conduct. Nor are the physical facilities and local services (health care, schools, public water supplies, sewage disposal, etc.) able to cope adequately with the increased population. The result is generally severe deterioration in life-style.[99] The question is then raised: What happens when the coal seam plays out 35 years later?* With rangeland destroyed, aquifers fouled or depleted, soil contaminated with particulate fallout, wildlife population

---

*This is the average lifetime of a U.S. coal field.[94]

decimated, what do the people turn to? To sociologists, this is a cause for real concern.

Along the way, what other changes occur that affect the lives of the people? Health effects and effects on the water supply have to be two major considerations. Rose et al.[100] have directed their attention to an assessment of the health "costs" of mining, transporting, and burning coal. Mining is the major source of accidents leading to disabling injuries ("black lung" is included in the data) and death; combustion is more likely to lead to "environmental deaths" (fatal heart and lung ailments from power plant emissions). Coal transport would be a minor factor, and should be negligible if large-scale slurry transport becomes a reality. Coal-mining fatalities are estimated to be 3 to 10 times as numerous as those involved in nuclear power,[100,101] and, of course, pneumoconiosis (black-lung disease) is unique to coal mining. A recent study carried out at the University of West Virginia Medical Center showed that of 150 coal miners claiming disabling progressive massive fibrosis (pneumoconiosis), all were also cigarette smokers.[102] Coal combustion produces oxides of sulfur and nitrogen, trace metals, and particulates. Electrostatic precipitators can remove up to about 99.8 weight percent of the particulates, but those that are not removed are the very fine particles that penetrate deeply into the lungs and have the most severe effects on health.

As noted earlier, modern deep mining techniques, particularly longwall mining, produce large amounts of dust and might be expected to lead to an increased incidence of black lung in the future. Mine safety regulations, however, have been tightened to the extent that no worker now entering the miner's trade should develop black lung by the time he or she retires.[100] Mine ventilation has been improved to the point where fires and explosions should become a thing of the past, but there will still be accidents—fewer fatalities, but a continuing high level of disabling physical injuries. Deep mining would be expected to result in about 80 percent of the mining accidents and fatalities between 1970 and 2000 while yielding only 40 percent of the coal output. The dollar "cost" of these accidents in deep mining would add nearly six times as much to the cost of coal, as would accidents in surface mining, but land reclamation, in the latter case, would overshadow all social costs of coal, accounting for nearly one-third of such costs overall. Subsidence prevention would be the major social cost of deep-mined coal, and AMD control would be a major factor in the Appalachia and Interior surface mines. Over this 30-year time span, the total social costs of coal are expected to amount to over $19 billion, nearly $2 per ton.[68]

A study reported by the Health and Safety Commission (U.K.) showed that the use of coal for electric power generation was far more hazardous than oil-, gas-, or nuclear-based power, and the greatest hazard occurs in the extraction phase.[103] The Office of Technology Assessment (OTA) estimates that tripling of U.S. coal production would lead to 370 fatal and 42,000 disabling injuries each year in the extraction of this fuel.[104] Quite apart from the personal-injury aspect, environmental effects would also be noted, but not necessarily in direct relation to the amount of coal produced.

Some unquantifiable social factors that are not tabulated include: reductions in property values near the mines due to dust, noise, and aesthetics; and damage to roads. In this regard, primarily because of land reclamation costs, the social costs of surface mining coal are higher than deep mining. This does not, however, take into account possible gaps in restitution to miners' families following accidents, keeping in mind that deep mining is far more hazardous than surface mining.

The most visible evidence of property damage is the ravishing of the land by mining activities. It was stated back in 1944 that "we [the United States] have ruined more good land in less time than any other nation in recorded history."[105] One of the major despoilers has been the coal industry. For every ton of coal mined, about 0.0001 acre of land is undermined and 1/4 ton of waste must be disposed of. From 1970 to 2000 alone, this should mean 2 billion tons of waste.[68] When one considers the land destroyed by subsidence and the mountainous slag heaps left by deep mining, more acres have probably been damaged by deep mining than by surface mining up to this point in history.[19] In addition to this, many slag heaps and mines have been burning out of control for many years, in some cases causing accumulations of carbon monoxide in buildings.

In January 1980 it was reported that probably 200 coal mines were burning throughout the country, threatening human life and property. One of them, in Centralia, Pennsylvania, has been burning continuously since 1963, after being ignited by rubbish burned by local residents. About $4 million has been spent to excavate the pit and to pump water, clay, sand, and fly ash in, without avail. A $12 to $20 million ditching program was rejected for economic reasons, and now a $6.1 million slurry-flushing program has been proposed.[106] In addition to the underground mine fires, over 450 refuse banks were burning in seven states in 1969.

In rebuttal to environmentalists' complaints about the large area of land destroyed by strip mining, Bagge states: "The 'vast' strip mining of the West envisioned by the media is in fact only vast in the volume of coal. The maximum land disturbed in the year 2000 would be an area roughly equal to the acreage inside the beltway surrounding Washington, D.C.—out of more than 315,000 square miles in Wyoming, Montana, and North Dakota."[107]

Anyone who has seen a strip-mined area must be convinced that steps have to be taken to restore the land to some degree of its previous condition. The main question concerns the extent of reclamation. The Bureau of Mines has referred to "basic reclamation" as grading, revegetation, and drainage control, and cited an average cost of $500 per acre.[68] The Bureau went further, to indicate a cost of $1100 per acre to return the land to productive use, with some specific projects ranging as high as $15,000 per acre. In West Germany and Great Britain, where surface mining is allowed only in relatively flat terrain, reclamation costs (returned to full agricultural productivity) have ranged from $3000 to $4500 per acre.[68] The Council on Environmental Quality and the Soil Conservation Service have agreed on $2000 per acre, but this does not accommodate the added costs of reclaiming land in hilly terrain.

To meet the cumulative goal of 12 billion tons of surface-mined coal from 1970 to 2000, up to 1.5 million acres will be stripped. At a $4000 per acre reclamation cost, the total outlay will amount to $6 billion or $0.50 per ton.* In the American West, where yields sometimes range between 40,000 and 80,000 tons per acre, but where water for reclamation is in very short supply, it is difficult to say how to assess the cost, let alone the feasibility of reclamation.

It is generally felt that if the evapotranspiration rate is not too high, areas receiving 20 or more inches of rainfall annually can be "rehabilitated" or at least have a "high potential for reclamation" if they are managed well. In the Four Corners region of the Southwest,† where rainfall is scarce, $5000 per acre reclamation expenditures have resulted in some very sparse vegetation, where only 1 to 20 percent coverage is normal. It is generally conceded that where rainfall is less than 5 in. per year, the land can never be reclaimed. New Mexico has land areas bereft of vegetation for five centuries. These were sites of Indian turquoise mines dug in the fifteenth century.[108] Land receiving 5 to 10 in. of rain per year may require centuries to recover.‡

The drought of 1976–1977 accentuates the problem of extracting the western coal, but visions of the millions of dollars worth of coal lying beneath the 30 acres of grazing land needed to support a single cow[52] make it likely that coal experts will find a way of getting the coal to market short of suffering open warfare with environmentalists, hydrologists, ranchers, and farmers.

SOCIAL COSTS OF COAL COMBUSTION

Coal has rightly been called the "ugly duckling" of fuels. It is dirty and dangerous to mine, ravishes the land and the waters, and threatens the safety of those living or working in its shadow. It is bulky and expensive to transport and poisons the environment when burned. It has little to recommend it except that it is there, in our own backyard, and in quantity, at a time when the supplies of its more desirable fossil competitors are dwindling rapidly.

Many people, when considering the nuclear alternative, tend to shunt nukes aside as impossibly hazardous. It remains to be seen where these same people will be making their stand when it comes down to the time to place their blessings on coal. Apart from the problems of extracting it, what are the problems with this "ugly duckling" as a fuel?

Aside from particulates that contribute heavily to coal's environmental notoriety, sulfur is undoubtedly the chief offender. Sulfur is present in coal as pyrites (mainly FeS_2), sulfates, elemental sulfur, and organic sulfur. Organic sulfur accounts for 30 to 70 percent of the total sulfur in most coals.[109] In one coal it has been measured at 11 weight percent, on a moisture-free basis,

---

*Based on observed 1970 averages for Appalachian fields: 5.7-ft seam thickness; 8200-tpa yield.
†Where Utah, Colorado, Arizona, and New Mexico meet.
‡About 40 percent of the western coal lands fall in these latter two categories (i.e., under 10 in. of rainfall).

although it will more normally range from $0+$ to 10 percent by weight. The organic sulfur components have remained unidentified because of the masking effect of large amounts of polymeric and aromatic matrices present in coals. Recently, Casagrande and Siefert[110] conducted a study of peat, as a precursor of coal, in an attempt to characterize the organic sulfur constituents.

In its natural environment, peat (which can be thought of as a highly organic soil) is combined with a complex living mixture of plants, litter, water, burrowing animals, and microorganisms. Earlier studies of inorganic soils have disclosed the presence of two forms of organic sulfur: (1) those with carbon–sulfur linkages, and (2) those with carbon–oxygen–sulfur linkages ("ester sulfates").[111] It is now felt that the latter constitute a major part of the organic sulfur in coal.

The functional part of coal comprises largely polycyclic aromatic hydrocarbons (PAHs),* organic oxygen, nitrogen, sulfur, coalification-modified residues of biological compounds, organically bound metals, and some aliphatic hydrocarbons. Mineral matter present contains an assortment of elements virtually blanketing the periodic table.[114]

The condensed PAHs that make up coal occur as lamellae with molecular weights ranging between 2000 and 12,000 (far lower than molecular weights—in the millions—typical of petroleum-based tars and residual fuel oils). Therefore, coal is potentially a more reactive feedstock for various conversion processes. The oxygen-, nitrogen-, and sulfur-containing functional groups have considerable influence on this reactivity and the resulting products. Coal liquids formed may, as a result, contain various contaminants including phenols, aromatic nitrogen compounds, and catalyst-poisoning sulfur compounds.

Oxygen in coal is generally incorporated in phenoxy and carbonyl groups, but some is present in heterocyclics and, with lignite and subbituminous coals, as carboxyl groups. The oxygen is also involved in the binding of metal ions. In conversion processes, it may lead to the formation of phenols. Nitrogen may be present in coal up to a level of 5 percent by weight, but more typically will fall in the 1 to 2 percent range. During combustion or conversion, the nitrogen in coal is probably a major source of NO_x compounds[114]; another source is atmospheric nitrogen.

The contaminants in coal can be a serious detriment to the expanded adoption of this fossil fuel as our major energy source. A study by the Bureau of Mines, which included 70 percent of the coals used for electric power generation, has shown that only about 14 percent were capable of meeting existing Environmental Protection Agency (EPA) air pollution standards applying to new power plants.[54] The major contaminants were found to be organic and pyritic sulfur, ash, and trace elements.

To avoid exceeding air pollution standards, several practices are followed in stationary plants: (1) low-sulfur coal is used; (2) the coal is highly beneficiated or desulfurized (or may be solvent-refined at some time in the future); (3) elec-

---

*PAHs are reported to cause a high incidence of colon cancer in human beings,[112] gene mutations in bacteria, and cancer in lower animals.[113]

trostatic precipitators and stack gas scrubbers are used to remove particulates and gaseous pollutants, respectively, from the flue gas; or (4) the flue gas is diluted and dispersed over a wide area by discharging it through tall stacks.

Solvent refining and beneficiation are "before-the-fact" methods that can provide clean fuels, but they are high in capital and operating costs, as well as in lead times for designing, producing, and installing the equipment. Solvent refining (Fig. 4.2) comprises essentially de-ashing and desulfurizing pulverized coal by noncatalytic treatment with hydrogen and coal-derived solvent(s). The coal can be recovered as a low-sulfur, low-melting ($\sim 375°F$) solid. Its suitability as an electric power-generating fuel is not yet fully established, although commercial test burns have been completed successfully, as of this writing. Sulfur and NO_x emissions were well below state standards for Georgia, where trials were run. At this time SRC does not appear to be economically competitive except for intermittent use in old coal-fired plants not equipped with pollution control devices.[116]* Beneficiation has been discussed in the section on coal preparation.

Desulfurization

Coal lends itself to cleaning at various stages: before, during and after burning. The "during" option is best exemplified by fluid-bed combustion (FBC), which is discussed later. The "after" option implies the use of flue-gas desulfurization (FGD) scrubbers. The simplest form of precleaning probably involves crushing and pneumatic separation, which deposits the heavy, sulfur-laden "hutch" in a dust-collector system.[117]

A new desulfurization process—*Sulf-X*— is receiving considerable attention because it is cheap, effective, and versatile.[118,119] It is a closed-cycle process based on a gas–solid reaction between SO_2 and a proprietary sulfide of iron, which can be mixed with stack gases as an atomized slurry. The unstable prod-

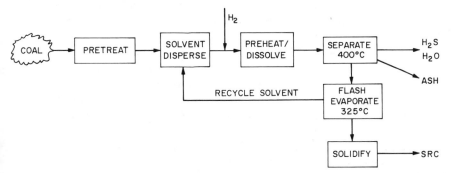

Figure 4.2   Process for solvent refining coal. (Adapted from reference 114.)

*SRC is expected to increase the cost of coal-generated electricity by 30 percent.[115]

uct precipitates out. It is then reconstituted to form the original iron sulfide, and the benign solid residue is readily disposed of as landfill.[119]

The economics of Sulf-X are aided by the fact that iron sulfide is an abundant waste product in pyrite mine tailings and can be extracted from acid mine drainage for less than the cost of neutralizing the acid for disposal. The desulfurization process is said to be 90 to 98 percent effective in removing SO_2; hence it should be able to meet the EPA standards even with high-sulfur, low-heat content coal. Also, the Sulf-X process uses only about 1.5 to 3 percent of the boiler's energy as compared to 6 to 15 percent for other regenerable processes. Capital and operating costs should be only one-half to two-thirds those of competitive systems that are adaptable only to postcombustion operation.[118] A 100-MW demonstration unit has been proposed.[119]

Another proposal for desulfurizing coal is magnetic filtration, a method used in the kaolin and clay products industries for mineral beneficiation.[120] Macroscopic pyrite occurs in coal as veins, in geologic lenses, as nodules, and in aggregates. Coal is diamagnetic,* whereas pyrites are paramagnetic.† As a result, the two can be separated readily if the pyrites can be liberated from the coal by fine grinding.‡ As a modification of this process, the pyrites and ash can be coated with a thin magnetic skin by exposure to iron carbonyl vapor. Experimentally, pyrites have been removed up to 93 percent. Separator operating cost has been estimated at 37¢ per ton for a 500-tph plant, including capital amortization over a 10-year period. This should be compared with about $1.25 per ton for mechanical coal cleaning.

The major limitation of magnetic filtration is that it removes very little chemically bound sulfur. This can best be done by air oxidation at 284 to 392°F (140 to 200°C) up to 1000 psig.§ Oxidation removes more than 90 percent of the pyritic sulfur and up to 40 percent of the organic sulfur, with a fuel loss of less than 10 percent and at a total cost of $3.50 to $5.00 per ton. Oxidation by an ammonia/oxygen system is yet another option under investigation.¶

There is, in the offing, a new and promising development aimed at desulfurizing coal "before the fact" at possibly one-half the cost of scrubbing. Military radar research by General Electric disclosed that sulfur contained in coal absorbs radiation much more effectively than the surrounding carbon molecules, so when pulverized coal is bombarded with microwaves, the sulfur is converted to the gaseous state and can be condensed and collected as elemental sulfur, without harm to the coal. The only requirements are that the coal be pulverized and be passed through a microwave generator.[121]

Precipitation of particulates and scrubbing of SO_x from flue gas are "after-the-fact" methods that pose their own unique problems, but these problems are

---

*It is repelled by a magnet and tends to orient itself normal to magnetic lines of force.
†They align themselves parallel to a magnetic field.
‡Method under development by H. H. Murray (Purdue University).
§DOE process.
¶Kennecott Copper Corp. method.

real and urgent. In 1976, a consortium of electric utilities was forced to abandon plans to construct a 3 million kW, coal-fired power station at Kaiparowits, Utah, billed as the world's largest. The quantity of contaminants in the flue gas became a key issue. The plant would burn 1000 tph of coal and release 300 tpd of contaminants into the surrounding area.[122]

The fly ash particles from a typical power station may comprise in excess of 10 percent of the mass of the coal burned.[123] In 1975, U.S. utilities burned over 400 million tons of coal. Assuming a 10 percent ash content for the coal and an average retention of 30 percent in the furnace, about 28 to 30 million tons of fly ash could have been emitted. A 600-MW_e station will emit about 1200 lb per minute, or over 800 tpd of fly ash, without electrostatic precipitators.[123] The coal-fired power plant at Farmington, New Mexico (Four Corners) has been said to emit 10^{23} submicron particles per day, which contaminate 100,000 mi^2 of a four-state region.[124]

About three-fourths of the effluent gas from burning pulverized coal comprises free nitrogen. Carbon dioxide is present in the range 12 to 15 percent. The harmful pollutants are present at relatively low levels: SO_2 (0.002 to 0.2 percent), SO_3 (1 to 30 ppm), and NO_x (100 to 1500 ppm). The balance of the effluent gas consists of oxygen (4 to 5 percent), from excess air used for combustion and from air leakage in the furnace; and water (4 to 12 percent), from the moisture content of the coal and the humidity of the combustion air.[125] It is now felt that there is substantial evidence to the effect that the primary health hazard from sulfur is not the SO_2 emissions as such but rather the resulting sulfuric acid and the sulfate particulates.[126]

Electrostatic Precipitation

Fly ash is a fine,* airborne, mineral residue of coal combustion from power plants. It comprises essentially silica, alumina,† and iron oxide matrices, but eight other metals have also been detected in fly ash, including cadmium, selenium, lead, and beryllium. The emitted elements which pose the most hazard to living systems are those that are both highly toxic and accessible. Many toxic elements are either insoluble (e.g., barium) or rare (e.g., indium, zirconium, gallium)[129] and need be of little concern unless present in an unusually high concentration. The fine fly ash particles adsorb sulfur and chlorine preferentially. The water solubility and deliquescence of the sulfates and chlorides formed produce excessive haze- and cloud-forming nuclei in the tail of the flue-gas plume. This is considered a major environmental hazard.[130]

Mixed organic substances are also adsorbed on the surface of fly ash passed through electrostatic precipitators in stacks of power plant boilers. Polynuclear aromatic hydrocarbons have been identified on the fly ash. Tests have shown

*The finest being about 2 μm.
†The average alumina content of U.S. coal fly ash is 22 percent. One process is being directed at recovering the aluminum from fly ash collected in power plant precipitators.[127,128]

these particulates to be mutagenic to bacteria.* They may also prove to be carcinogenic.[131] The bulk of the fly ash (now about 60 million tpy) is being captured, some by electrostatic precipitators; nonetheless, some gets through. It is estimated that about 2.6 million tons were released to the atmosphere in 1974.[132]† For a thorough review of electrostatic precipitation of fly ash, see White (references 123, 125, 134, 135).

Recent studies by the National Oceanic and Atmospheric Administration (NOAA) indicate that electrostatic precipitators remove 99.8 percent of the flue gas particulates, but the negatively charged particles getting into the atmosphere constitute a potential new menace by adding to the electrical fields of the atmosphere. About 800 ft downwind of the stacks, NOAA has recorded an electrical charge equivalent to that in a thunderstorm. The electrical plume has been detected over 50 mi downwind. They calculate that the 10,000 electrostatic precipitators now in service throughout the world could simultaneously generate a total electric charge equivalent to 40 thunderstorms.‡ The research team has expressed some concern over the effect these electrical plumes may have on living organisms as well as on the weather.[136]

The resistivity of the dust being collected is a major factor in electrostatic precipitator performance. As resistivities exceed 10^{10} ohm-cm, performance of the fly ash depends on the chemical composition of the ash, ambient flue gas temperature, moisture and sulfur trioxide content of the ash, and other factors. Most of the sulfur in the flue gas is oxidized to SO_2, but about 1 percent is converted to SO_3.§ The SO_3 content will be greater for higher-sulfur coals, but there is no one-to-one correspondence between sulfur content and resistivity. Under certain conditions, most of the SO_3 will condense on the fly ash particles and free-SO_3 vapor in the flue gas will be reduced correspondingly. Experience has shown that low-sulfur coals do not produce efficient operation of precipitators due to the high resistivity of the fly ash, although furnace conditions (particularly flue-gas temperature) can have an equally significant effect on resistivity. Sodium can be added to the coal feed to help compensate for low sulfur when a precipitator is used.[125] One cleaning process under development uses a high-intensity ionizer to charge the high-resistivity particles, thus making it easier to knock them off the collector plates.[138]

---

*The finest particulates are the most important biologically, inasmuch as they are the most difficult to remove electrostatically and in the nasal passages, their atmospheric residence time is the highest, and they are deposited most deeply in the lungs.

†Based on assumptions that U.S. power plants burned 600 million tons of 11-percent-ash coal, with 80 percent of the ash being fly ash, of which about 5 percent was released to the atmosphere. Actually, in 1976, only about 445 million tons of coal was burned in power plants,[133] hence the released-fly-ash figure should probably be closer to 2.0 million tons.

‡About 1500 thunderstorms exist globally at any one time.

§Conversion to sulfates may rise to as high as 5 percent, essentially developing within a few kilometers of the emission source. Conversions as high as 20 percent may be encountered with oil-fired plants.[137]

Acid Rain

Electrostatic precipitators remove, primarily, particulates. The flue gas may still contain large quantities of sulfur dioxide, which must also be removed to meet air pollution standards.

Within three or four days, some of the SO_2 from the flue gas will be converted to sulfuric acid, which is deposited eventually as acid rain, normally within a few hundred miles of the source.[139] More than 90 percent of such deposits occur in the northern hemisphere. The total annual natural contribution of sulfur (as H_2S from decaying vegetation and SO_2 from volcanoes) is estimated to be equivalent to 200 million tons of sulfates (or 4 million tons of hydrogen ions). Our worldwide SO_2 emissions from the combustion of fossil fuels is about 100 to 150 million tons per year, the equivalent of nearly 200 million tons of sulfuric acid.

Acid rain is formed essentially in accordance with the following basic reactions[140]*:

$$2SO_2 + O_2 \rightarrow 2SO_3 \tag{7}$$

$$SO_3 + H_2O \rightarrow 2H^+ + SO_4{}^{2-} \tag{8}$$

$$2NO + O_2 \rightarrow 2NO_2 \tag{9}$$

$$3NO_2 + H_2O \rightarrow 2H^+ + 2NO_3{}^- + NO \tag{10}$$

$$4NO_2 + 2H_2O + O_2 \rightarrow 4H^+ + 4NO_3{}^- \tag{11}$$

Normal rainwater, in equilibrium with atmospheric carbon dioxide,

$$H_2O + CO_2 \rightarrow H_2CO_3 \tag{12}$$

has a pH of about 5.6 to 5.8, due to the carbonic acid formed. The lowest recorded rainfall pH† was 2.4, in Pitlochry, Scotland, on April 10, 1974.[141]‡ In recent years, values in the western hemisphere have ranged from 2+ to 6, but yearly averages have been between pH 4 and 5, so the normal acid content of rainfall in the western hemisphere is considerably higher than could occur in nature with water saturated with carbonic acid. Values as low as pH 4.5 were recorded in our Northeast as early as 1955–1956.[140]

---

*Sulfur dioxide is formed by the combustion of sulfur-containing fossil fuels. Oxides of nitrogen (NO_x = NO and NO_2) are produced by high-temperature combustion in the presence of nitrogen.

†The pH is the negative log of the hydrogen ion concentration. A pH of 7 is neutral. Values above this, to a limiting value of 14, are alkaline, while values below 7 (down to a limiting value of zero) would be acid, with the lower pHs being the more acidic.

‡The acid equivalent of vinegar.

Addition of 4 million tons of hydrogen ions to the hemisphere's annual 2×10^{14}-ton rainfall would increase the acidity a full pH unit.[119]* Most of this acid would fall in the northeastern portion of the United States, eastern Canada, and northern Europe. Fish kills are already occurring in Norway and Sweden due to acid waters falling from industrialized regions south and west of those countries. North American fish kills are less well documented, but in northern Ontario and rural New Hampshire the pH of the rain, in the early 1970s was no more than 4.1 and the forest runoff frequently fell below 5.[139]

In the case of plant life, the acid leaches essential nutrients—calcium, magnesium, and potassium—from ecosystems (plants and soil) that are already naturally deficient. Loss of calcium, particularly, is damaging to timber growth.† It has been estimated that in Sweden alone, recovery of this loss of timber growth would make emission controls in western Europe economically viable.

Stack-Gas Scrubbing

Stack-gas scrubbers have been touted by environmentalists as the answer to emissions control, but they have had little to say about the economics, efficiency, reliability, and even the environmental impact of the devices. Most scrubbers are barely out of the development stage. Little is yet known of their economics or reliability. For a simple description of types of scrubbers used, see reference 143.

Commercial and developmental scrubbers vary considerably in their design and operation and in the reagents or absorbents they employ. Nearly 90 percent use lime, limestone, or dolomite absorbents; others combine lime with magnesia, use ammonia or ammonium salts, buffered sodium phosphate, sodium sulfite, activated carbon, citrates, aqueous carbonate, and so forth.[108] The absorbent determines the nature of the resulting sludge and its environmental impact.

The "conventional" scrubber (if there is such a thing), using lime [$Ca(OH)_2$] or limestone ($CaCO_3$), produces a voluminous wet sludge (basically $CaSO_4$) that is difficult to cope with. One ton of coal will yield about 8 or 9 ft^3 of sludge (or about 80,000 ft^3 per day for a 1000-MW$_e$ generating plant using high-sulfur coal).[128] At one power plant in Pennsylvania, the sludge is being pumped 7 mi into a holding basin behind an earth-fill dam. Over 25 years it will fill the valley to a depth of 150 ft and a length of 5 mi! There is some concern that the sludge may resist bacterial attack and, in the end, release hydrogen sulfide to the atmosphere.[126]

Other scrubbers employ reagents that produce a regenerable system yielding dilute SO_2 (for use in sulfuric acid manufacture), free sulfur, or dry magnesium

---

*It is estimated that the annual emissions in the United States alone are now 26 million tonnes of SO_2 and 22 million tonnes of NO_x.[141]

†The acid rain from emissions has reportedly reduced photosynthesis in the forests by 10 percent. This loss equates to the energy output of fifteen 1000-MW nuclear reactors.[142]

oxide reagent and SO_2 for conversion to sulfuric acid. Gases other than SO_2 can be scrubbed from stack-gas effluents as needed.

A recent development calls for the selective homogeneous gas-phase reduction of NO by ammonia.* Nitric oxide (NO) is formed during combustion by thermal fixation of atmospheric nitrogen ("thermal NO_x") and by oxidation of organic nitrogen compounds present in the fuel ("fuel NO_x"). The relatively safe nitric oxide is slowly converted in the atmosphere to the highly poisonous, smog-producing nitrogen dioxide (NO_2).[144]

The $deNO_x$ process simply injects ammonia into the flue gas at the proper temperature,† and the nitric oxide is reduced rapidly in the gas phase to form nitrogen and water:

$$6NO + 4NH_3 \rightarrow 5N_2 + 6H_2O \qquad (13)$$

with up to 70 percent efficiency. Trace amounts (1 to 2 percent) of N_2O (nitrous oxide) are also formed, but this gas is generally regarded as harmless. Hydrogen cyanide (HCN) can be formed only if hydrocarbons are present in the injection region, but not if hydrogen is also added. At most, the concentration of HCN would be less than that present in automobile exhaust.[133]

Scrubbing equipment typically costs in the neighborhood of $100 or more per kilowatt of power generated, making it a real factor in capital outlay for the power station. In addition, it will consume hundreds of thousands of tons of the scrubbing reagent each year, besides leaving a mountain of sludge for disposal.‡ Furthermore, the scrubbers typically consume up to 5 percent of the power output of the plant, in part because the scrubbed stack gas must be passed through mist eliminators, which cool the effluent gas to a point where it must then be reheated to about 175°F before being discharged to the atmosphere.

Waste in operation is not the only cost of scrubbing SO_2 from flue gases. The Department of Energy estimates that capital expenditures for equipment, through 1990, would cost $32 billion for EPA's full-scrubbing option (across-the-board 85 percent reduction of SO_2) or $19 billion for the DOE partial-scrubbing option (reduction on a sliding scale).[145] Adoption of the latter option was announced in May 1979. This regulation is expected to clear the air of acid rain by the year 2000. It applies to all power plant construction begun after September 18, 1978. By 1995, $770 billion in capital will have been expended on new plants, with 4.5 percent ($35 billion) going for pollution controls alone.[146] In 1977, all U.S. pollution control spending (for air, water, and solid waste) amounted to $37.5 billion.[147]

All in all, stack-gas scrubbers leave much to be desired, but they are better than nothing when one considers the potential hazards of burning increasing

---

*Exxon Thermal Denox Process (sometimes written $deNO_x$).

†Temperature is critical to the success of the process.

‡With the increasing use of coal, it has been estimated that the production of sludge in the year 2000 could be sufficient to cover the city of San Francisco to a depth of 9 ft.[132]

amounts of coal over the years. Air pollution cannot be dismissed lightly when one remembers that during the great London fog of December 1952, there were approximately 3500 to 4000 fatalities in excess of the number that would normally have been expected.[137]

Fluidized-Bed Combustion

Fluidized-bed combustion (FBC) of coal was pioneered by the United Kingdom's National Coal Board (NCB) and is now attracting considerable attention throughout the industrialized world as dwindling supplies of other fossil fuels bring coal to the fore. Fluidized-bed combustion promises to be an efficient, clean, and economical means of burning this dirtiest of fuels once the technology is sufficiently developed.[148]

Basically, the principle of FBC is simple. The combustor or furnace is partially filled with ash and/or pulverized limestone which is kept bubbling like a fluid by blowing air up through a porous bottom plate (the air distributor grid). This fluidized bed circulates around water-filled coils. The bed is heated to a temperature of about 1600°F using a clean fuel such as propane. Crushed coal (about 1 percent by weight of the limestone) is added and is ignited by the hot limestone. The propane is then shut off and the burning coal continues to heat the water in the furnace coils as additional fuel coal is fed in.[149] Fluidized-bed combustion has several unique features[149-151]:

1   The furnaces are more compact* and efficient than conventional combustors; hence construction† and operating costs are lower.

2   Use of a limestone bed permits SO_2 removal in situ, with over 90 percent efficiency.‡

3   High heat-transfer coefficients permit lower operating temperatures (1550 to 1650°F), reducing NO emissions by 80 percent and minimizing slag and clinker formation, thus lowering maintenance costs.

4   Fuel particles do not agglomerate because of the low combustion temperature; hence they burn more thoroughly and distribute heat better.

5   The combustors are potentially very flexible in that they can be designed to burn oil and gas as well as all ranks of coal, thus reducing dependency on a given fuel in a time of short supply.

6   In theory, they can be built for large and small electrical power generators, small boilers and drying kilns, gas turbines, waste incinerators, and so forth.

---

*Typically one-half to two-thirds the size of conventional boilers.
†Shop rather than field construction may be possible because of the smaller size.
‡EPA emission standards announced in May 1979 call for at least a 90 percent reduction in emissions for plants using high-sulfur coal, and 70 percent (minimum) reductions with low-sulfur coal.[146]

Detractors of FBC claim that they have severe problems with boiler-tube and floor-bed corrosion, and there is a still unresolved question of the practicality of scale-up to anything beyond a moderate industrial size.[150] These questions must be answered in the near future.

It has been noted above that SO_2 emissions can be controlled and NO_x can be reduced by FBC, but what of other pollutants—trace elements, organic compounds, and particulates? The work to date in these areas is very limited, but enough has been done to suggest possible ranges of contaminants that might be encountered in the effluent. A "worst-case analysis" indicates that beryllium, arsenic, uranium, lead, chromium, vanadium, and chlorine may be vaporized and emitted from coal-fired FBC boilers in hazardous concentrations without being captured by particulate collectors. The study also suggests that enrichment on particulates too fine (below 7 μm) to be removed from the flue gas may occur in the case of lead, chromium, selenium, bromine, and mercury. Preliminary particulate analysis shows that the diameter of the mass median of flue gas particulates is about 7 μm, suggesting that fine particles may indeed prove a problem with FBC, but this is not an unusual situation for conventional combustion systems.[151]

It is obvious that much remains to be done in the area of FBC technology. Numerous facilities are under construction or in operation to begin establishing burner operating parameters, waste disposal, regeneration and recycling, emission control procedures, and special problem areas. Many should be in service as we enter the 1980s.

Dispersion

A favorite, relatively low cost procedure for utilities to employ in meeting air quality standards is to emit flue gases through very tall stacks. The basic principle is that the higher the effluent is discharged above the surrounding terrain, the more dilute it will be when it comes back to earth.[152] Power station stacks are typically on the order of 200 to 300 ft high. The new generation of stacks may exceed 1000-ft heights.

This procedure eases the problem with one's immediate neighbors, but it does not reduce the contamination of the environment as a whole, unless electrostatic precipitators or scrubbers are also incorporated in the installation. Nearby conditions may be improved by preventing the plume from being trapped by local geographic features, but someone is still going to have to suffer the effects of the emissions. They will still contribute to acid precipitation in someone else's backyard, forest, or country, because "the impacts of our actions do not stop at national borders."[153]

Contaminants in Coal

As noted earlier, one concern of environmentalists and public health officials is the possible effects on human beings of discharge of various trace elements into the atmosphere due to burning of coal. Table 4.24 gives typical elemental analysis of coal and solvent-refined coal (SRC), by way of suggesting one possible solution to this problem through the burning of the refined product. Very little is actually known about the possible dangers of these elements in power station effluents, because their presence has not been monitored on a routine basis. Some pollutants are undoubtedly accumulated in the residual ash, where they may be subject to leaching into aquifers. Some may be adsorbed on fly ash and be collected in electrostatic precipitators. But some of these elements must end up in the air we breathe.

Cost of Cleanup

Table 4.25 gives an indication of the capital and operating costs of pollution control for a coal-fired 1000-MW$_e$ power station Although the data are pre-1975, it is still clear that they are significant factors in our energy costs. Table 4.26 shows industrial pollution control costs for 1970 and 1975. In this span of five years, the costs nearly doubled, because of inflation and increasingly stringent environmental regulations. On all counts, coal is the most environmentally damaging of the four main fuels for electric power generation

Table 4.24 Typical Elemental Analysis of Coal and SRC

	Coal				Coal		
Element	Mean, ppm	Standard Devia-tion, ppm	SRC, ppm	Element	Mean, ppm	Standard Devia-tion, ppm	SRC, ppm
Aluminum	12,900	4,500	65	Lead	34.8	43.7	—
Antimony	1.3	1.3	—	Magnesium	500	400	196
Arsenic	14.0	17.7	1.0	Manganese	49.4	40.2	1.7
Beryllium	1.6	0.8	—	Mercury	0.2	0.2	—
Boron	102.2	54.6	—	Molybdenum	7.5	6.0	—
Bromine	15.4	5.9	6.2	Nickel	21.1	12.4	—
Cadmium	2.5	7.6	—	Phosphorus	71.1	72.8	—
Calcium	7,700	5,500	—	Potassium	1,600	600	—
Chlorine	1,400	1,400	167	Selenium	2.1	1.1	—
Chromium	13.8	7.3	—	Silicon	24,900	8,000	—
Cobalt	9.6	7.3	0.8	Sodium	500	400	2.6
Copper	15.2	8.1	—	Tin	4.8	6.2	—
Fluorine	60.9	21.0	—	Titanium	700	200	96
Gallium	3.1	1.1	—	Zinc	272.2	694.2	—
Germanium	6.6	6.7	—	Zirconium	72.5	57.8	—
Iron	19,200	7,900		Moisture	9.0%	5.0%	0.0%
				Ash	11.4%	2.9%	0.05%

Source: Reference 114.

Table 4.25 Costs of Emissions Control—1000 MW$_e$^{a}

Equipment	Pollutant	Removal Efficiency, %	Cost, $/kW Capital	Cost, $/kW Operating^b
Electrostatic precipitator	Particulates	99	8–12	0.2
Two-stage scrubber	SO$_2$	85	40–60	0.8
Mechanical-draft cooling tower	Heat	n.a.	10–12	0.8
Total costs			54–84	1.8

Source: Reference 154.

^{a}Coal-fired electric power station.

^{b}Annual.

Table 4.26 Industrial Pollution Control Costs

	Billions of Dollars 1970	Billions of Dollars 1975
Air	0.5	4.7
Water	3.1	5.8
Solid waste	5.7	7.8
Total	9.3	18.3

Source: Reference 155.

(coal, oil, natural gas, and nuclear). It creates far more emissions, it requires more water, it creates more waste, and it requires more land area.[156]

In the 1960s the accepted means of regulating emissions from fossil fuel combustion was to establish sulfur emission standards. More recently, there has been some support for a sulfur emissions tax intended to provide a financial incentive for polluters to desist. The tax rate would possibly begin at 5¢ per lb of emitted sulfur per ton of coal, and rise in steps to 20¢.

In 1973, steam electric coal shipped to utilities averaged 2.9 percent sulfur by weight*; however, coal from the major supply regions, western Kentucky and Illinois, exceeded even this. An economic study[157] showed that in order to use surface-mined, high-sulfur coal from Illinois and meet sulfur emission requirements at no cost premium, the allowable cost of stack-gas scrubbing would be 22 to 24¢ per million Btu, where the midrange cost of scrubbing is actually

*Equal to 58 lb of sulfur per ton or $2.90 per ton tax at the 5¢ rate.

58¢. In markets traditionally served by Appalachian producers—Pennsylvania, New York, New Jersey—utilities would find it up to $15 per ton cheaper to use low-sulfur underground Appalachian coal than it would to burn lower-energy, low-sulfur western coal. The effect of the suggested emissions tax would be a decrease in emissions, but with a significant increase in energy costs. The alternative would be a potentially even more costly increased dependence on foreign oil.

The environmental/social impact of increased coal usage for electricity generation is shown in Table 4.27 for both deep- and surface-mined coal. Because of rapidly escalating coal prices, it is fruitless to attempt to quantify the future effect of social costs on coal prices. The average U.S. price of bituminous coal went up about 150 percent between 1972 and 1976.[159] It is hard to visualize social costs of coal increasing at a higher rate, so the percentage contribution to coal prices is likely to diminish somewhat, if anything. The social costs of surface-mined coal appear to be somewhat higher than for deep-mined coal, but the overall cost of the former is appreciably less because of economics of extraction.

But this is only part of the coal problem. There is a growing store of evidence that combustion of coal can "foul the nest" even more than was previously thought, even to the extent that radioactive elements in coal-combustion wastes are so long-lived that after 500 years the toxicity of these wastes will exceed that of all fission wastes with the exception of recyclable plutonium 239.[160] The dose

Table 4.27 Environmental and Social Impact of 1000-MW_e Electric Power System—Coal-Fired[a]

Environmental Effect	Deep-Mined	Surface-Mined
Air emissions, tons	383×10^3 (5)[b]	383×10^3 (5)[b]
Water discharges		
Heat	7.33×10^3 tons $\Big\}$ (5) 3.05×10^{13} Btu	40.5×10^3 tons $\Big\}$ (5) 3.05×10^{13} Btu
Solid waste, tons	602×10^3 (3)	3267×10^3 (5)
Land use, acres	29.4×10^3 (3)	34.3×10^3 (5)
Occupational Health		
Deaths	4.00	2.64
Workdays lost	8.77×10^3	3.09×10^3
General	Mine accidents; subsidence in urban areas	Landslides

Source: Reference 158.

[a]At 0.75 load factor, with generally prevailing environmental controls.

[b]Numbers in parentheses () indicate severity on the following scale: 5, serious; 4, significant; 3, moderate; 2, small; 1, negligible; 0, none.

rate of coal-consumption radioactive emissions, based on scanty data, might range considerably upward of 500 times that of fission fuels producing the same amount of electricity.[161] The most deadly of these pollutants could prove to be radon—a radioactive gas.[162]

Carbon Dioxide

Perhaps the most insidious effect of coal combustion could prove to be the release of carbon dioxide.* Much has been written recently concerning carbon dioxide in the atmosphere and projections of its effect on climate. Needless to say, the crucial questions remaining are: How did it get there? What is the rate of concentration change? How can it be controlled? And what will its effect be? Unfortunately, there seems to be no across-the-board agreement, although recent reports do appear finally to be converging on a common view.

Release of carbon dioxide is an inherent result of fossil fuel combustion. In the case of coal, the reaction is essentially

$$C + O_2 \rightarrow CO_2 \qquad 94 \times 10^6 \text{ cal/kg-mole} \qquad (14)$$

Sulfur and nitrogen oxides can be eliminated in theory, or reduced in practice, but this does not hold for CO_2, because the reaction shown is the source of heat in the combustion process. Coal is very inefficient in thermal output per mole of CO_2 formed in combustion. Methane yields over twice as much energy, and is closely followed by ethane and propane on this basis.[163]

In the past, it has been generally held that the observed increase in the CO_2 content of the atmosphere, over the past 100 years or more, has been due largely to the combustion of fossil fuels.† Holland claims that burning *all* the world's fossil fuel reserves (except for low-grade, uneconomical coals) would increase the release of CO_2 by a factor of about 10, but half of this would be absorbed by the oceans for a net increase of about five times. The process would be reversible after 1000 to 1500 years, when all recoverable fossil fuels would be consumed. His general view is that fossil fuel burning is merely "a pimple on a historical scale."[164]

The growing feeling now is that the increase in CO_2 content of the atmosphere is due in approximately equal measure to the worldwide destruction of forest cover and the oxidation (decay) of humus.[165] Instead of forests being a vast sink for excess CO_2, as previously supposed, the biota is apparently serving as a net source. It is believed that the biota could release to the atmosphere up to 3.6 times as much carbon dioxide as combustion of fossil fuels does.‡

The amount of carbon tied up in humus is estimated by various investigators to range from 700 to 3000 $\times$ 10^{15} g§ vs. possibly 830 $\times$ 10^{15} g for living ter-

*Each ton of burning coal emits about 3 tons of CO_2.
†Manufacture of cement is also a significant contributor to atmospheric CO_2.[164]
‡About 18 $\times$ 10^{15} g carbon per year vs. 5 $\times$ 10^{15} g, respectively.[6]
§1 $\times$ 10^{15} g equals 1 Gte (1 billion or 10^9 tonnes).

restrial biota. Together, their carbon reservoirs are perhaps two to three times that of the atmosphere. The concentration of carbon in the tropics is high because of the large mass of tropical forests, but the temperate and boreal zones have high humus concentrations because of the slow rate of decay in the colder climates.[166] The balance between input and output of carbon in these large reservoirs largely will determine the net flow of carbon dioxide into the atmosphere at any given time. Figure 4.3 shows various estimates of the balance of carbon released from the biota.

Some experts have been so bold as to suggest that the net effect of increased CO_2 might be global cooling, resulting from some complex, but unspecified, feedback phenomena.[172] There are those who argue that increasing temperatures due to CO_2 in the atmosphere could even benefit humankind by the poleward expansion of agricultural zones, accelerated photosynthesis, and a faster hydrologic cycle.[173] Unfortunately, all speculation on the effects of CO_2 are based on limited data obtained in restricted geographical areas and extrapolated to a global scale. In short, as of now, the estimates and hypotheses are all largely intuitive judgment.[174]

Many climate variables must be monitored regularly if we are to gain an understanding of the effects of atmospheric CO_2. Among these variables are: solar flux, cloud cover, surface albedo, marine and surface temperatures, polar ice thickness, water vapor, CO_2, tropospheric aerosols, and perhaps 15 or 20 other variables. Our society has been at work altering at least three of these: albedo, aerosols, and CO_2. The range of variations in these areas can perhaps

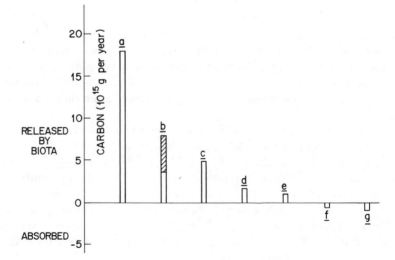

Figure 4.3 Annual release of carbon dioxide from biota (including humus). One hundred-year mean from carbon isotopes. Pre-1950. [(a), Woodwell et al. (maximum), (b), Woodwell et al. (range), (c), Adams et al., (d), Stuiver, (e), Bolin, (f), Oeschger et al., (g), Machta (references 166 to 171).]

best be illustrated by a sampling of expert opinions on the subject of CO_2 content of the environment and its effects.

Bolin (University of Stockholm) claims that the release of CO_2 by terrestrial biota is equal to only about 10 to 35 percent of the fossil fuel contribution. Woodwell (Marine Biological Laboratory—Woods Hole) and his colleagues estimate the figure at 80 to 160 percent, and attribute it to reforestation, but indicate that there is no evidence that the biota is a net sink. Keeling (University of California at San Diego) and Machta (NOAA) claim that terrestrial biota is a sink. Wong (Institute of Ocean Science) states that the biotic contribution is no more than 30 percent.[174]

Between A.D. 900 and 1900, the forest cover in western Europe was reduced from 90 percent to 20 percent, a situation that had occurred earlier in history throughout the eastern Mediterranean area. In our New England states, the forest cover gradually decreased until 1900 (from about 2.5×10^{15} g of carbon to 0.9×10^{15} g), at which time the abandonment of farms led to a reversal of the trend. The standing forest crop is probably less than half today (about 1.2×10^{15} g) what it was originally.[166] The most alarming change is certainly taking place in the Amazon Basin, the largest concentration of biomass remaining in the world. Here, in certain sections, there is an annual net decrease in forest cover of 0.6 to 1.5 percent,[165] which means that, at the present rate, the forest cover there may be reduced by half within the next 50 to 100 years.* Once the forests are cut, even intensively managed regrowth will be much smaller than the original stand in terms of plant mass.

The total marine plant mass† is only about 0.2 percent of the terrestrial plant mass. About 92 percent of the latter occurs in forested areas. Even so, approximately 32 percent of the net primary biomass production occurs in marine environments, three-fourths of it in the open oceans.[174] Converting tropical rain forests to cultivated land would probably result in about a 70 percent reduction in net primary biomass production in the area involved. Conversion of temperate forests or savannas to cultivated land would cause a 48 or 28 percent reduction, respectively.[175] There is considerable evidence that higher concentrations of carbon dioxide in the atmosphere stimulate photosynthesis. However, the amount of increase in plant growth due to the now higher carbon dioxide concentration would be so small as to be an insignificant factor when one considers the release of CO_2 from the biota and combustion of fossil fuels.[166]

There is, of course, a seasonal variation in the carbon dioxide content of the atmosphere due to seasonal changes in photosynthesis, respiration, and human-produced release of CO_2. The peak occurs in late winter or spring; the minimum is attained in late summer or fall.[166] Stuiver has used carbon isotope ratios to estimate the 100-year (1850–1950)‡ release of carbon from the biota.[168] Fossil fuels contain no ^{14}C and are richer in ^{12}C than in ^{13}C. The biota does contain ^{14}C and is also rich in ^{12}C relative to the atmosphere. Fossil fuel combustion

---

*Using the "rule of 70."

†As carbon, assuming a carbon content of 45 percent of the dry plant matter.

‡During which period the total terrestrial biomass decreased by 7 percent.

dilutes the ^{14}C content of the atmosphere (the Suess effect). The Suess effect, combined with the $^{12}C:^{13}C$ ratio, was used by Stuiver in his estimates, which show a total annual release from the biota of 1.2×10^{15} g, and from fossil fuels of 0.6×10^{15} g, over the 100-year period.

In keeping with the current annual rate of clearing tropical forests (0.5 to 1.0 percent), Woodwell et al.[166] estimate the most likely range for total world release from the biota at present is 4 to 8×10^{15} g carbon per year, which is approximately equal to the release rate from fossil fuels. Holland reported that the terrestrial biosphere has a carbon cycle time of about 10 years.[164]

Although the oceans provide the major reservoir for carbon dioxide in the form of calcium carbonate/bicarbonate, they are unable to accomodate all the excess from the atmosphere because of the low mixing rate. A better understanding is needed of the flux of carbon into the oceans to account for the balance, because current speculations on oceanic mixing show a limited capacity for the oceans to absorb CO_2 on this scale in a short time frame.[166]

Rising levels of CO_2 in the atmosphere correlate well with increasing temperatures of South Pacific waters. This "Southern Oscillation" extends over thousands of miles and lasts 6 or 7 years.[176] As the atmosphere warms up due to the greenhouse effect, marine microorganisms accelerate release of more CO_2 from the oceans, thus compounding the problem.

Of the approximately 4.5×10^{15} g (5 billion tons)* of CO_2 added to the atmosphere each year, the U.S. contribution is about 30 percent, or 1.35×10^{15} g (1.5 billion tpy).[177] A National Academy of Sciences report suggests that the peak CO_2 atmospheric concentration between 2150 and 2200 might be four to eight times the pre-Industrial Revolution level.[178,179] The temperature rise resulting from the greenhouse effect could be disastrous in its impact on agriculture, precipitation, and coastal flooding. Broecker (Lamont Doherty Geological Observatory) estimates that only a 4°F (2.2°C) difference in temperature existed between the last glacial age (16,000 to 20,000 years ago) and the last warming period (3,000 to 6,000 years ago). The greenhouse effect from our fossil fuel burning could have a similar effect. A SRI International study suggests that by the year 2025, the atmospheric CO_2 will have doubled and the average temperature will have risen by 5°F (2.8°C), leading to melting of the polar ice. If further atmospheric testing over the next 5 to 10 years shows the greenhouse effect to be as serious as feared, the only solution may prove to be cessation of our continuing economic growth, which is presently tied to the consumption of fossil fuels,[180] and restrictions on the reduction of forest cover. The results on our social structure would be incalculable. It is unlikely that most individuals would be ready emotionally to accept this hard solution.

SUMMARY

The coal option has been reviewed by Landsberg.[181] In President Carter's National Energy Plan (NEP-1), the role of coal was to serve as the "swing fuel,"

*1×10^{15} g equals 1.1×10^{9} tons (1.1 billion tons). Both units appear in various publications.

taking up the slack in supply left by gas and oil curtailments or shortages. With the Administration's stated ultimate goal of reducing oil imports to 7 million bpd (or possibly even 6 million bpd), the target for coal output had been set at 1250 million tpy in 1985. To attain this lofty goal, we would have to count heavily on coal from the Rocky Mountains and the Northern Plains states.*

The critical competitive edge of western coal is the high-volume mechanized output of low-sulfur fuel. Unfortunately for our energy future, the Administration energy plan advocated the use of scrubbers to reduce sulfur in *all* new utility and industrial coal-burning facilities, not just those employing high-sulfur coal. Since scrubbers add about 25 percent to the cost of a new power plant,[48] this policy placed an economic burden on western coal that had to be added to the already imposing transport, social, and water allocation burdens borne by this fuel. Fortunately, this regulation has been modified somewhat, but scrubbing is still required on even low-sulfur coals.

The aim of tripling coal use in nonutility areas (thus accounting for 38 percent of our coal consumption in 1985) may well be thwarted by the resulting higher cost of using western coal. Furthermore, the lower heat content of western coal† would require that oil imports be increased to 8.5 million bpd (from 7 million bpd), or that coal production rise to an even less attainable goal—1300 million tpy—to compensate for the lower heating value. If the Administration relies on supply/demand market pressures to solve these problems, Landsberg foresees an energy balance shortfall of as much as 3 million bpd of oil by 1985.[181]

Experts are now showing increased signs of pessimism regarding the realities of a "new age of coal," even to the extent of the prospects of doubling the output of coal by 1985 as envisioned by the Carter Administration. Another stumbling block is the projected rate of growth of energy consumption (5 percent to 1987, then 3.5 percent to 2000). The growth rate may go to 5.5 percent, according to at least one expert.[183] With the 1980 recession, triggered in part by recent OPEC price increases for crude oil, chances are good that the growth rate will actually be far below the estimates made in 1977—perhaps more like 2 percent or less.

All in all, it is easy to look at resources data and conclude that, with coal, and the technical feasibility of converting it to substitute natural gas or crude oil, we have no need to concern ourselves with energy shortages for hundreds of years. When one considers all the environmental, economic, technological, and logistical factors involved, however, it becomes increasingly questionable whether we truly want to convert to a new coal economy. The adverse environmental aspects of nuclear power have tended to blind many to the fact that coal, too, has its environmental shortcomings.

There are social costs that cannot be calculated equitably in dollars. There is the question of providing sufficient water to extract, prepare, and transport coal

---

*Only three big strip mines with a capacity of 50 million tpy were operating in Montana in 1974,[45] and five were operating in 1976.[182] Expansion of production would involve only seven new mines there, but they would account for about 11 percent of our total national output.[181]

†Averages 20 to 25 percent less than eastern coal.

and to convert it to fluid fuels. There is a question of the worsening greenhouse effect on our climate. There is no near-term economic problem with justifying desecration of pastures, forests, and tillable land through extraction of the underlying coal, but with an ever-increasing population, can we afford the long-term loss of this real estate at *any* monetary cost? It appears that we still need to have other energy options fully developed and ready to use on a large scale at the earliest possible date.

REFERENCES

1 R. Bojkov; reported by P. J. Bernstein, *Star-Ledger* (Newark, NJ) **63**(94), 39 (May 26, 1977).
2 B. Bolin, *Sci. Amer.* **223**, 125-132 (Sept. 1970).
3 H. Brown, "Energy in Our Future," in J. M. Hollander and M. K. Simmons, Eds., *Annual Review of Energy,* Vol. 1, Annual Reviews, Palo Alto, CA, 1976, p. 28.
4 D. Behrman, *Solar Energy—The Awakening Science,* Little, Brown, Boston, 1976, p. 141.
5 R. M. Rotty; reported by W. Lepkowski, *Chem. Eng. News* **55**(22), 10-14 (May 30, 1977).
6 C. F. Baes, Jr., *Amer. Sci.* **65**(3), 310-320 (May-June 1977).
7 E. Cook, *Man, Energy, Society,* Freeman, San Francisco, 1976, p. 75.
8 B. Tissot, *Recherche* **8**(77), 326-334 (Apr. 1977).
9 G. H. Cady, "Coal," in D. N. Lapedes, Ed., *McGraw-Hill Encyclopedia of Energy,* McGraw-Hill, New York, 1976, pp. 126-134.
10 G. H. Cady, "Peat," in *ibid.,* p. 546.
11 *ASTM Book of Standards,* Pt. 26, Designation D388, "Classification of Coals by Rank," Amer. Soc. Test. Mater., Philadelphia, 1978, pp. 220-224.
12 R. C. Binning and R. S. Hockett, "Heating Value," p. 364, in reference 9.
13 I. G. C. Dryden, Ed., *The Efficient Use of Energy,* IPC Science & Technology Press, London, 1975, p. 8.
14 *Energy Facts II,* Sci. Policy Res. Div., Libr. Congr., Serial H, U.S. GPO, Washington, DC, 1975, p. 110.
15 H. Enzer et al. (DOI, 1975); reported in reference 14, p. 97.
16 DOC (FPC, 1975); reported in reference 14, pp. 21, 96.
17 *Coal Facts: 1978-1979,* Natl. Coal Assoc., Washington, DC, p. 7.
18 World Energy Conf. Surv. Energy Resour. (1974); reported in reference 17, p. 71.
19 D. E. Carr, *Energy and the Earth Machine,* Norton, New York, 1976, p. 64.
20 Reference 17, p. 74.
21 Reference 14 (FPC data).
22 I. Ross, *Reader's Dig.* **110**(662), 153-160 (1977).
23 *Bus. Week* (No. 2487), 80-85 (June 13, 1977).
24 A. Epstein, *Phila. Inquirer* **296**(114), 1-A, 10-A (Apr. 24, 1977).
25 Reference 17, p. 72.
26 A. A. Bartlett, *Phys. Today* **29**(12), 9, 11 (Dec. 1976).
27 Reference 17, p. 82.
28 *Ibid.,* p. 9.
29 W. J. Lanouette, *Natl. Observer* (May 9, 1977), p. 4.
30 J. Arthur; reported in *Public Util. Fortn.* **98**(7), 36-39 (Sept. 23, 1976).
31 *Chem. Eng. News* **55**(7), 24, 26, 31 (Feb. 14, 1977).

32 G. Atwood, *Sci. Amer.* **233**(6), 23–29 (Dec. 1975).

33 N. J. W. Thrower, Ed., *Man's Domain: A Thematic Atlas of the World,* McGraw-Hill, New York, 1975, p. 31.

34 Reference 17, p. 85.

35 E. Faltermayer, *Fortune* **89**(6), 136–139, 244–252 (June 1974).

36 *Energy User News* **2**(24), 11 (June 20, 1977).

37 *Ibid.* **3**(22), 12 (May 29, 1978).

38 Reference 19, p. 58.

39 Reference 17, p. 82.

40 Reference 19, p. 60.

41 D. Clippinger, *Phila. Inquirer* **296**(128), A-1, A-10 (May 8, 1977).

42 *Courier-News* (Bridgewater, NJ) **94**(118), B-12 (Dec. 6, 1977).

43 Reference 17, p. 88.

44 Reference 19, p. 63.

45 E. Faltermayer, *Fortune* **89**(5), 215–219, 334–338 (May 1974).

46 J. D. Reilly, "Coal Mining," pp. 141–144, in reference 9.

47 R. A. Schmidt and G. R. Hill, "Coal: Energy Keystone," pp. 58, 59, in reference 3.

48 *New Sci.* **74**(1050), 274 (May 5, 1977).

49 Reference 17, p. 84.

50 Reference 19, p. 62.

51 *Civ. Eng.—ASCE* **47**(1), 43–47 (Jan. 1977).

52 E. Just, "Mining," pp. 448, 449, in reference 9.

53 *Bus. Week* (No. 2496), 114B (Aug. 15, 1977).

54 *Chem. Eng. News* **54**(45), 14 (Nov. 1, 1976).

55 J. J. Dowd, "Mining, Strip," pp. 453–455, in reference 9.

56 *Modern Energy Technology,* Vol. 1, Research and Education Assoc., New York, 1975, p. 11.

57 P. J. Phillips and P. P. DeRienzo, *Public Util. Fortn.* **99**(8), 29–33 (Apr. 14, 1977).

58 M. M. McCormack, "Protecting the Environment," pp. 49–59, in reference 9.

59 Reference 47, pp. 48, 49.

60 F. P. Aplan; reported in *Mech. Eng.* **99**(5), 22 (May 1977).

61 *Chem. Eng. News* **55**(25), 20 (June 20, 1977).

62 Reference 17, p. 80.

63 *Ibid.,* p. 93.

64 R. C. Longworth, *Phila. Inquirer* **296**(149), 10-D (May 29, 1977).

65 Manalytics, Inc., for EPRI; reported in *Chem. Eng. News* **55**(1), 14 (Jan. 3, 1977).

66 A. T. Yu, *Civ. Eng.—ASCE* **47**(1), 52–55 (Jan. 1977).

67 *Public Util. Fortn.* **99**(8), 73 (Apr. 14, 1977).

68 G. E. Dials and E. C. Moore, *Environment* **16**(7), 18–24, 30–37 (Sept. 1974).

69 W. Dempsey; reported in *Bus. Week* (No. 2483), 78 (May 16, 1977).

70 R. E. Briggs; reported in *ibid.* (No. 2485), 6 (May 30, 1977).

71 *U.S. News World Rep.* **82**(16), 75, 76 (Apr. 25, 1977).

72 NCA (1974); reported in reference 3, p. 53.

73 *Ibid.,* p. 52.

74 T. Roncalio; reported in reference 69.

75 *Chem. Eng. News* **57**(21), 18–20 (May 21, 1979).

76 M. Rieber et al., *The Coal Future,* Natl. Sci. Found., Washington, DC, 1975.

77   P. M. Boffey, *Science* **190**(4213), 446 (Oct. 31, 1975).

78   *Civ. Eng.—ASCE* **47**(2), 22 (Feb. 1977).

79   J. J. Josephson, *Environ. Sci. Technol.* **10**(12), 1086, 1087 (Nov. 1976).

80   *Chem. Eng. News* **54**(33), 21 (Aug. 9, 1976).

81   *Bus. Week* (No. 2451), 90D, 90H (Sept. 27, 1976).

82   E. Calvelli, *Mech. Eng.* **99**(6), 102 (June 1977).

83   J. Dooher et al., *ibid.* **98**(11), 36–41 (Nov. 1976).

84   M. K. Hubbert (1971); reported by T. M. Thomas, "Recent Trends in Energy Consumption and Supply," in G. J. S. Govett and M. H. Govett, Eds., *World Mineral Supplies: Assessment and Perspective,* Elsevier, New York, 1976, pp. 147–172.

85   BuMines (1973); reported by B. S. Cooper, "U.S. Policies and Politics," p. 38, in reference 9.

86   J. O'M. Bockris, *Energy—The Solar-Hydrogen Alternative,* Wiley, New York, 1975, p. 33.

87   *World Energy Supplies, 1969 to 1972,* Stat. Pap., Ser. J, No. 17, United Nations, New York, 1974, p. 31.

88   G. C. Gambs, "Energy Sources," pp. 250–253, in reference 9.

89   Reference 17, p. 66.

90   *Ibid.*, p. 70.

91   BuMines (1971); reported in reference 7, p. 327.

92   G. L. Decker; reported in *Mech. Eng.* **99**(6), 70 (June 1977).

93   G. S. Savitsky, *Energy User News* **2**(24), 13 (June 20, 1977).

94   *Environ. Sci. Technol.* **10**(13), 1200, 1202 (Dec. 1976).

95   Reference 19, p. 67.

96   A. Haas, Jr., *Bus. Soc. Rev.* (No. 12), 53–57 (Winter 1974–1975).

97   *Chem. Eng. News* **55**(14), 27, 28 (Apr. 4, 1977).

98   *Ibid.* **54**(42), 22 (Oct. 11, 1976).

99   Kohrs (1974); reported in B. Christiansen and T. H. Clack, Jr., *Science* **194**(4265), 578–584 (Nov. 5, 1976).

100 NAS (1975); reported in D. J. Rose et al., *Amer. Sci.* **64**(3), 291–299 (May–June 1976).

101 S. Baron, *Mech. Eng.* **96**(12), 12–20 (Dec. 1974).

102 W. K. C. Morgan (1980); reported in R. Kotulak, *Phila. Inquirer* **302**(169), 5-A (June 17, 1980).

103 *New Sci.* **77**(1096), 837 (Mar. 30, 1978).

104 *Chem. Eng. News* **57**(20), 27 (May 14, 1979).

105 Bennett (1944); reported by W. G. Ackermann, *EOS, Trans. Geophys. Union* **57**(10), 708–711 (Oct. 1976).

106 M. Walton, *Miami Herald* **70**(37), 30-A (Jan. 6, 1980).

107 C. E. Bagge, *Public Util. Fortn.* **98**(7), 36–39 (Sept. 23, 1976).

108 Reference 19, p. 66.

109 *Environ. Sci. Technol.* **9**(1), 18, 19 (Jan. 1975).

110 Kreulen (1952); reported by D. Casagrande and K. Siefert, *Science* **195**(4279), 675, 678 (Feb. 18, 1977).

111 Freney (1967); reported in *ibid.*

112 *New. Sci.* **72**(1021), 14 (Oct. 7, 1976).

113 *Chem. Eng. News* **55**(23), 21 (June 6, 1977).

114 D. W. Koppenaal and S. E. Manahan, *Environ. Sci. Technol.* **10**(12), 1104–1107 (Nov. 1976).

115 *Chem. Eng. News* **55**(28), 25 (July 11, 1977).

116 A. L. Hammond, *Science* **193**(4256), 873–875 (Sept. 3, 1976).

117 *Environ. Sci. Technol.* **11**(13), 1148, 1149 (Dec. 1977).

118 *Chem. Eng. News* **55**(22), 24, 25 (May 30, 1977).

119 *Public Util. Fortn.* **99**(11), 55, 56 (May 26, 1977).

120 *Chem. Eng. News* **55**(13), 26, 27 (Mar. 28, 1977).

121 D. Pothier, *Phila. Inquirer* **296**(114), 8-C (Apr. 24, 1977).

122 B. Gilbert, *Sports Illus.* **45**(25), 54–81 (Dec. 20–27, 1976).

123 H. J. White, *J. Air Pollut. Control Assoc.* **27**(1), 15–21 (Jan. 1977).

124 J. J. Devaney, *Phys. Today* **30**(4), 13, 15, 65 (Apr. 1977).

125 H. J. White, *J. Air Pollut. Control Assoc.* **27**(2), 114–120 (Feb. 1977).

126 A. L. Hammond, *Science* **194**(4261), 172, 173, 218–220 (Oct. 8, 1976).

127 *Chem. Eng. News* **54**(43), 33 (Oct. 18, 1976).

128 M. Kenward, *New Sci.* **74**(1050), 286, 287 (May 5, 1977).

129 J. M. Wood, *Recherche* **7**(70), 711–719 (Sept. 1976).

130 R. F. Pueschel, *Geophys. Res. Lett.* **3**(11), 651–653 (Nov. 1976).

131 C. E. Chrisp et al., *Science* **199**(4324), 36–38 (Jan. 6, 1978).

132 *Ind. Res.* **19**(11), 44, 46 (Nov. 1977).

133 Reference 17, p. 64.

134 H. J. White, *J. Air. Pollut. Control Assoc.* **27**(3), 206–217 (Mar. 1977).

135 *Ibid.* **27**(4), 308–318 (Apr. 1977).

136 R. E. Schmid, *Phila. Inquirer* **300**(173), 5-A (June 22, 1979).

137 J. Forrest and L. Newman, *Atmos. Environ.* **11**(5), 465–474 (1977).

138 *Bus. Week* (No. 2482), 92E (May 9, 1977).

139 I. C. T. Nisbet, *Technol. Rev.* **76**(4), 8, 9 (Feb. 1974).

140 A. J. Vermeulen, *Environ. Sci. Technol.* **12**(9), 1016–1021 (Sept. 1978).

141 L. R. Ember, *Chem. Eng. News* **57**(49), 15–17 (Dec. 3, 1979).

142 R. Peterson; reported in *Mech. Eng.* **98**(10), 18 (Oct. 1976).

143 M. Mullany, *Chem. Ind.* (No. 2), 58–61 (Jan. 15, 1977).

144 *Environ. Sci. Technol.* **11**(3), 226–228 (Mar. 1977).

145 L. J. Carter, *Science* **202**(4363), 30 (Oct. 6, 1978).

146 *Bulletin* (Phila.) **133**(44), 3 (May 25, 1979).

147 *Chem. Eng. News* **57**(11), 17 (Mar. 12, 1979).

148 K. W. Daykin and J. C. V. Rumsey, *Phys. Technol.* **6**(6), 245–250 (Nov. 1975).

149 *Bus. Week* (No. 2456), 90B, 90F, 90H (Nov. 1, 1976).

150 F. Calhoun, *Energy User News* **2**(30), 22, 23 (Aug. 1, 1977).

151 P. F. Fennelly et al., *Environ. Sci. Technol.* **11**(3), 244–248 (Mar. 1977).

152 J. J. Mutch, *J. Air Pollut. Control Assoc.* **27**(6), 567–571 (June 1977).

153 R. W. Peterson, *Chem. Eng. News* **55**(25), 39–42 (June 20, 1977).

154 Reference 56, pp. 42, 43.

155 *Ibid.*, p. 115.

156 C.-G. Ducret, "Environmental Aspects of Energy Conversion and Use," in I. M. Blair et al., Eds., *Aspects of Energy Conversion*, Pergamon, Oxford, 1976, p. 649.

157 A. Schlottmann and L. Abrams, *Rev. Econ. Stat.* **59**(1), 50–55 (Feb. 1977).

158 Counc. Environ. Qual., *Quality, Energy and the Environment*, U.S. GPO, Washington, DC, 1973, p. 14, Table 3.

159 Reference 17, p. 67.

160 J. J. Devaney, *Phys. Today* **29**(6), 78 (June 1976).

161 J. H. Ray, *ibid.*, pp. 77, 78.

162 A. L. Simon, *Energy Resources,* Pergamon, New York, 1975, p. 43.

163 J. H. Krenz, *Energy—Conversion and Utilization,* Allyn and Bacon, Boston, 1976, p. 36.

164 H. D. Holland, presented at Gen. Res. Colloq., Bell Laboratories, Holmdel, NJ, Oct. 14, 1977.

165 G. M. Woodwell, *Sci. Amer.* **238**(1), 34–43 (Jan. 1978).

166 G. M. Woodwell et al., *Science* **199**(4325), 141–146 (Jan. 13, 1978).

167 J. A. S. Adams et al., in G. M. Woodwell and E. V. Pecan, Eds., *Carbon and the Biosphere,* AEC Tech. Inf. Cent. Washington, DC, 1973.

168 M. Stuiver, *Science* **199**(4326), 253–258 (Jan. 20, 1978).

169 B. Bolin, *ibid.*, **196,** 613 (1977).

170 H. Oeschger et al., *Tellus* **27,** 168 (1975).

171 L. Machta, in reference 167.

172 P. H. Abelson, *Science* **197**(4307), 941 (Sept. 2, 1977).

173 J. Alker, letter in *Amer. Sci.* **65**(5), 530 (Sept.–Oct. 1977).

174 R. A. Kerr, *Science,* **197**(4311), 1352, 1353 (Sept. 30, 1977).

175 R. H. Whitaker and G. E. Likens; reported in reference 165.

176 K. Hanson; reported in *Chem. Eng. News* **55**(41), 30 (Oct. 10, 1977).

177 *Ibid.* **55**(49), 4 (Dec. 5, 1977).

178 *Technol. Rev.* **80**(1), 18–20 (Oct.–Nov. 1977).

179 *Phys. Today* **30**(10), 17, 18 (Oct. 1977).

180 W. Lepkowski, *Chem. Eng. News* **55**(42), 26–30 (Oct. 17, 1977).

181 H. H. Landsberg, *Science* **197**(4298), 9 (July 1, 1977).

182 B. Christiansen and T. H. Clack, Jr., *Science* **194**(4265), 578–584 (Nov. 5, 1976).

183 J. F. Hogerton; reported in *Energy User News* **2**(26), 6 (July 4, 1977).

5 COAL-DERIVED SUBSTITUTE FUELS

There is no technical question that we can gasify coal. An economic and political decision is needed to get on with the job.

J. CROWLEY

There has developed in the energy field the unfortunate and confusing use of a term that applies both to the subject of this chapter and to the quite different subject of conservation. To energy specialists, "conversion" means not only the production of substitute fluid fuels* from coal, peat, oil shale, tar sands, and so forth, but it also refers to the conversion of gas- and oil-fired boilers to coal-fired units, to conserve the less plentiful fuels. Throughout this chapter and the following two chapters (dealing with tar sands and oil shales, and biomass- and refuse-derived fuels), "conversion" will take on the former meaning.

As noted in Chapters 2 and 3, fluid fossil fuel supplies are on the wane. For example, the most credible estimates now available place remaining recoverable, conventional U.S. oil resources at 156 to 377 billion bbl.[1]† Unfortunately, these fluids have become essential to our present way of life, particularly in the residential and transportation sectors of our economy.

The particular fossil fuel apparently closest to exhaustion, with regard to the U.S. market, is natural gas. The depletion of this fuel has been stimulated, in large measure, by the fact that it is not only the cleanest and most desirable fuel, but at the same time, its price has been federally regulated at rates lower

---

*The writer will use the term "fuels" broadly to mean fuels to meet our energy needs, as well as materials to fuel our chemical feedstock requirements.

†Remaining as of December 31, 1975. Higher figure is based on 60 percent of oil originally in place [National Petroleum Council (NPC) data] reduced by cumulative production to date noted. The lower figure is a compromise between the latest U.S. Geological Survey (USGS) and oil company estimates, assuming 32 percent recovery. The USGS estimate of oil originally in place (440 billion bbl) is about 45 percent of NPC estimate (810.4 billion bbl), and USGS claims approximately half of ultimately available domestic oil has already been produced.[2]

than its competitors—petroleum and coal—on an equal-energy basis. Table 5.1 shows the steady decline in the growth rate of the U.S. gas industry since 1965, despite a continuing high demand for this fuel.

Recognizing the fact that fluid fossil fuel reserves are dwindling and demand for them continues to increase, the nation is faced with the question of how to meet our energy needs for the future. An American Society of Mechanical Engineers (ASME) Task Force on Energy Conversion Research recently completed a report on the subject.[4] This report identifies several broad areas for major energy research effort involving fossil (or fossil-like) fuels. They are:

1   Direct combustion of coal (but in compliance with necessary environmental restraints).
2   Synthesis of substitute pipeline and producer gases (syngases*) and liquid fuels (synoils, or synfuels when including syngases) from more plentiful raw materials.
3   Production of refuse-derived fuels (RDF).

It was felt by the Task Force that the latter two represented the largest blocks of "new" fluid energy available in the foreseeable future. The report was formulated basically on technological grounds, with only secondary regard for political or social factors.

The need, then, is to provide a means for supplying substitute fluid fuels derived from plentiful (or renewable) raw materials, and to do so at a competitive price and without sacrifice of quality. Our most readily available potential feedstock is coal, a fossil fuel whose use has been wavering since natural

Table 5.1   Average Annual Consumption Growth Rate of Domestic Natural Gas

	Average Annual Growth Rate, %
Pre-World War II (1935–1940)	6.7
Preembargo (1965–1970)	6.5
1971	3.6
1972	−0.1
1973	−0.2
1974	−3.7
1975	−6.5
1976	−1.4

Source: Reference 3.

*Not to be confused with syn gas, or synthesis gas (CO plus H_2).

fluid fossil fuels initially found their place in our economy. (See Chapter 4 for a more detailed account of coal.)

As early as the end of the seventeenth century, the embryonic technology was noted for converting this readily available solid fuel to cleaner and more versatile fluid forms. In 1670, an English minister noted that coal heated in a closed vessel produced "a spirit which issued out and caught fire at the flame of a candle." Conversion technology progressed to a point where manufactured gas ultimately was made available for lighting the streets of London in 1815. By 1834, water gas was made in France, by the reaction of coal and steam, and at about the same time, the French extracted oil from shale.[5] Early coal-based gases were poisonous and foul-smelling due to the presence of hydrogen cyanide and hydrogen sulfide, but it was found that they could be cleaned by scrubbing with lime and iron oxide.

Approaching World War II, Germany recognized the need for developing a secure domestic source of fluid hydrocarbon fuels to sustain her war effort. As a result, her scientists and engineers developed the Bergius process, which called for hydrogenation and catalytic hydrorefining (both at 660 to 930°F and 3000 to 10,000 psi), of her ample coal. This process produced a 35 to 50 percent aromatic blend gasoline and a paraffinic–naphthenic mixture. Tar acids (phenols) and tar bases (organic nitrogen compounds) were by-products of the hydrogenation stage, in the amounts of 5 to 10 weight percent of the mineral-ash-free (m.a.f.) coal.[6] This process provided the bulk of her fuel for the duration of the war and beyond, but by 1958, the process had become uneconomic, because of the availability of cheap Mideast oil, and was abandoned.

In the meantime, South Africa, realizing her isolation from oil and gas sources, introduced the SASOL coal conversion process (with technical assistance from U.S. sources). This method produces a wide range of hydrocarbons and has helped the country maintain its balance of payments with the outside world and retain its independence from OPEC, despite the premium prices of the resulting products.

Gasification and liquefaction of coal have proven successful technically speaking, but this fact is overshadowed by the generally poor economics of the processes. Only when foreign oil prices rise sufficiently will this situation be reversed. There is a question as to how long the Organization of Petroleum Exporting Countries (OPEC) would then be able to offer its sole resource at a price greater than the cost of replacing it with a competitive product.[7] OPEC is in a position, however, where it could drop its prices abruptly to discourage a budding conversion industry.

CONVERSION OPTIONS

There are a number of options open to us for producing substitute fluid fuels from fossil source materials. Domestic gasification and liquefaction of coal and retorting and distilling of oil shale offer the brightest hopes, because both raw materials are plentiful and the basic technology is understood. Actually, oil shale conversion has an edge over coal conversion in that the H:C ratio of oil shale is higher (1.5 to 1.6 vs. 0.5 to 1.0 for coal),[8] thus requiring less drastic conversion treatment. Tar sands are a promising possibility in Canada, but our tar sands reserves are miniscule compared to our northern neighbor's. Peat, of which we have large reserves, is another prospective candidate as a source for substitute natural gas.

Basically, the goal of any synthetic fuel process is to increase the hydrogen-to-carbon ratio (the carbon efficiency) and reduce mineral, sulfur, and nitrogen contents, to enable the product(s) to meet emission standards. Increasing the H:C ratio can be accomplished by either adding hydrogen or extracting carbon. All this, or course, must be accomplished at a cost competitive overall with natural fuels.[5] The atomic hydrogen-to-carbon ratios and energy contents of a wide variety of fuels are given in Table 5.2

Ideally, the starting material for synthesis of substitute fuels should have an H:C ratio as close as possible to the desired product and should contain a minimum amount of oxygen. The extent of deviation from these criteria determines the degree of processing, hence the plant complexity and cost, required to achieve the ultimate conversion. The cost of capital may make it cost-effective to start with more expensive, higher H:C ratio raw materials, such as oil shale (1.5 to 1.6) or tar sands (1.4 to 1.5) rather than with cheaper, low-H:C coal (0.5 to 1.0).[8] Table 5.3 notes potential fossil raw materials for synthetic or substitute fuels—their characteristics and current major uses.

In light of the impending natural gas shortage, there are currently various fuel options for major users—electric utilities and industrial plants[9]:

1   Conversion from gas-fired boilers to low-sulfur or No. 6 fuel oil.
2   Conversion to No. 2 fuel oil (to permit retention of existing boilers).
3   Conversion from gas-fired boilers to direct combustion of coal.
4   Coal gasification.

The first option means scrapping existing boiler facilities and going to a more costly and dirtier fuel. The second is acceptable only for limited-term emergency use, because the fuel is more expensive and requires boiler modifications. The third option, conversion to coal, has been tagged "utterly ridiculous," as it calls for entire new plants, with coal handling and storage facilities, new boilers, and extensive pollution control equipment. Furthermore, coal is not acceptable, in any case, for use in the Los Angeles basin and certain other areas of the country.

The most serious constraints to the expanded employment of coal are the environmental issues associated with its extraction, preparation, and combustion. Sulfur and mineral ash make coal an inherently dirty fuel. Consequently, its expanded use has been strongly inhibited by the failure of scrubbing to measure up environmentally and economically, and the questionable acceptability of tall stacks as a means of keeping the environment suitably free of pollutants.[10]

Table 5.2 Hydrogen-to-Carbon Ratio and Heating Value of Hydrocarbon Fuels

Fuel Type	Atomic H:C[a]	Gross Heating Value, Btu/lb
Gaseous		
High Btu		
Methane	4.0	23,940
Natural gas	3.5–4.0	21,060–21,420
Intermediate Btu		
Hydrogen	—	61,020
Coke-oven gas	4.9	17,280
Water gas (H_2/CO)	2.0	7,740
Low Btu		
Producer gas[b]	1.2	2,160
Liquid		
Methanol	4.0	9,540
Gasoline	2.0–2.2	20,160–20,520
No. 2 fuel oil	1.7–1.9	19,260–19,800
No. 6 fuel oil	1.3–1.6	18,000–18,900
Crude shale oil	1.6	18,540–18,720
Bitumen[c]	1.4–1.5	17,640–18,000
Solid		
Kerogen[d]	1.5	18,000
Lignite	0.8	7,020–9,720
Subbituminous coal	0.8	9,900
Bituminous coal	0.5–0.9	12,060–15,840
Anthracite	0.3	15,840
Low-temperature coke	0.4	14,760
High-temperature coke	0.06	14,400

Source: Reference 8. [Cokes-Spiers (1962); coals, Given (1975); shale oil, kerogen, bitumen, Cameron Engrs. (1975); others, Perry et al. (1963), Linden (1965).]

[a]Mineral-free carbon; total organic hydrogen.
[b]Nitrogen-diluted, from bituminous coal.
[c]Athabasca tar sands.
[d]Green River oil shale.

The only long-term option that makes sense for fossil-fuel-burning power plants is the fourth—coal gasification—which permits use of existing power generating stations, provides low-pollution fuel, and makes use of a relatively abundant raw material.

Inherently, coal gasification plus gas combustion is less efficient than direct burning of coal, by about 30 percent.[11] Another source claims an overall effi-

Table 5.3 Potential Fossil Raw Materials for Substitute Fluid Fuels

	Characteristics and Major Uses
Peat	Coal precursor. Partially decomposed marshland growth. Low heating value prohibits shipment as fuel. Potential uses: fuel for on-site power generation and raw material for syngas. Primary current use: soil conditioner.
Lignite	Coal of recent origin. Low heating value prohibits long-distance shipment. Major use: minemouth power generation.
Subbituminous coal	Intermediate between lignite and bituminous coal. Low-sulfur, medium-heating-value western coal. Economical power generation fuel for large-scale use as far east as western Ohio.
Bituminous coal	High-volatile, high-sulfur, high-heat-content "soft" coal. For coking, power generation, and general industrial use.
Anthracite	Essentially pure-carbon, low-sulfur, low-nitrogen, high-heating-value "hard" coal. Major use: home heating fuel.
Shale oil	Insoluble hydrocarbon (kerogen) associated with fine-grained sedimentary rock. No current commercial uses in U.S. Potential source of syncrude and chemical feedstocks.
Tar or oil sands	Soluble hydrocarbon (bitumen) associated with minerals (predominantly sand). Under commercial development in Canada as source of syncrude.

ciency for the process of about 60 percent.[12] Furthermore, gasification technology places a drain on water supplies in the arid western plains where coal is most plentiful.* The necessary gasification facilities also add to the already large amount of land disturbed by mining operations. Nonetheless, compared to coal combustion, the resulting product is cleaner and probably results in reduced air pollution overall.

CONVERSION CHEMISTRY AND TECHNOLOGY

Unlike most other countries, we are blessed by[13]: substantial oil and gas production capability, large recoverable resources of coal and oil shale, developing conversion technology, a nuclear power industry and the raw materials to sup-

*USGS estimates that 37 to 150 gal of water is required to produce 10^6 Btu of SNG, depending on the specific process.[10]

port it, and an established energy research and development program. Nonetheless, we cannot yet deliver alternative fluid fuels in commercial quantities.

The basic constituents of fossil fuels are, of course, hydrogen and carbon. The atomic ratio of the two has considerable bearing on the physical form of the fuel, as well as its heating value per unit weight, as suggested by the two limiting forms of fossil fuels as found in nature: methane and anthracite. The former has a H:C ratio of four (3.5 to 4.0 for actual natural gas) and a heating value of 24,000 Btu per lb (21,000+ Btu for natural gas), as compared to 0.3 H:C and 15,800 Btu per lb for anthracite. In between, there are numerous liquids and solids intermediate in both properties.[8] The basis of coal conversion processes (CCPs) is using water (steam), heat, and pressure, together, to release hydrogen from coal, and then catalytically combining the hydrogen with carbon atoms in coal to produce hydrocarbons ranging from methane to asphalt.[14]

Fluids fuels have become essential to our economy in at least three areas: transportation, space heating, and peak shaving in electric power generation. Yet we are becoming increasingly dependent on foreign oil to meet these needs. The major candidates for replacing foreign oil are coal-based synfuels.

There are several basic systems for liquefying coal, some with many potential commercial variations: (1) add hydrogen; (2) reject carbon; (3) tear down the coal to individual carbon atoms, and build them up again into the desired compounds; (4) extract hydrocarbons with solvent and hydrocrack the extract; and (5) gasify the coal, convert to methanol, and then to gasoline. Stated another way, these variants reduce to pyrolysis, indirect liquefaction, and direct hydroliquefaction. Options 3 and 5 are probably the most speculative, on the basis of economics. Option 1, (direct liquefaction) is the most probable major source of coal-derived liquid fuels. Pyrolysis (option 2) could become significant if the low liquid yields can be tolerated and an outlet can be found for the large volumes of by-product char.[15] Option 3 (indirect liquefaction) is the only commercial liquefaction process operational in the world today.

There is no commercial domestic liquefaction plant in operation at this time, and if the timing of the EDS commercialization process is any indication, up to 30 years may be required to develop a new technology, from conception to commerical operation. Fortunately, three process developments (SRC-II, H-Coal, and EDS) are far enough along now—close to the large-pilot-plant stage—that commercial operation could be a reality by about 1995.[15] The Institute of Gas Technology has gone so far as to suggest that the United States could phase out crude oil and natural gas imports by the year 2000, provided that massive federal support is forthcoming in relation to direct financial involvement, tax restructuring, streamlining of regulatory processes, water allocation, and so forth.[13] This schedule, however, does not seem credible to this writer, because of inescapable bureaucratic delays and political maneuvering, logistics and environmental problems, and massive labor and materiel needs.

The most logical way to achieve conversion of coal is to remove carbon as the gaseous oxides and/or to add hydrogen to the molecule, using water as the primary source of hydrogen. Simply stated, the desired overall reaction would be[16]

$$2C + 2H_2O \rightarrow CH_4 + CO_2 \qquad -2765 \text{ kcal/kg-mole} \qquad (1)$$

similar to bacterial conversion, in nature, of cellulose to methane and carbon dioxide. Expressed more explicitly, the reaction for coal is[17]

$$CH_{0.8} + 0.8H_2O \rightarrow 0.6CH_4 + 0.4CO_2 \qquad (2)*$$

with 40 percent of the carbon being rejected as CO_2—just sufficient to eliminate oxygen from the added water vapor (steam) and to generate the necessary hydrogen.

It should be remembered that coal is the least desirable fossil fuel to use as a starting raw material, because it has to be processed more extensively to attain the required H:C ratio. The complexity of the process and of the plant (and therefore the cost) obviously is greater than if one were to start with tar sands or shale or crude oil.

In practice, conversion to high-heating-value gas (largely methane) by the *gasification* process occurs in three basic stages[16,17]:

$$C + 1/2O_2 \text{ (pure)} \rightarrow CO \qquad +26,637 \text{ kcal/kg-mole} \qquad (3)$$

Combustion of some of the coal (Equation 3) provides thermal energy necessary for the *formation of the process gas* (Equation 4).

$$C + H_2O \xrightarrow[\text{(1900–2500°F)}]{\Delta} CO + H_2 \qquad -32,457 \text{ kcal/kg-mole} \qquad (4)$$

The *water-gas shift* reaction is controlled to provide a hydrogen-to-carbon monoxide molar ratio of 3.

$$CO + H_2O \rightarrow CO_2 + H_2 \qquad +7,838 \text{ kcal/kg-mole} \qquad (5)$$

This sets up the final, *catalytic methanation* stage (Equation 6). Acid-gas cleanup removes H_2S and CO_2 contaminants to prevent fouling of the catalysts[16]:

$$CO + 3H_2 \xrightarrow{\text{cat.}} H_2O + CH_4 + \text{heat} \qquad (6)$$

Water is removed in a dehydration step. The heat produced in Equation 6 is too low quality (low temperature) to use in the process; thus the efficiency is reduced to 50 to 55 percent overall.

---

*$CH_{0.8}$ represents the approximate combination of carbon and hydrogen in coal. Equation 1 ignores the presence of hydrogen in coal. Actual analysis has disclosed the molecular formula of a specific coal (Clifty Creek No. 6, high-volatile bituminous coal) to be $C_{0.54} H_{0.45} S_{0.01} N_{0.01}$,[18] or a combination of $CH_{0.83}$). Thus it can be seen that the H:C ratio cannot really be generalized.

One of the anomalies of the gasification process is that the formation of methane is favored by low temperatures, but the reaction rate at low temperatures is too slow to be of commercial interest. Formation of methane is inhibited by temperatures above 480°F (250°C); therefore, processes now under development produce synthesis gas in a high-temperature stage (Equation 4) and then subsequently convert the synthesis gas catalytically to methane in a low-temperature stage (Equation 6).[16]

A more efficient conversion technology (typical of the newer processes under development) employs *hydrogasification* to produce a substantial quantity of methane directly by reaction with hydrogen-rich gas, using partial combustion of char as a source of process heat:

$$C\ (char) + O_2 \rightarrow CO_2 + heat \qquad (7)$$

and char and steam as the hydrogen source:

$$C\ (char) + H_2O + heat \rightarrow CO + H_2 \qquad (8)$$

The overall *hydrogasification* reaction is[17]

$$2CH_{0.8}\ (coal) + 1.2H_2 \rightarrow CH_4 + C\ (char) + heat \qquad (9)$$

The high-temperature thermal energy released by Equation 9 is used to produce hydrogen from the char and steam (Equation 8). Also, the hydrogasification process has a lower oxygen demand than gasification with methanation (Equations 1 to 6), making the former process more efficient overall (65 to 70 percent process efficiency).[17]

Purification or cleanup of the raw gas is simpler than stack-gas cleanup of coal combustion fuel gas, because the producer gas affords a smaller volume per unit of energy generated, and because the sulfur is in the form of hydrogen sulfide rather than sulfur dioxide. For industrial use and for electric power generation close to the generation site, the carbon dioxide removal and methanation stages can be omitted. For more detailed accounts of the general gasification and hydrogasification technologies, see references 16 and 17.

Another conversion technology that shows considerable promise is *pyrolysis* or thermal cracking, which consists basically of heating the coal in an inert atmosphere. Coal begins to decompose at 660 to 750°F (350 to 400°C), forming ultimately a hydrogen-rich volatile fraction and a carbon-rich residue or char. This destructive distillation continues up to a temperature of about 1740°F (950°C). If this temperature is maintained for an extended period, the resulting residue will be almost pure carbon, with a structure similar to graphite. The relative amounts of hydrogen-rich gases and liquids will depend on the coal type and the heating method.[8] As an example, pyrolysis might yield gas, oil, and char in the proportions (based on heat content) of 17, 19, and 56 percent, respectively, with the remaining 8 percent lost in the overall process. Much of the thermal energy for the process is provided by the hot synthesis gas from the char gasification and by the recycled hot char itself.[17] One problem with the pyrolysis process is that over half of the energy may be retained in solid form (although there are numerous variants of the process).[16]

There is now evidence that the first step in the pyrolysis of coal is expansion of the softened coal by the formation of gaseous products. This is then followed by the dissolution of the organic material. The presence of a hydrogen-donor solvent greatly enhances the ultimate yield. Several percent of fusinite remain unconverted in the end, depending on rank and maceral content of the coal. Syncrudes from coal typically have a high nitrogen content,* which may make it difficult to employ them as chemical feedstocks. Catalytic refining operations (cracking and hydrocracking) require essentially nitrogen-free feeds, since many nitrogen compounds are basic and react with the acidic surface sites, thus poisoning the catalysts. Petroleum feeds usually are catalytically hydrodenitrogenated before seeing the hydrocracking catalyst. The same treatment of coal syncrudes would require much more severe pretreatment conditions because of the higher nitrogen content.[19]

The fourth major technique of coal conversion is solvation or liquefaction. Liquefaction of coal by existing technology is fundamentally more complex than gasification, because it entails subtle rearrangements of the chemical structure of coal to insert the additional hydrogen. The route is circuitous. Basically, the coal is gasified and then the gaseous products are used to synthesize the liquids. The process is inherently inefficient because it requires high temperatures, and it breaks all C—C bonds in the coal and then rejoins some of them to form low-molecular-weight hydrocarbon fractions.[17]

Current liquefaction technology calls for separating the coal into undissolved solid coal and clean coal liquid fractions in a separator operating at 500°F and 300 psi, exposing the liquid fraction (at 825°F and 3500 psi)† to hydrogen-donor solvents (produced in the process) or to hydrogen generated by treating the undissolved coal in a gasifier (at 1800 to 3500°F and 300 psi). The resulting liquids are then destructively distilled and the vapors selectively condensed to form the desired liquids—generally, SNG, gasoline, diesel fuel, and fuel oil.[20]

A major cost factor in liquefaction is provision of the necessary hydrogen. To make coal-derived syncrude a suitable replacement for heavy refinery oil as boiler feed, 5 to 10 weight percent of hydrogen (based on the weight of the coal) must be added. Perhaps one-fourth of the product energy value is consumed in providing the necessary hydrogen.[21] About two-thirds of the cost of syncrude would be due to cost of capital for the facilities.[22] Furthermore, refining product to gasoline would consume about 10 percent of the amount of energy delivered to the consumer.[21] It is expected that high H:C liquids derived from coal might cost at least twice as much as similar liquids derived from imported crude oil, at 1976 prices.

---

*As high as a few weight percent.
†Higher temperatures destroy the hydrocarbons.

Coal can also be gasified directly in situ—*underground coal gasification* (UCG)—by partial combustion of the coal seam in place.[23] This technology will be discussed in a later section of this chapter.

GENERAL CONSIDERATIONS

Coal gasification technology was in commercial use long before natural gas was discovered and exploited commercially. It is capable of producing gas in grades ranging over a decade in energy content: 100 Btu per scf by UCG; 100 to 250 Btu by gasification with air; 250 to 550 Btu by gasification with oxygen, followed by cleanup; and 950+ Btu by gasification with oxygen, cleanup, and catalytic methanation.[24]

Processes producing low-heating-value gas will generally be located close to the industry or power plant slated to use the fuel, whereas facilities yielding 950+ Btu per scf gas are more likely to be situated at the mine, with product being pipelined into the normal gas distribution network.[25]

The low- and medium-heating-value product gases are mixtures of H_2, CO, CO_2, H_2S, CH_4, N_2 (and other inerts), H_2O, NH_3, and small amounts of various hydrocarbons and other impurities. With removal of impurities, the low-energy gas becomes suitable for power generation and industrial use, but the presence of toxic CO and the low energy content make it unsuitable for residential use. If the gasification plant is sufficiently close to an industrial consumer, intermediate-heating-value gas can be cleaned up for use in the manufacture of chemicals (e.g., ammonia) and fuels (methane, methanol).[25] High-heating-value gas (SNG) results when producer gas is methanated to yield about 93 percent methane. The conversion (from coal to SNG) can be accomplished at a thermal efficiency of about 56 percent, or nearly 72 percent overall, if the by-products are included. Approximately 18 percent of the coal input is used to fuel the process.[26]

The higher the heating value of the product gas, the more costly the conversion. High-Btu gas enjoys a reasonably secure market because most residential and commerical users, and many small industries, cannot adapt readily to other fuels, and because considerable capital is invested in gas transmission and distribution facilities.[8]*

The gas composition and heating value depend on the temperature, pressure, and residence time in the gasifier (and whether or not the methanation option is employed). Another important factor is the gasifying medium used. Oxygen-steam typically yields about double the heating content of product gasified with an air-steam medium.[25] Many conversion options are available. See reference 28 for some of the possibilities. Of all the methods available, only the Wellman-Galusha gasifier is employed commercially in the United States at this writing. It produces a low-Btu gas in a moving-bed,

---

*The U.S. transmission and distribution network measures 918,000 mi and is worth billions of dollars.[27]

countercurrent gasifier. Figure 5.1 illustrates a typical fuel gas cleanup system employed to remove impurities from the raw product gas prior to methanation or residential use.

SPECIFIC COAL CONVERSION PROCESSES

An effort will be made here to describe briefly some of the specific coal conversion processes (CCPs) currently under development or processes commercial in various parts of the world. It is not possible to categorize all these processes unequivocally according to the four process categories previously described, because many represent major modifications of the four, and others combine two technologies. Still others represent different, one-of-a-kind methods. If this portion of the chapter seems overly long, the writer rationalizes his treatment on the grounds that among the methods discussed, one or more is almost certain to

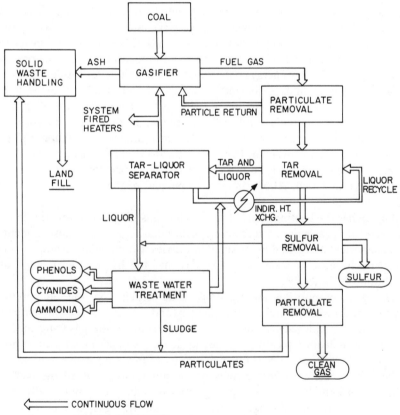

Figure 5.1   Fuel gas cleanup system. (Adapted from reference 25.)

be the source of an important (although perhaps small) segment of our energy budget in the foreseeable future. At this time, it cannot be said which method it will be.

Each coal gasification process involves up to five unit processes[29]:

1 Gasification (formation of process or synthesis gas).

2 Purification (removal of contaminants from synthesis gas stream).

3 Water-gas-shift conversion (provision of proper ratio of CO to H_2 to produce methane).

4 Gas cooling and cleanup (separation from raw gas of particulates, tars, oils, and water).

5 Catalytic methanation (conversion of CO/H_2 to methane).

Gasification Processes

The commercial *Lurgi* process is based on one of the earlier technologies. The overall process is shown in a simplified version in Fig. 5.2. It can be air-blown or oxygen fed. The former option produces a low-Btu gas for local use. The latter provides a medium-Btu producer gas which can be upgraded to a pipeline grade gas (950+ Btu per scf) by methanation. Table 5.4 shows the potential output of a Lurgi plant. Based on the output of pipeline gas (SNG), the thermal efficiency is 56.2 percent. The overall efficiency (including by-products) is about 71.6 percent,[26] while the combined-cycle* efficiency should reach nearly 94 percent.[10]

Lurgi is the only fully commercial process deemed applicable to the manufacture of pipeline gas. The *Koppers-Totzek, Wellman-Galusha,* and *Winkler* technologies all operate at pressures only slightly over atmospheric, hence have high oxygen demand and low conversion efficiencies compared to the Lurgi process.[17] Furthermore, the Wellman-Galusha method is presently limited in size to a maximum coal feed of 85 tpd.[10]

Lurgi

The Lurgi process feeds lignite or subbituminous coal ($> 1/4$ in.) through a lockhopper system into a pressurized (up to 450 psi) fixed-bed gasifier, into which steam and oxygen or air are fed. Because of the slow gas flow rate and resulting long retention time inherent in the fixed-bed gasifier, gas production rate with the Lurgi is probably only about 12 percent that of the newer (but commercially untried) fluidized-bed or entrained-bed systems. In preliminary tests, midwestern caking coals required 50 to 70 percent more steam and 45 to 50 percent more oxygen, and allowed feed rates only 70 to 75 percent those of noncaking coals. Fine grades of coal cause channeling and excessive particulate carryover in the gas stream.[17] A modified *Slagging Lurgi* process offers promise of greater efficiency and the ability to use caking coals effectively.[30]

---

*Combined-cycle plants employ the waste heat from large gas turbines to generate steam for conventional steam turbines, rather than venting the heat to the environment.

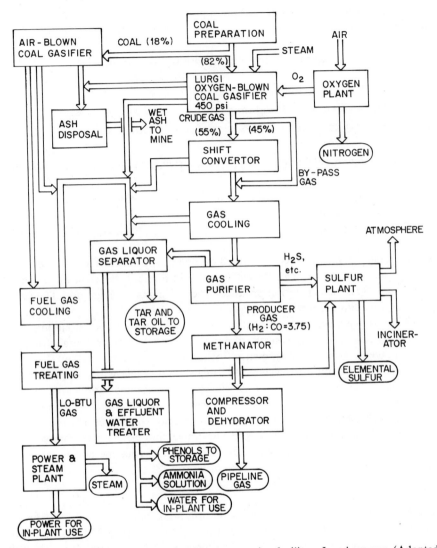

Figure 5.2 Simplified flow diagram of coal conversion facility—*Lurgi* process. (Adapted from reference 26.)

Modified Lurgi

Although Lurgi has been demonstrated commercially, some of the newer *modified-Lurgi* technologies (e.g., HYGAS, CO_2 Acceptor, Synthane, and BI-GAS), have more to offer for the long haul. These latter processes have four features in common[17]:

1 They use continuous fluid-bed or entrained-bed gasifiers at pressures at least equal to the Lurgi process.

Table 5.4 Potential Output of Lurgi Plant

	Energy, 10^6 Btu/hr	Product, Common Units
Input		
Coal	20,400	n.a.
Output		
SNG	11,470	785×10^6 scf/day
Tar	1,510	239,250 gpd
Tar oil	840	157,370 gpd
Phenols	160	n.a.
Naphtha[a]	370	74,900 gpd
Ammonia	210	260 tpd
Sulfur	60	167 ltpd
Total output	14,620	—
Thermal efficiency		
Based on SNG	56.2%	
Overall (including by-products)	71.6%	

Source: Reference 26.
[a]Primarily BTX.

2 They form more methane directly by initial hydrogasification, aided by more effective contacting of raw or lightly treated coal with hot, hydrogen-rich synthesis gas at elevated pressure.

3 They use less (or no) oxygen in the production of synthesis gas from the residual char remaining from the primary gasification stage.

4 They all use a final catalytic methanation step to produce pipeline-grade gas. (The methane content will vary substantially with the process.)

HYGAS Process This Institute of Gas Technology process is presently under development. It consists essentially of combining coal with hydrogen to form methane.[30] It normally uses lignite or subbituminous coal feed, but caking-type bituminous coal can be used by pretreating with air at atmospheric pressure and 700 to 800°F (371 to 427°C) in a fluidized-bed reactor. The coal is slurried with by-product light oil and pumped into the pressurized (1000 to 1500 psi) reactor, where the oil is recovered and recycled. (This method of transferring solids into a high-pressure environment is probably superior to the use of lockhoppers.[17]) A flowchart of the process is shown in Fig. 5.3.

The short retention time and the presence of activated carbon in the early stage of gasification speed up the process considerably. Hydrogen-rich synthesis gas is produced from steam and residual coal char in an oxygen gasifier or, optionally, a method such as the *Steam-Iron* process can be used to form hydrogen by reacting steam with reduced iron oxide.[17] The purified producer gas can be

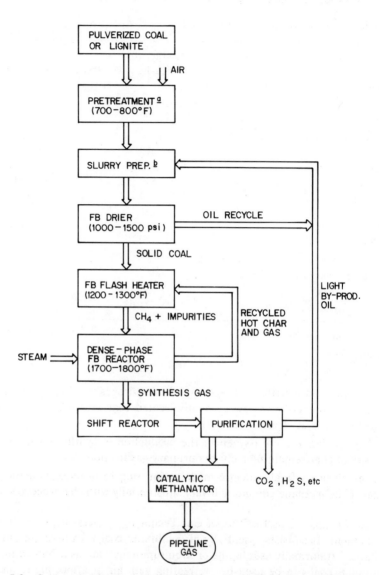

Figure 5.3 Coal conversion by hydrogasification plus methanation—HYGAS process. (*a*) Can be omitted with noncaking grades; (*b*) medium for transferring coal to high-pressure reactor. (Adapted from reference 17; reproduced with permission from the *Annual Review of Energy,* Vol. 1; © 1976 by Annual Reviews, Inc.)

methanated to pipeline gas. The Steam-Iron option* (Fuel Gas Associates) promised a 10 percent cheaper product than the original HYGAS process. The higher capital cost was to be offset by the lower operating costs and the reduced purification needed because of the high-grade hydrogen produced.[32]

CO_2 Acceptor Process   This Conoco Coal Development Company process is another method under development. Pulverized reactive coal (lignite or sub-bituminous) is fed into a fluidized-bed gasifier at 1500°F (816°C) and 150 psi. Hot dolomite (the acceptor) is also fed in to provide some of the process heat and to react with the CO_2 and H_2S. The spent acceptor is transferred continuously to a regenerator at 1900°F (1038°C), where it is calcined. Cyclones separate particulates from the gas. After purification of the raw gas, the product gas, which has a $H_2:CO$ ratio of about 3.2, can be methanated directly, without shifting. Methanation and dehydration produce pipeline grade gas.[17]

Like the Steam-Iron modification of the HYGAS process, the CO_2 Acceptor process avoids the need for an oxygen plant. But it is a complex method requiring operation within a rather restrictive range of conditions; hence it is perhaps less certain of success as a commercial process than some of its competitors. It also calls for the use of input coal grades that are basically clean enough to be burned directly, in most areas.[31] It would not be economically viable if used with bituminous coal, because of the lower reactivity of this rank.[33]

Synthane Process   The distinguishing feature of this Department of Energy (DOE) process is the pretreatment of caking coal in a fluid-bed reactor at 800°F (427°C) and at full gasification pressure in the presence of steam and oxygen. The gases produced are added to those from the main gasifier, where the temperature is raised to 1800°F (982°C) and the bulk of the steam and oxygen (about 88 percent) is introduced. The resulting char (about 30 percent of the input coal carbon) is fed to the power plant. In the overall process, the $H_2:CO$ ratio is raised from 1.7 to 3.0.[17] The raw gas from the shift converter can be purified, methanated, and dehydrated to produce pipeline gas.

Table 5.5 shows typical compositions of raw producer gases produced by SASOL-I variations (essentially the fixed-bed Lurgi process compared with a fluid-bed variation similar to Synthane or HYGAS). The compositional differences suggest the impact of altering process parameters. Although similar to HYGAS, the Synthane process is inherently simpler, in that it employs a single reactor stage.[31] It also uses any rank of U.S. coal.

BI-GAS Process   This Bituminous Coal Research process is the other major modified-Lurgi technology. It uses an entrained-bed system to permit all grades of pulverized coal to be heated rapidly to 1700°F (923°C) by hot gas from a lower reactor. (Steam is fed into the upper reactor along with the coal.) Devolatilization of the coal produces methane and char during a residence time of a few seconds. The char is fed to the lower section of the gasifier, where it

*This option, in an experimental facility, failed to live up to its expectations and has been shut down. Hydrogen output rate was much lower than expected.[31]

Table 5.5   Typical Product Mixes from Coal Conversion Processes (SASOL-I)

		Percent	
		Fixed-Bed	Fluid-Bed
Light gas	C_1, C_2	7.6	20.0[a]
LPG	C_3, C_4	10.0	23.0
Gasoline	C_5–C_{12}	22.5	39.0
Diesel fuel	C_{13}–C_{18}	15.0	5.0
Heavy oil	C_{19+}	41.0	6.0
Oxygenated compounds		3.9	7.0
Aromatic, % of gasoline		0	5

Source: Reference 34. Reprinted with permission from G. A. Mills, *Chemtech* **7**(7), 418–423 (July 1977). Copyright by the American Chemical Society.
[a]Half methane.

produces additional gas with steam and oxygen and causes slagging of the ash. After shifting, the gas is purified, methanated, and dried to produce pipeline gas.[17] This process is looked upon as a "brute-force" method, with a high degree of uncertainty of success as a commercial process. On the other hand, it could prove to be a "sleeper,"[31] since it does produce larger quantities of methane directly from the coal.

Hydrane Process   This is a hydrogasification method (DOE) which reacts all grades of U.S. coal directly with "pure" (95.6 percent) hydrogen[33] in a free-fall zone at a pressure in excess of 1000 psig. The intermediate char formed reacts with steam and oxygen to produce synthesis gas for hydrogen formation. The raw producer gas contains as much as 60 to 73 percent methane.[17, 33] The process differs basically from the Synthane process in that the bulk of the methane (about 95 percent) is produced by direct reaction of hydrogen and char rather than through an intermediate synthesis gas.[33]

ATGAS Process   This Applied Technology Corporation process injects any grade U.S. coal into a molten iron bath [at 2600°F (1430°C) and 5 psig] to react with steam and oxygen and form a synthesis gas for methanation to pipeline gas. The sulfur from the coal is removed in the slag.[17]

U-Gas Process   DOE funded this development by IGT and Memphis Light, Gas & Water to produce a 300-Btu-per-scf gas for industrial use, using caking, high-sulfur eastern bituminous coal feed. A demonstration plant will produce about 175 million scf per day (at the rate of about 62,500 scf per ton). The carbon-to-gas efficiency is expected to be more than 95 percent. The coal is gasified by a steam-oxygen medium in a single-stage fluidized bed, at about 1900°F (1038°C). An ash agglomeration technique avoids slagging, and entrained coal fines are recycled to the gasifier.[35]

Exxon Catalytic Methanation Process This technology is unique in that it combines gasification, shift, and methanation in a simultaneous catalytic operation in a single reactor. Minus-8-mesh Illinois bituminous coal is sprayed with alkali (e.g., KOH) catalyst solution to a level of 10 to 20 percent. The catalyzed coal is passed through lockhoppers into the reactor, where it combines with a preheated steam-recycled-CO/H_2 fluidizing medium.[36] A simplified flow diagram is shown in Fig. 5.4.

KilnGas Process This development by Allis-Chalmers Corp. and 11 utility companies injects air and steam into a tumbling bed of coal. (The tumbling action prevents agglomeration of caking coals.) Hot gases dry the coal and heat it to volatilization temperature. Fuel gas is drawn off for direct utility use.[37]

GEGAS Process This General Electric technology is a gasification process tailored to in-house power generation by the small industrial user. Instead of employing lockhoppers or slurries to feed the coal under pressure, it calls for extruding 6-in.-diameter "logs," formed from particulate coal and by-product tar, through a gastight seal into a fixed-bed, pressurized gasifier.[38]* Once inside the gasifier, the logs are broken up. This process differs from most gasification processes in that it uses all U.S. grades of coal without pretreatment, and in a size that normally could be carried out, unreacted, in the effluent stream.[40] Inert diluents (e.g., silicon carbide and coal ash) are used to reduce swelling and caking of the coal.[17] The product gas has a heating value of 160 Btu per scf.[41]

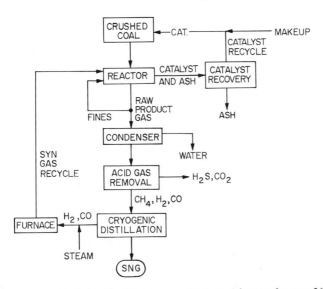

Figure 5.4 Exxon catalytic gasification process. (Adapted from reference 36; reprinted with permission of the copyright owner, The American Chemical Society.)

*The tar binder is obtained by washing and quenching the raw producer gas.[39]

Composite Process

The *COGAS* process (FMC and others) is a composite system of pyrolysis and gasification. It is a thermally efficient method which produces gas and liquids in nearly equal amounts. The process requires four reactors for the char and oil step (pyrolysis) alone, but it makes up for part of this high capital outlay by eliminating the need for an expensive oxygen plant.[30]

Subbituminous and/or bituminous coal is fed to the series of reactors at a pressure of 50 psig, with temperatures increasing as the volatiles are driven off. The residual char remaining after separation of the initial pyrolysis products is reacted with steam, using the combination of char in air as an indirect source of the required heat. Pyrolysis gas (about 35 percent methane) is combined with synthesis gas from the gasifier and the product is then upgraded to pipeline gas by the usual procedures.[17] The oil is hydrotreated to provide a syncrude. This could be the Achilles heel of the process, as no natural market exists for the unrefined synthetic oil.[30] A marketing outlet must be established to make the process economically viable, or further processing must be carried out to ready the syncrude for a conventional refinery.

Liquefaction Processes

COED

The *COED* process (Char Oil Energy Development) is a direct pyrolysis method. The $1/8$-in. dried coal feed is heated to successively higher temperatures in a series of fluid-bed reactors at 6 to 10 psig, beginning at an initial temperature of 600°F (315°C), and increasing to 1600°F (870°C) in the final stage. Combustion of char in oxygen provides the necessary heat.

Oils are separated from the gases, and the gases are cleaned and reformed to produce hydrogen. The filtered syncrude is hydrotreated at 750°F (400°C) and 1500 to 3000 psig to remove sulfur, nitrogen, and oxygen. Typical nongaseous products from a ton of coal are: 1.04 bbl (43.7 gal) syncrude plus 1177 lb residual char.[42] Various options exist to permit feeding product gases to a power plant or using them for chemical feedstocks.

TOSCOAL

The *TOSCOAL* process (The Oil Shale Corp.) is another direct-pyrolysis or carbonization method of conversion. Crushed dried coal is preheated by hot fuel gases and is then fed into a rotating drum containing hot ceramic balls that heat the coal to 800 to 1000°F (427 to 538°C). The char produced is about half the weight of the input coal and contains about 80 percent of its heating value. The vapors are cooled and the condensed liquids are fractionated into gas oil, naphtha, and residuum. One ton of Wyodak* coal typically yields 21.8 gal of

---

*Wyoming–Dakotas formation subbituminous coal.

16,000-Btu-per-lb oil when carbonized at 970°F (520°C) or 10.5 gal when carbonized at 800°F (427°C). An outlet must be found for the large amount of char produced in order to make the process commercially viable.[42]

Synthol

The *Synthol* process (Bureau of Mines) is a direct solvent-extraction method. Dried pulverized coal is slurried in product oil and fed to a fixed-bed catalytic reactor with hydrogen added to desulfurize the coal. The slurry is held at 850°F (454°C) and 2000 to 4000 psig for up to 14 min, at which time the liquid is separated from the gas phase and centrifuged to remove solids. The gas is cleaned of H_2S and ammonia and recycled with fresh hydrogen to the reactor. Western Kentucky coal with a 4.6 percent sulfur content will typically yield 3 bbl (126 gal) of 17,700 Btu-per-lb oil per ton of feed coal [mineral-ash free (m.a.f.)].[42]

H-Coal

The *H-Coal* process (Hydrocarbon Research Inc.) is another direct solvent-extraction technique. Dried minus-40-mesh coal is slurried in coal-derived oil, mixed with hydrogen, and fed into an ebullated-bed catalytic reactor at 850°F (454°C) and 3000 to 3500 psig (Fig. 5.5), where it is held for 10 to 30 min. The gas is separated, cleaned, and recycled with new hydrogen and coal-oil slurry to another reactor.

The liquid is cleaned of particulates, stripped, and fractionated into naphtha, middle oil, and fuel oil. The H_2S-containing liquid is treated in a Claus unit to recover elemental sulfur. The H-Coal conversion efficiency is reported to be 90 to 94 percent. One ton of m.a.f. coal will yield about 3 bbl (126 gal) of C_4-to-975°F (524°C) distillate. This is the only demonstration process that produces both refinery feedstock and an environmentally acceptable fuel oil.[42]

The H-Coal process, together with the *Exxon Donor Solvent* (EDS) process, have been designated by two panels of the National Research Council* as deserving of early expansion to demonstration units and adoption as standards of reference for the performance of competing systems.[44] DOE has awarded Exxon, together with affiliated Carter Oil, EPRI, and Phillips Petroleum, a $240 million contract for a demonstration plant producing low-sulfur oil, liquids for blending gasoline, and utility fuel.[45]

Exxon Donor Solvent

The *EDS* process developed from industrial research initiated in 1966. A recycled H-donor solvent and a hydrogen plant provide the hydrogen to liquefy the coal noncatalytically at moderate temperature and pressure. The spent solvent is regenerated catalytically by hydrotreating, but the catalyst sees only uncontaminated materials.[22] See Fig. 5.6 for a schematic of the process.

*Ad Hoc Panels on Low-Btu Gasification of Coal and on Liquefaction of Coal.

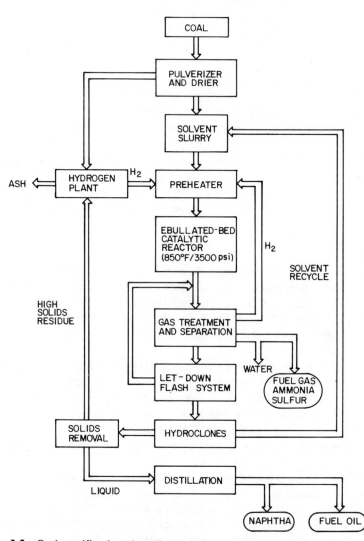

Figure 5.5 Coal gasification by direct hydrogenation process—H-Coal method. [Adapted from reference 43; reprinted with permission from *Environ. Sci. Technol.* **11**(2), 122, 123 (Feb. 1977); copyright by The American Chemical Society.]

The process yields vary with the type of feed, as shown in Table 5.6. The naphtha can be reformed to a high-octane gasoline (RON = 101),[46] with a yield of about 2.5 to 3.0 bpt.[47]

Coalcon

Coalcon (Union Carbide Corp. and Chemical Construction Corp.) is yet another direct solvent-extraction process that once appeared to be very promising. It

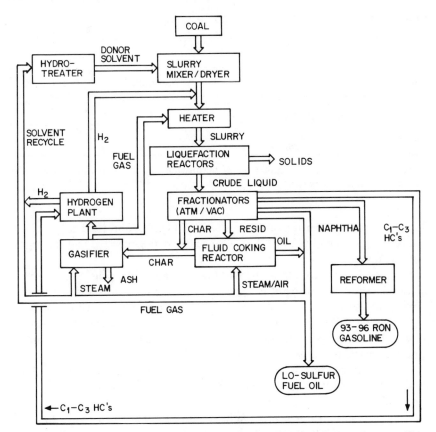

Figure 5.6 Exxon Donor Solvent (EDS) process. (Adapted from reference 46; reprinted with permission of the copyright owner, The American Chemical Society.)

was slated for construction of a demonstration plant, partially funded by the federal government, but it was plagued from the start by managerial and technical problems.* These led to the project's demise on June 15, 1977.[49]

Solvent-Refined Coal

Solvent-Refined Coal (SRC) (Pittsburgh and Midway Coal Co.) is another direct solvent-extraction process, and the only current conversion method capable of removing tightly bound organic sulfur while reducing ash content.[50] Dried coal, pulverized to 80 weight percent minus-200 mesh, is slurried in coal-based solvents [boiling range 550 to 800°F (288 to 427°C)] at a solvent-to-coal weight ratio of 1.5 to 4. The slurry is pumped to a preheater together with 1000 to

---

*The chief of which was caking of the gasifier nozzles by high-sulfur eastern coals. This was solved technically by injecting a mixture of "reduced" charcoal, hydrogen, and oxygen, but with an exorbitant cost overrun.[48]

Table 5.6 Effect of Feed Coal Type on EDS Product Blend

Feed Coal		Products	
Type	C Content, %	Naphtha, %	Fuel Oil, %
IL—bituminous	70	18	27
WY—subbituminous	68	19	19
ND—lignite	64	17	15

Source: Reference 46. Reprinted with permission of the copyright owner, The American Chemical Society.

20,000 scf hydrogen per ton of slurry, at 850°F (454°C) and 1000 psig, and held for a residence time of 0.2 to 1.8 hr. The coal is depolymerized and the organic matter is dissolved.

The gas is flashed off at 625°F (329°C) and 995 psig, scrubbed and separated, and the hydrogen is recycled. The slurry from the flash vessel is filtered and the solids (35 to 55 weight percent undissolved carbon and 5 to 8 weight percent sulfur) are used to heat the Claus unit for recovery of elemental sulfur. The filtrate is vacuum-flash-distilled, the solvent is recycled, and the light hydrocarbons are diverted to the by-product stream.[21]

The bottom fraction is SRC, having a solidification temperature of 300 to 400°F (149 to 204°C), depending on the residual solvent content. It can be transported as a hot liquid (above 400°F). Made from western Kentucky coal (7.1 percent ash and 3.4 percent sulfur), the SRC has been found to have a heating value of 16,000 Btu per lb, 0.1 percent ash, and 0.8 percent sulfur. The SRC can also be hydrotreated catalytically (*Hybrid* process) to produce a syncrude together with other liquids.[42]

The capital cost of a 15,000-tpd (input) SRC plant would be about $1 billion (1976 dollars), which means that the product probably would not be competitive with direct combustion of low-sulfur coal or with high-sulfur coal with stack-gas scrubbers. However, scrubbers are expected to become more expensive, whereas SRC is expected to become cheaper with time. At present, the proper role for SRC would appear to be for combustion in old coal-fired boilers for intermittent service—in other words, systems not economical to equip with pollution control facilities.[24]

Two nontechnical problems facing SRC are: (1) it must be classified as a solid fuel to meet EPA regulations, and (2) it normally would be classified as a high-tariff, manufactured product rather than as coal, which would subject it to prohibitive shipping costs.[24] On the other hand, most states allow the cost of the premium fuel to be passed on to consumers as a fuel adjustment charge, whereas scrubbing can be added to electricity rates only on the basis of ROI.[50]

A remaining technical and economic problem involves the necessity of properly assessing the optimum hydrogen plant capacity to handle the variety of coal feeds. The hydrogen plant represents about 40 percent of the capital investment, and no satisfactory means has yet been established for predetermining the hydrogen requirements for each grade of coal.[50]

SRC-II

A modification of this process—*SRC-II*—has been developed by Gulf under the sponsorship of DOE. A coal-derived solvent and hydrogen form a slurry with coal and convert it to a liquid stream containing unreacted coal and mineral residues and a raw gas stream. Pipeline gas (PLG) and propane and other light hydrocarbons are recovered together with free sulfur. But the main product is a synthetic liquid fuel obtained by distillation of the liquid stream, together with by-product naphtha, usable as a feed for gasoline or petrochemicals.[51] See Fig. 5.7 for a flowchart of the process.

Others

Dow Chemical Process This is a new technology for liquefying coal, which employs a novel *emulsion catalyst system*. The molybdenum disulfide catalyst is dissolved in water and then emulsified in oil. This emulsion is combined with the coal-in-oil slurry feed, and on heating, the water vaporizes, leaving behind a finely dispersed metal salt residue (1-μm particles). Because of the efficiency of catalyst dispersion, catalyst cost is only about $2 per ton of coal treated.[14] The process is illustrated in Fig. 5.8. For each ton of coal feed, it yields 1404 lb low-sulfur fuel and 436 lb residue. The composition of the fuel is typically 8.8 percent methane, 18.7 percent LPG, 18.5 percent naphtha (C_6 to 400°F), 47.9 percent distillate (400 to 975°F), and 6.1 percent fuel oil (975+ °F).[52]

Exxon estimates the cost of producing liquids from coal by its EDS process at $20 to $25 per bbl. Gulf estimates $21 per bbl for SRC-II liquid produced

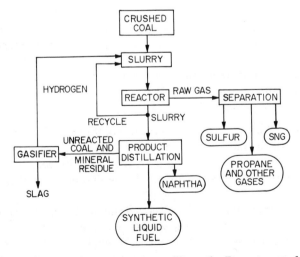

Figure 5.7 SRC-II liquefaction process. (From the Department of Energy.)

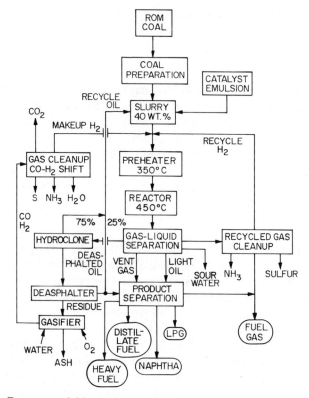

Figure 5.8 Dow expendable-catalyst coal liquefaction process. (Adapted from reference 52; reprinted with permission of the copyright owner, The American Chemical Society.)

commercially, and Hydrocarbon Research Inc. figures $19 to 30 per bbl, the exact number hinging on how plant and financing costs are determined. Dow refuses to put a number on the cost of its emulsion-catalyst liquid product until it has been scaled up further, but it claims that its process should cost 20 percent less to set up and should benefit from 10 percent greater operating efficiency than the competing technologies.[14]

Fischer-Tropsch Process   This is a derivative of the basic Bergius process mentioned earlier. It was modified by the South Africans and became the basis of the SASOL process. It uses Lurgi gasifiers (Fig. 5.9); in fact, the Lurgi process is essentially the SASOL process or the Fischer-Tropsch synthesis with methanators.

The gasifiers are charged with pulverized (minus-$3/8$ in.) noncaking coal and oxygen. The product gases are purified and shifted to a $H_2:CO$ ratio of 1.8. The synthesis gas and recycled gases are passed through a fixed catalyst bed at 430 to 490°F (221 to 254°C) and 360 psig to produce straight-chain, high boiling

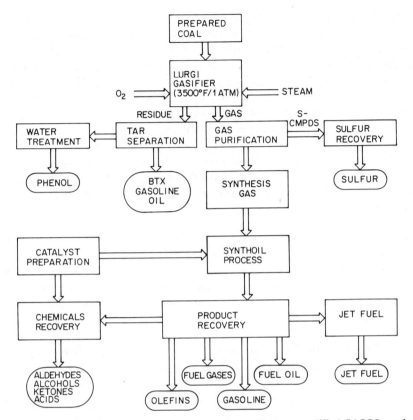

Figure 5.9 Gasification-synthesis process—Pullman-Kellogg modified-SASOL method. (Adapted from reference 53; reprinted with permission of the copyright owner, The American Chemical Society.)

hydrocarbons; medium-boiling oils; diesel oil; LPG; and oxygenated compounds.

Fresh synthesis gas, fixed-catalyst-bed tail gas, and tail gas from the fluid-bed synthesis operation are reformed with steam, and the reformed gas is fed to the fluid bed together with synthesis gas at 600 to 625°F (316 to 329°C) and 330 psig. The resulting products are mainly C_1–C_4 hydrocarbons and gasoline, with minor amounts of medium- and high-boiling liquids. Substantial quantities of oxygenated compounds and aromatics are also formed.[42] (Table 5.5 showed typical commerical outputs from *SASOL-I* plants equipped with fluid-bed and fixed-bed synthesizers.)

Mobil Process   This is a third-generation process that starts with methanol, made from CO and H_2, products common to all coal gasification processes. High-octane (93 to 96 RON) gasoline is produced directly from methanol feed using an aluminosilicate zeolite catalyst.[34]

The *Fischer-Tropsch* and *Mobil* (methanol-to-gasoline) processes are simple, relatively low-pressure, conventional catalytic processes, as opposed to the complex, high-pressure, direct hydrogenation processes. Their primary disadvantage lies in the fact that they have about twice the gas-volume demand of the latter processes. High-hydrogen syn gas costs almost as much as hydrogen, but second-generation technology conceivably could produce syn gas for less, and thus introduce downstream economies over direct hydrogenation, such as the H-Coal process. The slagging moving-bed and other gasifiers are potential sources of low-cost syn gas.[54] One of the difficulties with direct hydrogenation is that the inherent variability of coals—even within a single seam or a particular mine—makes it difficult to predetermine the hydrogen capacity needs of a given conversion facility, yet the hydrogen plant may represent 40 percent of the capital investment.

Miscellaneous Conversion Processes

The other coal conversion processes are numerous and developed to such a limited extent as to scarcely warrant a detailed discussion. Some of them will be mentioned very briefly here, since they ultimately may offer major technological breakthroughs that will become important in the future:

Texaco   Gasifies all U.S. coals by burning incompletely in oxygen. Characterized by low thermal efficiency and high product cost.[17,30]

Gulf (Gulf General Atomic and Stone and Webster)   All U.S. coals are dissolved and solution-hydrogasified to produce gas ready for purification. Heat is supplied by nuclear reactor.[17]

Citgo Hydropyrolysis (Cities Service Research and Development)   Noncatalytic, single-stage hydrogenation of dry coal at high pressure. The residence time of 100 to 3000 msec produces up to 16 percent light, aromatic liquids (about 94 percent benzene) and high-Btu gas with a conversion efficiency of 50 to 80 percent.[55]

Flash Pyrolysis (Occidental Petroleum)   Pyrolyzes all grades U.S. coal in a single pass through an entrained-flow reactor at atmospheric pressure, with a coal residence time of fractions of a second. Similar to the slower (minutes) COED process. Produces gas, liquids, and large amounts of char.[8,17]

Synthoil (Foster-Wheeler)   High-temperature, high-pressure reaction between hydrogen and coal suspended in fluid. Sulfur is removed.[56] Gasoline production can be maximized with iron powder catalyst.[57]

ADL Extractive Coking (Arthur D. Little, Inc.)   Relatively easy, noncatalytic liquid extraction, similar to Synthoil and H-Coal, but more efficient utilization of hydrogen. Low pressure permits use of low-cost commercial equipment.[57]

Molten Salt (Rockwell International)   Gasification in a high-pressure molten sodium carbonate bath contained in an alumina reaction vessel. Salt catalyzes reaction to form high-methane product.[17]

Agglomerating Ash (Battelle-Union Carbide)   Bituminous coal or char feed combusted with air in fluid-bed reactor, and steam-gasified at 100 psi in a separate fluid bed. Produces synthesis gas.[17]

Liquid-Phase Methanation (LPM) (Chem Systems Inc.)   Desulfurized synthesis gas flows through a reactor with an inert liquid that picks up heat in a fluidized catalyst bed. Produces methane in a single pass.[17] The *LPM/S* variation combines the shift and methanation steps in a common reactor for greater efficiency.[58]

Consolidated Synthetic Gas (CSG)   Also known as CO_2 *Acceptor* (described earlier).

High-Temperature Nuclear Reactor (HTR) (West Germany)   A nuclear reactor supplies process heat (at 950°C, in quantity >3000 MW). Uses coal feed only; no oxygen required. Overall efficiency (electric and gas production) about 64 percent vs. about 50 percent for conventional gasifier.[59]

Others

The nation's largest coal gasification test facility, developed by Combustion Engineering, Inc. (C-E), was placed on stream on October 10, 1977. It is a two-stage, atmospheric-pressure, entrainment-type gasifier, producing a minimum amount of solid waste.[60]

Other routes have been pursued in synthesizing gasoline commercially. During World War II, Japan had facilities in North Korea for a coal conversion technique that was quite different (and far more expensive than current technology). They first produced coke from the coal, then made calcium carbide from the coke in an electric furnace. The next steps in order were: acetylene, acetaldehyde, butyraldehyde, octanol, and finally octane. In Taiwan, starchy root vegetables were fermented to butanol and then butyraldehyde was formed to provide octanol and finally octane.[61]

Supercritical Solvent and Antisolvent Processes

Work with supercritical solvents, and even antisolvents, is under way in an effort to improve coal conversion efficiencies. Use of certain solvents* (e.g. toluene or *p*-cresol) at supercritical temperatures and pressures permits extraction of heavy coal components at temperatures well below their normal boiling points, by raising their volatility as much as four orders.[62]

Autoclave tests (Catalytic, Inc.) have shown that toluene extracts about 21 percent of the coal and yields about 68 percent char. Solvent can be almost completely recovered (99 percent). A 9:1 *p*-cresol-to-water mixture gives a higher extraction rate, but chemical complexing seems to preclude further use of this system. Similar technology is under development in the U.K.[63] An economic analysis showed that the total capital investment (including interest and initial

---

*The most effective solvents are those having critical temperatures between about 600 and 850°F. Preferred coals are noncaking and highly volatile (e.g., Wyodak subbituminous).

catalyst cost, but excluding land cost) would be nearly $400 million for a 10,000-tpd (dry coal) capacity. The corresponding coal–liquid product cost would be $6.14 per million Btu.* Improved extraction rates would drop the product costs significantly.[64] Exxon's EDS process yields liquids projected to cost $5.00 per million Btu, equivalent on an energy basis to $29 per bbl of crude oil.† Both SRC-II and H-Coal products would be within 10 percent of the EDS product cost—or about twice the cost of imported crude delivered to Gulf Coast refineries in the fourth quarter of 1978.[65]

Kerr-McGee is investigating the critical solvent de-ashing of coal for the SRC process. The feed is mixed with the solvent under critical conditions and is then allowed to settle. Ash and undissolved coal form the heavy, settled phase. The extract and solvent are withdrawn from the top and heated to decrease the density of the solvent, and allow inverse solubility to effect separation of the dissolved coal. The de-ashed coal is removed from the bottom of a second settler. Solvent is stripped from residues of both settlers and is recycled.[66]

A C-E Lummus process under development uses an *antisolvent* to enhance agglomeration of small mineral particles in a coal–liquid solution and allow them to settle out. The dissolved SRC is contained in the overflow. Antisolvent recovery rates are expected to be over 99 percent.[62]

UNDERGROUND COAL GASIFICATION

Underground coal gasification (UCG) or controlled combustion of coal in situ has been developed out of the necessity of avoiding the hazards of mining, reducing environmental impact, opening up deep coal seams to extraction, and increasing recovery rates. The possibility was first suggested over a century ago (1868). Over the past 50 years, development effort has been concentrated in the U.S.S.R., peaking in the mid-1960s. By the early 1970s, there was little or no work going on in the field, because of poor production quality and quantity, but more important, because of the economic facts of life. The products of UCG could not compete with cheap natural gas and crude oil. About 1972, however, the Bureau of Mines undertook some experimental work, which has been reported in reference 67.

Underground coal gasification entails three basic operations: *pregasification* (providing access to the seam and linking the blast inlet and gas offtake through the seam), *gasification* (input of gasification medium, ignition, maintaining flame front, and process control), and *utilization* (on-site generation of electricity or PLG, or nonenergy uses).[67]

The three fundamental approaches to pregasification involve *shaft* methods, requiring underground labor to create large-diameter openings; *shaftless* methods, using multiple boreholes driven from the surface; and *combination* methods, employing both shaft and shaftless technologies. The shaft technology

---

*Assuming a 21.2 percent toluene extraction yield with Wyodak coal.
†Based on a 24,000-tpd plant built in 1978, and using Illinois No. 6 bituminous coal.

includes the *chamber* or *warehouse* technique, which is characterized by a low recovery rate and poor product uniformity; *borehole producer,* which requires extensive underground work; and the *stream* technique, which is now the favored method for dealing with steeply pitched seams. The shaftless methods include the *blind borehole* technique, in which product quality deteriorates as gasification progresses, and the *percolation* or *filtration* technique, which is difficult to control because of the parallel flows involved.[67] Only two, the stream and percolation technologies, will be described further, because they represent the most successful ones for steeply sloping and horizontal seams, respectively.[68]

The *stream* method requires considerable underground work, but it offers the most effective means of gasifying steeply sloping coal seams. Parallel inclined galleries* are constructed and connected at the base by horizontal "fire drifts." The seam is ignited in the fire drift, and the flame front progresses up the face of the inclined gallery, normal to the gas flow. Mineral ash and roof material fall into the void space.[67]

The *percolation* method comprises drilling a pattern of boreholes into the coal seam and carrying out gasification between pairs in such a way as to ensure a continuous supply of product gas. Lignite beds have some natural porosity, but higher ranks of coal generally call for provision of linkage paths.[67]

Linking is usually achieved in one of four ways: *hydraulic fracturing* ("hydrofracking"), by injecting high-pressure water into natural breaks to increase porosity†; *electro-linkage,* involving application of voltage (about 200 V) to electrodes at base of boreholes, to dry out the seams and then pyrolyze them, leaving a permeable path of coke between air inlet and gas offtake; *pneumatic linkage,* using high-pressure air (about 70 atm) injected at one borehole, to expand openings to the adjoining borehole‡; and *reverse combustion* (countercurrent combustion), injecting air at one borehole and igniting the seam at the other, so that the flame front propagates toward the injection hole, counter to the air flow.[69]§ Unfortunately, linking measures usually open only narrow linear channels rather than planar or broad area paths.[68]

The reverse combustion procedure is generally recommended for shrinking or mildly swelling coals. Use of domestic eastern coals for UCG by current technology is difficult because of the swelling tendencies of these bituminous coals. On heating, they form a plastic mass that plugs channels, large and small. The most promising locations for UCG facilities appear to be Wyoming and Texas.[68]

Pyrolysis, char combustion, and gasification (partial combustion) occur in different zones of the linkage path. Very small quantities of paraffinic hydrocarbons, oxides of carbon, and water are evolved well below 570°F (300°C). Small amounts of oily liquids form slightly above 570°F (300°C). The

---

*Approximately 60 yd long and 100 yd apart. The panel may contain 12,000 tons of coal.
†Sand is sometimes included to prop cracks open when pressure is released.
‡Effective with lignite, but very slow with higher ranks.
§When linkage at 2 atm is established, air flow is reversed.

bulk of the tars is given off at 660 to 1020°F (350 to 550°C), together with moderate amounts of combustible gases. Most of the hydrocarbon gases are formed between 840 and 930°F (450 and 500°C), and above 1020°F (550°C); the residue of semicoke produces a gas mixture which is predominantly hydrogen, methane, and carbon oxides. When low air pressures are employed to minimize leakage, very little methane is produced, and most of that is produced in the pyrolysis stage.[68]

Thin (<3 ft thick) coal seams of poor-quality, high-ash coal that cannot be mined economically also do not lend themselves to UCG. Thin seams lose too much heat to adjoining strata.[69] Environmentally speaking, UCG is not as innocuous as it might seem on the surface. Many unsightly drilling rigs must be erected and much acreage must be disturbed to support a single facility. Furthermore, the product gas is generally dirtier and lower in heating value than the "town gas" produced by carbonizing coal in retorts. The use of a variable-quality raw material and an underground "retort" constantly growing in size make for a very difficult quality control problem.[69]

There is some concern that phenols and aromatic hydrocarbon by-products might pollute nearby aquifers, but no such problem has been detected in the U.S.S.R. Some tests suggest that the unburned coal absorbs such emissions, much like activated charcoal.[23]

A joint West German–Belgian operation contemplates gasifying coal seams at a depth of at least 1000 m, too deep for normal mining, considering the low grade of the coal. The hot gases will be used to feed on-site electric power generating stations,[70] since the heating value is too low to allow transmitting it economically over distances greater than 10 mi.[71]

Normally, over 75 percent of the coal in the seam is consumed in underground coal gasification, and 50 to 70 percent of the heating value of the consumed coal is recovered in the form of product gas.[68] On an energy basis, this means that 38 to 52 percent of the coal in-place is recovered.

Different sources show little agreement on the composition of the product gas, because of process and raw material variables, but a few things are fairly certain. In air-fed gasification, the raw product gas may contain about two-thirds nitrogen, with virtually all but 2 or 3 percent of the rest being carbon oxides and hydrogen. The heating value of the gas will range from about 50 to 140 Btu per scf. Eliminating the nitrogen by using an oxygen gasification medium nearly doubles the heating value (180 to 250 Btu per scf) and makes it possible to methanate the product to a pipeline gas (PLG). The maximum heating value results from using an air-steam gasifying medium (250 to 280 Btu per scf).[67]

In the production phase, a number of problems require attention before UCG can become a commercial reality. *Combustion control* is paramount. Good gas quality and yields are dependent on minimizing production of CO_2 and water, and completely consuming the free oxygen in the gasification medium. As the void volume in the linkage path increases, the medium experiences poorer contact with the coal. Combustion control is also affected by loss of heat through adjoining strata and roof collapse. *Roof collapse* also causes surface subsidence, gas leakage, and groundwater seepage, with accompanying loss of thermal energy. *Linking* seldom produces uniform permeability. Fracturing increases leakage, disturbs adjoining strata, and frequently short-cuts areas which cannot then be extracted. Boreholes give more uniform link-ing. *Leakage* is highly dependent on geologic conditions, being particularly prevalent in faulted zones, shallow seams, and beds contiguous with porous sedimentary layers. *Groundwater control* problems result largely from roof col-lapse.

Efficient, economical utilization of the product gas, of course, is the real key to acceptance of UCG. The alternatives, all based on utilization at the mine-mouth, because of the relatively high cost of transmitting low-energy gas, are: electric power generation, upgrading to pipeline gas (by purification, shift con-version, and methanation), process heating, conversion to liquid fuels via Fischer-Tropsch synthesis (equivalent to the SASOL technology), and direct utilization of low-Btu gas (to manufacture ammonia, methanol, and so forth).

Practical power generating plants fueled with UCG product gas are prob-ably limited to 100- to 500-MW maximum capacity due to the size of the coal field required to support them. A high load factor would also have to be maintained because of the relative lack of flexibility in the output of the UCG operation. The greatest need for electric power is in the industrial east, the region most remote from the more favorable sites for underground coal gasifica-tion (Wyoming and Texas). The economic feasibility of minemouth electric power generation with UCG gas depends heavily on geographical location, geo-logic conditions, cleanup requirements, cost of fill material,* and other factors.

Upgrading to pipeline quality gas calls for gas cleanup at the surface or, at the very least, exclusion of air (nitrogen) from the gasification environment by use of oxygen or steam/oxygen as the gasifying agent. Since the alternatives to this option (aboveground coal gasification, light-oil gasification, or importation of LNG) are all expensive approaches, UCG may offer a viable answer, even though capital costs would be high, as they would be in the case of conversion to liquids.

Efficient use of UCG gas for process heating would require a large number of manufacturing plants grouped closely together, and a constant supply of gas would have to be assured.

Best estimates in the U.K. suggest that on an energy basis, deep-mined coal costs 1.25 to 4 times as much as surface-mined coal, and UCG would cost about twice again as much, with the drilling of boreholes accounting for about 60 per-cent of the total.[69] On a strictly cost basis, it would be cheaper to mine the coal, but this is not the sole consideration, as suggested at the beginning of this sec-tion.

PEAT—RUDIMENTARY COAL

Results of a study begun in 1974 by the Institute of Gas Technology (IGT) and the Minnesota Department of Natural Resources suggest that the United States

---

*To backfill the void space in the mine and prevent subsidence.

has another significant, yet unheralded fossil source of substitute fluid fuels—*peat* ("coal in the making"). Peat is a dark brown or black, partially decomposed residuum of marshland growth. Our peat reserves are second only to those of the U.S.S.R. and, on an energy basis, amount to more than our combined uranium, natural gas, and petroleum reserves. Our recoverable reserves of peat are estimated by IGT at about 1443 Q. Of this, about half, or 685 Q (240 billion boe), is located in the lower 48, mainly in the north-central tier.[72]

Considerable interest is also being shown in a much smaller deposit in eastern North Carolina. A Research Triangle Institute study estimates the latter deposits at over 400 million tons (about 4.2 Q).* The deposit comprises a 5- to 6-ft layer of woody peat covering about half of a 372,000-acre tract deemed desirable as farmland.[73] Table 5.7 shows estimated recoverable fossil reserves in the lower 48.

The IGT suggests that as many as 196 peat conversion plants could be supported in Alaska alone, with minimal environmental impact. Since peat generally occurs at the surface, in deposits averaging 7 ft thick, it can be extracted ("harvested")† readily without significant alteration of the earth's contour and without removal of vast amounts of overburden.

One of two methods is commonly used to mine or harvest peat. The first

Table 5.7   Recoverable Fossil Reserves of Lower 48 States (IGT Estimates)

Type	Quads (10^{15} Btu)[a]
Anthracite	100
Lignite	200
Crude oil	200
Natural gas	280
Oil shale	430
Peat	685[b]
Subbituminous coal	1640
Bituminous coal	3100

Source: Reference 72. Reprinted with permission of the copyright owner, The American Chemical Society.

[a]Data not precise; scaled from graph.
[b]Equal to about 65.9 billion tons (at about 50% moisture content). Other sources have indicated estimates of U.S. reserves (air-dried basis) as low as 13.8 million tons.

*Based on a heating value of 5200 Btu per lb (10.4 million Btu per ton) at 50 percent moisture content, or slightly less than 40 percent that of bituminous coal.
†So called because it can only be done in summer months.

calls for cutting the peat ("sod cutting") into bricks or cylinders and allowing it to lie on the surface of the bog to dry for several weeks. The moisture content drops typically from 90 percent to 30 to 40 percent. In the second method, "milling," a machine shreds the peat on the surface, and it is left to dry for three days, to a moisture content of about 55 percent.[74] The removal of peat is expected to uncover a biologically more productive soil, more suitable for cultivation of forests and possibly even some food crops.

Since peat has an average moisture content of 30 to 40 percent after field drying[75] (but as high as 55 percent)[74] and is generally associated with ample groundwater, it may ultimately be mined hydraulically. A major technological breakthrough has been the development of a technique for mechanical dewatering of the peat. Solar dewatering has also been explored, together with use of some of the conversion plant energy output. The latter proved uneconomical.[72]

Gasification appears to be technically and economically feasible according to tests carried out in a demonstration unit by IGT, DOE, and Minnesota Gas Co. (Minnegasco) since 1976. The projected output of a commercial plant is shown in Table 5.8. Only single-stage hydrogasification is needed since, with peat, four times as much carbon is converted directly to hydrocarbons, as in the case of lignite feed, and this is accomplished at a lower hydrogen partial pressure.[72] Perhaps one of the more serious drawbacks to the industrial utilization of peat is that, in the northern tier at least, harvesting, is limited to the summer months; hence large quantities of peat must be stored for drying to provide a uniform energy supply.[74]

ENVIRONMENTAL EFFECTS

Approximately 176 sites have been identified in the United States as having sufficient coal and water to support a large coal-to-substitute natural gas (SNG) facility of 250×10^6 scf per day capacity (each) for 20 years. This requires pro-

Table 5.8 Projected Output of Peat Conversion Plants

Product	Quantity
SNG[a]	250×10^6 scfd
Liquids[b]	29,000 bpd
Ammonia	465 tpd
Sulfur	48 tpd
Capital cost[c]	$900 million

Source: Reference 72. Reprinted with permission of the copyright owner, The American Chemical Society.

[a]At 20-yr average cost of about $2.00 per 1000 scf.
[b]Mostly benzene, naphthalene, and phenols.
[c]1976 dollars.

viding (for each site): 15,000 tpd of coal, 15,000 tpd of water, and 3000 tpd of oxygen. It also entails disposal of 2500 tpd of solid wastes from the coal pretreater and gasifier. High-sulfur eastern coal would also yield as much as 600 tpd of by-product sulfur. About 110×10^9 Btu per day of waste heat and large amounts of waste gases would be released to the atmosphere. Some gasification processes would require 3×10^6 gpd (12.5×10^3 tpd) of coolant and scrubber water, in addition to the input water consumed in the processing—water that is not abundant in arid coal-producing regions. This scrubber water would pick up over 1000 ppm each of phenols, ammonia, COS, and large quantities of suspended particulates, thiocyanates, and cyanides.[76]

As noted in Chapter 4, coal contains polycyclic aromatic hydrocarbons (PAHs), which are known carcinogens. It is reasoned that, in coal, these compounds are relatively stable, but conversion conditions, particularly hydrogenation, liquefy the coal and liberate the PAHs and other complex coal-based compounds, which could have unknown effects on the environment and on human health.[77]

There is considerable concern that the products of gasification and, particularly, liquefaction are highly contaminated with these carcinogenic polycyclic aromatic hydrocarbons and probably various organometallic compounds of undesignated compositions.[76] Indeed, Calvin estimates that a 25,000-tpd coal liquefaction plant will produce, each day, liquid hydrocarbon fuels containing 200,000 lb of PAHs and 10 lb of benzo(α)pyrene, one of the more carcinogenic of aromatic chemicals. Natural crude oil contains 1 to 5 ppm of the latter chemical, but coal-derived syncrude contains 10 to 100 ppm.[78] The chemical nature of coal, and the conversion conditions, are also conducive to the formation of various organometallics.[76] Among the organometallics believed to be formable in synfuel processing are: metal porphyrins and carbonyls, metallocenes or π complexes, arene carbonyls, metal alkyls, organohydrides, and metal chelates.[76] Unless aromatic chemicals and organometallics can be removed economically, conversion of coal could prove to be an environmental disaster.

Western coal, which contains less sulfur than eastern coal, contains radioactive atoms at a level one to two orders higher than eastern coals. Although these atoms can be removed from the combustion fuel gas by electrostatic precipitators, they are then concentrated in the fly ash and must be disposed of in an environmentally acceptable manner. Preliminary testing gives no indication of a health hazard from these radionuclides released by burning western coal, but investigators are awaiting results on potential biomagnification in the food chain and the implications of using and disposing of the fly ash.[79] About 60 million tons of fly ash is now being produced each year, and with increased use of coal, both in direct combustion and in conversion, this figure may rise to 1 billion-plus tons by the year 2000,[80] so the concern about radionuclides is certainly well founded.

Environmental concerns with coal-derived synfuels do not end with the processing. Combustion of these fuels is estimated to release to the atmosphere nearly 50 percent more CO_2 than coal itself, at an equal energy level. Table 5.9 compares CO_2 release of various fossil fuels. Mention was made in Chapter 4 of the potentially hazardous materials adsorbed on fly ash. Much remains to be learned about where these various pollutants end up in the case of coal-derived synfuels—in their by-products, in the various waste streams that result, or in the synfuels themselves—and what their impact is on health and the environment.

A coal hydrogenation plant, operated with great care for six years during the 1960s, reported an incidence of skin cancer in workers which was 16 to 37 times that in the normal population. Syncrude from Oak Ridge National Laboratory (ORNL) was also found to contain mutagenically active material, hence should be regarded as a "presumptive carcinogen."[82]* All these factors add up to the need for extreme caution.

Table 5.10 lists some of the pollutants from the overall gasification process which generates the particular effluent. A 250-million-cfd gasification plant is estimated to emit: 300 to 450 ltpd of sulfur (mainly H_2S), 100 to 150 tpd of ammonia, up to 2 tpd hydrogen cyanide, 10 to 70 tpd of phenols, 50 to 300 tpd of benzene, and up to 400 tpd of oils and tars.[17] For the most part, these materials would be collected for sale, to offset some of the cost of fuel production, but some could find their way into the environment. The numbers will vary widely with the particular process and with the specific feedstock. It will be remembered that Fig. 5.1 illustrated a typical procedure used to clean up syngas. In another procedure described in the literature, the tar and heavy oils are removed by quenching and cooling. Ammonia, light oils, phenols, and dust are removed in a venturi scrubber or by direct water wash. Acid gases (CO_2, H_2S) are removed by physical absorption in methanol or chemical reaction with an amine solution, and H_2S is converted to elemental sulfur. Following

Table 5.9 Release of Carbon Dioxide by Combustion of Fossil and Coal-Derived Substitute Fuels

	10^{15} g Carbon/ 100 Quads Energy
Coal-derived synfuels[a]	3.4
Coal	2.5
Oil	2.0
Natural gas	1.45

Source: Reference 81. Reprinted with permission of the copyright owner, The American Chemical Society.

[a]Production and combustion.

*A substance shown to cause cancer in one or more species of laboratory animals, but not as yet in human beings.

Table 5.10   Effluents from Coal Gasification

Stage	Effluents
Coal preparation	Dust
	Particulates
Devolatilization	Sulfur dioxide
	Organics
	Naphthalene
	Anthracene
	Pyridine
	Phenols
	Cresols
	Trace metals
Gasification	Pitch
	Tar
	Phenols
	Aromatics
	Hydrogen sulfide
	Hydrogen cyanide
	Benzene
Purification–methanation	Sulfur compounds
	Ammonia

Source: Reference 83.

methanation, water is removed by a glycol wash, and then the clean, dry gas is compressed to pipeline pressure.[16]

The process water stream may contain phenols, cresols, light aromatics such as benzene, oils, tars, ammonia, sulfur compounds, traces of cyanide, coal, char, ash, and so forth. The raw gas stream is largely contaminated with SO_2, H_2S, NO_x, and CO_2. Typically, a cleaned SNG stream might contain less than 0.1 percent CO, 0+ percent H_2S, and 5 percent hydrogen. Particulates will arise from the various operations involving coal, char, ash, and lime.[16] Again, some of the by-products are potential commercial materials; hence there is an economic incentive, and perhaps even a financial necessity, to collect them and thus keep them out of the environment.

ECONOMICS

Reduced volumes of gas extraction and lower rates of interstate transmission, in the face of much higher drilling rates over the past seven years or so, should signal the nature of our natural gas problem, and the situation is similar for crude oil. Barring the discovery of large, new reserves of unconventional gas, and development of the technology for extracting and exploiting them, our reserves of natural gas are on an irreversible decline. Our domestic crude oil situation is in even greater jeopardy, because we lack any reasonable expectation of *major* new discoveries.

There can be little doubt that our success in producing synfuels in significant quantities, and at a reasonable price competitive with foreign oil, could have a major impact on future OPEC pricing policies. Significant advances in the technology and economics of gasifying and liquefying our large coal (or oil shale) resources could cause OPEC oil ministers to reassess their positions. The big question is: What are the prospects of accomplishing this monumental task? If this goal is achieved, the political problems may be next.

Historically, the politics of energy have been very ephemeral—changing with each administration, and even within the term of office of a given administration. In the 1976 election campaign, candidate Carter established a cardinal principle that stated: "Prices should generally reflect the true replacement cost of energy. ... We are only cheating ourselves if we make energy artificially cheap and use more than we can really afford."[84] This principle has stood up during his administration—to a degree—with the phasing in of federal price decontrol of domestic natural gas and crude oil. This policy has resulted in higher fluid fuel prices, but it has provided some degree of encouragement for synfuel research and development programs.

In 1978, the Administration toyed with a further proposal to force oil refiners to begin introduction of 5 or 10 percent of synthetic liquids into their refinery output by 1985 or 1990. The higher costs of the synfuels would have been averaged into the costs of their standard products if this intent had prevailed. Substitute gases were not considered specifically in this proposal.[85]

In the interim, during the debate over such issues, valuable lead times have been eroded, delaying the initiation and consummation of our synfuel efforts. But the first major step has been taken. On June 16, 1980, congressional negotiators agreed on the basic features of an energy bill that would, in part, provide $20 billion over a five-year period for loans and price guarantees for the production of synfuels.[86]

Eddinger[87] claims that from the day the technical concept of a process is born, until the commercial production facility is brought on stream, 15 to 20 years may pass. At the present time, we are probably no further down that scale than possibly 10 years, with regard to the most advanced of the technologies discussed earlier, and the 5 to 10 years remaining will not see production at a full commercial scale.

Required lead times already preclude meeting DOE's initial schedule of having the first commercial gasification facility on stream by 1978 or 1979. To meet the goal of 200 such plants built between 1981 and 2000 would entail a 17 percent annual growth rate for the coal conversion industry![88]

The major factor remaining at this point is the sheer magnitude or scale of the operations required for synfuels to have a significant bearing on our energy budget. It has been estimated by the Ford Foundation that domestic coal conversion plants could produce as much as 2 to 3 Q of synfuels per year by 2000.[89]

Other estimates have ranged from 0.5 million bpd of oil equivalent (or 1 Q per year) in 1987 and 2 million bpdoe (or 4 Q per year) in 1992,[90] to 10 quads (maximum) of total synfuels,[89] or 4 tcf (quads) of syngas alone,[91] depending on the scenario used. Production of 2 to 3 Q of all coal-derived synfuels would be a major technical undertaking in itself, requiring a feed volume of about 700 million tons of coal (4 to 5 Q).[89]

To attain commercial-scale production of synfuels would be a technological, logistic, and economic challenge. The following factors must be considered: overall capital needs, coal requirements (including consideration of conversion losses), conversion facilities, water demands, labor commitments, product storage, and product price.

Capital

It has been estimated by one source that it would require a total investment of about $100 billion (or $14 billion per year over 10 years) to provide the facilities necessary to ensure this level of commercial gasification capability.[92] A further delay of five years might mean about 60 percent escalation of capital costs because of inflation. Yet another source, the Office of Technology Assessment (OTA), reported in 1977 that it should cost no less than $120 billion to set up a synthetic fluid-from-coal industry in the United States. This assumes 80 commercial conversion plants at $1.5 billion each. The price of liquid product was estimated at $25 to $40 per boe.

The tightness of this quantity of speculative investment capital on the money market would undoubtedly create a shortage of capital to produce other "necessities" for our way of life and to finance other necessary growth in our economy.

Capital costs for the most advanced gasification methods (HYGAS, Lurgi, CO_2 Acceptor, and COED) would probably fall in the range $2500 to $5000 per million Btu (or per 1000 scf) per day capacity. Coal mine fitting costs would be about $600 to $1200 per million Btu per day capacity.[61] The financial carrying costs of plant have been estimated to contribute 25 to 45 percent of the price of the gas produced.[12]

An economic analysis of coal gasification shows the main cost factors in the overall process to be: steam and utilities, 23.3 percent; acid-gas removal, 14.2 percent; oxygen plant, 8.4 percent; and coal (combined preparation, storage, and handling) and sulfur recovery, each 7.2 percent. The actual gasification (conversion) stage amounts to only 6.5 percent of the total cost.[34]

There are those who contend that "big is better," but it might be smarter for industrial plants or even small utilities to put their money into by-product coke ovens, to produce high-Btu coke-oven gas and then use the coke in small gasifiers to generate low- or intermediate-Btu gas for local use in processing or central heating. Low-Btu gasification facilities could be built for only about $4 million, in part because no oxygen plant would be required. Medium-Btu facilities would be more costly because of the need for an oxygen plant.[93]

The nature and magnitude of the conversion process are what make capital requirements so high. The reaction vessels are massive in order to operate at the temperatures and pressures required for conversion of coal, and the corrosive nature of some of the products accounts for the expensive special alloys called for.

Hydrogen sulfide released in the process is one of the primary culprits. It becomes increasingly corrosive to carbon steel at temperatures over 450 to 550°F (230 to 290°C). In combination with hydrogen, at temperatures above 800°F (427°C), precipitation of carbides occurs along grain boundaries. A high chromium content is required to offset this, so 300 series austenitic stainless steels are normally used. At the higher temperatures, clad or weld-overlay corrosion resistant steels are employed.[93]

Hydrogen at temperatures of about 450°F (230°C) and pressures over 200 psi causes decarburization of carbon steel and forms methane at internal interstices, thus producing blisters. Molychrome alloys suppress this tendency.[93]

In SRC plants, chlorine concentrations over 100 ppm and 1.5 percent water in the dissolver section may cause stress-corrosion cracking. It is not unusual for hydrogenators to require wall thicknesses over 12 in. to withstand the 850°F (455°C) temperatures and 4000 psig pressures often encountered.[93]

Coal Requirements

To produce the coal required for a commercial, 4-Q (tcf) per year syngas industry, about 50 huge new coal mines would have to be opened, each equal to the largest coal mine in operation now.[91] The largest domestic mine produced 10.2 million tpy in 1976. The next three produced 9.3, 7.4, and 6.5 million tpy, respectively.[94] The mine opening and development costs would be perhaps 15 to 25 percent as much as the costs for the corresponding conversion facilities. Each conversion plant would require about 6 million tpy of coal feed, a total of about half our 1977 coal production. In addition, of course, the electric utilities industry plans to increase its coal consumption 50 percent by 2000, to compensate for growth and for phasing out of gas- and oil-fired boilers.

Conversion Facilities

To attain 4-tcf annual production level for syngas, 50 new gasification plants, each with a daily capacity of 250 million scf, would have to be erected.[91] Considering the materiel factors noted above, this would represent a massive undertaking for the metals and other associated industries, as well as the energy industry.

Water

Since many of the conversion facilities would have to be built at the minemouth, for economic reasons, water demands could become critical, because the bulk of the large mines are surface mines located in arid or semiarid regions. It has been estimated by one source that the 50 SNG plants would consume water at a level equivalent to 20 percent of the total projected water consumption of *all* domestic electric utilities in 1980.[91] Estimated in different terms by another source, conversion of coal to yield 10 billion Btu of syngas requires 1.1 acre-ft* of water, just for the gasification process, compared with 4.5 acre-ft for an energy-equivalent coal-fired power plant with scrubbers, and 9.0 acre-ft for an equivalent nuclear power plant.[95]

Labor

It would require about 1000 trained miners plus support people to operate each of the supporting coal mines once they are opened. Furthermore, each conversion facility would require a 5000-worker construction force. Upon completion, 1000 chemical engineers would be required to maintain each plant on stream.[91] The United States has only about 25,000 competent engineers for this line of work, most of whom are already gainfully employed in other endeavors, so engineering labor demands alone could scuttle such a program.[61]

Product Storage

Natural gas wells operate continuously for long periods at reasonably steady, predictable rates and with little loss. Conversion facilities, on the other hand, are nothing more than chemical plants, which typically operate at 75 to 90 percent of nameplate capacity. Furthermore, these conversion plants are plagued with a raw material that is inherently variable in its composition and reactivity. Also, the product output mix is highly dependent on process variables. It will undoubtedly be necessary to provide product storage facilities to tide the plant over shutdowns and to act as a "surge bin" for varying output rates. This adds to the capital needs of the conversion facilities.

Product Price

Even the most optimistic observer expects the cost of substitute gases to be about twice that of domestic natural gas. One reason is the conversion efficiencies. Natural gas can be extracted with little or no loss. Conversion efficiencies of other fluid fuels are shown in Table 5.11.

The Ford Foundation estimates the price of medium-Btu syngas from coal to be $4 to $5 per million Btu, and high-Btu syngas (SNG) to be $5 to $6 compared to about $2.60 for domestic interstate natural gas. Coal liquids are estimated at $35+ per bbl using the gasification-synthesis technology, and $25+ using hydrogenation (with or without donor-solvent) technologies.[97] The actual prices will also depend somewhat on the value of the by-products.

*An acre-ft of water is 325,850 gal.[95]

Table 5.11 Conversion Efficiencies (Coal-to-Fluid Fuels)

	Btu/scf	Conversion Efficiency, %[a]
Gaseous product		
Natural gas (for reference)	1027	—
Synthetic pipeline gas	1000	60
Coal gas, low-temperature	780	80
Coke-oven gas	540	95
Water gas	300	95
Producer gas	130	70
Liquid fuel, Btu/bbl		
Syncrude	5,600,000	60
Methanol	2,720,000	65

Source: Reference 96. From *Man, Energy, Society* by Earl Cook. W. H. Freeman and Company. Copyright © 1976.

[a]Percentage of chemical or latent-heat energy in coal that can be obtained by burning the conversion product.

Peat, of course, is a potential alternative to coal as a feed for conversion to synfuels. With peat, the situation would be little different than with coal, except that the process of gasification of peat is more direct than coal, so the plant might be somewhat less complex and the selling price of the gas somewhat lower, but the difference would be relatively modest.[88]

SUMMARY

There are some basic considerations that should be used to guide us in the decision to convert coal to fluids or to use it directly as a fuel. Synfuels generally have little to offer for base or even average-load electric power generation, except for those cases where the transportation costs of coal from the mine to the power plant would be excessive, or where existing fossil fluid power plants cannot be assured continuing fuel supplies. It would be cheaper to build large, modern, efficient coal-fired plants, complete with the latest pollution control devices, rather than suffer the capital outlays and the energy losses inherent in coal-conversion options. The synthetics might best be used as fuels for transportation, home heating, and peak-power generation, where the convenience, transportability, storability, and quality of synfuels offer advantages. Even then, the synfuels should be produced from low-grade feeds that cannot reasonably be used for direct combustion in large power plants.[54]

The primary initial outlets for coal liquids are expected to be electric utilities, to replace phased-out petroleum-based liquids in existing power plants. The large pilot plants for the H-Coal and EDS processes are to be followed by a larger demonstration plant for SRC-II scheduled to come on stream in 1983. By 1990, commercial facilities should be able to meet about 30 percent (450,000 bpd)* of the 1978 demand (1.5 million bpd) for oil by U.S. electric utilities.[98]

Economics is, obviously, one of the key issues in the conversion of coal and peat to fluid synfuels. The switch of even a modest portion of our energy budget to these fuels would require tying up a very significant portion of our nation's available capital. The undertaking would also occupy a very large segment of our technical labor pool. The water requirements and questions regarding toxicological problems of the conversion processes, the wastes, and the resulting products pose real environmental issues which demand careful attention.

In the meantime, political issues appear to be the major stumbling blocks. Our large oil companies would, perhaps, have sufficient size to press the development of the technology to a commercial scale. In some cases, this would mean merely completing projects they are already sponsoring. They might even be able to provide capital for getting commercial synfuel production off the ground. The problem is that there are those in government who still feel these companies should be prohibited from entering into energy projects outside the petroleum arena. The political uncertainty of this issue had virtually paralyzed progress, although the pressure has now eased somewhat with failure of the divestiture issue. The federal support now being proposed under the aegis of the Synthetic Fuels Corporation promises to be crucial toward ending the stalemate.

At present, it looks highly improbable that a synfuels-from-coal industry will make sufficient impact on our energy budget to rescue us from our dilemma in the foreseeable future. The scope of the synfuels undertaking is too large to more than relieve a little of the pressure, without making some other portion of our economy suffer severe hardships. Nonetheless, our coal reserves are ample to provide, theoretically, sufficient syngas to meet our needs for many years. For example, based on 1977 domestic natural gas production levels, the economically recoverable reserves of Pittsburgh No. 8 coal (19 billion tons by BuMines estimate) underlying Pennsylvania, West Virginia, and Ohio would be sufficient to provide a 20-year supply of SNG. Our natural gas reserves are estimated at a 12-year supply.[99]

There is no consensus as to the probability of establishing a viable coal conversion industry, even among the industrialists responsible for the development. A survey reported in 1977 disclosed that these business managers felt it would take up to 10 years for low-Btu gas, 10 to 20 years for high-Btu gas, and 10 to 25 years for syncrude production to reach a commercial level. Even then, there was little agreement as to what these levels might be. Production of low-Btu gas was

---

*About 200,000 bpd from SRC-II, 150,000 bpd from H-Coal, and 100,000 bpd from EDS.

thought by some to be headed for about 20 percent of our natural gas demand (although some opted for an estimate as low as 6 percent). Better agreement was reached on coal liquids—5 to 10 percent of our nation's energy budget by 1995.[100] These estimates are more optimistic than those of the Ford Foundation[89] and other, more recent sources.

It is obvious that the synfuel option will only partially meet our future needs. Even then, our past record of coal strikes must leave some doubts in our minds as to the stability of supply of such coal-derived synfuels for the long haul.

REFERENCES

1 J. D. Parent and H. R. Linden, "A Survey of United States and Total World Production, Proved Reserves and Remaining Recoverable Resources of Fossil Fuels and Uranium, as of December 31, 1975," Inst. Gas Technol., Chicago, Jan. 1977, p. 13.

2 S. B. Alpert and R. M. Lundberg, "Clean Liquids and Gaseous Fuels from Coal for Electric Power," in J. M. Hollander and M. K. Simmons, Eds., *Annual Review of Energy*, Annual Reviews, Palo Alto, CA, 1976, pp. 87-99.

3 *Coal Facts: 1978-1979*, Natl. Coal Assoc., Washington, DC, p. 58 (DOE data).

4 "Research Report: Energy Conversion Research," Amer. Soc. Mech. Eng., New York; condensed by G. P. Cooper, *Mech. Eng.* **99**(11), 22-28 (Nov. 1977).

5 O. H. Hammond and R. E. Baron, *Amer. Sci.* **64**(4), 404-417 (July-Aug. 1976).

6 J. H. Field, "Bergius Process," in D. N. Lapedes, Ed., *McGraw-Hill Encyclopedia of Energy*, McGraw-Hill, New York, 1976, p. 112.

7 *Public Util. Fortn.* **98**(9), 69-73 (Oct. 21, 1976).

8 D. B. Anthony and J. B. Howard, *AIChE J.* **22**(44), 625-656 (July 1976).

9 C. A. Stokes; reported in *Chem. Eng. News* **54**(37), 13 (Sept. 6, 1976).

10 D. A. Tillman, *Environ. Sci. Technol.* **10**(1), 34-38 (Jan. 1976).

11 IGT study; reported in *Energy User News* **2**(41), 16 (Oct. 17, 1977).

12 S. M. Dix, *Energy: A Critical Decision for the United States Economy*, Energy Education Publ., Grand Rapids, MI, 1977, p. 118.

13 *Energy User News* **2**(41), 16 (Oct. 17, 1977).

14 *Bus. Week* (No. 2553), 132D, 132H (Sept. 25, 1978).

15 L. E. Swabb, Jr., *Science* **199**(4329), 619-622 (Feb. 10, 1978).

16 E. H. Thorndike, *Energy and Environment: A Primer for Scientists and Engineers*, Addison-Wesley, Reading MA, 1976, pp. 94-96.

17 H. R. Linden et al., "Production of High-Btu Gas from Coal," pp. 65-85, in reference 2.

18 D. L. Hagen, "Methanol: Its Synthesis, Use as a Fuel, Economics, and Hazards," NP-21727, Natl. Tech. Inf. Serv., Springfield, VA, Dec. 1976, p. I-6.

19 B. C. Gates, *Chemtech* **9**(2), 97-102 (Feb. 1979).

20 N. P. Cochran, *Sci. Amer.* **234**(5), 24-29 (May 1976).

21 *Technol. Rev.* **79**(1), 20 (Oct.-Nov. 1976).

22 A. L. Hammond, *Science* **193**(4256), 873-875 (Sept. 3, 1976).

23 T. H. Maugh II, *ibid.* **198**(4322), 1132-1134 (Dec. 16, 1977).

24 Bodle and Vyas (1973); reported by F. D. Cooper, "Coal Liquefaction," pp. 138-141, in reference 6.

25 R. B. Engdahl and R. E. Barrett, "Fuels and Their Utilization," in A. C. Stern, Ed., *Air Pollution*, Vol. 4, 3rd ed., Academic, New York, 1977, pp. 380–424.

26 J. L. Hatten, *Mech. Eng.* **97**(7), 31–35 (July 1975).

27 L. E. Smartt, *Public Util. Fortn.* **100**(9), 4, 7 (Oct. 27, 1977).

28 F. C. Schora, "Substitute Natural Gas (SNG)," pp. 652–654, in reference 6.

29 J. M. Talty, *Environ. Sci. Technol.* **12**(8), 890–894 (Aug. 1978).

30 A. L. Hammond, *Science* **193**(4255), 750–753 (Aug. 27, 1976).

31 *Chem. Eng. News* **56**(31), 23 (July 31, 1978).

32 V. D. Chase, *Popul. Sci.* **210**(1), 91–94 (Jan. 1977).

33 R. C. Weast, "Coal Gasification," pp. 134–138, in reference 6.

34 G. A. Mills, *Chemtech* **7**(7), 418–423 (July 1977).

35 *Chem. Eng. News* **57**(6), 20, 24 (Feb. 5, 1979).

36 *Ibid.* **56**(47), 26, 27 (Nov. 20, 1978).

37 *Ibid.* **56**(19), 29 (May 8, 1978).

38 *Bus. Week* (No. 2482), 44L (May 9, 1977).

39 R. Gorman, *Popul. Sci.* **211**(2), 102–105 (Aug. 1977).

40 *Mach. Des.* **49**(12), 18 (May 26, 1977).

41 *Sci. News* **111**(18), 280 (Apr. 30, 1977).

42 F. D. Cooper, "Coal Liquefaction," pp. 138–141, in reference 6.

43 *Environ. Sci. Technol.* **11**(2), 122, 123 (Feb. 1977).

44 G. S. Schatz, *News Rep.* (NAS) **28**(1), 1, 6, 7 (Jan. 1978).

45 *Energy User News* **2**(34), 7 (Aug. 29, 1977).

46 *Chem. Eng. News* **55**(51), 26, 27 (Dec. 19, 1977).

47 *Public Util. Fortn.* **100**(6), 58 (Sept. 15, 1977).

48 *Bus. Week* (No. 2508), 84D, 84F (Nov. 7, 1977).

49 *Chem. Eng. News* **55**(35), 7, 8 (Aug. 29, 1977).

50 *Bus. Week* (No. 2500), 104B (Sept. 12, 1977).

51 *Public Util Fortn.* **102**(9), 70, 71 (Oct. 26, 1978).

52 *Chem. Eng. News* **56**(39), 43, 44 (Sept. 25, 1978).

53 *Ibid.* **54**(30), 24, 25 (July 19, 1976).

54 R. Shinnar, *Chemtech* **8**(11), 686–693 (Nov. 1978).

55 *Chem. Eng. News* **54**(37), 33, 34 (Sept. 6, 1976).

56 *Ibid.* **54**(27), 16 (June 28, 1976).

57 *Ibid.* **54**(28), 30 (July 5, 1976).

58 *Ibid.* **54**(38), 22 (Sept. 13, 1976).

59 K. H. van Heek; reported in *Chem. Eng. News* **54**(45), 16, 17 (Nov. 1, 1976).

60 *J. Air Pollut. Control Assoc.* **27**(12), 1210, 1211 (Dec. 1977).

61 D. F. Othmer, *Mech. Eng.* **99**(11), 29–35 (Nov. 1977).

62 *Chem. Eng. News* **56**(47), 28, 30 (Nov. 20, 1978).

63 G. O. Davies, *Chem. Ind.* (No. 2), 560–566 (Aug. 5, 1978).

64 R. R. Maddocks; reported in reference 62.

65 L. E. Swabb, Jr.; reported in reference 62.

66 A. H. Knebel; reported in reference 62.

67 R. M. Nadkarni et al., *Chemtech* **4**(4), 230–237 (Apr. 1974).

68 D. W. Gregg and T. F. Edgar, *AIChE J.* **24**(5), 753–781 (Sept. 1978).

69 P. N. Thompson, *Endeavor* (New Ser.) **2**(2), 93–97 (1978).

70 *Chem. Eng. News* **54**(42), 22 (Oct. 11, 1977).

71 M. Kenward, *New Sci.* **71**(1014), 396 (Aug. 19, 1976).

72 IGT; reported in *Chem. Eng. News* **55**(45), 39, 40 (Nov. 7, 1977).

73 L. J. Carter, *Science* **199**(4324), 33, 34 (Jan. 6. 1978).

74 E. W. Cook; letter in *Chem. Eng. News* **56**(6), 2 (Feb. 6, 1978).

75 C. H. Fuchsman; letter in *ibid.* **55**(50), 4 (Dec. 12, 1977).

76 D. W. Koppenaal and S. E. Manahan, *Environ. Sci. Technol.* **10**(12), 1104–1107 (Nov. 1976).

77 *Bus. Week* (No. 2420), 36E (Feb. 23, 1976).

78 M. Calvin, *Chem. Eng. News* **56**(12), 30–36 (Mar. 20, 1978).

79 C. E. Styron and B. Robinson, *ibid.* **55**(46), 6 (Nov. 14, 1977).

80 *Ind. Res.* **19**(11), 44, 46 (Nov. 1977).

81 Counc. Environ. Qual.; reported in *Chem. Eng. News* **57**(29), 6 (July 16, 1979).

82 B. Commoner, *Chemtech* **7**(2), 76–82 (Feb. 1977).

83 C. A. Zraket, "Energy Resources for the Year 2000 and Beyond," PB-247413, DOC, Washington, DC, Mar. 1975, p. 24.

84 Quoted in G. S. Savitsky, *Energy User News* **3**(12), 18 (Mar. 20, 1978).

85 *Energy User News* **3**(11), 18 (Mar. 13, 1978).

86 C. Haas, *Phila. Inquirer* **302**(169), 1-A (June 17, 1980).

87 R. T. Eddinger, *Chemtech* **7**(9), 556–558 (Sept. 1977).

88 M. Kenward, *New Sci.* **71**(1008), 84, 85 (July 8, 1976).

89 *Energy: The Next Twenty Years,* H. H. Landsberg, Chmn., Ford Foundation, Ballinger, Cambridge, MA, 1979, p. 310.

90 Congressional conference (1980); reported in *Phila. Inquirer* **302**(172), 9-A (June 20, 1980).

91 J. Sudbury; reported in *Energy User News* **2**(46), 20 (Nov. 21, 1977).

92 C. A. Stokes, *Energy* **3**(3), 16–18 (Summer 1978).

93 J. B. O'Hara, *Met. Prog.* **110**(6), 33–38 (Nov. 1976).

94 *Keystone Coal Industry Manual;* p. 86, in reference 3.

95 H. W. Welch, in J. E. Bailey, Ed., *Energy Systems: An Analysis for Engineers and Policy Makers,* Dekker, New York, 1978, p. 78.

96 E. Cook, *Man, Energy, Society,* Freeman, San Francisco, 1976, p. 126.

97 Reference 89, pp. 308, 309.

98 EPRI; reported in *Public Util. Fortn.* **102**(9), 27 (Oct. 26, 1978).

99 W. B. Carter; reported in *Chem. Eng. News* **56**(29), 7, 8 (July 17, 1978).

100 S. W. Herman et al., Eds., *Energy Futures: Industry and the New Technologies,* Ballinger, Cambridge, MA, 1977, Introduction.

6 TAR SANDS AND OIL SHALES

Many experts believe that the net energy yield from shale will be negligible.

W. CLARK

Tar Sands

Tar sand is sand saturated with a crude hydrocarbon substance too viscous to be recovered in its natural state by pumping from a well. The term "tar" is something of a misnomer, since in petroleum refining, tar is a residue from a thermal process. "Bitumen" is a more correct term, since this properly implies a carbon disulfide–soluble oil. The bitumen in tar sands has an average molecular weight of about 650, and roughly half is distillable without cracking.[1]

Tar sands ("oil sands") apparently started as pools of crude oil lying beneath an ancient sea. When the land rose and the water drained away, the release of overlying pressure permitted the oil to migrate upward into the still-wet seabed. The heterogeneity of the seabed accounts for the present variability of tar sands. In some deposits (notably much of the McMurray Formation in Alberta, Canada), the moisture from the seafloor maintained a thin film of water and seashore clays around each grain of sand, thus separating the mineral sand from the oily coating. The oil, in turn, prevented the water from evaporating. Exposure of the coated sands to the atmosphere allowed the lighter hydrocarbons to evaporate, leaving behind heavy, carbon-rich hydrocarbons (a form of bitumen), together with sulfur, metal ions, and other impurities (Fig. 6.1).[2]

The Athabasca tar sands have a typical void volume of about 35 percent, mainly connate water, with the balance air, although methane has been found in some test borings. A typical composition would be 83 weight percent sand and the remaining 17 weight percent, bitumen combined with water. This combined 17 percent shows up with astonishing regularity, although the actual bitumen-to-water ratio varies over a considerable range. The hydrocarbon is ap-

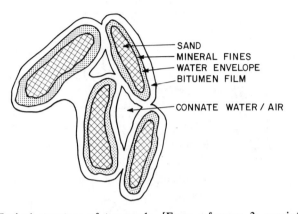

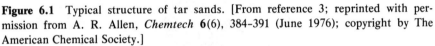

Figure 6.1 Typical structure of tar sands. [From reference 3; reprinted with permission from A. R. Allen, *Chemtech* **6**(6), 384–391 (June 1976); copyright by The American Chemical Society.]

proximately 83 percent carbon, with the other major elements being hydrogen (about 10 percent), sulfur (about 5 percent), and nitrogen (0.5 percent). Vanadium and nickel are generally present at levels of about 250 and 90 ppm, respectively. The hydrocarbon component is essentially 47 percent oils, 32 percent resins, and 20 percent asphaltenes. It has a heating value of 17,690 to 17,910 Btu per lb (average 17,800 Btu) on the basis of organics. Over 99 percent of the mineral content is quartz and clays.[1]

These tar sand deposits were first reported in 1788 by Peter Pond, an itinerant New England fur trader, as "a sticky substance oozing from the river banks which the indians used to waterproof their canoes." Scientific interest in investigating the tar sands was first shown by the Canadian government in 1890, but it then required 67 years more to develop a large-scale commercial extraction technology. This work was largely spearheaded by Karl A. Clark of the Research Council of Alberta, who developed the hot-water extraction process.

At elevated temperatures, bitumen flows and will float on water. At room temperature or below, it is denser than water and does not flow. The bitumen from the east-central portion of Alberta (near Lloydminster) is relatively low in viscosity, owing, in part, to the reservoir temperature. (See Fig. 6.2 for the geographical locations of Alberta tar sand deposits.) The viscosity of the Lloydminster bitumen is about 300 times that of normal crude oil, and the deposits just north (Cold Lake) and in the western part of the province (Peace River) are 1000 times more viscous than the Lloydminster bitumen. The Lower Cretaceous period deposits in the northeastern part of the province (Athabasca and Wabasca) are 10 times again more viscous.[4]*

*Some reports speak loosely of Athabasca tar sands as *all* such deposits occurring in Alberta Province (an entity perhaps more appropriately referred to as the Alberta oil sands deposits). Actually, the Athabasca sands are particular Alberta deposits, as seen in Fig. 6.2. The precise meanings encountered in specific literature references are seldom clear.

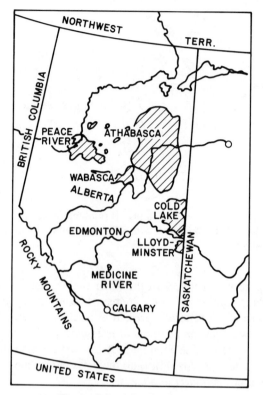

Figure 6.2   Alberta oil sands.

RESOURCES

Over geologic time, thick layers of boulders, gravel, sand, clay, and muskeg
have been deposited on top of the tar sands. These sands may now lie up to 2600
ft below the surface,* in layers 100 or more feet thick. The Athabasca deposits†
of Alberta, Canada—the world's largest—represent about 88 percent of
Canada's Lower Cretaceous in-place reserves of tar sands. (The Triassic
deposits on Melville Island, N.W.T., are not yet fully assessed, but are unquestionably much smaller.) Canada's Lower Cretaceous sands alone are generally
estimated at about 711 billion boe of in-place reserves.[1] These reserves may actually be as high as 1000 billion boe, according to a broad sampling of
estimates. These deposits are spread over at least 12,500 mi² and have an
average pay thickness of 150 ft. Although some of these deposits are very deep,

---

*Less than 10 percent of Canada's deposits lie within 150 ft of the surface, where they can be open-
pit-mined.[4]
†This excludes the Bluesky-Gething and Grand Rapids deposits, often included as "Athabasca" tar
sands.

those currently being worked commercially are "surface" deposits, up to only 150 ft deep. About 90 percent of Canada's tar sands lie 150 to 700 ft deep, where other extraction technologies must be employed.[2]

Most of the Wabasca, Cold Lake, and Peace River sands, totaling an estimated 260 billion bbl of oil in-place, are too deep for open-pit mining (i.e., they have an excessively high overburden ratio).* They are not generally considered to be economical to work[5] until such time as presently unproven extraction technologies are demonstrated to be economically viable. The Cold Lake reserves alone have been estimated at about 160 billion bbl, accessible only by in situ extraction.[4,6]

Venezuela, endowed with the world's second largest deposits of tar sands, has in-place reserves estimated from as low as 200 billion boe[1] to as high as 700 billion boe.[7] They cover a 53- by 373-mi (85- by 600-km) area extending along a roughly east-west, discontinuous band north of and parallel to the Orinoco River. Overall recoverability is expected to be poorer than from the Canadian deposits (10 percent vs. about 25 percent), with only 70 billion bbl of raw oil† being recoverable from the higher reserves estimate of 700 billion boe.[7]

The total U.S. in-place reserves (>30 billion boe)[4]‡ are less than 10 percent of Venezuela's, with nearly 90 percent being located in the Uinta Basin of Utah, closely associated with our western oil shale reserves. The balance is scattered through California, Texas, Kentucky, Missouri, and Kansas, with small amounts in New Mexico and Louisiana.§ The major U.S. deposits range in thickness up to 300 ft and in depth below grade to 2000 ft. Domestic tar sands have a very low sulfur content (about 0.5 percent vs. 4 to 6 percent for both Canadian and Venezuelan oil).[11] Most of the sulfur comes out in the processing, primarily in process water as dissolved organic compounds or as hydrogen sulfide.[4,12] Our Uinta sands lack the intervening water film between the sand and the bitumen, thus making separation more difficult.[4] A similar situation exists with some Canadian deposits.

The extractable or proven reserves are, of course, much smaller than those in place, because of the effects of deposit thickness, deposit accessibility (geographical location and overburden ratio), bitumen content and viscosity, and recovery rate (mining or in situ extraction, presence or absence of intervening water film, etc.). It has been estimated that Canadian tar sands are potentially convertible into five times the amount of crude in all the known natural petroleum deposits in North America.[2]

---

*Ratio of overburden thickness to tar sand deposit or pay thickness.

†This raw oil will yield only about 70 percent (49 billion bbl) of a syncrude.[7]

‡Averitt estimates about 8.3 percent of these in-place reserves (or 2.5 billion bbl) are proven reserves, and half (15 billion bbl) are ultimately recoverable.[8]

§There is also a pool of about 160 billion bbl of oil too viscous to pump, lying beneath 42 counties near the junction of Kansas, Missouri, and Oklahoma.[9] Southern California also has about 20 billion bbl of heavy crude about 1200 to 5000 ft deep which requires enhanced recovery methods for extraction. The problem is that for steam recovery, 1 bbl of crude must be consumed to generate steam to recover 3 bbl. Prior to recent OPEC price increases, such recovery was not economical.[10]

Unfortunately for the United States, the technological development of our large oil shale deposits lags well behind Canadian tar sand technology, and we have poor prospects of benefiting directly from the hydrocarbon output of their tar sands. At best, by the 1990s Canadian industry will probably provide only about 1 million bpd of tar sand syncrude or about 30 percent of their own domestic crude oil needs, while conventional petroleum supplies will be on the decline.[4] A new, large-scale syncrude plant* would have to be constructed each year just to keep up with Canada's increasing demand for oil.[2]

The United States can possibly benefit indirectly, however, from Canada's development of tar sands in one or more of three ways: (1) development of a new major source of oil in Canada should reduce somewhat the worldwide competition for dwindling supplies of conventional petroleum, (2) extraction and conversion technology developed for the Canadian industry could be licensed for the development of our own modest tar sand resources, and (3) there might be some technological spin-off of value to our own shale oil industry.

EXTRACTION

Many problems face those charged with extracting the bitumen from tar sands. Not the least of these problems is the muskeg overlying much of the Alberta deposits. The muskeg, a semifloating mass of partially decayed vegetation and scraggly spruce, jack pine, and tamarack trees, may constitute up to an 18- or 20-ft-thick layer on top of the mineral overburden. This surface is impassable in the summer, swallowing up land vehicles that attempt to traverse it. In the winter, it becomes frozen so hard as to virtually defy fracturing. To mine the tar sands, it is deemed best to first drain the muskeg for at least two years and then remove the remaining vegetation while it is frozen.[4] Once past this hurdle, one is faced with handling a sticky, abrasive, semiplastic mass at temperatures that may rise into the 90s or plunge as low as $-60°F$ $(-51°C)$. It undoubtedly constitutes the most difficult-to-handle material ever mined on such a large scale. Table 6.1 lists the major technologies under development for recovering bitumen (or syncrude) from tar sands. They will be discussed at greater length in the following pages.

General

To be adaptable to surface mining techniques, tar sand deposits must lie within about 150 ft of the surface. To feed an economically viable extraction plant, it is necessary to ensure a steady delivery of ore on a round-the-clock, year-round basis, despite the fact that at high summer temperatures tar sands stick to everything, clogging cooling systems, external vehicle controls, and conveyors, and the bitumen dissolves natural rubber vehicle tires and conveyor belts.[4] At

*It takes five to seven years to design and build a plant of this sort,[2] which will be described in the following pages.

Table 6.1 Bitumen Recovery Technologies

Mining and surface extraction
   Hot-water separation
   Hot-water extraction
   Direct fluid coking
   Anhydrous solvent extraction
   Cold water extraction
   Sand reduction
   Spherical agglomeration
   Microbial extraction
   Guardian extraction
   Petroleaching

In situ extraction
   Thermal (fireflooding)
      Forward combustion
      Reverse combustion
      Reverse combustion followed by forward combustion
      COFCAW
   Steam injection/in situ combustion
   Emulsion steam drive
   Steam stimulation
   Huff-and-puff
   Continuous steam injection
   Cyclic steam pressurization/depressurization
   Electrical resistance heating
   Atomic fracturing

Heavy-oil extraction
   Sol-Frac
   Steam injection

---

cold, winter working temperatures [down to $-45°F$ ($-43°C$)], the 100-lb teeth of bucket-wheel excavators glow red hot and wear out in one shift.[2] The solutions to these problems appear to be explosive fracturing of the deposits in the summer, redesign of the bucket teeth with abrasion- and heat-resistant alloys, and use of oil-resistant rubber recipes.[4]

Because of the fact that a typical tar sands recovery process (Syncrude) may comprise a train of about 41 in-line plants, operating on a round-the-clock basis,[11] it is essential that precautions be taken to ensure that no process disruptions occur. Redundancy is not economically practical in most cases; therefore, mining and aboveground extraction operations are generally designed with a time-surge coupling to avoid a disastrous domino effect that would shut down the entire facility for a minor failure in an individual stage. In the case of the GCOS process, a 30- to 40-min coupling, in the form of short-term storage bins and surge piles, takes up the slack and offsets mismatching and breakdowns of the individual in-line steps.[3]

The pioneer and present leader in the development of tar sands mining and extraction technology is Great Canadian Oil Sands Ltd. (GCOS), an organization 96 percent owned by Sun Company, Inc. (Philadelphia). Production of synthetic crude oil (SCO)* at the Fort McMurray GCOS plant is now about 50,000 bpd. To produce this much syncrude, GCOS has to remove 33,000 m^3 of overburden and mine and dispose of 100,000 tons of spent tar sand daily—this, in the face of the exceedingly poor handling characteristics of the ore.[4] The overburden ratio at the GCOS site is about 0.4, but open-pit mining of ratios as high as 2.5 to 3.5 has been discussed.[1]

On the basis of capacity, this operation of GCOS is little more than a pilot scale of the Syncrude Canada Ltd. facility,† a new joint project of Imperial Oil Ltd. (31 percent), Canada-Cities Service Ltd. (22 percent), Gulf Oil Canada Ltd. (17 percent), Petro-Canada (the national oil company, 15 percent), Alberta Syncrude Equity (provincial agency, 10 percent), and Ontario Energy Company (provincial agency, 5 percent).[13]‡ This facility, now under construction about 7 km north of the GCOS site, is expected to produce SCO at an initial rate of 52,000 bpd and reach 130,000 bpd by 1981.

Imperial Oil Ltd. (an Exxon subsidiary) has requested approval to build a $4 billion plant of similar capacity and expects to produce 120,000 to 145,000 bpd by 1985. This facility is to be erected at Cold Lake, near the site of a 13-year-old pilot plant which has been producing 5000 bpd of heavy oil by steam injection.[6]

The method adopted for getting specific hydrocarbon deposits out of the ground will depend largely on the depth to the top of the tar sand bed. Table 6.2 shows expected ore recovery rates as a function of depth and extraction technique.

Table 6.2   Recovery Rates for Tar Sands

Method	Depth to Top of Seam, ft	Recovery, % of Ore
Open-pit mining	0–150	85–90+
Underground mining	>300	50–60
In situ extraction	>600	20–30

Source: Reference 3.

*A product that feels like a 5W motor oil and is about 80 percent refined.[2]
†The processes are, however, quite different, as will be seen below.
‡The three government agencies have a 30 percent interest, replacing Atlantic Richfield Canada Ltd. (ARCAN), who pulled out because of escalating costs. One report has the consortium spending $2.5 billion for construction, including the power plant and the 266-mi Edmonton pipeline.

Surface Retorting Processes

The basic GCOS process, which was developed by Clark in 1923, has been in operation on a small scale since the early 1960s. It is a *hot-water separation* process, comprising three major stages: conditioning (mixing or pulping), separation, and scavenging.[1] The tar sands are mined and reduced in size by "ablation." The resulting "pulp" is mixed mechanically, reacted with any required chemicals (pH control agents, wetting agents, organic diluents, etc.), and then heated to process temperature. This *conditioning* stage is primarily a heat-transfer means. Heating is accomplished with open steam introduced in a horizontal rotating drum. Water is added to form a 60 to 85 percent solids pulp at 180 to 200°F (82 to 93°C). The pulp is then screened and mixed with additional water.

The pulp is pumped to the *separation* cell, consisting of two settlers, one on top of the other. The top settler floats the bitumen, while the bottom one settles the sand, most of which is discharged as tailings. The bitumen is removed as froth. The middlings stream from the separation cells is mainly water, with some fines and bitumen particles. Some of this stream is recycled to the conditioning drum. The balance of the middlings, known as the drag stream, is fed to the *scavenger* cells (or froth flotation cells), where the pH is adjusted to 8.0 to 8.5 with monovalent bases.*

The basic separation procedure typically will yield about 2 percent minerals in the bitumen stream and 15 to 20 percent bitumen in the tailings stream. When 50,000 bpcd of crude bitumen is fed through the GCOS coker to recover synthetic crude oil, the nonvolatile hydrocarbon molecules are broken down to distillable products—naphtha, 36 percent; kerosene, 24 percent; and gas oil, 40 percent. The by-product coke and the aromatic and olefinic by-products are used as plant fuel, and the gases are used to manufacture hydrogen. The distillates are then treated with the hydrogen to remove sulfur. The product output of the recovery plant is approximately: 2250 tpcd of coke, 270 ltpcd of sulfur, and 39,000 bpcd of SCO.[1] Approximately 92 to 93 percent recovery of the in-place bitumen is claimed.[4]

The primary differences between the GCOS and Syncrude operations lie in the approach to mining. The former employs large, custom-designed bucket-wheel excavators, hydraulic or bucket-ladder dredges, and supersized drag lines. The Syncrude process uses a larger number of smaller and cheaper units—essentially scrapers, trucks, and power shovels. The large, custom-designed, bucket-wheel excavators used by GCOS have to be funded and ordered far in advance of startup; Syncrude's scrapers are virtually off-the-shelf items and can be paid for over the life of the project.[1] Syncrude also has proposed a somewhat different bitumen upgrading system, shown in Fig. 6.3.

---

*Polyvalent cations tend to flocculate clays and thus raise the middlings consistency in the separation.

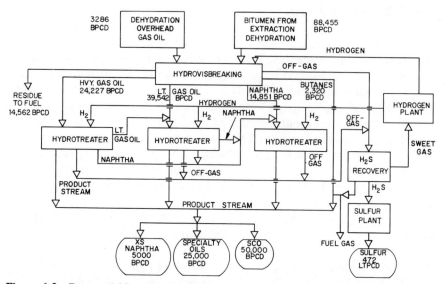

Figure 6.3 Proposed bitumen upgrading system (Syncrude Canada Ltd.). bpcd, bbl per calendar day; ltpcd, long tons per calendar day. (Adapted from reference 1.)

The *hot-water extraction* process (Bureau of Mines) differs from hot-water separation (GCOS) in that a solvent is also used.[1] Fuel oil (20 to 25 volume percent aromatic) is added in the conditioning stage at a weight ratio of 1 part oil to 3 parts bitumen feed. A typical froth composition from Edna, California, feed would be about 73 weight percent bitumen, 6 percent minerals, and 21 percent water. The final product is similar to that obtained from the cold water extraction process to be discussed later. Demonstrated bitumen recovery rates are 96 percent for Vernal, Utah, ore and 90 percent for Sunnyside, Utah, ore. Recovery rates would be expected to fall below 50 percent if polyvalent cations were to be present in the connate water or in wetting agents employed, owing to the formation of a gelatinous slime. The Edna deposits are unusually high in iron and calcium salts in the connate water, but prewashing with hot water reduces the polyvalent cation concentration sufficiently to avoid slime formation.[1]

The *direct fluid coking* process (Mines Branch, Canadian Department of Mines and Technical Surveys) feeds tar sands to a fluid-bed coker or still, heated to 900°F (480°C) with hot, clean sand, and distills off the volatiles. The weight ratio of sand to oil in the coker may be as high as 35.[1] The residuum is then thermally cracked by transferring the coked solids (sands encased in coke) to a fluid-bed burner or regenerator at 1400°F (760°C), to burn off the coke. About 20 to 40 percent of the clean sand is rejected and the balance is recirculated to the coker. Vapors from the coker are condensed in the product receiver, and receiver off-gases are used to fluidize clean, hot sand recycled to the coker. The receiver condensate is largely a heavy crude.

This is a simple, direct process involving technology similar to that of fluid catalytic cracking. Scale-up should pose no problem. High mineral fines have no adverse effect on the process, but the large amount of sands circulated may lead to excessive erosion of equipment, and much thermal energy is lost in dumping clean sand at 1400°F (760°C). Heating tar sands to this temperature requires about 240 Btu per lb, and since the heating value of 1 lb of tar sand (12 percent bitumen) is about 2100 Btu, the process heat loss from this source alone amounts to about 10 percent of the bitumen in the tar sand feed.[1]

The *anhydrous solvent extraction* process (Cities Service Athabasca Ltd.) comprises four major steps: mixing, draining, solvent recovery from solids, and solvent recovery from product. In the *mixer* step, raw tar sand is mixed with light, recycled hydrocarbon solvent and small amounts of bitumen, water, and minerals. The solvent-to-bitumen ratio is about 0.5.

The *drainage* step is a three-stage, countercurrent wash, requiring a 30-min (nominal) settling and drainage time for each stage. Solvent is drained from the bed of sand remaining from each extraction step until the interstitial pore space is emptied. Excess mineral fines may lead to bed plugging.

The first of the final two steps—solvent stripping from solids and from product stream—is the key to the economic viability of the process. In *solvent stripping*, a theoretical solvent recovery of 93 percent is possible but, in practice, up to nearly one-fifth of the solvent may be lost in the process and must be made up from the bitumen product stream. In the worst-case situation, net bitumen recovery could fall to as low as 19 percent.[1] A situation in which only a fifth of the bitumen could be recovered would be economically intolerable.

The *cold-water extraction* process (Mines Branch) actually uses a combination of cold water* and kerosene.† The pH is adjusted to 9.0 to 9.5 by the addition of soda ash.‡ A wetting agent is also added. This mixture is disintegrated in a pebble mill, and the sand that settles out is transferred to a raked classifier. The effluent from the disintegrator, diluted with additional water, is pumped to the thickener together with the overflow of oil and sand from the classifier. A separation is then effected to provide kerosene for recycle. The crude product stream is typically 73 percent oil (1:1 bitumen to kerosene), 25 percent water, and 2 percent minerals.

The *sand reduction* process is similar to the Mines Branch cold water separation process, without the solvent, but it is aimed primarily at providing feed for fluid coking. Tar sand is combined with water§ at 70°F (21°C) and fed to a screw conveyor that transports the pulp to a water-filled settling vessel, where it is screened through an immersed rotary screen. The bitumen agglomerates and is retained on the 20-mesh screen and is withdrawn as oil product. About 80 percent of the sand feed passes through and goes to the waste

---

*About two to three times the weight of tar sands.
†Equal parts by weight of solvent and bitumen.
‡About 1.5 lb per ton of tar sands.
§Ratio of 0.75 to 3.0 tons of water per ton of tar sand. The lower ratio is preferred.

stream. The nominal oil stream composition is 58 weight percent oil (bitumen), 27 percent minerals, and 15 percent water.[1]

The *spherical agglomeration* process (National Research Council, Ottawa) is similar to the sand reduction process. The tar sand is mixed with water and the pulp is ball-milled. Dense agglomerates of bitumen form and are separated for upgrading. Their approximate composition is 75 to 87 weight percent bitumen, 12 to 25 percent sand, and 1 to 5 percent water.[1]

A quite different proposition is *microbial extraction* (University of Western Ontario), a technique that could result in considerable energy conservation over *hot-water extraction.* Microbes (bacteria and fungi) have been known to build on hydrocarbon substrates (e.g., kerosene and crude oil) as primary sources of carbon. About 80 cultures capable of using hydrocarbons as their food have been isolated. Two major effects have been produced: formation of a surface slick of floating bitumen droplets released from the sands, and concentration of the bitumen on sand nodules.[14] Much work remains to be done.

Some thought has been given recently in the United States to technology for extracting tar sands. Athabasca tar sands have been envisioned as the first target, but U.S. tar sands could follow if world oil prices were to rise sufficiently. Using the *Guardian* process (Guardian Chemical Co.) the Athabasca sands should yield 1 bbl per ton of ore, at an estimated cost of $6.00 to $7.50 per bbl. About two years would be required to erect a 100,000-bpd plant upon completion of negotiations.[15]

The Guardian process consists of using a rotating mixer to combine hot water and a small amount of a proprietary chemical with the tar sand.[15] The solid and liquid streams can then be separated by any one of the methods already described. The product from this process is a semiliquid tar requiring transport by hot pipelining or by rail. After cracking, it can be pipelined cold. According to Guardian, GCOS experiences a recovery rate of 85 to 88 percent vs. over 94 percent for the Guardian process. Furthermore, the latter process uses only about one-third the energy of the GCOS process. The leaner Utah deposits would yield only about 0.5 bbl per ton, probably sufficient to be of interest at the present world prices of oil.

Economical extraction of the Venezuelan tar sands is doubtful at this time. Although these deposits are the world's richest (several barrels per ton), they occur at depths down to 20,000 ft. *Petroleaching* is envisioned as one likely extraction technology.[15] Although this method has not been fully described, the name would suggest that it comprises extraction with light hydrocarbon liquids, probably recycled from the upgraded product stream.

In Situ Processes

As noted earlier, approximately 90 percent of Canada's known tar sand deposits lie more than 150 ft below ground level and must therefore be extracted by notably less efficient in situ technologies. Such processes require employing a "driver" substance in the formation to force the bitumen from the injection point to the production well(s). To contain the drive, the overburden must be present in a sufficiently thick and impervious layer to retain the necessary pressure and prevent uncontrolled escape of fluids.[2] This generally implies that the top of the deposit must be at least 600 ft below grade level, although the exact depth required will vary with the composition and geological features of the formation and the extraction technique employed. At depths between open-pit and in situ extraction depths (150 to 600 or 700 ft), there is a region in which deep or underground mining can be employed. The extraction cost will be high, and the recovery rate, using room-and-pillar (pillar-and-stall) technology, is only about 50 to 60 percent at best. About 15 percent of Canada's tar sand resources lie in this depth range, compared to 60 percent where extraction must be by in situ means. In situ technologies are yet unproven on a commercial scale, but much hangs in the balance and considerable development effort is under way. Potential in situ methods are described below.

The *thermal* (or *fireflooding*) process has many variations, but basically it entails igniting the bitumen at the base of a well driven into the deep formation, and sustaining combustion (or partial combustion) by injecting air. The heat reduces viscosity and makes the bitumen more mobile. The formation temperature rises to over 650°F (340°C) and causes cracking of the bitumen to coke and distillates,[2] so upgraded crude is recovered rather than bitumen.[1]

There are two major variations of fireflooding (Fig. 6.4). *Forward combustion* means that the deposit is ignited near the air injection well, and the flame front is propagated forward toward the production (or collection) well(s). Since the hydrocarbons released are driven into a relatively cold zone of the deposit, the viscosity of the recovered liquid is somewhat limited. Burning out of the deposit is complete, however, and the formation is swept clean of bitumen.[1]

In the *reverse combustion* process, the deposit is ignited at the production well and the flame front advances toward the air injection well, counter to the air flow. This process leaves some of the hydrocarbon in the formation; hence the recovery rate is poorer than with forward combustion. On the other hand, the combustion products are forced into the heated zone; therefore, they are in no way viscosity-limited. A wide range of organic components is present.[1] Reverse combustion technology failed in early tests in the United States because of channeling due to the high degree of heterogeneity of the tar sand formation.[4]

Reverse combustion followed by forward combustion was more successful. It required less injected air and proved easier to control. It is estimated that 50 percent recovery of in-place bitumen can be achieved at a process consumption rate of about 10 percent.[4] This work is being carried out by DOE's Laramie Energy Research Center (LERC).

The *COFCAW* process is a combination of forward combustion and waterflooding. It is being developed by Amoco Canada Petroleum Ltd. Four production wells are drilled at the corners of a square and an injection well is drilled at the center. Combustion is initiated at the center well, and air and water are injected to force the liquid bitumen to the production wells. The feasibility of this process has been proven in a 1000-bpd pilot facility using

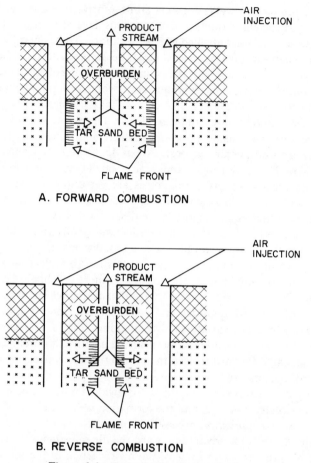

Figure 6.4 Basic fireflooding methods.

Athabasca tar sand feed. Recovery rate for the method has been estimated at 50 percent or more, with consumption of 4 to 5 percent of the bitumen in the process.[4]

A combine comprising AOSTRA (Alberta Oil Sands Technology and Research Authority), BP Canada Ltd., Hudson's Bay Oil and Gas Ltd., and PanCanadian Petroleum Ltd. has been investigating *steam injection with in situ combustion.* It should provide a higher recovery rate than steam injection alone.[4]

Emulsion-steam drive is a relatively low-temperature process in which a "driver" is used to sweep the bitumen toward the production wells. This driver may be a nonionic surfactant or caustic in water or steam or a hydrocarbon diluent. In any case, the bitumen is emulsified for high mobility. Injection wells are located at the four corners and the production well at the center of a 4-acre square. The sweep efficiency using caustic has been estimated at 70 to 100 percent. A 25 to 30 weight percent bitumen-in-water emulsion is recovered. The fuel requirement with a steam driver is about 16 percent of the bitumen recovered. The efficiency will be still less if there are heat losses upward into the overburden or if lower grades of ore are used, requiring the heating of larger amounts of sand.[1]

Steam stimulation in various guises is being looked at quite carefully. Exploitation of deep deposits by this method was first proposed by Imperial Oil Ltd. using feed from their Cold Lake lease. Their process calls for *intermittent steam stimulation*, also known as the *huff-and-puff* process. The "huff" stage comprises injecting steam at 660°F (350°C) and 2000 psi into deep (500-m) wells for a four- to six-week period. The "puff" stage means subsequent pumping out of the bitumen for a period of up to six months. When the output stream becomes merely hot water, the cycle is repeated.[2,4]

Imperial has proposed a 160,000-bpd-capacity facility costing in excess of $4 billion, over half of which would be for a plant to upgrade the bitumen to SCO suitable as feed for existing conventional refineries. Ground breaking is scheduled for 1981, and startup for 1985.

This project would require drilling about 10,000 wells over 20 years, with 1000 to 2500 in operation at any one time. The production life of a given well would be about five to six years or 9 to 10 huff-and-puff cycles. Energy for steam production in a huff-and-puff facility of this size would amount to about half the output of the current GCOS plant, but SCO production cost is expected to be even less than for the new Syncrude facility.[4] Norcen Energy Resources Ltd. (Cold Lake) is investigating a process similar to Imperial's huff-and-puff.

Murphy Oil Ltd. is evaluating a *continuous steam injection* process, and under AOSTRA sponsorship, Shell Canada Resources Ltd. (Peace River) is developing a *cyclic steam pressurization/depressurization* process. The Peace River deposits are unique in that they are situated atop a water-saturated zone, which provides a channel for the steam-driven bitumen between the wells. To use the same technology in other tar sand formations would require development of an acceptable means for interconnecting the wells underground in the absence of the Peace River's water-saturated layer. Numac Oil and Gas Ltd. is investigating *hydraulic fracturing* technology for tar sands.[4]

Petro-Canada Exploration Ltd., Canada–Cities Service, and Imperial are studying *electrical resistance heating* as a means of freeing bitumen from the sands. An alternating current is applied to raise the temperature enough to extract the bitumen with a steam flush. Such a method should require no fracturing of the formation.[4] *Atomic fracturing* has been proposed but is untried as yet. The original intent was to detonate a 9-kt nuclear device at a depth of 1250 ft, creating a 230-ft-diameter cavity which would contain up to 50 percent of the atomic energy as useful thermal energy for driving the oil.[1]

Heavy-Oil Extraction

As for the very heavy oil found at the confluence of the Kansas, Missouri, and Oklahoma borders, it is not properly designated either bitumen (from tar sands) or shale oil. It is mentioned here only because it is a fluid, fossil hydrocarbon, unpumpable and otherwise unclassified. The Bureau of Mines has developed a potential process—*Sol-Frac*—for extracting it. This comprises explosive fracturing (nonnuclear) and solvent extraction.[16] Nothing is known as to recovery rates or economics because the field has not yet been commercially exploited and no pilot plant has been constructed.

Steam injection is another extraction technique under investigation for heavy-oil recovery.

ENVIRONMENTAL AND SOCIAL FACTORS

Development of tar sands as a major energy source brings to light some imposing environmental and social problems. At the Fort McMurray location of GCOS, the population of the area rose from 1000 to 28,000 between 1964 and 1978. Despite the existence of much open land, the lack of central planning with the involvement of local residents led to severe overcrowding in town, and sanitation, educational, and other facilities became overtaxed. Prices became exorbitant as greedy speculators sought to make a killing. Divisions arose between the old and the new residents. In short, everything seemed to lead to a decline in the quality of life. It is quite apparent that a major need in future communities associated with tar sands development will be dispersal of population.[4] Generally speaking, the social problems are similar to those encountered in the open-pit mining of western U.S. coal (see Chapter 4).

From the viewpoint of environmentalists, the major concern appears to be with the tailings ponds. They can become loaded, in short order, with toxic organic chemicals and metals, and become covered with a thin surface film of bitumen. Since these ponds will be among the larger bodies of water in the area,* their appearance along a major flyway of migratory waterfowl could prove disastrous.† It should be noted that bitumen is particularly toxic to fowl. Spillover or leakage into surface streams could also cause severe pollution of public water supplies. Even in situ extraction processes might poison underground aquifers.[4]

Earth-fill dams up to 300 ft high have been constructed to contain these waters, but the available soil has very poor load-bearing properties.[3] Even if the waste process water can be retained adequately, the entire operation poses yet another serious water problem—a potential water shortage for processing purposes. Effort is going on toward determining the possibility of water reuse.[4]

Still another environmental problem involves the swell factor resulting from fluffing up of the solids during mining and extraction. Even with removal of 10 to 12 weight percent bitumen, the swell factor of the mineral component is such that overburden and tailings from an open-pit mining area will fluff up to 70[3] to

---

*Syncrude will ultimately mine an area of 2800 hectares (6900 acres) and require a tailings pond covering 3000 hectares (7400 acres).[4]

†The Syncrude pond is especially poorly located from the standpoint of migratory waterfowl.

100^9 ft above the original grade level. Local regulations prevent placing such soil outside the operator's lease area and, of course, it is impractical to spread it on top of recoverable reserves.[3] Again, the environmental aspects of extracting bitumen from tar sands are at least similar, in many respects, to those associated with western coal.

Upgrading

These various extraction processes yield a wide variety of hydrocarbon products ranging from gases (by in situ extraction) through distillable products (naphtha, kerosene, and gas oil by the GCOS process) to bitumen (by most extraction processes) and semiliquid tars unsuitable for cold pipelining (Guardian process). Many of these products must be upgraded before being fed to conventional crude oil refineries. Tar sand bitumen, the usual raw product, contains large quantities of asphaltenes, polynuclear aromatics linked with alkyl chains. These normally have high molecular weights and low $H:C$ ratios and contain rather high concentrations of vanadium, nickel, and other metals. Such characteristics have prevented these bitumens from being processed catalytically, because of catalyst poisoning and excessive hydrogen consumption.[17]

Upgrading usually entails thermal cracking and distillation at temperatures in excess of 900°F (480°C), a procedure that generally consumes about one-fourth of the energy value of the product. Considerable hydrogen may also be consumed in side reactions produced at these process conditions.

Chiyoda Chemical Engineering and Construction Company has developed a new catalytic process for upgrading these high-asphaltene bitumens without poisoning their proprietary catalysts. A single pass of asphaltenes with molecular weights of 1000 to 20,000 yields a product similar to solvent-deasphalted oil. The aromatic character of the asphaltenes is virtually unchanged, but the alkyl chains linking the aromatic condensed rings are selectively broken to produce oil-soluble, low-molecular-weight fractions, with the $H:C$ ratio virtually unchanged. The sulfur and vanadium contents are reduced in the process.[17]

ECONOMICS

There are those who believe that tar sands could compete with OPEC oil at prices paid in 1977.[18] Most others acknowledge the necessity for government support to make ends meet.[19]* The main reason for the marginal economics of a tar sands extraction operation is the high capital investment required. The Syncrude plant has been on stream since 1978 following a capital outlay of about

---

*In accordance with an agreement between Syncrude and the Canadian government, the latter payed an import compensation to a domestic refiner or users purchasing SCO from Syncrude at the world price of conventional oil (about $15.50 per bbl in Canada in the first quarter of 1978), despite the price of $11.75 per bbl current then for conventional Canadian crude.

$2.5 billion, including power station, pipeline, and extraction facility. Their estimated cost of producing SCO is about $9.50 per bbl (including operating expenses, depreciation, and continuing capital expenditures), at an average production rate of 125,000 bpd.[4]

The capital cost for Syncrude then becomes $20,000 per bpd capacity (vs. $6000 for GCOS).[13] Other plants may run as high as $24,000 to $32,000* compared to about $325 to $350 per bpd capacity for pumping conventional Saudi Arabian crude out of the ground.[4,13] Product from GCOS sold initially for $12.30 per bbl and went up to $13.30 in July 1978.[4]

A tar sands plant of 200,000-bpd capacity would require about 30,000 worker-years to build, placing tremendous pressures on available skilled labor resources. Canadian tar sand resources could possibly support 15 such plants for a period of 25 years, but the most likely production rate for the year 2000 is estimated to be only 800,000 bpd.[20]†

Another factor affecting the overall cost of liquid fuels from tar sands is the character of the products formed. Some processes provide a crude bitumen, some a semiliquid, tarlike product, and others an essentially 80-percent-refined crude ready to mix with other refined petroleum products or to feed to a conventional oil refinery. The necessity of hot pipelining a semiliquid tar to the refinery can add materially to the end costs. On-site cracking to provide products suitable for cold pipelining will also entail additional capital outlays. These are significant economic considerations that cannot be ignored.

The uncertainties of rising costs, world oil prices, high taxes, and stifling governmental regulations, combined with the sheer magnitude of operations, have led more than one potential investor to back out. It has become increasingly evident that government support, or at least relaxation of some government regulations, will be necessary to ensure the continued development of tar sands technology. There can be no doubt that projections of future world petroleum prices will play a dominant role in the direction government policies will take. The hazard is that bureaucratic delay in responding to such indicators will stretch out interminably the action necessary to get the needed policies into operation. And delay seems to go hand in hand with escalating capital costs.

Oil Shale

Most people look on oil shale and tar sands extraction and conversion technologies as something new. In a sense they are right, inasmuch as the commercial development of these fossil hydrocarbons has, as yet, made little or no impact on the energy scene. A note appeared in an American Chemical Society (ACS)

---

*Shell Canada Resources Ltd. facility, proposed for production in the second or third quarter of 1985.

†It should be noted that the Syncrude process will consume 18,000 bpd, cutting the net output 14 percent from 130,000 bpd to 112,000 bpd.[13]

publication in 1927, which read: "Shale oil obtained in the operation of the Federal Government's experimental oil shale plant near Rulison, Colorado, is now available for distribution by the Bureau of Mines to laboratories that might be interested in conducting tests with such oils."[21] An announcement in another ACS publication appeared 50 years later. It stated that "products from shale oil may be available as early as 1981."[22] The progress in half a century has been virtually nil, except for the refinement of some experimental processes and the issuance of hundreds of patents largely related to extraction technology.

In 1944, Congress passed the Synthetic Liquid Fuels Act, to encourage development of synthetic or substitute fluid fuels, since discovery rates of petroleum and natural gas were declining in the face of booming consumption. This was followed by a brief flurry in synthetic liquid fuels activity, which came to a rather abrupt halt by 1953, with the realization that the Middle East oil finds were significant, and the belief that they would provide "all the oil that modern ingenuity could consume." Now, only a quarter century later, with a global decline again in natural gas and petroleum discoveries relative to consumption, concern with alternative sources of fluid fuels has finally become essential. History has always had its dreamers, so interest in coal, oil shale, and tar sand conversion has never come to a complete standstill, but domestic commercialization has never been realized on more than a local scale.

The time finally seems to have arrived for the development of oil shale and tar sand conversion industries, as well as coal conversion. The prospects of a new Middle East–scale energy bonanza are virtually nil, unless one believes in real long shots, such as geopressured* natural gas reservoirs. Nonetheless, substitute hydrocarbon fluids from fossil fuels should not yet be considered a sure thing commercially. There are many problems to be overcome. These problems will be the subjects of many of the following pages.

DESCRIPTION AND OCCURRENCE

Mineral formations, especially sedimentary deposits, contain many solid hydrocarbons, including tar, asphalt, wax, bitumen, and kerogen. Oil shale is neither "oil" nor "shale" but a combination of *kerogen*,[23] a complex organic compound of indefinite composition, and marlstone rock.† The kerogen, a cross-linked, high-molecular-weight, combustible organic material, decomposes to a crude oil on heating above 900°F (480°C).[25] The sedimentary rock of oil shale ranges in geologic age from Recent to Precambrian. It occurs in many parts of the world, where it may be known variously as *tasmanite* (an impure coal), *tripolite* (a diatomaceous earth), *kukersite* (algal-based Ordovician shale), *ichthyol* (bituminous schist), or *torbanite* (kerosene shale).[23]

---

*Also called geopressurized natural gas. "Geopressured" appears to have become the more fashionable term.

†Small amounts of soluble bitumen may be present in some "shales," but insoluble kerogen (molecular weight 1000 to 10,000 plus) is the common hydrocarbon.[24]

The kerogen was formed by slow decomposition under anaerobic conditions of aquatic organisms, spores, pollen grains, and vegetable matter. It is the valuable portion of oil shale. It was formed under relatively low geologic temperatures, hence might be considered an embryonic petroleum. There are basically three types of kerogen: (1) *humic*—derived from higher organic materials, which generally yield aromatic hydrocarbons; (2) *algal*—derived from algae, which are normally aliphatic; and (3) *mixed*—which is a combination of the two. Kerogen has proven difficult to analyze, in part because it is insoluble, relatively unreactive with most reagents, and degrades to yield complex secondary products.[26] Despite its uncertain and variable composition, it might be generalized as approximately 75 percent carbon, 15 percent oxygen, and 10 percent hydrogen. It also contains lesser, but significant amounts of sulfur (up to nearly 9 percent in some grades) and nitrogen (up to 3 percent).[23]

The distribution of kerogen in the inorganic matrix is basically interparticle. Only a small amount is either directly or chemically bonded to the matrix. Porphyrins in the kerogen may be chelated with some of the associated inorganics. Common kerogen is a highly naphthenic (alicyclic) polymer with associated, randomly distributed aromatic, and nitrogen and sulfur heterocyclic ring systems. The typical molecule resembles coal in that there is definite evidence of a benzenoid structure, although it is less pronounced than in the case of coal.[23]

The heating value of the richer shales deposited in our West during the Eocene epoch (50 million years ago) ranges around 9500 to 9600 cal per gram (based on organic content).[24] These western shales are concentrated in the Green River Formation, covering about 16 million acres. The commercial quality ores, defined as yielding at least 25 gpt (gal syncrude per ton ore) and occurring in beds at least 10 ft thick, underlie a total of about 11 million acres. See Fig. 6.5 for the geographical range of these shales. It should be noted that nearly all are found within a 125-mile radius of the juncture of Wyoming, Utah, and Colorado.

Available data on the distribution of oil shales show a wide disparity in estimates of recoverable oil and the effect of deposit richness on recoverability. In the Green River Formation, localized deposits may range in richness between 7 and 77 gpt, based on the Fischer assay [retorting at 932°F (500°C)].[27]

By far the richest deposits in the United States lie in the Piceance Creek Basin in Colorado, over an area of 960,000 acres. They are covered by 1000 ft of overburden, but they are about 2000 ft thick and average 25 gpt. The kerogen here is associated with *nahcolite* ($NaHCO_3$) and *dawsonite* [$NaAlCO_3(OH)_2$], potential sources of soda ash and aluminum respectively. In the Green River Basin of Wyoming, *trona* ($Na_2CO_3 \cdot NaHCO_3 \cdot 2H_2O$) is interbedded with the oil shale. It is mined for soda ash.[23] These three minerals contribute to the natural alkalinity of oil shale.*

Less-organic-rich shales formed during the earlier Devonian period (350 million years ago) and in the Mississippian epoch underlie a large portion of the

---

*Spent shale from the TOSCO process has a pH of about 9.[28]

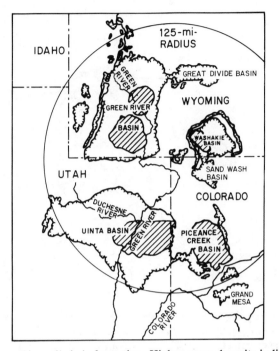

Figure 6.5 Green River oil shale formation. Higher-assay deposits indicated by cross-hatching.

eastern U.S. land area in at least 28 states.[23] They stretch from Texas to New York State, and from Alabama to the Canadian border, and cover an area of 160 million acres (of the total U.S. land area of 2270 million acres). Ninety-eight percent of these reserves occur in Ohio, Kentucky, Tennessee, and Indiana.[29] The geographical range of eastern shales is shown in Fig. 6.6. These deposits, although extensive, are not as well defined or as rich in organic materials as the Eocene shales of the west. This eastern shale has a heating value (based on the organic content) of about 8500 to 8900 cal per gram.[31] Eastern shales are lower in carbonate minerals than western shales (typically 0.5 weight percent vs. 15 percent, respectively),[32] hence should pose a less severe environmental problem.

The eastern shales are lower in hydrogen content and convert to trapped coke under conditions of the standard Fischer assay. Potential hydrocarbon yields, as extracted by normal retorting technology, are therefore indicated as low;[32] hence these ores have generally been considered to be economically impractical to exploit. Recent developments, however, indicate that this is probably not so. Two technologies for exploiting the Antrim shales of lower Michigan and the other Devonian Basin deposits appear to be within our grasp. These will be discussed in subsequent pages.

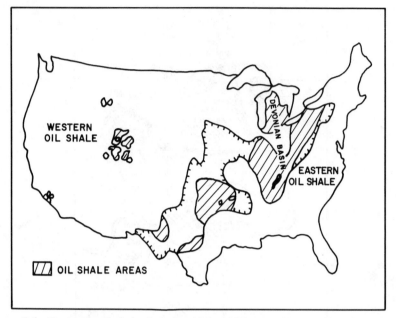

Figure 6.6 Domestic oil shale deposits. (Adapted from reference 30; reprinted with permission of the copyright owner, The American Chemical Society.)

RESOURCES AND RESERVES

Oil shale deposits are found in large quantities in the lower 48 states and in Alaska, but the bulk of the richer deposits (those assaying over 15 gpt) are found in the Green River Formation.[28] It is difficult to arrive at meaningful reserves and resources figures for shale oil, because only this formation has been probed to a significant extent, and these deposits have not yet been extracted on a commercial scale. The economics of recovery are paramount in determining proved reserves, yet they are heavily dependent on many imponderables: geographical location, depth below grade, recovery rate (in turn, dependent on the particular extraction method chosen), assay, deposit thickness, water availability, accessibility of transportation, labor supply, and the ever-present environmental and political factors. The Devonian and Mississippian deposits of the central and eastern United States are even more questionable, as they are so widely scattered, little explored, and comprise lean ores that do not yield to conventional assay methods.

The premium-grade deposits—the ones most commercially exploitable—exist primarily in the Piceance Creek Basin of Colorado (83 percent), with the balance nearly equally distributed between the Uinta Basin of Utah (9 percent), and the Green River, Washakie, and Sand Wash* basins of Wyoming

*Reference 34 reported the Sand Wash Basin as lying in Wyoming. Actually, it is primarily in Colorado, south of the Washakie Basin.

(8 percent).[34] All are in the Green River Formation. Recovery rates of 50 percent may be attained on a localized basis, but this is probably too high for an average recovery rate. Most of these reserves exist on federally leased lands (78 percent), in beds 30 or more feet thick and less than 1500 ft below grade,* and comprise ores assaying at 30 gpt. These deposits represent about 170 billion boe of recoverable shale oil. If lower-yield (average 25 gpt), thinner deposits (minimum 10 ft) are included, the proved recoverable reserves rise to 730 billion boe—a quantity twice as great as the entire Middle East petroleum reserves.[34]

More generally, the estimates of proved recoverable reserves of U.S. shale oil have ranged from 34 billion to over 1000 billion boe, depending on the assumptions made relative to geographical and geological accessibility, deposit thickness, assay, recovery rate, and so forth.[35] Unfortunately, the poorer the prospects of commercial recovery from a given grade of ore (less well defined, less accessible, and low-assay deposits), the larger our shale oil resources appear to be. Poorly defined, 15-gpt deposits in the Piceance Creek Basin contain three-fourths of that Basin's resources, which, in turn, are probably two-thirds of our total oil shale resources.[36]† The most credible figure for proved recoverable reserves is probably in the range 74 to 80 billion bbl. This compares with an FEA (1974) figure for proved recoverable reserves of domestic petroleum, amounting to 35.3 billion bbl. All but the most pessimistic and limiting estimates show shale oil output prospects to exceed conventional petroleum in the long haul.

The National Petroleum Council (NPC) assessed the current shale technology and economics, and projected little near-term production of shale oil. The Council felt that near-term prospects were much more promising for the Athabasca tar sands than for U.S. shale oil, from the standpoint of both accessibility and processability.[36]

One U.S. Geological Survey (USGS) report gave very high U.S. shale oil resource estimates (26 trillion bbl), but this included low-yield shales to a depth of 20,000 ft, hardly a promising commercial prospect. High-yield shale oil resources were estimated at 1300 billion bbl,[37] more in keeping with an earlier estimate of Averitt, also of USGS (1140 billion bbl).[38] Reference 34 suggests our in-place reserves may exceed 2000 billion boe in the Green River Formation alone.

Variations in the assessment of oil and gas potentials of the Devonian and Mississippian shales are extremely wide. Strippable deposits in Indiana and Kentucky alone are estimated to contain 18.6 billion boe—equivalent to about 130 tcf‡ on gasification. The methane yield of the entire Devonian and Mississippian proved recoverable reserves have been placed at 8000 to 16,000 tcf.[39] The Brooks Range in Alaska appears to have very rich (130-gpt assay), thin (4 ft) deposits of *tasmanite* from the Mesozoic era, but the in-place reserves have not been fully assessed.[23]

---

*Most U.S. coal mines are considerably shallower.
†These leaner ores add to the economic and environmental problems by calling for moving and heating more ore and disposing of larger volumes of spent shale for a given yield of kerogen.
‡Or approximately 130 quads of energy.

One thing is certain—the United States has the world's largest known reserves of oil shale.* It is quite apparent that no one really knows the true assessment of this resource. Part of the trouble is that recoverability is so intimately tied to assay and accessibility of the ore, as well as the extraction technology employed to recover the hydrocarbon, but more important, to the existing socioeconomic and political climates.

The eastern and Alaskan oil shale resources also have not been fully assessed. Some sources claim that the former alone represent as little as 2 percent of U.S. oil shale reserves,[27] whereas others evaluate them at 2 trillion bbl, about the same as the Green River reserves.[33]

Shale Gas

One "exotic" or unconventional source of fuel/feedstocks that has been probed only superficially, is natural gas in the Devonian (brown) shale formation. The deposits cover an area of about 163,000 mi^2 in 13 eastern states—the same general area in which oil shale occurs. The gas-bearing formation ranges in thickness from a few feet to over 8000 ft.[40] There is a wide range of opinions as to the magnitude of this resource. The Department of Energy believes that a "conservative estimate" of the gas in place would be 2400 tcf with 10 percent (240 tcf or quads) potentially recoverable.[41] The Office of Technology Assessment (OTA) sees it differently. It claims only about 1 tcf per year will be recoverable—for a period of about 15 to 25 years (5 percent of our current annual consumption of natural gas).[40]

At the present time, there are approximately 10,000 gas wells in the brown shale formation. About 69,000 new wells would be necessary to maintain output at 1 tcf per year.[40] Small-scale production has been going on for over a century, but the production rate has been low because these shales yield only a very slow flow rate (200,000 cfd vs. 2 to 5 $\times$ 10^6 cfd for normal sandstone natural gas wells). Technology must be developed for less expensive well drilling methods and for more effective fracturing of the formation. The current development of this resource through the offices of the Eastern Gas Shales Project (DOE), has been under way only since about January 1976.[41]

"New" interstate gas sold for about $1.46 per 1000 scf at the end of 1977. It is felt that a price of $2 to $3 per 1000 scf would be necessary to stimulate Devonian shale gas development sufficiently. At this writing, the price of interstate gas is about $2.60, so it remains to be seen how the shale gas project will develop.

EXTRACTION TECHNOLOGY

Recovery of syncrude from oil shale comprises, basically, opening up the ore bed (fracturing in place or mining and crushing) and applying heat to the ore to

*Virtually all North American deposits are within our borders.

release the organic material from the mineral matrix. The following pages will describe the major proposals in greater detail.

Table 6.3 makes a broad-brush comparison of the three basic technologies: open-pit mining, underground room-and-pillar mining with surface retorting, and in situ fracturing and extraction. The option of the basic extraction technique for any given oil shale deposit hinges primarily on the depth of the ore bed below ground surface, the bed thickness, and the assay of the deposit. A firm of mining engineers has come up with a set of specific guidelines for Devonian shale: (1) do not mine if the overburden is over 200 ft thick or the seam is less than 10 ft thick, (2) do not mine if the overburden ratio is greater than 2.5, and (3) do not mine shale containing less than 10 percent organic carbon.[29] Similar guidelines, but with different trigger numbers, would have to be applied to other ores, depending on depth, assay, thickness, extraction process, accessibility, and so forth.

A major problem in dealing with oil shale is that (on average) a ton of shale yields less than a barrel of oil,[28] hence massive quantities of spent shale must be disposed of—quantities that are about 30 percent greater in volume than their in-place volume due to "fluffing"* and are highly alkaline—making land reclamation and protection of aquifers extremely difficult. Some of the related oil shale statistics are shown in Table 6.4.

Since the recoverable shale oil in the Piceance Basin has been estimated at 20 billion bbl, this means that over the lifetime of this resource, 33.4 billion tonnes (36.8 billion tons) or 14.6 billion m^3 (515.5 billion ft^3 or 19.1 billion yd^3) of spent shale must be disposed of, and one-fourth of this residue will be highly alkaline oxides (mainly calcium and magnesium), a fact that does not bode well

Table 6.3 Comparison of Shale Oil Recovery Methods

			Oil Shale Resource	
			Oil, 10^9 bbl	
Recovery Method	Retorting	Grade, gpt	In Place	Recoverable
Underground mining, room-and-pillar[a]	Surface[a]	30	100	54
Open-pit mining	Surface	20	700	660
Mining rubblization[b]	In situ[b]	20	700	350

Source: Reference 42.

[a]Capital cost $1.0 to $1.5 billion per 100,000 bpd output.
[b]Capital cost $0.6 billion per 100,000 bpd output.

*Expansion of ore and overburden during mining and processing.

Table 6.4   Shale Oil Extraction Statistics (Estimated)

U.S. oil consumption rate (1978)	19.4×10^6 bpd
Assuming shale oil production rate	1.0×10^6 bpd
Mine requirements	
Number	20
Capacity (each)	62,500 tpd
Water requirements	160,000 acre-ft/yr
	or 7×10^9 ft^3/yr
Shale mined	
Weight	1.25×10^6 tpd
Volume	525,000 m^3/da
	or 18.5×10^6 ft^3/da
	or 685,000 yd^3/da

Source: Scaled up from data in reference 43.

for the aquifers or the soil.[7] The matter of process water will be considered in greater detail later in the chapter.

Surface Retorting Methods

The major extraction technologies call for retorting the raw shale in surface facilities. Three methods are available to provide heat for surface retorting: (1) residual coke left on the spent shale can be burned, (2) externally produced hot gases can be circulated through the retort, or (3) hot solids recycled through the system can be used as the heat-transfer medium. At least 10 different variations of these three processes are presently undergoing development.[27]

The earliest surface retorting extraction of shale oil on a large commercial scale (up to 780,000 bbl in 1952), took place in Scotland (Scottish Oil Ltd.) and continued from 1860 to 1963, when the high-grade reserves finally played out.[23] The extraction was accomplished by the *Pumpherston* process, in which the shale gases, after scrubbing, are burned to provide the heat for retorting. The waste heat goes to a boiler where it generates steam, which is fed into the bottom of the retort to sweep the oil out to the condenser.

Three more advanced retorting methods have been developed and tested on a scale large enough to give some indication of their commercial potential. They are the TOSCO-II, the gas-combustion retorting, and the Union Oil counter-current internal combustion technologies.

The *TOSCO-II* (The Oil Shale Corp.) process is characterized by the co-current flow of externally heated ceramic balls and cold crushed ore in a rotary-kiln retort. The balls, which are heated in an external heat generator fueled by residual carbon retained on the spent shale, serves as the heat-transfer medium. The process affords a high throughput rate without the need for massive equipment. The off-gas has a high heating value since it is not diluted with combustion products.[23] It is, however, a rather complex system, and requires considerable process water to limit dusting of the spent shale and to condense the liquid products.[44]

The *gas-combustion* process (Bureau of Mines) has been demonstrated in several pilot plants since the late 1940s. The largest unit processed 150 tpd. The crushed oil shale is fed into the top of a vertical retort and falls (countercurrent to the retorting gases) successively through four zones: preheating (and product cooling), retorting, combustion, and cooling (or heat recovery). The raw product gas is cooled by the incoming shale feed, condensing the oil as a mist. The vapor stream exits into a demister to separate out the oil. Some of the low-heating-value (100 Btu per scf) product gas is recycled as fuel for the retort. The thermal efficiency and throughput of the process are high, and oil yields are good. An added bonus is the fact that no process water is needed to condense the oil stream.[23,44]

The *countercurrent internal-combustion* process (Union Oil Co.) feeds the crushed shale upward, by means of a rock pump, through a funnel-shaped retorting vessel, from which the spent shale spills over to a discharge chute. The flame front, heated by organic residues in the spent shale and air fed from the top, progresses downward, counter to the ore feed, and fluid products, condensed by the cool shale feed, flow downward and are collected from near the bottom of the cone of the retort. The retort accepts virtually any size ore and processes it efficiently without requiring coolant water.[44]

A batch-feed modification of this retort, known as the *N–T–U* process (for Nevada–Texas–Utah), operates in similar fashion. A large unit was built and used by the Bureau of Mines in the late 1940s.[23]

The *Petrosix* process (Petrobrás—The Brazilian National Oil Co.) injects externally heated recycle gases into the bottom of a vertical retort while shale feed moves by gravity through feed preheat (product cooling), retorting, and heat-recovery zones. Oil is carried off at the top as a gas-borne mist stream, which is then separated in cyclones and electrostatic precipitators. External heating of the retorting gas eliminates clinker formation by permitting good temperature control.[44] The product gas has high heating value and is suitable for petrochemical manufacture.[23] Aside from the external gas heating, the process is similar to the gas combustion process.[44]

The *refractory heat carrier* process (U.S.S.R.) employs hot spent shale ash as the heat-transfer medium to heat the crushed shale feed. The shale is pyrolyzed in a semicoker at 660 to 970°F (350 to 520°C). Indications are that the equipment can be scaled up to handle feeds in excess of 1000 tpd.[23]

The IGT *hydrotreatment* process hydrotreats the kerogen in the retort, resulting in a yield of up to 115 percent of the Fischer assay. Using Green River shales, typically 95 percent conversion of the kerogen in place can be achieved.[30] By varying the operating conditions, production of oil boiling below 730°F (388°C) can be optimized [using a middle-zone retort temperature below 1200°F (650°C)], or gas production can be favored (at a retort temperature above 1200°F). High temperatures generally favor a wider range of product compositions.

Perhaps the process of greatest potential interest to environmentalists is the *multimineral* process (Superior Oil Co.). It reportedly integrates room-and-pillar mining, crushing to extract naturally occurring nahcolite (sodium bicarbonate), surface retorting in a closed-system retort to extract oil, leaching of spent shale to produce aluminum hydroxide or alumina and soda ash, and returning the dewatered spent shale to the mine. Extraction of the three inorganic chemicals makes this latter method of waste disposal possible, because the volume of the spent shale is reduced in the process by 50 percent rather than being increased by about one-third due to fluffing, as in more conventional processes.[27] It may be necessary to select ores of appropriately high mineral content to make this method function well and economically.[33]

There are three variations of the *Paraho kiln* method of processing oil shale. The first employs product gas to heat the retort internally. The second introduces product gas heated externally, to combust the shale. The third method feeds fuel and char to an external heater and employs the hot solids to generate heat for the pyrolyzer.

Surface retorting of an average product mix of rich (10 to 70 gpt) Green River Formation shales has typically yielded product mixes of about 70 weight percent oil, 10 percent gases and light oils, and 20 percent coke. Yields from most potential commercial processes range from 75 to 100 percent of the Fischer assay.[27] Although oil is the major product, the gases and coke cannot be ignored in the economic analysis. By appropriate alterations in the operating parameters, it may be possible to control the product mix as desired.

In Situ Methods

The need to minimize water consumed in conversion, washing spent shale, lowering alkalinity, and laying dust would suggest that in situ processing could provide a viable answer. The presumed lower yields, however, seem to more than offset the advantages, but in situ retorting has not yet been carried out on a large enough scale to provide a high degree of certainty as to recovery rates.[27]

In situ retorting is considered most effective in extracting oil from deep deposits that would be too expensive to mine by room-and-pillar methods. Conventionally, wells would be drilled and the ore formation fractured to allow air or hot gases to circulate through the rubble. Retorting would then be carried out in much the same manner as underground coal gasification (see Chapter 5).

Alternatively, *modified in situ retorting* (also known as *stoke mining*) calls for 15 to 25 percent of the shale to be mined out to form an underground room, and the mined ore is retorted above ground. The balance is "rubblized" with explosives, allowing the room roof to collapse. The rubblized section is then sealed, and ignition is started in the loose ore. Air and steam are injected at ceiling level and the oil formed from the kerogen flows into sumps in the chamber floor and is drawn off. Unless fracturing is carried out with care, the flame front may advance through an overly porous channel, and conversion in the bed as a whole will be incomplete. Tests show that 60 to 70 percent conversion can probably be achieved by modified in situ processing, a figure that is comparable to surface retorting when all energy expenditures are considered.[33]

The in situ concept has not yet been developed on a commercial scale. Occidental Petroleum Co. and Ashland Oil, Inc., have begun construction of a $440 million, 57,000-bpd commercial facility of this type, which is scheduled for completion late in 1983. Production costs have been estimated at $8 to $11 per bbl, suggesting that the operation should be economically viable. This modified in situ process is best carried out in thick deposits relatively close to the surface. Various other companies are also investigating in situ methods, in hopes that one technique will be found that will prove to be universally applicable to all western shale deposits.

A modified in situ process designed to produce 1 million bpd of shale oil would require the handling of about 200 million tpy of shale, compared with up to 1 billion tons of material to be handled in an equivalent surface retorting process. The latter amount is thought to be the upper limit of shale that can be handled each year in light of the nation's other mining needs. (About 600 million tons of coal were produced domestically in 1977.) The modified in situ process would require only a third of the labor force called for in a deep mining and surface retorting operation, and would require only 25 to 33 percent of the water, perhaps the strongest argument for this technique. Furthermore, spent shale disposal would be reduced by at least 80 percent, and more if Superior Oil's method of recovering nahcolite ($NaHCO_3$), alumina (Al_2O_3) or aluminum hydroxide [$Al(OH)_3$], and soda ash (Na_2CO_3) from the mined fraction proves feasible.[33] It should be noted, however, that unlike conventional in situ processing, *stoke mining* requires an aboveground retorting facility to extract oil from the 20 percent ore fraction that is mined, thus adding to capital costs.

Yet another in situ extraction method is *leaching*, in which very hot water is used to leach $NaHCO_3$ from the mineral matrix. Superheated steam is then injected and passes through the voids left by the leached bicarbonate. The kerogen is retorted by the steam without the need to move rock.

Aluminum, as mixed carbonates with sodium, may be extractable from some deposits. Sufficient heat may even be developed to produce alumina (Al_2O_3) or even elemental aluminum. Sulfur, sodium bicarbonate, and alumina would normally be yielded as by-products.[18] This method has the advantage that it avoids fluffing of the spent shale, recovers commercially valuable byproducts, and leaves the spent bed free of inorganic salts that might otherwise be leached into aquifers.

Although little work has been done on *bacterial treatment* of oil shale, it is known that sulfur-oxidizing bacteria (*thiobacillus*) will produce sulfuric acid from organic sulfur. This acid should disintegrate the mineral matrix and increase shale porosity. It is also possible that the kerogen would be degraded to lower-molecular-weight, more useful forms.[45] It has been reasoned that injection of a thiobacillus into oil shale might break the bonds between kerogen and the mineral component and result in the formation of a mixture of an acidic-oil gunk and minerals. The hydrocarbon could then be separated more readily with solvents or water, at moderate temperatures, and processed to a syncrude.[46]

Extraction of Devonian Shale

As noted earlier, eastern shales, when heated at 932°F (500°C), in accordance with the Fischer assay conditions, produce low yields (less than 15 gpt vs. 30 to 50 gpt for many Green River deposits). Actually, it has been found that the organic contents of the two shales are comparable (about 13.5 percent as carbon),[32] but Devonian shale is lower in kerogen. Hydrogen-rich kerogen (such as the Piceance Creek deposits) converts readily to oil on thermal retorting, but hydrogen-lean kerogen (Devonian) turns to coke which remains locked in the mineral matrix.[33]

The relatively thin Antrim beds in Michigan, although deep (about 1200 to 1440 ft),[29] can apparently be extracted efficiently by *in situ combustion with air injection*. Yields of oil and gas appear to be comparable, on an overall energy basis, to those of western shales, but fracturing technology must be further developed to make the process economically viable.[33]

A second conversion technology, suitable for the relatively shallow, but thick, hydrogen-lean shale beds, involves strip mining followed by the IGT-developed surface *hydroretorting* method described earlier. Coking is reduced and the output of medium-boiling fluids is increased, so that less refining is required. The actual yield is 2½ times that predicted by the Fischer assay. A price projection of $15 to $20 per bbl has been made.[33]

Refining

Figure 6.7 illustrates a shale oil refinery operation based on *hydrofining*.* The conversion of shale-based SCO to useful fuels is easier than coal liquefaction, because the net H:C ratio of shale oil is closer to the desired fuels than is coal, but once the raw shale oil is produced, problems are not at an end. The kerogen-based synthetic crude (SCO) is not the same as petroleum-based crude. For one thing, SCO contains as much as 10 times the amount of nitrogen found in conventional crude (about 0.25 percent), causing engine knock with gasoline derived from it; and such fuel deteriorates rapidly on storage. Nitrogen present in shale oil is in the form of homologs of pyridine, quinoline, pyrrole, and benzonitrile.[25]

Furthermore, sulfur content is high (up to 1 percent), fouling refinery catalysts and polluting the air. The sulfur compounds are disulfides and thiophene-type structures. Petroleum wax content is also relatively high, contributing to the fouling of engines. Oxygen is present primarily as phenol homologs.[25]

---

*A fixed-bed catalytic process to desulfurize and hydrogenate raw hydrocarbon feedstock.

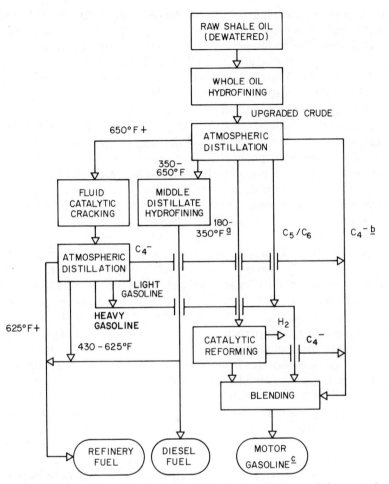

Figure 6.7 Shale oil refinery based on hydrofining. *a,* Heavy naphtha—equal to or better than Arabian light crude (higher in naphthenes, lower in paraffins); *b,* butanes and lighter; *c,* high octane. (Adapted from reference 47; reprinted with permission of the copyright owner, The American Chemical Society.)

Initial tests suggest that the major pollutants can probably be eliminated by treating the oil with hydrogen, with possible recovery of ammonia and elemental sulfur as by-products.[33]

ENVIRONMENTAL AND SOCIAL FACTORS

In a surface retorting facility, disposal of spent shale accounts for about 40 percent of the water needs. Work at the Paraho Development Corporation facility in Anvil Points, Colorado, shows that it may be possible to compress spent shale into a strong, dense, cementlike material, using the small amount of carbonaceous residue as both lubricant and "glue" for the process. The compressed product is very stable and impermeable to water, suggesting that there would be minimal leaching of the alkali into aquifers and streams. Uncompressed spent shale from a 100,000-bpd plant would, in the course of 20 years, occupy a volume 300 m (985 ft) deep by 3 km^2 (1.16 mi^2).[33]

Eastern shales present a somewhat different problem from the western shales, in that their mineral carbonate content is about a thirtieth that of western shale (0.5 vs. 15 percent).[32] The eastern shale is largely illite (a mineral containing iron, aluminum, potassium, magnesium, and silicon) with very little carbonate to pose alkali pollution problems. The eastern shales also are less endothermic than western shales, thus reducing process energy requirements. Eastern shales produce higher yields of gas than do western shales (as much as 1000 scf per ton), with a heating value of between 450 and 500 Btu per scf indicated by one test.[30]

Oil shale mining operations will inevitably alter groundwater patterns. To support a 1 million-bpd shale oil industry in the Green River Formation would consume about one-half the established water supply in an area where average annual rainfall is only 12 to 15 in.[28] The influx of additional people* into the area to support the industry would make further demands on the water supply. Several volumes of water are required to produce one volume of oil,† and additional amounts are needed to leach the spent shale and to reclaim the land.[48] Removal of the needed water from the river would increase the salinity of the lower Colorado River by about 2 percent and make noticeable changes in the entire Colorado River basin, which affects 8 percent of our total land area.[48]

In the end, the rate of extraction of shale oil probably will be more dependent on water supply than on shale reserves or the ability to mine, transport, and dispose of waste.[28] A study has shown that pipelining water from the Mississippi River to the Green River Formation would be a cost-effective way of providing the necessary water,[51] but some environmentalists fear that the pain would merely be shifted to another site.

Colony Development Corp. conducted 11 revegetation studies on spent shale between 1965 and 1973. They found that spent shale initially would not support growth as effectively as native soil; however, they claimed that with proper treatment, shale could be converted into an acceptable growing medium. The major problems with spent shale are the low fertility due to nutrient deficiency, excess soluble salts, structural defects of the soil that lead to poor infiltration, and overheating due to the low albedo of the dark shale.[34] These problems are, of course, amplified by the shortage of water.

The practicality of reclaiming the shale is somewhat suspect, inasmuch as independent observers found only three grasses and two nonwoody shrubs had

*Doubling of the population is estimated (from 100,000 initially to 200,000).[48]
†Variously estimated at 3:1[48] or 1.4 to 4.5:1.[49] One source claims that 1.14 acre-ft (371,470 gal) of water is required to produce 10^{10} Btu of shale oil.[50]

flourished on the shale, and then only with heavy watering, fertilizing, and mulching. None of this growth would be suitable forage for the large herds of migratory mule deer native to the area. Furthermore, there is no commercial source of large quantities of four-winged flatbush[28] or other seeds necessary for revegetation.

In situ processing would somewhat ameliorate the environmental impact of shale oil recovery by reducing waste disposal and land reclamation problems. It would also decrease water consumption by one-third to one-half that required for mining and aboveground retorting (Table 6.5). Water requirements of the two general modes of mining oil shale are open to question. Another source[25] shows open-pit mining requiring twice as much water as deep mining (average 300×10^3 acre-ft per year vs. 150 acre-ft per year), half of it going to disposal of spent shale. One thing is certain—in situ extraction requires much less water. Bureau of Mines and Occidental Petroleum experiments suggest possible shale oil extraction rates of 53 to 70 percent by in situ techniques vs. up to 100 percent by surface retorting, but room-and-pillar mining makes only 62 to 65 percent of the in-place hydrocarbons accessible, so overall yield may be poorer.[25]

Land reclamation is not the only environmental factor to be considered in dealing with oil shale. The magnitude of the task of merely moving the shale throughout the mining, extraction, and disposal processes is almost beyond comprehension. It can perhaps best be visualized by realizing that to support a 1-million-bpd shale oil surface extraction effort, the tonnage of earth and rocks moved would be equivalent to that excavated in constructing the Panama Canal—and this would have to be repeated every 9 months for the working life of the deposits![28] The discouraging fact is that the shale oil would then supply only about 5 percent of our petroleum demand (based on 1978 consumption). Even with modern excavating equipment and transport technology, this would be a massive, continuing undertaking, particularly in light of the fact that it would result in such a small portion of our energy needs.

Quite apart from the matter of water and logistics problems, a shale oil

Table 6.5 Environmental and Social Impacts of Shale Oil Recovery Methods[a]

Process	Population Influx	Spent Shale Disposal, 10^9 tpy	Water Requirements, 10^3 acre-ft/yr
Aboveground extraction			
Underground mining	130,000	0.5	176
Open-pit mining	140,000	1.2	173
In situ extraction	100,000	0.23	96

Source: Reference 42.

[a] Assumes production rate of 1 million bpd.

industry would add particulates to the air from the mining and conversion operations and from disposal of spent shale. It would also introduce sulfur from the retorting operation. Around the White River valley, adjoining the Piceance Creek Basin, air pollution standards are said to be so stringent that they are already exceeded by emissions from decaying native vegetation and from shale.[52] As suggested earlier, in situ extraction technology offers some promise of reduced environmental impact, but in the words of one observer: "in situ recovery may simply hide the environmental problems underground," without really eliminating them.[46]

Like the products of coal liquefaction, shale oil has been regarded as containing presumptive carcinogens. This is based on investigations of syncrude from ORNL, which was found to include mutagenically active material, and of Scottish shale oil produced in the late nineteenth century.[53] Carcinogens may also be present in the spent shale and in the water distillate from the retorting process.[24] This has not been confirmed conclusively.

Perhaps one of the more disturbing aspects of shale oil is the effect of its extraction and combustion on the CO_2 content of the atmosphere. This is particularly true in the case of the western shales, which are associated with carbonate minerals: nahcolite, calcite, and dawsonite. High-temperature retorting processes, in most of which the retort heat is provided by burning shale ore in the retort, produce temperatures of 700 to 1100°C, sufficient to calcine the carbonate minerals and release CO_2 to the atmosphere. Low-temperature processes (500°C) normally employ an external heat source and do not calcine the carbonate minerals. (Eastern Devonian shales have few such carbonates associated with them, hence pose little or no problem.)[54]

The retorting process, the nature of the associated minerals, and the ore assay all affect the amount of CO_2 produced. Initial calculations indicate that extracting and burning shales from the Green River Formation would probably release 1.25 to 4 times the amount of CO_2 produced by burning an energy-equivalent amount of bituminous coal, 1.5 to 5 times the amount for conventional oil, and 2 to 7 times as much as natural gas.[54] There is already much concern about the impact on our air and our climate of increased coal utilization. With substitution of shale-based synoils, the problem would be worsened further.

In addition, the oil shale industry would bring to bear the same societal disruptions associated with massive coal mining. On the overall scale of things, there is the probability that shale oil production would be even more disruptive of the environment and society than would coal production.[48]

ECONOMICS

The practicality of oil shale recovery and conversion to commercial products competitive with conventional petroleum products hinges largely on the going price of OPEC crude. For the first quarter of 1978, the dockside value of imported crude averaged \$14.40, up 3.4 percent (from \$13.92) over the corresponding period of 1977, but showing a slight downward trend from December 1977 (from \$14.53).[55] In the second half of 1979, the price went up (about 42 percent since the end of 1978), with a new price of about \$20 per bbl.[56] Further increases, to as high as \$34.72 per bbl for Libyan light crude, went into effect in January 1980.[57] With projections of continuing increases in OPEC prices, there should come a time when oil shale development becomes economically practical. The crucial question revolves about the uncertainties as to what the critical figure is for the price of crude. There seems to be a general feeling that the break-even point might be about \$20 per bbl for conventional crude. When crude sold for \$18 to \$20 per bbl, government financial participation or assistance in the form of tax incentives was almost mandatory.[33] Using the rich Piceance Creek Basin deposits, the breakeven figure might be shaved to as low as \$16 per bbl, because of their high assay.[58] These numbers are highly speculative, because they depend so much on the political climate and on the timing—particularly with respect to inflation and its effect on the availability and cost of capital. It must be remembered that long lead times are required to make a substantial entry into any such new energy venture. Another factor in the economic picture that cannot be neglected is the effect of the cost of the competitive coal conversion option, but there is reason to believe that syncrude from shale oil could cost up to \$6 per bbl less than coal-derived syncrude.[59]

The hydrogen-to-carbon mole ratio of shale oil is intermediate between conventional petroleum and coal carbonization products; hence shale oil should be the preferred raw material (over coal) for fluid fossil fuels. Less drastic, hence less costly processing is called for to produce equivalent substitute hydrocarbon fluids from shale.[23]

One development project (Rio Blanco Oil Shale Project—Gulf Oil Corp. and Standard Oil Co. of Indiana) is under way in northwestern Colorado. Production is expected to go as high as 1500 bpd during the five-year period of the project. The director of the venture felt confident that a syncrude price of \$13 to \$14 per bbl should be in the cards, provided that the producer's costs do not outstrip escalating oil prices. It is expected that the unrefined syncrude output might be mixed directly with fuel oil from conventional refining, for burning at generating stations,[60] thereby avoiding an additional \$1 to \$4 per bbl refining cost.[61] One of the more optimistic cost figures for shale-based syncrude was Paraho's \$11.50 per bbl.[33] The Institute of Gas Technology estimated that Devonian shale could be extracted and processed to sell at \$15 to \$20 per bbl.[32] Table 6.6 gives one estimate of costs to produce syncrude from oil shale. The present inflation rate and OPEC pricing now make these data suspect, although the relative figures for the two techniques are probably valid.

The governmental action deemed most likely to stimulate appropriate industrial interest in the development of syncrude would be loan guarantees and price protection, whether the crude be derived from coal or shale. This action is expected from the Synthetic Fuels Corporation, a new government body. Without such support, there is some question whether the shale oil industry can hope to get

Table 6.6   Shale Oil Production Cost (Estimated)

Extraction Method	Investment Cost, $/bpd Capacity	Operating Cost, $/bbl	Production Cost,[a,b] $/bbl
Modified in situ	5,000–7,000	3.50–5.00	8.00–11.00
Surface retorting	14,000–23,000	4.00–5.00	16.00–25.00

Source: Reference 62.

[a] Oil produced may be prerefined to quality equivalent to Arabian crudes at additional cost of $1 to $4 per bbl.

[b] At 15% discounted cash flow, 100% equity.

off the ground. The present level of inflation makes the economic situation especially precarious.

The major deterrents to shale oil development (all of which involve economics directly or indirectly) have been identified as[25]:

1   Uncertain economic climate
   (a)   A three-year inflationary increase in 50,000-bpd plant cost from $0.4 to $1.0 billion, for example.
   (b)   OPEC capability to flood market with cheaper oil.
   (c)   Uncertainty over federal economic incentives.
2   Regulatory "morass"—the need to streamline and consolidate agency approvals for construction and operation.*
3   Technological problems—unexpectedly low measured yields from in situ processes and other problems with scale-up.
4   Lack of a specific national energy policy, supported by the executive and legislative branches of the government—inconsistency, shortsightedness, and lack of specific economic incentives.
5   Potential shortages—water, skilled labor, construction equipment and materials, oil-rich deposits, adequate transport, and so forth.
6   Overly strict environmental standards—many contradictory, unstable, and ill-defined requirements.

Various possible steps to eliminate the first deterrent have been cited.[25] Tax incentives could be granted (e.g., accelerated depreciation or increased income tax credits). A floor could be placed on the price of imported oil, thus preventing flooding of the market with cheap OPEC oil. The government could guarantee to purchase shale oil at a set price that would allow the producers to make a reasonable profit. Low-cost loans or loan guarantees might be granted to pro-

*This was presumably resolved by the Carter Administration late in the third quarter of 1979.

ducers. These or other economic levers could be, but have not been, applied to stimulate shale oil development.

It should be remembered that producers must spend $100 to $200 million before they even know the answers to their technological questions.[63] Such a sum could be expended before it was determined that a process would not be technically possible or economically viable. Most business managers and stockholders are unlikely to hold still for such a gamble, especially in light of present regulatory and economic uncertainties. There are exceptions, however, as noted earlier, for example, in the case of the $440 million joint venture of Occidental Petroleum and Ashland Oil, where the goal is merely to see *if* commercial production of oil shale is possible.[52]

The costs of new, previously unproven processes, as a general rule, tend to double when put into operation. The costs of synfuels are highly sensitive to the cost of the needed capital and to the stream factor.* Oil refineries operate at a stream factor of about 0.91 and synthetic plants should be similar, but if major operating problems were to be encountered in practice, the stream factor might fall to 0.5, nearly doubling the cost of synthetics.[51]

SUMMARY

It must be apparent from the immediately preceding pages and from the previous chapters that there is an urgent need for new major sources of primary energy, especially for fluid fuels for the transportation, residential, and commercial sectors of our economy. On the surface, coal and shale oil would appear to be ideal sources—available in large quantities and technologically convertible to fluids. On closer examination, we have seen that both sources suffer from similar deficiencies—largely economic, ecologic, and logistic.

The best guess up until recently was that we might see a handful of commercial shale oil conversion plants in operation by 1985; however, this now seems highly improbable, in light of the dearth of activity toward this end and the long lead times required to bring such events to pass. If the logjam were to be broken, it is estimated that the maximum number of aboveground retorting plants that could ultimately be supported would be about 20, with a total capacity of about 1 million bpd. This would represent only about 5.2 percent of our 1978 consumption of oil (19.4 million bpd). A few would-be producers feel the figure could possibly reach up to 10 percent. The main constraints to reaching this level of production would be lack of water and shortage of capital.[25]

There is the feeling that if in situ technology could be proven, it would raise the production potential to several million bpd, but in no case could the maximum be attained before the end of the century. Recovery rates and effects of in situ retorting on aquifers are still very much in doubt.

In any case, recovery of hydrocarbons from such organic-lean ores as oil shale and tar sands introduces some imposing problems of providing investment

*Actual output divided by nameplate capacity, assuming continuous operation.

capital and large amounts of skilled labor to transport and process huge quantities of the raw material. None of this is made easier by the remoteness from adequate labor pools, the limited reserves of qualified engineers and heavy machine operators, and the shortage of needed, dependable water supplies.

Many recent articles have neglected or made light of the economic problems of extracting shale oil. The remoteness and low organic content of the deposits, the capital requirements, the transport and water demands, and the various environmental issues, however, make the economic viability of this energy source questionable.

The oil shale and tar sands industries are necessarily highly capital-intensive. Many potential producers already have vast sums tied up in oil and gas exploration, nuclear energy, coal mining, and so forth, and place a much lower priority on the oil shale option, which, until recently, has received little more than lip service from the federal government. The day this option actually receives tax incentives or price supports from the federal government will be the day some hard decisions will be made in the industry. Until then, our synthetic fluid fuels program can be expected to remain pretty much in limbo. There is always the threat, of course, that if synthetic fuels are subsidized in some form or other, the incentive to conserve fuel may be diminished and consumption of our cheaper fuels may remain at an excessively high level.[51]

Another view on the future prospects of shale oil was expressed at a recent symposium and drew almost universal support from those in attendance. Simply stated, it was the feeling that shale, as a source of hydrocarbon fuels and feedstocks, would never get off the ground, in large measure because of the imposing environmental problems and the expected low net yields of useful products.[64]

All discussions of our energy resources and our conversion options seem to ignore totally one aspect of extraction that could well prove to be crucial to future generations. There is much talk about recovery rates, purely from the viewpoints of production and of extraction costs. No thought seems to be given to the fact that, for example, room-and-pillar mining dooms 30 to 50 percent of the resource to be forever locked up in the bowels of the earth, with little or no prospect of recovering it later, even at escalated prices. It would appear that if we are to make the most of our limited fossil fuel resources, we must restrict our efforts to those technologies that remove essentially *all* of the hydrocarbon on the first pass.

REFERENCES

1 F. W. Camp, "Tar Sands," in Kirk-Othmer, *Encyclopedia of Chemical Technology*, 2nd ed., Vol. 19, Interscience, New York, 1969, pp. 682–732.

2 C. E. Maurer, *Popul. Sci.* **210**(5), 80–83, 170 (May 1977).

3 A. R. Allen, *Chemtech* **6**(6), 384–391 (June 1976).

4 T. H. Maugh II, *Science* **199**(4330), 756–760 (Feb. 17, 1978).

5 *New Sci.* **77**(1087), 219 (Jan. 26, 1978).

6 *Energy User News* **2**(47), 21 (Nov. 21, 1977).

7 M. K. Hubbert, "Outlook for Fuel Reserves," in D. N. Lapedes, Ed., *McGraw-Hill Encyclopedia of Energy.* McGraw-Hill, New York, 1976, pp. 11-23.

8 P. Averitt, "Coal Resources of the United States," USGS Bull. 1412, U.S. GPO, Washington, D.C., 1975.

9 D. E. Carr, *Energy and the Earth Machine.* Norton, New York, 1976, p. 80.

10 R. Yates, *Phila. Inquirer* **301**(162), 17-A (Dec. 9, 1979).

11 G. R. Gray, "Oil Sand," pp. 533-535, in reference 7.

12 R. Schutte; reported in *Chem. Eng. News* **55**(23), 35, 36 (June 6, 1977).

13 P. Nulty, *Fortune* **97**(10), 72-78 (May 22, 1978).

14 *Chem. Eng. News* **55**(23), 35, 36 (June 6, 1977).

15 E. Pearson, *Energy User News* **3**(17), 12 (Apr. 24, 1978).

16 Reference 9, pp. 80, 81.

17 *Chem. Eng. News* **57**(28), 22-24 (July 29, 1979).

18 D. F. Othmer, *Mech. Eng.* **99**(11), 29-35 (Nov. 1977).

19 *Bus. Week* (No. 2529), 44B (Apr. 10, 1978).

20 C. L. Wilson, *Energy: Global Prospects 1985-2000.* WAES (MIT), McGraw-Hill, New York, 1977, pp. 218, 219.

21 *Ind. Eng. Chem.* **19**, 388 (1927).

22 *Chem. Eng. News* **55**(30), 14 (July 25, 1977).

23 R. E. Gustafson, "Shale Oil," pp. 1-20, in reference 1.

24 R. Gallois, *New Sci.* **77**(1091), 490-493 (Feb. 23, 1978).

25 S. W. Herman et al., *Energy Futures.* Ballinger, Cambridge, MA, 1977, pp. 357-399.

26 R. P. Philp; reported in *Chem. Eng. News* **55**(24), 20 (June 13, 1977).

27 D. L. Klass, *Chemtech* **5**(8), 499-510 (Aug. 1975).

28 W. D. Metz, *Science* **184**(4143), 1271-1275 (June 21, 1974).

29 B. S. Lee, *Energy User News* **3**(8), 22, 23, (Mar. 6, 1978).

30 *Chem. Eng. News* **54**(43), 36, 38, 40 (Oct. 18, 1976).

31 J. W. Smith and H. B. Jensen, "Oil Shale," pp. 535-541, in reference 7.

32 *Chem. Eng. News* **55**(24), 19, 20 (June 13, 1977).

33 T. H. Maugh II, *Science* **198**(4321), 1023-1027 (Dec. 9, 1977).

34 C. H. Prien, "Survey of Oil-Shale Research in the Last Three Decades," in T. F. Yen and G. V. Chilingarian, Eds., *Oil Shale.* Elsevier, New York, 1976, pp. 235-267.

35 S. Rattien and D. Eaton, "Oil Shale: The Prospects and Problems of an Emerging Energy Industry," in J. M. Hollander and M. K. Simmons, Eds., *Annual Review of Energy.* Vol. 1, Annual Reviews, Palo Alto, CA, 1976, pp. 183-212.

36 "U.S. Energy Outlook—A Report of the National Petroleum Council's Committee on U.S. Energy Outlook," NPC, Washington, DC, Dec. 1973.

37 W. C. Culbertson and J. K. Pitman, in D. A. Brobst and W. P. Pratt, Eds., *U.S. Mineral Resources.* USGS Prof. Pap. 820, U.S. GPO, Washington, DC, 1973.

38 P. Averitt, "Coal Resources of the United States, Jan. 1, 1967," USGS Bull. 1275, U.S. GPO, Washington, DC, 1969.

39 T. F. Yen and G. V. Chilingarian, "Introduction to Oil Shales," pp. 1-12, in reference 34.

40 J. P. Smith, *Courier-News* (Bridgewater, NJ) **94**(118), B-12 (Dec. 6, 1977).

41 J. Fisher, *Energy User News* **2**(28), 18 (July 18, 1976).

42 A. E. Lewis and A. J. Rothman, "Rubble in situ Extraction (RISE)," Lawrence Livermore Lab., UCRL-51768, Mar. 1975.

43 *Energy Supply, Demand/Need, and the Gaps Between, An Overview,* Vol. 1, (MIT), Natl. Tech. Inf. Serv., Springfield, VA, Mar. 1976, pp. 30, 31.

44 G. U. Dineen, "Retorting Technology of Oil Shales," pp. 181–198, in reference 34.

45 J. D. Keenan, "Biochemical Sources of Fuels," in D. M. Considine, Ed., *Energy Technology Handbook,* McGraw-Hill, New York, 1977, pp. 4-5 to 4-11.

46 C. H. Prien; reported in reference 9, p. 79.

47 R. F. Sullivan; reported in *Chem. Eng. News* **56**(2), 33, 34 (Jan. 9, 1978).

48 E. H. Thorndike, *Energy and Environment: A Primer for Scientists and Engineers,* Addison-Wesley, Reading, MA, 1976, pp. 167, 168.

49 M. M. McCormack, "Protecting the Environment," pp. 50, 51, in reference 7.

50 H. W. Welch, in J. E. Bailey, Ed., *Energy Systems: An Analysis for Engineers and Policy Makers,* Dekker, New York, 1978, p. 98.

51 O. H. Hammond and R. E. Baron, *Amer. Sci.* **64** (4), 407–417 (July–Aug. 1976).

52 *Bus. Week* (No. 2501), 39, 40 (Sept. 19, 1977).

53 B. Commoner, *Chemtech* **7**(2), 76–82 (Feb. 1977).

54 E. T. Sundquist and G. A. Miller, *Science* **208**, 740–741 (May 16, 1980).

55 *Energy User News* **3**(18), 15 (May 1, 1978).

56 J. R. Emshwiller, *Phila. Inquirer* **301**(25), 1-A (July 25, 1979).

57 UPI; reported in *Naples* (FL) *Daily News* **57**(138), 1 (Jan. 3, 1980).

58 R. Meeker; reported by W. Chapman, *Phila. Inquirer* **295**(59), 1-A, 8-A (Aug. 28, 1976).

59 G. A. Rial; reported in *Chem. Eng. News* **55**(30), 14 (July 25, 1977).

60 A. McCue, *Energy User News* **2**(26), **12** (July 11, 1977).

51 J. E. Swearingen; reported in *Bus. Week* (No. 2531), 76–79, 84, 88 (Apr. 24, 1978).

62 Reference 20, p. 220.

63 *Bus. Week* (No. 2425), 36, 37 (Mar. 29, 1976).

64 A. Kaufman, "Two Views of Resource Scarcity: Engineering vs. Economic," presented at Natl. Symp. Crit. Strategic Mater., Amer. Chem. Soc., Washington, DC, June 5–7, 1978.

7 HYDROCARBONS FROM REFUSE AND BIOMASS

Energy from the sun is free ... in large supply ... and is subject neither to monopolistic nor nationalistic dislocations.

A. H. BROWN

REFUSE

Although concern over solid wastes goes back to our early history, published literature acknowledging the potential of waste for energy generation largely dates back only to the recent Middle East oil embargo. Even then, most publications quote data on gross amounts of solid waste generated, without regard for recoverability.[1] Published literature on hydrocarbon (methane) recovery from municipal solid wastes dates from at least the early 1940s, with the thrust being fuel rather than hydrocarbon feedstocks.

An estimate of the average domestic component of municipal solid waste (MSW) generated in the early 1970s was 3.22 lb per capita-day, according to one reference.[2] (Most estimates range upward of 3 lb, approaching a near-term projection of 3.5 lb.) Nearly a third of the MSW is paper; yard waste comprises nearly a fifth, and is closely followed by food waste. The remaining 32 percent is made up of glass, metals, wood, plastics, rubber, leather, textiles, and miscellany, according to EPA data (1974).[2]

Estimates of total MSW seem to range from about 110 to 175 million tpy, with the most likely average falling between 140 and 165 million tpy. The combustible or organic fraction of MSW (hence the "manmade ore" potentially available as a source of hydrocarbons) is estimated at about 65 percent by weight,[3] with values ranging from as low as 52 to as high as 80 percent, because of the wide-ranging nature of lifestyles, environments, and so forth, in the United States. It is this organic fraction that is the major concern of this chapter. The inorganic and metallic portions are looked upon primarily as potentially recoverable components whose resale value should enhance the economics of refuse-derived fuel (RDF) or of potential feedstocks.

On average, U.S. municipalities send about 300 million tons per year of refuse to dumps and incinerators*—according to Bureau of Mines estimates. Of this, over half would be suitable for reclaiming energy and/or materials.[5] It has been estimated that if all the combustible U.S. waste could be burned with 40 percent recovery of the thermal energy released, 2.4 million typical midwestern homes could be heated each year.[6] Collection of all widely dispersed solid wastes for combustion in centralized conversion plants serving the entire population is not remotely attainable. It is perhaps more pertinent to note that combustible solid wastes from all Standard Metropolitan Statistical Areas (representing 70 percent of the U.S. population) could generate 1 Q of energy,[7] or about 1.4 percent of our current energy budget. If this seems like an inconsequential amount,† it must be considered that with the rise in our petroleum imports, the decline in our domestic petroleum and natural gas reserves, and the added pressures placed on all our fossil fuels by the lack of a national commitment to nuclear power, we must take advantage of all such "inconsequential" energy options regardless of their size.

Urban and Industrial Wastes

Solid waste disposal is a growing problem in the United States, particularly around large metropolitan centers, where land is expensive and scarce and air and water pollution problems abound. It is a situation that can be eased while helping to solve our energy dilemma.

Municipal Solid Wastes (MSW)

The intelligent management of refuse from our society is important for two major, energy-related reasons: (1) combustible wastes can be recycled or can be converted to provide heat or to form useful hydrocarbons to supplant partially our nonrenewable fuel resources, and (2) recycled metals and glass can be converted to products with greatly reduced consumption of process energy (Tables 7.1 and 7.2). The latter not only conserves energy but also reduces consumption of virgin metal resources. A third, and possibly equally important result of recycling, is the reduction in the load placed on our shrinking solid waste landfill sites.

Solid waste disposal was estimated to cost the United States $6 billion annually (in 1973)—about $25 per ton to collect waste in urban areas, and about $10 per ton to process and deposit it in landfills.[11] (These figures vary considerably with geographical location.) Transport distances to licensed landfill sites are increasing rapidly in most parts of the country as sites become full or are more tightly regulated by environmental authorities. By 1985, the national

---

*Of all the U.S. municipal solid waste disposed of in 1970, 77 percent was disposed of in open dumps, 13 percent in landfill sites, and most of the balance (10 percent) was incinerated.[4]

†Actually it is equivalent to about 50 percent of the energy generated by all our hydroelectric generators.[8]

Table 7.1 Energy and Resource Conservation Aspects of Recycling

	Typical Energy Saved by Recycling, %[a]	Contribution of Recycled Materials to U.S. Resource Base, %[b]
Copper	87	40
Aluminum	96	25
Lead	63	49
Zinc	63	14
Paper	70	20
Stainless steel	n.a.	30
Gold	n.a.	35
Silver	n.a.	23
Textiles	n.a.	13
Rubber	n.a.	4

Source: Adapted from reference 9. Copyrighted by and reproduced with the permission of the National Association of Recycling Industries (NARI).

[a] EPA, AEC, and others.
[b] BuMines.

average disposal cost per ton is expected to rise to $12, while collection and disposal combined will be $40 to $50 or more per ton.[11]

One source estimates that there is enough energy potentially recoverable from our MSW to provide approximately 3 percent of the nation's total energy needs.[12] A typical ton of refuse will yield the energy equivalent of 1 bbl of oil or $1/3$ ton of coal.[13] By any reckoning, these represent significant quantities of energy.

Before an evaluation can be made of the feasibility of a program to recover energy or feedstocks from waste, it is necessary to assess the quantity and quality of the resource. Any attempt to produce figures for "typical" U.S. municipal solid waste compositions are fraught with uncertainties. The ghetto area of a large city is apt to be high in demolition debris—wood, rocks, bricks, glass, dirt, and so forth. The city as a whole will be low in yard debris—grass clippings, leaves, tree trimmings, and the like—while suburban areas of the same metropolitan region will have a still different waste makeup. Variations will also occur due to geographical and climatic differences. Some published data are based on ovendry weight and others are "as received" (20 to 25 percent moisture—some estimate higher). Some compilations combine household or residential ("domestic") waste with commercial refuse. Others do not specify the particular source of urban refuse. Perhaps one of the more definitive compilations of the breakdown of the combustible portion of typical U.S. municipal solid waste is that of the Bureau of Mines (Table 7.3). The composition of the noncombustible portion (from yet another source) is approximately half water,

Table 7.2 Energy Savings from Recycling vs. Use of Primary Materials, 1976

Material	Tons Recycled	Energy Content, Primary Materials, 10^6 Btu/ton	Energy Content, Recycled Materials, 10^6 Btu/ton	Energy Saved By Recycling, 10^6 Btu/ton	Total Energy Saved, 10^{12} Btu	Savings, 10^3 boe
Aluminum	1,433,000[a]	175.4[b]	6.8[b]	168.5	242.0	41,710
Copper	1,423,600[a]	46.2[b]	5.9[b]	40.3	57.3	9,880
Iron/steel	46,111,500[c]	14.6[b]	5.7[b]	9.2	425.6	73,340
Lead	670,000[a]	8.7[d]	3.2[e]	5.5	3.7	636
Paper	10,158,000[f]	20.3[g]	8.6[g]	14.4	146.0	25,156
Rubber	125,000[h]	31.2[h]	9.1[h]	22.1	2.8	475
Zinc	179,400[a]	19.7[d]	7.8[e]	11.8	2.1	366
Total					879.5	151,563

Source: Reference 10. Copyrighted by and reproduced with the permission of the National Association of Recycling Industries (NARI).

[a]BuMines.
[b]ORNL.
[c]BuMines/A. D. Little Corp.
[d]Battelle Memorial Institute.
[e]A. D. Little Corp.
[f]BuCensus.
[g]EPA.
[h]Industry estimates.

Table 7.3 Combustible Part of U.S. MSW (Approximate)[a]

	Weight Percent	
	Composition	Paper Fraction
Paper	50.2	
Corrugated cardboard		26.7
Newsprint		18.0
Other paperboard		13.3
Packaging		12.6
Office paper		12.2
Magazines/books		7.6
Tissue paper/towels		5.3
Other nonpackaging		2.9
Paper plates/cups		1.4
Yard debris	17.8	
Food	16.8	
Plastics	5.2	
Wood	4.6	
Leather/rubber	3.4	
Textiles	2.0	
Total	100.0	100.0

Source: Reference 14.

[a]Residential and commercial.

one-fifth to one-fourth ferrous metals, one-seventh glass, and the balance is a mixture of metals and minerals.[15] Data on the quantity of the resource base are very uncertain at best.

Table 7.4 shows the proximate analysis of three different combustible mixes from municipal solid waste vs. coal. The sulfur content of MSW is appreciably lower than coal, but the chlorine content is higher. The lower heating value of the waste tells us that it would be necessary to burn more combustible waste than coal to produce an equal quantity of heat. This, of course, will change the relationships of coal and MSW pollutants from those noted, because comparisons should be made on an equal-energy basis. Ultimate analysis of the total MSW has been made in Table 7.5.

There are a number of advantages in employing MSW as a source of fuel[18]:

1 It reduces the consumption of nonrenewable fossil fuel resources.
2 The volume of solid residues for landfill disposal is reduced.
3 Atmospheric and waterborne pollutants are decreased.
4 Sulfur oxide emissions are decreased.
5 Use is made of an available, renewable, increasing resource.

Table 7.4   Proximate Analyses of Coal and Combustible Portion of MSW

	Coal,[b] wt %	Combustible Portion MSW,[a] wt %		
		c	d	e
Moisture	3.5–7.0	12.4	5.0	6.9
Ash	9–12	9.0	11.2	21.7
Sulfur	1–2.5	0.20	0.28	0.28
Chlorine	<0.1	0.45	1.24	0.64
Btu/lb[f]	10,000–12,600	7500	8800	7100

Source: Reference 16.
[a] Partially dried basis.
[b] BuMines data.
[c] Light-gage plastics and paper.
[d] Heavy-gage plastics, paper, leather, rubber, and wood.
[e] Yard waste and putrescibles.
[f] As received.

Table 7.5   Analysis of Domestic Municipal Solid Waste (Estimated)

	Weight Percent
Moisture	28.3
Carbon	25.6
Hydrogen, net[a]	3.4
Oxygen	21.2
Nitrogen	0.6
Sulfur	0.1
Metals	20.8
Heating value (max.), Btu/lb	4450

Sources: Reference 17.
[a] Total hydrogen less 1/8 oxygen.

It has been noted, however, that paper as products is more valuable than paper as energy[19]; hence judgment must be exercised to balance all factors of composition, process costs, transport costs, marketability of product, energy value, and so forth.

Nonwood Industrial Wastes

Before entering into a discussion of industrial waste, it is important to understand the meanings of some terms relating to the subject. *Home scrap* is residual material generated in primary production and returnable directly to the production process (e.g., excess metal from an ingot, and thermoplastic sprues and runners). *Prompt industrial scrap* (also known as *new scrap*) is waste generated in a manufacturing or fabricating process (lathe turnings, punch press scrap, drillings, etc.). *Old scrap* is that which is generated from products that have reached the end of their useful lives. Home scrap generally will be returned directly to the process, whereas new scrap, which usually can be separated readily from other plant waste, is more likely to be sold for salvage. Old scrap most commonly ends up as a component of urban waste.[20] Metals in the form of pigments, tetraethyl lead, foil, nonprecious metal platings, and so forth, have little likelihood of eventual recovery because of their disperse nature and their low unit value.

Plastics pose a problem for recycling because of economic and technical restraints. There are at least 40 different types of plastics and generally a variety of classes or variations in each type. Many are so incompatible as to discourage attempts at blending them. The commercial technology for separating these plastics by chemical type does not exist at this time, and hand sorting is seldom economically feasible. Progress is being made in both separation and end-use technology, but the lack of compatibility makes it appear that most recycled plastics will come from either home or new scrap rather than old scrap.

In specific cases, poly(vinyl chloride) has been extracted mechanically from scrap wire and cable,[21] polyester soft drink bottles have been sorted and recycled into polyester fabrics and other forms for noncritical uses, and chlorinated polyethylene has been employed as a compatibilizer for a heterogeneous mixture of thermoplastics.[22] Thermosets pose a difficult recycling problem, inasmuch as they cannot be substituted for virgin material because of their lack of moldability. Some thermoset runners and sprues, however, have been ground and blended into virgin molding material at levels up to 10 weight percent, for use as fillers or extenders.[23,24]

A market exists for recycled plastics, inasmuch as scrap plastics traditionally have been valued at 15 to 25 percent of the cost of virgin material. The practicality is in doubt, however, once the product has gotten past the fabricator's plant, because of the collection, sorting, identification, and blending problems that face the recycler. At this point, it generally will be most practical to include such plastics in the combustible part of industrial or municipal solid waste for pyrolysis to hydrocarbon feedstocks or for combustion as boiler fuel.

The major combustible component of urban wastes is paperstock. As noted earlier, paper is more valuable as paper than as fuel, but this must be tempered by a consideration of the composition of specific input waste and the difficulty of sorting it. In the case of industrial waste generated in a given plant, sorting can often be carried out readily in the daily course of normal housekeeping. The composition will be highly dependent on the plant product mix. If the source (the manufacturing facility) is situated geographically convenient to a market for recyclable paperstock, there is little doubt that such should be treated as a fiber feedstock and not as fuel. In a heterogeneous industrial waste, where paperstock is not necessarily a predominant component of the waste, or in municipal solid wastes, the economics of sorting may make it more practical to burn or pyrolyze the paperstock. Although paper consumption has more than doubled since 1950, the amount of paper recycled has increased very little over this time frame, largely owing to technological, institutional, and economic factors.[20] It is a generally accepted statistic that at our present level of paper disposal, each one of us discards the equivalent of eight trees per year,[25] hence there is much to be gained by recycling paper more effectively than at present.

Industrial wastes are estimated to amount to approximately 5 percent of our total national (dry) waste stream, with little likelihood of a significant change through 1980 (Table 7.6). Figure 7.1 indicates a typical cycle for industrial scrap. The total amounts of solid waste are significant, because each American is responsible for generating nearly a ton of waste per year.[27]

Sewage Sludge

Currently, sewage sludge contributes only a very minor portion of our energy budget.* The methane produced by anaerobic digestion of sludge is used largely for operating the wastewater treatment facilities—for space and sludge heating† and for the generation of electricity.

There are over 3200 sewage treatment plants in the United States employing anaerobic digesters, which will be discussed later. Assuming 1 ton of active solids per million gpd of flow and yields of 12 scf of 566 Btu gas per pound of solids, these plants are estimated to produce gas equivalent to only about 47 $\times$ 10^{12} Btu annually.[31]

The primary goal of sludge disposal is to reduce its volume with minimum consumption of added fuel while meeting all environmental regulations. Generation of fuel (or energy) is becoming an increasingly desirable added benefit.

Landfills

Some writers have used the term "unnatural" gas and others have used "dump-ground" gas to signify swamp gas formed biologically from organic materials buried in landfills (or "reverse mines"). To date, production of this gas has been virtually a cottage industry, having negligible impact on our energy budget. The output has been compressed and used locally for cooking, space heating, and so forth, but at this writing, it has been produced on only a very limited commercial scale (about 1.2 $\times$ 10^{12} Btu in 1976[8] or 2 to 4 billion scf). The potential maximum recoverable landfill gas is estimated at 3 Q in 1985 and 5.6 Q in 2000.[32]

The withdrawal of this gas serves a useful purpose in that it reduces the fire and explosion hazard existing in some landfill sites. If the gas were not put to gainful use, the landfill gas would probably have to be vented or flared to pre-

---

*Each person accounts for an average of 3.75 lb per day of sewage sludge solids.[28]
†Digesters are generally operated at temperatures ranging from 95 to 118°F (35 to 58°C).[29,30]

Table 7.6 Energy Contents of Waste Streams

	Waste Stream[a,b]				Specific Energy Content, Btu/lb	Total Energy, Quads	
	1971		1980			1971	1980
	10⁶ tpy (dry)	Percent	10⁶ tpy (dry)	Percent			
Agricultural crop/food processing	390	44	390	37	5,500	4.29	4.29
Animal manure	200	23	266	25	3,500[c,d]	1.40	1.86
Urban refuse	129	15	222	21	4,500[e]	1.16	2.00
Logging/wood manufacturing residues	55	6	59	5	6,000[c]	0.66	0.71
Miscellaneous industrial wastes	44	5	55	5	10,000[c]	0.88	1.00
Miscellaneous organic wastes	50	6	60	6	7,000[c]	0.70	0.84
Municipal sewage solids	12	1	14	1	2,000[c]	0.05	0.06
Total	800	100	1,066[f]	100	5,193[g] 5,071	9.14	10.76

Source: Reference 26.

[a] L. L. Anderson (1972).
[b] A. Poole (1975?).
[c] *Combustion* (Feb. 1977).
[d] R. D. Smith.
[e] H. W. Schultz et al. (Jan. 1976).
[f] Total shown in reference 26 has been corrected.
[g] Average: 1971, 5193; 1980, 5071 Btu/lb.

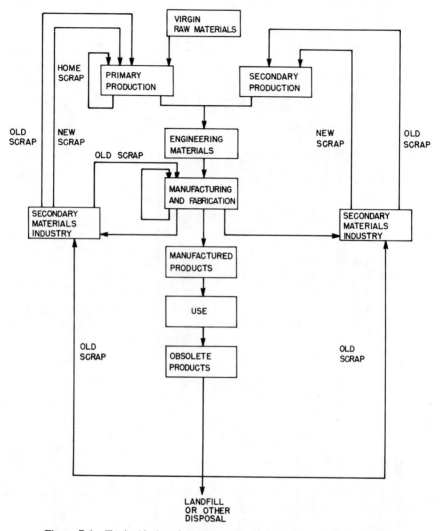

Figure 7.1 Typical industrial scrap cycle. (Adapted from reference 20.)

vent dangerous accumulation and lateral migration below ground. According to EPA, buried refuse will generate 0.45 ft³ of methane for each pound of land-filled waste.* Production begins after a two-year burial period, and should continue for about 20 years.[34]

In 1969, a flash fire and explosion occurred due to lateral migration of "unnatural" gas from a landfill to a North Carolina National Guard Armory 30

---

*Theoretically, 1 lb of cellulose will yield 7 ft³ of methane. Practically, the figure is closer to 4.5 ft³.[33]

ft away. Several persons were killed and more than a score injured, some totally disabled. A 1976 "unnatural" gas explosion in a Japanese irrigation tunnel under construction killed four workmen, and migrating gas forced closing of two Richmond, Virginia, schools.[35] Early in the 1970s, a collection and flaring system had to be installed at a Palos Verdes, California, landfill site to handle gas migrating into a residential area.[36] Methane is lighter than air and can be produced by biochemical decomposition of organic materials by specific bacteria. This occurs in man-made landfills as well as in natural sites such as peat bogs, marshlands, tar pits, and other hydrocarbon deposits. If upward movement to the atmosphere is impeded, lateral migration will occur along paths of least resistance for distances as great as several hundred yards. It is felt that in sanitary landfills, where layers of solid waste are separated by layers of compacted clays and silt, the tendency to migrate laterally may be aggravated. Freezing of subsurface water or capping the surface with impervious paving can also be contributing factors.[35]

The most effective ways to control lateral migration of landfill gas are to introduce impervious perimeter barriers down to bedrock or to the top of the continuous groundwater table, or to install low-flow, forced ventilation with a collector system. Care must be taken to avoid methane concentrations in the explosive 4 to 14 percent range,[35] a concentration that also coincides with what is considered to be a useful lower level for the recovered gas.[36]

Pacific Gas and Electric Company undertook to drill 18 producing wells in the Mountain View, California (San Francisco area), sanitary landfill site to demonstrate that the gas can be collected, purified, and distributed through established pipeline systems for use of its regular customers. The 700-acre area covered by the wells receives 2500 tpd of refuse[37] and is expected to produce about 1 million cfd of raw gas which is reduced to 600,000 cfd when purified to pipeline quality. It is estimated by EPA that as much as 5 million cfd of raw gas could be produced from the entire landfill area for a period of 8 to 10 years.[35]

Similar projects are also under way in Palos Verdes (Los Angeles County)[31,33,35] and New York City.[37] The projected annual energy yield from these latter projects is approximately 1.2×10^{12} Btu, only about two-thirds of 1 percent of the current total energy yield of MSW. Typically, the energy content of landfill gas ranges from 300 to 700 Btu per scf,[35] with the average nominally 500 to 550 Btu.[32] Table 7.7 shows a representative landfill gas composition.

Several options exist for utilizing the product gas[38]:

1 The raw landfill gas can be injected into an existing pipeline distribution network.

2 It can be delivered locally direct to an interruptible gas consumer.

3 It can be used to generate electric power on site, either before or after upgrading to pipeline quality.

4 The landfill gas can be purified and upgraded to pipeline quality gas for normal distribution.

5 Methanol can be produced on site from landfill gas feed.

6 The gas can be liquefied on site.

7 The gas can be vented or flared.

The first two options offer the lowest capital cost, since the landfill gas (approximately 50:50 methane to carbon dioxide) can be sold as is. The Los Angeles project is an example of the third option. The last option is unthinkable in these times of dwindling gas supplies unless the landfill gas supply is so small as not to warrant one of the other options. The three remaining possibilities call for large capital outlays, thus requiring long-term assurance of a continuing high-volume supply of feed. At best, exploitation of landfill gas is a risky, capital-intensive undertaking because of the variations in landfill construction, composition, and management.[31]

At the Palos Verdes site, the major problems have proved to be leaks in the collection system, lower-than-expected yields, and corrosion. The original system (1975) consisted of five major stages[36]:

1 Gas collection—vacuum gathering from eight wells spaced over 40 acres, through perforated vinyl pipe inserted into 36-in.-diameter holes backfilled with gravel and capped with concrete and earth.

2 Compression of raw gas.

3 Pretreatment—through mixed beds of molecular sieves and activated carbon to remove water, heavier constituents, and trace contaminants.

4 Carbon dioxide removal—through molecular sieves.

5 Compression of upgraded gas—to 180 psig and cooling to 100°F.

It was found that the contaminants accumulated in the pretreatment section, hydrolyzed, and decomposed to form corrosive sulfurous and hydrochloric

Table 7.7 Typical Landfill Gas Composition (Palos Verdes, California)

Component	Mole Percent
Methane	50–56
Carbon dioxide	40–45
Nitrogen	0.5–1
Trace components	0.5–1
Oxygen	0–0.1
Hydrogen sulfide	nil
Ethane	0
Heating value (raw gas), Btu/scf	500–550

Source: Reference 36. Reprinted with permission of the copyright owner, The American Chemical Society.

acids. The pretreatment section was then changed in (1977) to incorporate separate dehydration, contaminant removal, and acid-adsorbing stages. The system has now stabilized at about 550,000 cfd, with a greater than 90 percent on-stream time. Collection has now proven to be the rate-limiting stage.

Forest Wastes

Forests constitute the largest part of the earth's biomass. About one-third (754 million acres) of the total land area in the 50 states (2271 million acres) is covered by forests, of which two-thirds are commercial forests. In 1970, they held a standing crop of 706 billion ft^3, with 11.4 billion dry tons of merchantable bolewood. The average cellulose content of this bolewood is 45 percent (5.1 billion dry tons). Annually, these commercial forests produce 315 million dry tons of bolewood and an equal amount of tops, branches, and foliage, a considerable portion of which falls and ends up as decaying forest litter. The average annual yield per acre of commercial forest land is about 0.62 tpa of bolewood and total growth is 1.3 to 1.6 tpa (excluding roots).[39]

The conversion of this biomass to energy and to various feedstocks has been well demonstrated on a rudimentary scale. At this stage in its development, however, conversion technology leaves something to be desired. Biomass conversion to date has largely been involved with combustion for local heating and power generation use and with waste recycle.

Despite high overall productivity, the efficiency of conversion to biomass in nature is very low. Only a fraction of 1 percent of the solar energy received by the earth is used in the photosynthesis process.[40]*

Photosynthesis

There are two broad classes of plants with different metabolisms and climatic dependencies. One mechanism is the *C-3 metabolism*, characterized by wheat, barley, and commercial tree species. It involves the photosynthetic reaction of a five-carbon compound with carbon dioxide to form a six-carbon compound which then cleaves to form two stable three-carbon compounds. The other mechanism—*C-4*, characterized by sugarcane and corn—results in the formation of four-carbon compounds which are then decarboxylated to stable three-carbon compounds. The C-4 mechanism, which functions best in warm climates [77 to 105°F (25 to 40°C)], is about twice as efficient as the C-3 mechanism. The end result of photosynthesis is living matter with an empirical formula calculated to approximate

$$H_{2960}C_{1480}O_{1480}N_{16}P_{1.8}S.^{41}†$$

Water appears to be one of the more important elements of photosynthesis, but only in that it seems to control the plant's ability to absorb carbon dioxide.[42]

*Up to possibly 3 percent for some tropical rain forests.[40]
†Weighted average with woody plants predominant.

Turgidity of the leaf, due to the water content, opens the stomatal apertures in the leaf's surface, thus providing easy access for carbon dioxide.[43]

It is not the intent of this particular publication to dwell on the details of photosynthesis or the various cycles—carbon dioxide, oxygen, nitrogen, water, energy—that play roles in it. An informative series of articles on these subjects can be found in the September 1970 issue of *Scientific American*. Many other source references are also available. (See, for example, reference 44).

Figure 7.2 shows the availability of energy for photosynthesis, while Table 7.8 gives some typical photosynthesis efficiency estimates. There is some support for the thought that the average overall conversion of cultivated systems could be made as high as 5 percent through proper cultivation practices. This would be about an order-of-magnitude improvement on most natural ecosystems in the United States.[47]

The anatomy of wood is very complex and variable. Like other fibrous plant tissue, it comprises long, hollow, tubular cells having, in the case of wood, an aspect ratio of about 1000:1. The fiber walls are layered, with the middle layer

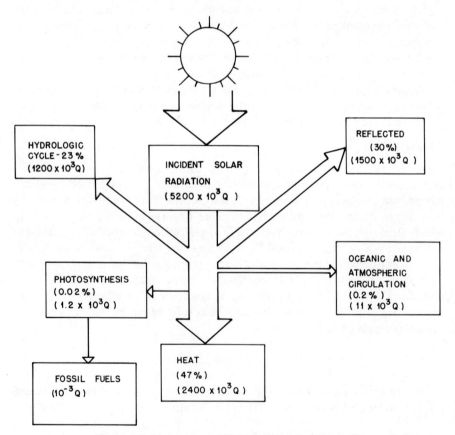

Figure 7.2   Solar energy. (Adapted from reference 45.)

Table 7.8 Photosynthetic Solar Energy Conversion by Plants

	Age, yr	Location	Solar Energy Conversion[a] (est.), %
Alfalfa[b]	<1	U.S. Midwest	0.29
Corn			
Mature silage	<1	U.S. Midwest	0.41
Stalks and ears	<1	U.S.	0.44–0.69
General crops	<1	U.S.	0.28–0.85
Sugarcane	n.a.	Louisiana, Florida	1.2
Slash pine			
Crown and bole	20+	Southeastern U.S.	0.24–0.30
Conifers	20 ± 2	U.K.	0.37
Sycamore	5	Georgia	0.64

Source: Reference 46. Reprinted with permission from G. C. Szego and C. C. Kemp, *Chemtech* **3**(5), 275–284 (May 1973). Copyright by the American Chemical Society.

[a] Assumes average annual insolation of 1300 Btu/ft²-day.

[b] Three cuttings per season.

largely determining wall thickness. The fibers of softwoods (gymnosperms) are typically about twice as long as hardwood (angiosperm) fibers. Because of their greater rigidity, wood fibers cannot be spun and woven as cotton fibers can. They can, however, form a wet web of matted fibers, bonded together through hydrogen bridges between hydroxyl groups of adjacent fibers.[48]

Silviculture

The net photosynthetic productivity of the earth has been estimated at 155 to 180 billion tpy.[40,49] Of this, forests account for about 42 percent and croplands 6 percent. Table 7.9 shows data on U.S. forest-based materials. Although the net productivity exceeds the annual consumption of fossil fuels on an equivalent energy basis, relatively little of the forest biomass is actually used as fuel. Most of our wood-based fuel (up to 1.1 Q) is used by wood-based industries, while another 2 Q of potential wood fuel remains available but unused.[49]

Ultimately, combustion for heat should not prove to be the most effective way of using wood. Its high moisture content drastically reduces its already low heat content. The same limitations also reduce wood's value as an alternative source of fluid fuels or chemical feedstocks. The value of forest residues as fuel is economically unattractive, in part because of the logistics of collecting them and delivering them to the consumer. At the present state of our harvesting and conversion technologies, however, we probably actually have no other choices at this time for utilization of our forest resources.

Trees are less demanding of fertility and overall soil quality than are field crops; hence they can be grown on land that is unsuited to field crop

Table 7.9 Forest-Based Materials, United States, 1970

	Billion ft^3	Million Tons^a
Consumption	15.1	241
Domestic production^b	14.0	225
Imports	2.4	38
Exports	1.4	21
Domestic growth^c	18.6	298

Source: Adapted from reference 50.

^{a}Average about 16 tons per 1000 ft^3 (approximately 15 for softwoods; 18 for hardwoods).
^{b}Projections: 1985, 17.5 bcf (280 million tons); 2000, 20 bcf (330 million tons).
^{c}On 495 million acres of commercial forest.

cultivation.[44] Between the two world wars, large amounts of marginal cropland reverted to forest cultivation. By the year 2000, the demand for softwood is expected to increase 40 to 50 percent and hardwood by 90 percent (over 1970 figures). Forest productivity has been increasing despite a relatively stable acreage dedicated to it.

It is felt that intensive cultivation practices should be capable of doubling the current productivity with no increase in forest land.[51] (See Table 7.10 for projected productivity.) This would entail introduction of the most up-to-date silvicultural technologies:[53]

1   Increased fertilization levels.
2   Improved irrigation and drainage.
3   Increase in stands of more valuable and faster-growing species at the expense of less valuable, slower-growing species.
4   Intensification of reforestation program.

Table 7.10 Projected U.S. Forest Productivity for about 2025^a

	Billion ft^3	Million Tons
Softwoods	19	290
Hardwoods	11	200
Total	30	490

Source: Reference 52.

^{a}Assumes: 5-in.-diameter (min.) bole, above 1-ft-high stump, up to 4-in.-diameter top. This represents about 50% of biomass. Complete harvest of stump, top, and large branches could increase yield by 25%.

5 Increased use of genetically improved growing stock.

6 Improved weeding efforts to favor more desirable species in mixed stands.

7 Increased thinning effort to concentrate growth on harvestable stems.

8 Intensified forest protection against fire, insects, and disease.

These technological improvements, however, must be reinforced by clarification or solution of a number of socioeconomic and institutional situations if the right climate for improvement is to be provided[54]:

1 Small woodlot owners, of whom there are about 1 million,[55] generally have little interest in forest management.

2 Environmental groups have a deep but somewhat ill-defined concern toward intensive forestry.

3 Burdensome real estate taxes on standing timber generally are applied annually to the same multiyear crop.

4 Environmental regulations discourage small operators and siphon off capital needed for productive capacity.

5 Regulation of transport through the Jones Act has had a significant effect on costs and efficiency of transport, as has the deterioration of the U.S. railroad system.

6 Outdated and conflicting building codes have had a depressing effect on a major outlet for timber products.

All of these factors, and their effects—direct or indirect—on forest products for fuel and feedstocks must be assessed and dealt with to achieve maximum efficiency in making these renewable resources available in increasing and reliable quantities.

Use of forest residuals (stumps, roots, limbs, etc.) is a major potential source of additional wood, and ultimately could nearly treble productivity from the 1970 base when combined with the doubling achieved through intensive management.[56] A conventional, natural forest yields about 1.4 to 2.7 dry ton equivalents (dte) per acre-year. (Each dry ton equivalent of wood has a heating value of about 17 million Btu.)[55] At the present time, stumps, limbs, crowns, and diseased and decaying trees comprise a neglected resource base amounting to hundreds of millions of tons.

If one were to assume a forest maturity rate of 30 years and a 25 percent efficiency in converting fuelwood to electric power, a 100-MW generating plant (serving 50,000 people) would require collecting the forest residues from about 1,568,600 acres (634,800 hectares) to provide the fuel. This means that the per capita requirement would be about 31.4 acres (12.7 hectares) to supply the electrical needs of the community.[57] A program was mounted by the Department of Agriculture (DOA) in 1976 to attempt to exploit this reservoir.[55] This does pose some potential hazards, in that such thorough harvesting of forest lands might deplete the soil of essential nutrients, which are concentrated in the boles, bark, and branches of trees.[58] This would then be reflected ultimately in decreased yields.[53] Although one might think living biomass is the major form of storage for organic matter (and hence energy) on forest lands, the organic matter below ground and lying on the forest floor contains 1.7 times as much energy as does the living-plant biomass.[59] The intent of whole-tree utilization is to harvest this unused resource.

Even without whole-tree usage, more intensive use of wood by clear-cutting of forests, can create havoc with the ecology by[60]:

1 Increasing stream flow.
2 Reducing transpiration.
3 Greatly increasing concentrations of dissolved chemicals in streams and soil.
4 Increasing erosion and particulate transport.
5 Raising soil temperature and moisture content.
6 Accelerating decay of forest debris.
7 Increasing nitrification.
8 Reducing organic content of forest floor.
9 Altering albedo.
10 Lowering pH of drainage waters.

To offset the potential harmful effects of such changes, the U.S. Forest Service now recommends that clear-cutting of hardwood forests be carried out on a 110- to 120-year rotation. In the Northeast, this means[60]:

1 Limiting clear-cutting to fertile, gently sloping lands having high recuperative capacity.
2 Clear-cutting only small areas, to ensure availability of seed and to minimize stream flow and nutrient extraction effects.
3 Minimizing forest floor and stream channel disturbance.

Resources

Like municipal solid waste and other urban industrial wastes, forests are renewable resources. On the surface, compared with MSW, they would appear to constitute a resource more uniform in composition and more predictable, and therefore somewhat easier to monitor and to forecast. This does not necessarily conform with reality. Wood waste, indeed is quite uniform in composition—at about 50 percent carbon, 44 percent oxygen, and 6 percent hydrogen, on a moisture-free basis. This is reasonably close to the elemental compositions of major crop residues and the organic portion of typical municipal solid wastes. Unfortunately, however, the wood resource quantity has been difficult to assess with any degree of accuracy, in part because of archaic methods of classification and measurement and the inaccessibility of some major stands.[54]

Classification of Forest Lands   About one-third of all U.S. land area (or about 754 million acres) is classified as forest land. It ranges in elevation from sea level to as high as 12,000 ft, depending, in part, on latitude. It covers a wide diversity of climates, soils, and topographies. Of this forest land, about two-thirds, or 500 million acres, is classified as commercial forest land. The classification of U.S. forest lands is shown in Table 7.11. The arbitrary yield limit for "commercial" forest lands is at least 20 ft^3 per acre-year. Below this limit, forests are considered "noncommercial." Above it, even private forest lands are classified as commercial, although they may not constitute producing stands.[54] Of course, rather extensive stands of timber are scattered in remote, hard-to-reach areas of the lower 48 states and Alaska, where cutting is not cost-effective. Nonetheless, they are still considered commercial if they meet the yield criterion.[62]

Estimates of yield are based on archaic methods not applicable to pulp, fuel, or feedstock needs. They are aimed, ultimately, at assessing the commercial volume of wood for sawtimber usage. A few definitions are in order before embarking on a more detailed account of the subject of wood resources:

Growing stock (commercial volume; standing volume)   The wood in the main stems (boles) of growing tree(s) over 5 in. in diameter* at breast height, above a 1-ft stump, up to a top diameter of 4 in.† May be expressed in volume, board-feet, or mass.‡

Table 7.11   Extent of U.S. Forest

Classification	Million Acres	Percent
Total U.S. land area	2270	100
Forest land	754	33
Commercial forest land	500	22
Privately owned	*365*	*16*
Publicly owned	*135*	*6*
Noncommercial forest land	254	11
Low yield	*211*	*9*
Public parks and wilderness	*17*	*0.7*
Deferred for possible wilderness classification	*3*	*0.1*
Inaccessible Alaskan forests	*23*	*1*

Source: References 54 and 61.

*Inside the bark.

†Stump, crown, limbs, foliage, bark, and roots are excluded from official statistics.

‡Based on average densities of 27.4 and 32.8 lb per ft^3 for softwood (gymnosperms) and hardwood (angiosperms), respectively.

Log scales (log rules)   Measuring devices for estimating volume of lumber recoverable from logs. Volume tables may be required in conjunction with some scales.

Sawtimber   Trees sufficiently large to contain at least one sawlog, that is, a log for cutting into lumber.

Board-foot   Nominally, a piece of lumber 1 in. thick and 1 ft square. (Actual seasoned lumber will have a minimum cross section of slightly over 70 percent of nominal, owing to shrinkage on seasoning and/or finishing.)

Biomass   Total content of plant [tree(s)], including main stem(s) [bole(s)], branches, twigs, leaves, fruit, bark, and roots.

For application to the manufacture of pulp and paper or chemicals, or for conversion to energy, current methods of arriving at needed statistics on renewable wood resources are not appropriate. Assessments should be based on mass, and include logging and manufacturing residues, rather than being based on recoverable sawtimber volume alone.*

A *renewable resource* is considered as one whose supply can be restored or replenished when the original stock has been depleted. This has come to imply that renewable resources are of biological origin.[63] The renewability ratio† for materials of biological origin is generally considerably higher than for materials of geological origin.

In practice, *silviculture* is far less advanced than agriculture, because of less intensive management throughout history, coupled with the far greater life span of forest crops. A few of the 60 or so domestic species that are commercially important have such long growing cycles (e.g., Douglas fir and redwoods) that they can scarcely be considered renewable resources in a practical sense.[63] Therefore renewability must be recognized as being species-, time-, and depletion-rate dependent.

Renewable resources are of two types, *primary* and *secondary.* The first are the output of a biological production system designed for the deliberate purpose of supplying plant- or animal-derived materials to industry. Most forest resources fall in this category. The second are by-products or residues of biological production systems whose essential purpose is to produce food, feed, or other nonmaterial goods.[64]

Yields

Wood accounts for over ¼ billion tons per year of industrial materials, with a market value of about $6 billion (in 1974) at local delivery points. Cotton-

---

*Pulping machinery must, however, be charged on a volume basis.
†The ratio of stock-renewal rate to stock-depletion rate.

derived materials are a distant second at 3 million tons per year, but at a higher value-added level per ton ($2 billion market value for fiber products alone).[54]

In 1970, the average volume growth rate for U.S. commercial forests was estimated at 38 to 46 ft^3 per acre-year,* depending on the source,[54,55] with growth ranging from 23 ft^3 on public lands in the Rocky Mountains to 65 ft^3 for intensively managed forest-industry stands on the Pacific coast. It has been suggested that had optimal state-of-art forest management practices been employed throughout the industry, the new annual growth rate could have been increased to as much as 74 ft^3.[54]

Simple application of nitrogen fertilizer at five-year intervals could increase production of Douglas fir on average sites by 15 to 20 percent, and up to 30 percent increases could be attained in many southern pine forests. The yields of conifers in the Pacific Northwest could be raised 30 to 35 percent by control of the stand density through frequent thinning.[65]

Between now and the year 2020, it is expected that our total forest acreage will decline by about 4 percent, but the softwood component of this acreage will increase by about 8 percent. Even so, it is felt that intensive management and whole-tree usage practices could increase net recoverable growth by a factor of 3 over this time frame. Greater attention to reducing high mortality losses (due to fire, disease, and insects) might easily provide the largest single source of increased forest production. The current mortality loss of close to 4.5 billion ft^3 destroys a substantial portion of the total annual growth.[66]

Not all forest residues are used effectively; in fact, nearly 10 billion ft^3 of wood residues from forest production can be classified as "nonused" (Table 7.12). This volume of residues is the energy equivalent of about 2 Q.[67]

Socioeconomics

The United States has been a net importer of wood-based products over the past six decades, basically because of our inefficient domestic production and transportation systems. In our society, it has been more attractive economically to employ fossil-based resources. Our more productive forest lands are in the southeastern and northwestern parts of the country, whereas our major population centers are in the southwestern, northeastern, and upper-midwestern sectors. To bring these products to market, we need a low-cost transportation system comprising both long-haul railroads and intercoastal water transport.

Over the past 30 years, our railroad system has become the laughing stock of the modern world, and federally regulated rate structures have put the kiss of death on long-haul rail transport of wood. The once-flourishing alternative of intercoastal marine transport has been virtually wiped out by the Jones Act, which places restrictions on interstate transport of goods by foreign carriers. This act, meant to reduce competition for our higher-cost domestic carriers, particularly in the petroleum field, has succeeded in undermining the water transport of forest materials from Alaska to the lower 48 states. What started

---

*About 0.5 to 0.6 dte of fiber per acre-year.

Table 7.12 Nonused Residues from Forest Production, 1970 (Estimated)

	Billion ft^3
Logging residues	
From growing stock	1.6
From nongrowing stock	1.6
Growth less harvest^a	
Softwood	1.1
Hardwood	3.5
Logging residues (growth less harvest)	
Softwood	0.1
Hardwood	0.4
Manufacturing residues (bark and wood)	1.5
Total nonused residues	9.8^b

Source: Adapted from reference 67.

^{a} Potential fuel pool.

^{b} Equivalent to about 2 Q of energy.

out with the purpose of becoming a socioeconomically beneficial regulation could well turn out to be an economic disaster.[68]

Silviculture, involving natural forests, as practiced today, is too inefficient—technically and economically—to stir up much interest as a potential source of a significant portion of our energy budget. Energy plantations based on short-rotation forest cultivation appear more promising. These will be discussed later in this chapter under "Tree Plantations."

Farm Wastes

Farm wastes are sometimes referred to as *agwastes. Agricultural refuse* (AR) is another term often used. It generally includes crop wastes, animal wastes (manure), and forest residues.

Field Crops

Crop residues are widely and thinly dispersed and often difficult to collect. As a result, they are frequently burned in the fields or plowed back into the soil. The principal crop waste energy resources at this time are bagasse and cotton gin trash. In 1976, bagasse contributed 25 million Btu* or 63 percent of the sugar industry's energy needs. Cotton gin trash† made an essentially "local" energy contribution of 0.95 million Btu.[31]‡ The by-products of these and other crop

---

*Equal to 3.7 million tons. Another source claims that 3.225 million tons was used in this way.[69]

†Comprising leaf fragments, sticks, other plant parts, and linty material (motes).

‡The heating value of most field residues is approximately 7500 Btu per lb.[70]

wastes also have little or no international mobility, being used almost entirely within our borders, or even within a relatively narrow regional range.[71]

Perhaps the major factor differentiating field crops from forest crops is that the latter are perennials or year-round resources,* whereas most agricultural plants are annuals or seasonal resources. With a forest-based resource, it becomes easy to establish stability in personnel and equipment and in facility utilization and to minimize storage capacity requirements,[72] because harvesting can be carried on throughout most of the calendar year, in response to demand. Annuals, on the other hand, must be planted and harvested in accordance with a seasonal growth cycle controlled by weather, climate, and the metabolism of the plant species.[73]

For seasonal crops on an annual growth cycle,† capacity would have to be provided for storage of at least a year's harvest, to even out fuel or feedstock supply over the calendar year. Prior to storage, it might be necessary to dry, grind, or otherwise process the material to ensure preservation. Perennials can be left standing virtually until needed, but even when cut, wood stores well. Since the bulk of the potentially usable wastes is produced in association with harvesting, the waste supply follows the same cycle as the crops. In warmer latitudes, longer growing seasons may make it possible to harvest seasonal crops on a shortened cycle—two or three crops per year—thus evening out supply and reducing storage requirements. *Canopy culture* (cultivation under cover of a transparent plastic film) also allows multiple harvests throughout the year.

In any case, the capital tied up in stored resources and underutilized facilities can play a major role in determining the economic feasibility of an energy option based on agwaste. For power generation, it could become economically necessary to provide an alternate (such as wood waste, coal, or lignite) to the crop waste resource, to even out the energy supply. Such an option is not necessarily feasible in the case of the generation of feedstocks from these wastes.

The keys to the efficient utilization of crop residues are price and availability. The prices of a wide variety of plant residues from farms and forests (1975 to 1976) are in the range of $20 to $30 per bone-dry ton (bdt).‡ This includes the cost of collecting, storing, and transporting the waste.[72] It has been estimated that, assuming an 80 percent generation load factor, to feed power plants having a total 90,000-MW power generation capacity would require 430 million tons of plant residues. To attain this level of utilization, trade-offs would have to be made to divert wastes from other uses.

A regional analysis shows that approximately 58 percent of our agwaste is now returned to the soil, 22 percent is "nonused" or excess residue (hauled

---

*Kelp, algae, water hyacinth, grasses, and bamboo are other potential year-round resources.[72]

†Essentially a 5.5-month growing season in the major U.S. farm belt.[74]

‡Moisture contents (as is) vary considerably. Grass straw (10 percent moisture content) and forest residues (50 percent) represent essentially practical limits.[72] Major agricultural crop wastes are generally assumed to have a 30 percent moisture content.[75]

away at cost or burned in the fields), and 20 percent is used for fuel (5 percent), animal feed (5 percent), or is sold for nonfuel uses (10 percent).[76] National figures, if available, would probably show a somewhat similar pattern of crop waste utilization.

The return of waste to the soil has beneficial consequences that must be weighed against the benefits of diverting the material to fuel and feedstock needs. Many of a plant's nutrients end up in grain stubble, corn stover, and other unused portions of plants,[39] and plowing them into the soil provides needed elements for future crops. The carbon content of the soil must also be maintained at a level of about 1.5 percent (equivalent to an organic content of approximately 2.6 percent).[57] The residues also serve admirably as soil conditioners: enhancing the water-holding capacity of the soil, increasing microbial activity, aerating the soil, reducing erosion by wind and rain, and providing a source of carbon dioxide for photosynthesis. The agronomic effects of withholding this waste from the soil could have an overall deleterious affect on soil quality, leading to subsequent reduced crop yields,[74] especially in soils of low fertility or in areas subject to severe leaching. The long-term removal of forest litter (branches and leaves) from European forests has reduced productivity there by an estimated 20 percent.[39]

The major energy and feedstock component of plants (woody and otherwise) is *cellulose.* Wood ranges from about 40 to 50 percent in *alpha-cellulose* content, whereas field crops may be either crude or *holocellulose.** The *crude cellulose* content of field crops ranges from 20 percent for bluegrass to 91 percent for cotton fibers.[39] The average annual productivity of field crops in the United States (1972) was nearly 4.5 times that of commercial forests (based on yield of merchantable bolewood only). Based on bolewood, branches, and new foliage, the factor is reduced to about 1.5 to 2.0.[39]

U.S. fields produced about 794 million tons of crops in 1972, of which the cellulose content was approximately 318 million tons or 40 percent.[39] About 60 percent or 474 million tons of the total crop was in the form of residues.[77] Nearly a third of our domestic field-crop agwaste is from cereal straw. About a fifth comes from corn and sorghum stover. Approximately one-fourth of the agwaste is corncobs, fruit pits, and nut shells, combined with soybean hulls and stover. Sugarcane bagasse accounts for only a little over 1 percent of the total. If animal residues are included, the total agricultural residues amount to 805 million dry tons, about 41 percent animal and 59 percent field wastes.[77]†

Food (and allied products) production is the fourth largest "industrial" consumer of energy in the United States (after the primary metal industry, chemical and allied products, and petroleum refining).[78] The feasibility of agriculture as a source of fuel and feedstocks must lie in part, with the *efficiency*

---

*The carbohydrate fraction of extractive-free wood.

†Between 50 and 80 percent of the farm animals are in confinement, where their wastes can be collected.

of cultural energy utilization.* Most field crops have an efficiency of about 4 to 10:1,[79] vs. approximately 1.2:1 for forests.[80] Sugarcane stands out with a ratio of 22:1.[79]

Cellulose is available in both plant material and in waste. The global fixation of carbon dioxide is estimated at 7.3×10^{14} lb (365 billion tons) per annum. Approximately 6 percent (22 billion tpy) is converted to cellulose. Of this, only about 20 percent (4 billion tpy) is readily available for conversion to fuels and feedstocks. Most of the cellulose is too widely distributed for economical collection and transport to conversion facilities.[81]

On a local basis, some crop wastes may be available in sufficient quantity to warrant collection for conversion. A case in point is the 1 million tpy of grass straw residue within a 50-mi radius of Eugene, Oregon. It is a residue of the grass seed industry, and is normally burned because the high lignin content makes it indigestible as an animal feed. Air pollution regulations serve as a further incentive for converting this waste.[81]

Primary cultivation of field crops (on energy farms) is a tempting option for energy to supplant conventional petroleum-based fuels. A closer look, however, reveals that we discard more of the *net primary production* (NPP) than we use for food and animal feed. Agwaste would therefore appear to be a more logical source of energy provided that the energetics, agronomics, and economics are favorable.[79]

Renewability of biomass resources (based on field crops) for use as fuel and feedstocks is not a foregone conclusion. It hinges on a number of factors, among which are[82]:

1. Competition with food cultivation for land use (which depends heavily on growth rate of population).
2. Availability of water† and nutrients (phosphorus in particular).
3. Technical advances in photosynthetic productivity.

Animal Wastes

Since the turn of the century, the meat-packing industry has been complimented on its efficiency in using "everything but the squeal," but now promises to add a new dimension to its legendary efficiency by employing manure as a source of energy.

It has been estimated that fermentation of only one-eighth of India's cattle manure‡ could provide that country with much of its fertilizer needs§ while pro-

---

*The ratio of crop productivity (in terms of that portion used for energy) to investment of cultural energy (fuel, fertilizer, and so forth, input by farmer, but excluding solar energy).
†About 100 lb (12.5 gal) of water is required to produce 1 lb of plant material. On drying, about $\frac{1}{4}$ lb bone dry biomass is left.[83]
‡Total estimated at 800 million tons annually.[84]
§Taking the place of $1.5 billion worth of synthetic fertilizers annually (more than half the nation's foreign exchange).[83]

ducing much-needed energy. A daily input of 100 lb of manure to a low-cost fermenter could make a family of five or six, owning four or five animals on about 7.5 acres (3 hectares) of land, self-sufficient with regard to fuel and fertilizer.[83]

In the United States it has been found that collections from a cattle feedlot with a 100,000-head capacity could produce $1.8 million (1976 dollars) worth of ethylene each year.* Conversion plants would have to be built near the sites of large feedlots (existing largely in the western high plains states and Florida), since the raw material is heavy, bulky, and has a high water content. In part because of the water, it has only half the heating value of coal.[85†]

Nationwide, we have an annual renewable resource of nearly 2 billion tons of animal waste, about 8 to 10 times our urban waste (household and commercial combined). Unlike crop wastes, this is an essentially constant supply, "harvestable" throughout the year. This raw material, however, introduces a problem of wastewater treatment and disposal because of its high water content. Also, collection and storage of the methane gas product can lead to problems for the small (two- or three-farm) producer, because of capital requirements. Furthermore, the low efficiency of digestion at temperatures below 86°F (30°C) makes it necessary to add heat in colder climates.[57] *Mesophilic digestion,* used in the earlier treatment plants, takes place at about 95°F (35°C) and requires a retention time of 15 to 30 days. The newer *thermophilic* process employs a bacterial strain found in cattle manure. It is carried out at 131 to 140°F (55 to 60°C) with a retention time of no more than 10 days. There is preliminary evidence that the higher temperature may partially sterilize the sludge.[87]

Most processes for energy recovery produce methane gas from manure, but the Pittsburgh Energy Research Center (PERC) of the Bureau of Mines demonstrated a method capable of producing a heavy oil from manure, with a reported yield of about 96 percent of theoretical.[88] This and other processes will be described later in the chapter. If *all* this country's agricultural wastes (including animal wastes) could be collected and converted, they could produce the equivalent of about 35 percent of the nation's 1978 oil consumption, or 2.45 billion bbl of oil annually.[89] Needless to say, much of such waste is economically unavailable because of its extensive dispersal in the fields. Collection of these wastes cannot be institutionalized as municipal solid wastes can be; hence, except for the large cattle feedlots, manure conversion could turn out to be little more than a local cottage industry.

ENERGY FARMING

Plants, including trees, can be cultivated intensively for the specific purpose of providing sources of fuels and/or feedstocks. In such cases, the primary goal is

*Ethylene is the backbone of the petrochemical industry.
†The human output of feces is only about ½ lb per day, sufficient to produce 680 Btu of methane, or enough to boil about 3 pt of water.[86]

to produce the largest possible quantity of recoverable biomass per unit cost and unit land area, irrespective of plant species, food value, or worth as a consumer product. Lesser goals are to achieve these high yields on lands of less than optimum fertility, in a diversity of terrains and climates, without the need for costly and possibly time-consuming land rehabilitation operations between crop cycles.

Tree Plantations

Tree culture faces a particularly perplexing dilemma in that its normal products, lumber (sawtimber) and pulpwood (roundwood), are more valuable economically than "tall crude" or wood for fuel or feedstocks. Furthermore, as noted earlier, intensive cultivation of forest lands can lead to severe environmental consequences. It is quite apparent that hard decisions must be made on how best to utilize this and other biomass resources. These decisions must be based on resource availability, existing technology, the environmental (including soil) impacts, economics, and the social necessity or desirability of employing the option.

As noted before, to increase forest production, we must convert to stands of higher-yield, genetically improved growing stock, intensify reforestation, weed and thin effectively, and reduce mortality. Through such action it should be possible to double wood productivity within about 50 years. Whole-tree utilization, in addition to more intensive forest management, could make it possible to triple productivity in the same time frame.[66]

For best results, intensive management would imply greater emphasis on tree plantations in accessible regions of the country such as the Southeast, but also including more level, open terrain in the Northeast and along the Pacific coast. It has been found that species such as sycamore and hybrid poplar (cottonwood) can be cultivated efficiently on one- to three-year (*minirotation*) cycles.[56,89] These fast-growing species can be propagated by *cuttings* (clippings of stemwood) or can be regenerated by *coppicing* (stem-sprouting) on tree plantations. These would normally be harvested on a 10- to 20-year (*short-rotation*) cycle to yield 430 ft³ per acre-yr (30 m³ per hectare-yr)[90*] vs. 100 ft³ per acre-yr (7 m³ per hectare-yr) for natural, mixed forests harvested on a 60- to 120-year cycle. *Juvenile coppices* (on a one- to two-year cycle) will yield 6.7 tpa-yr oven dry (15 oven-dry tonnes per hectare-yr) or more of total biomass. More normal annual harvests for minirotation cycles (two to three years) of poplar and sycamore would be 2.7 tpa (6 te per hectare),[57] although 6 tpa (13.4 te per hectare) of sycamores have been produced on otherwise useless land in the Southeast on a four-year growth cycle.[92]

Since these species produce regrowth from the stumps, cultivation of the soil

---

*In a private communication, Anderson[91] explained that volume units here reflect a method of assessing short-rotation growth nondestructively. On harvesting, assessment will be on a mass basis. Minirotation growth is measured by cutting and weighing. A single volume-mass conversion factor is not necessarily assured.

is not necessary, and soil erosion is minimized. Harvesting would entail the use of machines that fell and chip the trees for efficient transport to the power plant or conversion facility. Use of this product as fuel for a 100-MW power plant to provide electricity for a community of 50,000 would call for a 373,000-acre (105,960-hectare) sycamore plantation operating on a two-year harvest cycle. This amounts to about 7.4 acres (3 hectares) per person, and compares with the 31.4 acres (12.7 hectares) required using forest harvest residues, as noted earlier.[57] The latter would possibly lead to long-term, irreparable depletion and erosion of the soil, hence does not appear to be an attractive option, although failure to use these residues would decrease near-term yields of energy.

The preferred strategy for tree plantations would be to employ marginal lands that would not normally compete with production of food crops and natural forests. This could, however, lead to low productivity, which could make this option economically unattractive.

Fuel Cultivation

Two of the more imposing problems facing proponents of agwaste utilization and energy farming are deterioration of the soil, and the world's population growth rate and its pressures on the use of land for growing food. This section is intended to shed some light on these and related problems.

Soil Loss

The United States is blessed with a relatively high percentage of arable land* compared to the world as a whole (25 percent vs. 11 percent).† Another 75 million acres or so could possibly be made arable by draining swamps, irrigating deserts, and grading land. This would be a very costly undertaking—in both money and energy—hence the gain might not warrant the cost.

Over recent years, the United States has lost arable cropland at a net annual rate of 1.25 million acres. Equivalent increases in cropland through irrigation, drainage, and so forth, have been more than offset by losses to highways, urbanization, and special land uses. Since 1945, arable croplands nearly equal in area to the state of Nebraska have been lost to cultivation by such means. Between 1949 and 1969, 15 percent (58 million acres) of U.S. cropland was withheld from production, and crop production rose 30 percent), so these land losses were compensated for by increased productivity. The energy cost of achieving this level of productivity, however, was significantly higher.

Much has been said about the damaging effects on the soil of strip mining. About 153,000 acres are directly disturbed each year, and another 450,000 to 750,000 acres are affected by acid mine drainage (AMD) and soil erosion due to mining activities. Little is said about the 32 *million* acres of land covered by highways and roads to date, or the 667,000 acres devoted to merely parking our

---

*Approximately 470 million acres.
†About 81 percent of our arable land is under cultivation.[93]

cars. Nor has it been widely publicized that the bulk of these highway and urbanization losses (probably two-thirds) occur to croplands, since most population centers have been built on waterways, where the land is richest, and most highways follow flat river basins.[93]

But this is only the tip of the iceberg. Over the first 200 years of this nation's history, it is estimated that one-third of our topsoil has been lost. By about 40 years ago, approximately 100 million acres had been lost "for practical cultivation" through erosion. Soil erosion had "seriously impoverished" about 200 million acres of croplands prior to 1940.[94]

Normal agriculture forms soil at the rate of 1 inch per century (about 1.5 tpa-yr), and the average annual loss of topsoil from croplands is presently estimated at about eight times this rate. Water runoff carries 4 billion tons per year of sediment into the waterways of the lower 48 states. Three-fourths of this comes from agricultural lands, and another one-fourth from logging, construction, and so forth. One-fourth ends up in the ocean, with the rest being deposited in reservoirs, rivers, and lakes. About 450 million yd^3 of these sediments must be dredged from rivers and harbors each year, at a cost of $250 million. The overall cost of sedimentation damage is $500 million annually. For the United States as a whole, an amount of soil loss equal to one-fourth (1 billion tons) of the soil erosion due to runoff is attributed to wind.[95] Together, water runoff and wind erosion transfer 5 billion tons per year (equivalent to a 7-in. depth of soil over 5 million acres) to our waterways and elsewhere.* This amounts to a cropland removal rate of about 12 tpa-yr.

About 64 percent of the nation's croplands stand in need of treatment for excessive soil erosion. This is particularly true with land used for row crops (e.g., corn, soybeans, and cotton) planted on slopes. Yields from these eroded soils are lowered because of low nitrogen and organic content, deteriorated soil structure, and reduced accessibility of water.

Despite the "Dust Bowl" situation of the southern Great Plains† which developed in the 1930s, and more recent reminders on a lesser scale in the winter of 1976–1977, all evidence indicates that the intervening expenditure of nearly $15 billion on soil conservation has not achieved the desired results. Actions of those most directly involved—the farmers—suggest that near-term crop yields and profits have been accorded first priority consideration over long-term soil conservation, under the economic pressures inherent in a boom-or-bust business. It should be obvious that farm soil erosion which allows 15 tons of topsoil a second‡ to be carried out the mouth of the Mississippi River cannot be continued indefinitely without a major decline in crop yields.[96] Yet indications are that contour plowing and terracing, strip-cropping, crop rotation, leaving agwaste on the soil, conversion of marginal erosion-prone cropland to pasture land, planting windbreaks, and practicing conservation tillage are all receiving short shrift. The result is that from more than one-third of all our cropland[97] to

---

*One acre-inch of soil weighs about 150 tons. The average topsoil depth on cropland is 7 to 8 in.[94]
†Large portions of Texas, New Mexico, Colorado, Oklahoma, and Kansas.
‡Nearly 500 million tons per year, via a single watershed.

as much as two-thirds of our wind-erosion susceptible cropland[98] is sustaining excessive soil losses leading inexorably to a decline in productivity. Indeed, nearly 85 percent of 283 farms surveyed in Wisconsin have been found to be losing over 5 tpa-yr of topsoil.[99]* Although 18 states have passed sedimentation control laws to protect the watersheds, all but two of these states have seen fit to exempt soil erosion due to farm activities.

To ensure our continued ability to provide food and animal feed for future generations in the face of a growing global population, increased crop yields must be realized, and doubly so if we contemplate energy farming on any significant scale. It has been suggested that perhaps the only answer to our soil erosion problem is to make future farm loans, crop insurance, disaster relief, and price supports contingent on compliance with adequate soil conservation practices.[96]

Biosynthetic Conversion

The prospects of growing biomass specifically for fuels and feedstocks is closely tied to the future of technology for improving and controlling photosynthetic conversion of solar energy. There is much effort being expended in this direction at the present time, although it has been stimulated largely by the world's increasing needs to meet food demands for the rising population.

It has been found that selective breeding of some plant varieties for high yields can lead to more disease- and pest-prone strains. The potential ecological effects of pesticides, herbicides, and excess nitrogenous fertilizers are also causes for concern, as are long-range predictions of worsening weather conditions for crop cultivation on a global scale.

The prospects for energy farming are somewhat nebulous at present. The use of marginal land—with its attendant lower yields—is essential if we are to continue to provide food to the world. Careful choice of plant species and efficient biosynthesis will be necessary if we are to add the burden of energy cultivation to food cultivation.

The *photosynthetically active radiation* (PAR) of the sun (400 to 700 nm) is only a fraction of the total available radiation. Taking into account all factors contributing to photosynthesis, the maximum theoretical efficiency for conversion by aquatic plants (unicellular algae) is about 12.3 percent, and the maximum possible efficiency for land plants is about 10 percent. When reduced by the energy wasted in plant respiration, the latter figure becomes about 6.6 percent—the maximum solar energy conversion efficiency.[100] The actual productivity is even lower, as shown in Table 7.13. Selected C-4 plants have an actual metabolic efficiency of 53 to 73 percent of the theoretical maximum, and C-3 plants range from about 32 to 44 percent. Much is said about the high yield of sugarcane because it grows year round, but Mississippi Delta sugarcane has only a seven-month growing season and its productivity is appreciably lower. There is reason to feel that technology could ultimately increase actual solar energy conversion efficiencies to 4 to 5 percent.[101]

*Over three times the rate of natural formation of soil.[93]

Table 7.13 Photosynthetic Productivities during Active Growing Season

	tpa-yr	Percent of Theoretical
Theoretical maximum photosynthetic productivity (annual average)[a]	117	
Maximum measured productivity		
C-4 plants		
Sugarcane	62	53
Napier grass	64	55
Sudan grass (sorghum)	83	71
Corn (*Zea mays*)	85	73
C-3 plants		
Sugar beets	51	44
Alfalfa	37	32
Chlorella[b]	46	39

Source: Reference 100.

[a] U.S. Southwest.

[b] A unicellular algae grown in tanks.

Canopy Culture

Assuming a yield of 30 dry tons of biomass per acre-yr, and conversion of 1 lb of biomass to 5 scf of pipeline-quality SNG, about 70 million acres would have to be under cultivation just to meet our current demand for natural gas. This is somewhat more than the cropland area dedicated to corn cultivation in 1973, but much of this corn land was in areas with too short a growing season for biomass yields of the level assumed above.[74] One possible technical solution to this dilemma would be canopy culture. Such a practice would conserve water and provide carbon dioxide enrichment, but some fear that it could introduce problems of heat dissipation.

We noted above that C-4 plants, in air and bright sunlight, are more efficient solar converters than are the best C-3 plants. The main difference is that the latter waste energy in photorespiration. This effect could perhaps be offset by cultivating C-3 plants in atmospheres low in oxygen (2 percent vs. ambient 20 percent) and high in carbon dioxide (0.13 percent vs. ambient 0.032 percent).* Table 7.14 shows the effect of raised CO_2 levels on productivity. There is also a remote possibility of imparting C-3 plants with C-4 metabolism characteristics by breeding.

The ability to enhance net photosynthetic production by elevating the carbon dioxide content of the growing environment suggests the possibility of a viable *canopy culture* (or *closed-cycle*) technology. Alfalfa could be harvested

---

*High CO_2 levels also produce dramatic increases in nitrogen fixation by legumes. About 0.1 percent increase in CO_2 level raises fixation of nitrogen by alfalfa by a factor of 5.[100]

Table 7.14   Effect of Raised Carbon Dioxide Levels[a] on Photosynthetic Rate

	Approximate Increase in Productivity Rate, %
Corn, grain, sorghum, sugarcane	33–67
Soybean, sugar beet	40–87
Cotton	100–150
Sunflower	100–160
Rice	80–238

Source: Adapted from reference 100.

[a] Factor of about 3.

up to 12 times a year in such a controlled climate,[100] but canopy culture seems more suited to cultivation of higher value-added specialty crops for feedstocks. Bassham[101] suggests that 1-km^2 by 300-m high canopies of transparent, inflatable plastic might be erected in the U.S. Southwest, where sunlight is plentiful and water is at a premium. The plants would have to be selected specifically for cultivation under a transparent plastic cover, where the greenhouse effect would cause severe heating. Carbon dioxide could be obtained from power plant flue gases (or from CO_2 gas wells), but resistance to gaseous contaminants from fuel combustion could prove to be a problem. This use of combustion gases, however, would have the added benefit that CO_2 discharges to the atmosphere from fossil fuel combustion would be reduced. The plants might also scrub the remaining SO_2 from partially treated flue gases.

Overall, the growing environment would be enriched with CO_2, water would be retained and recycled, photosynthesis rate hopefully would be enhanced, and arid land would be brought under cultivation. There are mixed feelings as to whether or not plant diseases and insect infestations would be controlled more readily. The scheme is far from reality, from the standpoints of engineering and economics, but it is conceptually feasible and deserving of further study. One of the major unanswered engineering questions involves selection of a suitable canopy material—transparent to PAR, durable, strong, low-cost, and so forth. Initial calculations suggest that about 4000 mi^2 (10,000 km^2)* under canopy culture could provide the biomass fuel to meet the complete electric power demands of the state of California in 1985. Perhaps the most attractive aspect of the proposal is that low-cost, arid land not conceivably suitable for normal agriculture could be used.

Plant Growth Regulators

One possible means of improving plant yields is the judicious application of *plant growth regulators,* organic chemicals that affect physiological processes

---

*About a 63-mi^2 area.

in plants. There are four different classes of plant-produced growth regulators (*plant hormones* or *phytohormones*):   *auxins*—cause enlargement of plant cells; *gibberellins*—stimulate cell division, cell enlargement, or both; *cytokinins*—stimulate cell division; and *inhibitors*—inhibit or retard physiological or biochemical processes.

There is much speculation as to the mechanisms whereby these plant hormones function, but there is no clear-cut agreement at present. Ethylene is constantly evolved by plants, and there is some feeling that this or other unsaturated gases may be involved in plant bioregulation.[102] (Acetylene gas was used as early as the mid-1930s to initiate flowering of pineapple plants in Hawaii.)

Sugarcane is a so-called short-day plant, which initiates flowering only within a narrow range of day lengths. Suppression of flowering by application of plant hormones increases the sugar content by as much as about 1.3 tpa. As little as 2 oz of gibberellins per acre can increase cane yield over 5 tpa and sugar yield by 0.2 to 0.5 tpa, while a ripener (glyphosine) raises sugar yield 10 to 20 percent, depending on other factors, such as soil and weather conditions and plant variety.

Other plants are also affected in various ways by these plant hormones—ease of harvesting, plant size, environmental or pest resistance, latex flow and yield (in rubber trees), nutrient uptake, and so forth. There is much to be learned about these compounds, yet they already represent 5 to 10 percent of the agrochemical market.[102] Adoption of biomass as a fuel and feedstock resource may well hinge on developments in this field in the next few years.

Societal Pressures

At present, the average American directly or indirectly consumes about 1 ton of grain per year, 93 percent of which is animal feed to provide our meat, milk, and eggs. Our eating habits have rubbed off on large portions of the United Kingdom, West Germany, Italy, U.S.S.R., and, particularly, Japan, which are consuming meats in much greater quantity than in years past. The United States and Canada supply about 80 percent of the grains imported by these nations.[102] In 1974, the United States contributed to this international agricultural market, 85 percent of the soybeans, 60 percent of the feed grain (corn and sorghum), 45 percent of the wheat, 30 percent of the cotton, and 24 percent of the rice.[103]

The concern is that with the global population expected to rise to 5 billion by about 1987[103] and 6 billion by 2000,[104] our croplands may no longer be capable of supplying sufficient grain exports for the world, let alone providing energy crops for domestic use. One source suggests that just supplying adequate food for a populace growing at the present rate of 2.2 percent per annum* is a hopeless undertaking. The less developed countries (LDCs) would have to begin looking to their own resources for food, since the United States and Canada

*Doubling time 32 years.

conceivably cannot continue to meet the demand. This means, for the LDCs, a necessary change from subsistence farming to market agriculture—a change that may not be achievable.

Ultimately, population/food pressures must also begin to affect the industrialized nations as well. The end result is bound to bring about increasing pressures on our deteriorating soils. Population control measures would take time—possibly 75 years—to revert to *zero population growth* (ZPG), because of the low average ages of the inhabitants of many of the world's faster-growing nations. Toward the end of this period, when we might hope for our energy farming to be at the peak of its development, the pressures on our soils for food/feed farming should be at their maximum. Seemingly, the best we can hope to do is to employ only marginal lands for energy farming—concentrating on intensively managed forests and on specific, high-yield specialty crops capable of growing on poor soils and in relatively dry climates.

Povich[105] notes that from the standpoint of fuel, 1 million acres of sugarcane would be equivalent to about 55 million bbl of oil. But at 1976 sugar prices, this would yield $3 billion worth of sugar on the market—sufficient to purchase 200 million bbl of oil at $15 per bbl, or 100 million bbl at the second quarter 1980 OPEC price of crude. Under these circumstances, could we afford to produce sugarcane-based oil?

Unit Operations

A number of unit operations are essential in farming[106]:

1 Preparation of the site (clearing, grading, plowing, harrowing).
2 Sowing or planting.
3 Cultivation (weeding).
4 Cutting and harvesting product(s).
5 Gathering and transporting product(s).
6 Subdivision (shredding, grinding).
7 Conversion.
8 Packaging and distributing product(s).

Annual crops such as corn require all of these measures, but the first two stages can be omitted with plants that coppice. Even coppicing is not necessarily indefinitely repeatable, however, since after four such crops, eucalyptus growth, for example, loses vigor and yields decline. Stumps must then be pulled and new growth has to be planted. Algae require no sowing and planting, but the high costs of site preparation, pumping facilities, and so forth, more than offset the sowing.[106]

Basically, site preparation can entail drastic grading, terracing, and drainage on a vast scale, to prevent severe soil erosion. This can prove to be a very costly operation. Planting of windbreaks to minimize wind erosion may not be too practical on marginal land in arid climates, especially since maximum yields depend on high incident light intensity throughout. Yet protection of the soil is essential if our ability to provide high farm productivity is to continue.

Corn, a plant offering high yields, is used both as human food and as animal feed. Approximately 7 weight percent ends up as human food; the balance of the plant (excluding the roots) is used as cattle silage. If we were to adopt corn as a fuel and feedstock resource we would probably find it necessary to reduce meat consumption rather drastically. Adoption of any other plant for fuel and feedstocks would undoubtedly call for similar trade-offs in availability for food, feed, or fibers, sacrifices that society may not be able to afford if life is to be sustained.

In the cultivation of sugarcane, fertilizer is the major element of cost (approximately one-third in Brazil), because of the high cost of ammonia from natural gas. Biological nitrogen fixation in the sugarcane field could reduce this cost.[107] In Hawaii, the nitrogen requirement for sugarcane averages about 360 lb per acre per 20- to 24-month growing season (with a range of 300 to 450 lb).[105]

The fuel yield from sugarcane is about 320 million Btu per acre-yr (at 8000 Btu per lb). Bagasse is not a surplus commodity because of the high energy requirements of sugar production. In fact, the boiler fuel used is a combination of bagasse and a considerable quantity of oil.* Approximately half of the power generated is for irrigation, and the other half is used in the factory. The residual ash from the combustion of bagasse is physically unsuitable as fertilizer or soil conditioner and is generally disposed of as landfill. Sugar production also requires about 1 ton of water for every pound of sugar produced.[105]

Mississippi Delta sugarcane fields require no irrigation, but their yields are appreciably lower (25 tpa-yr of cane or 7.5 tpa-yr vs. 28.5 tpa-yr in Hawaii, of dry combustibles) due to the 7-month growing season. Excess bagasse is sometimes available, because power is not needed for irrigation.[105]

In energy farming, large amounts of water would be required for optimum yields; hence the success of such a venture could depend on water availability. The limitations of phosphorus resources could be another limiting factor in energy farming. If all our fuel were to be derived from energy farming by 2070, our nonrenewable phosphorus resources would last only 117 years.[105]

Nitrogen, too, is an important factor in any consideration of biomass, because it is also essential to plant growth. Unlike phosphorus, however, it is available—at a price in energy. Wild plants consume about 2.5 percent of photosynthetic energy (directly or indirectly) in fixing atmospheric nitrogen. Reducing nitrogen in nitrates, already fixed in the plant decay cycle, requires expenditure of another 5 percent of photosynthetic energy. In the manufacture of synthetic fertilizer, we expend 2 percent of our entire global energy budget in chemical fixation of nitrogen, and we duplicate this amount of energy in merely

*For example, about 1 bbl of oil for each 3 tons of bagasse.

distributing the resulting product.[108] The fundamental reaction in nitrogen fixation is

$$\tfrac{1}{2} N_2 + \tfrac{3}{2} H_2O \rightarrow NH_3 + \tfrac{3}{4} O_2 \tag{1}$$

with ammonia representing nitrogen in the highest energy form. Equation 1 consumes about 81.2 kcal (340 kJ) per mole of energy, which means the energy equivalent of 0.11 mole of glucose is consumed in producing 1 mole of ammonia (assuming theoretical energy transfer).[107]

Our synthesis of ammonia began with the direct combination of nitrogen and hydrogen by various high-pressure, catalytic processes:*

$$\tfrac{1}{2} N_2 + \tfrac{3}{2} H_2 \rightarrow NH_3 \tag{2}$$

Prior to World War II, most processes produced the required hydrogen by reacting coal (or coke) with steam, but current technology calls for *reforming* with natural gas and water:

$$CH_4 + H_2O \rightarrow CO + 3H_2 \tag{3}$$

and then undergoing the *water-gas shift:*

$$CO + H_2O \rightarrow CO_2 + H_2 \tag{4}$$

Together, these reactions require considerable energy input (39.5 kcal per mole). Air is introduced in the secondary reformer (equation 4) to produce a raw synthesis gas with a mole ratio of 3:1 H/N. The synthesis gas is purified by absorption of carbon oxides, to prevent poisoning of the ammonia synthesis catalyst, and the purified gas is methanated. The purified H_2/N_2 synthesis gas is then compressed and catalytically reacted to form anhydrous ammonia.[109]

The overall synthetic process is four and a half times as energy efficient as biofixation, but this advantage is reduced to a factor of about three and a half by the energy cost of distribution of the synthetic product.[108] This advantage is reduced still further through the loss, to air and water, of about 50 percent of the applied fertilizer by accelerated denitrification, leaching, and runoff. Net delivery of usable fixed nitrogen by the synthetic route is no better than by the natural biosynthetic route, because the plants must then expend energy re-reducing the synthetic fertilizer, which is bacterially oxidized to nitrate before being taken up by the plant.[108]

Of the total 1160 million tons of nitrogen fixed naturally each year, over 80 percent is the result of recycling. Somewhat less than 15 percent is biofixed, and the balance is fixed by atmospheric reactions—lightning, forest fires, ozonization.[108] If we had to depend solely on natural sources, much more marginal land

*Haber-Bosch, Claude, and Casale processes, to name only three.

would have to be brought under cultivation, and much energy and money would have to be expended to clear, grade, and irrigate it. Even then, erosion would be a constant threat to these new, substandard croplands.

Despite the fact that each unit mass of fertilizer nitrogen delivered and used costs the fossil fuel equivalent of six and a half times that mass of glucose, plus the plant energy equivalent of six times that mass of glucose to reduce the nitrate, there are good and just reasons for continuing to synthesize fertilizer[108]:

1 Neither biofixation nor decay can be increased sufficiently to keep up with our food and feed needs.

2 The benefit-to-cost ratios of fertilizers (about 3:1) provide ample short-term justification.

3 High yields, made possible only with fertilizers, reduce the amount of cropland required to provide our agricultural needs.

There is doubt that fertilizer manufacturing efficiency can be raised about 38 percent, but there is some hope that assimilation can be enhanced beyond the present 50 percent figure. If all sewage nitrogen were to be returned to the fields, only about 10 percent of our fertilizer needs would be met, and delivery costs would offset most of this.[108] The use of natural gas to manufacture fertilizer (at the rate of 30 scf for 1 lb of nitrogen) could well be leading to a long-term disaster. In this form, nitrogen is a nonrenewable long-term resource, requiring about one-third of all the fossil energy we devote to agriculture.[110] Global production of chemically fixed nitrogen fertilizer increased by two orders (441,000 to 44,100,000 tpy) between 1905 and 1974. Since then, it increased further by 25 percent, to 55,125,000 tpy in 1978 (estimated). Biologically fixed nitrogen appears to have stabilized at about 165 to 193 million tpy.[110]

Specific Plants

There are numerous plants that have been cited as potential hydrocarbon sources of raw materials for specific fuels or feedstocks. Calvin calls sugarcane (*Saccharum officinarum* L.) the "most efficient solar energy device we have to-day on a large scale."[111] It is a voracious feeder, exhibiting rapid growth over a 20- to 24-month growing cycle. It produces clean cane at the rate of 90 to 100 tpa per crop (or 60 tpa-yr). It is about 30 percent dry combustible material.* As noted earlier, it has high water (1 ton of water per lb of sugar) and fertilizer requirements.[105] Energy needs for producing raw sugar from the sugarcane are also high, but typically about two-thirds of the requirement is obtained from burning bagasse.† Industrywide, about 6.45 billion lb of bagasse‡ was consumed in this way in 1976.[112]

Bagasse is the lignocellulosic stalk, comprising pith, ring, and epidermis,

---

*Half sugar, half fiber.
†In Florida (1976), representing one-third of our domestic sugar production.
‡About 40 percent of production.

from which the raw sugar is extracted by milling. Its heating value is low, owing in part to its moisture content of over 50 percent. A new Canadian conversion process first extracts the pith, which contains 92 percent of the sugar, leaving the ring and epidermis in a more favorable form for use in pulp, paper, and woodlike products.[113] Any improvement in the value of by-products will aid in making the biomass conversion concept economically viable.

Among other potential agricultural crops for biomass are sorghum (including Sudan grass), alfalfa, corn, Napier grass, sugar beet, and kenaf, as well as other plants that yield particular hydrocarbons (oils, rubber latex, resins, etc.) potentially useful as feedstocks. In general, milder climates produce higher biomass yields—California sugar beets producing about 18.7 tpa-yr (dry basis) to 14.3 for Washington State, and Florida seaweeds yielding 32.6 tpa-yr (dry) to 14.7 for Massachusetts. Among the more prolific producing plants are Napier grass, seaweed, sugarcane, sorghum, and sugar beets.[106]

Guayule   This plant (pronounced wy-oo'-lee) is *Parthenium argentatum* Gray, a member of the sunflower family, with the demonstrated ability to replace natural rubber (*Hevea brasiliensis*).[114] There are two basic harvesting technologies for such plants—harvesting the whole plant product or tapping latex from the plants, as with hevea for natural rubber latex. The latter yields *reduced organic materials* (extractable by acetone and benzene) at an average 10 ± 5 weight percent of the latex.[115]

Guayule is not tapped, as in the case of natural rubber, because the latex is in discrete cells, each of which has to be ruptured to extract the latex. Natural rubber latex, however, is in tubules in the bark of hevea. Guayule is just one of about 2000 known plant species that have yielded a potentially commercial rubber. There is actually no detectable chemical or physical difference between guayule rubber and hevea rubber. Both are isoprene polymers $[(C_5H_8)_x]$, with a cis configuration, and both have regular repeating isoprene units. Molecular weights and lengths are also approximately the same.[116] Guayule rubber may, however, require different amounts of curing chemicals than hevea.[115]

In 1910, about 10 percent of the world's natural rubber came from guayule, with the balance based on hevea. In the late 1920s, 1500 tons of rubber were extracted from 8000 acres of guayule in California, but the price of hevea dropped to 2¢ per lb during the depression and guayule could not compete.[114]

Guayule exists in numerous strains which offer the opportunity for selection and genetic improvement, and it grows well under plantation conditions, where it can be harvested efficiently. Hevea cultivation is restricted to tropical rain forest climates within 10 degrees of the equator. Guayule, on the other hand, can be grown where rainfall and irrigation water are in short supply. It is estimated that about 5 million acres in California, Arizona, New Mexico, and Texas are suitable for its culture. This land would probably be marginal for cultivation of virtually any other crop without irrigation and fertilization.[114]

Hevea was imported at the rate of 719,000 tons in 1974 (over $500 million) and contributed to our balance-of-payments problems. Cultivation of domestic guayule could reduce somewhat this dependency on foreign raw materials while providing employment for American Indians.* Harvest of 10 tons of guayule would yield about 1 ton of deresinated rubber.†

Guayule grows from seeds to 2-ft-high bushy perennial shrubs. The native habitat is desert regions of the Southwest, at altitudes of 4000 to 7000 ft, where the temperatures range between 0 and 120°F (-18 to 49°C). The growth slows below 60°F, and the plants become semidormant below 40°F or during prolonged drought. Presently, native stands are scattered over an area of 130,000 mi^2 of Mexican and Texas deserts. Guayule plants typically are from 10 percent to as much as 20 percent dry rubber after four years of growth.[117]‡ The seeds, which proliferate readily, particularly in summer and fall, can be stored for several decades.[114]§

For commercial production, 11 to 25 in. annual rainfall is necessary. Heavier annual rainfall produces excess vegetation at the expense of rubber. Rainfall below 14 in. annually requires supplemental irrigation for plantation culture, although native growth seems to thrive on 9 in. or less. The optimum rainfall for long-rotation (four- to eight-year) crops is 16 to 18 in. annually.[114]

Highest yields seem to result from irrigation, perhaps because unevenly spaced stress (drought) periods are required to produce rubber. With irrigation it can be harvested after three years. Guayule requires no fertilization and does not deplete the soil.[114] In the native state, guayule is very resistant to disease and insect pests, but in plantation culture, like many other plants, it is susceptible to the diseases and pests that invade cotton and lettuce. The grasshopper is a major pest.[118]

Normally, the entire shrub—roots and all—is harvested, because one-third of the rubber occurs in the roots.¶ On the other hand, cutting 2 in. above ground level—*pollarding*—results in such heavy resprouting of new shrubs that one-year-old pollarded bushes are as fully developed as two-year-old seedlings. The limit to the number of pollarded harvests is not yet known. The shrub must be kept intact and extracted within a few days, inasmuch as the rubber contains no natural antioxidant. About 95 percent of the rubber is extractable.[113]

Prior to World War II, hevea plantations were producing annual yields of about 200 lb of rubber per acre. By 1974, yields had increased by a factor of six. During World War II, guayule rubber was being produced at levels of about 900 lb per acre. Present indications are that it might be possible to increase plantation-grown guayule yields to 1200 or 1500 lb per acre or higher, but the ultimate maximum yield for hevea might be 3000 to 6000 lb per acre, employing high-yield clones treated with chemical stimulants.[114] The strides made by the

---

*Over 500,000 acres of Indian reservation arid and semiarid land could be brought under cultivation.[116]

†Resin is present as a contaminant.

‡Native Mexican guayule shrubs are about 10 to 17 percent rubber.

§Seeds stockpiled since the 1950s are expected to have a 40 percent germination rate.

¶Tap roots extend to a length of more than 20 ft, while an extensive secondary root system ingests water from a large area.[118]

Rubber Research Institute of Malaysia in the cultivation of hevea have far outstripped the advances made in guayule cultivation, because of the lack of incentive in the latter case. Nonetheless, laboratory experiments have produced a 2.2- to 2.6-fold increase in yield by treatment of guayule seedlings with 2-(3,4-dichlorophenoxy)triethylamine.[119] It has been estimated that western hemisphere guayule researchers are at least a decade away from a practical alternative source of natural rubber.[120]

Economics is bound to be a major factor in the future of guayule. Wages in the Far East, where over 85 percent of the world's natural rubber is produced,* are between 20 and 30¢ per hr, while U.S. farm wages are 10 times that and still rising. Hevea is, however, among the world's most labor-intensive crops, whereas guayule can be harvested mechanically. Furthermore, guayule is harvested during a slack period for most California farm workers. Other economic uncertainties lie in actual yields, extraction costs, the market value of guayule by-products,† and the future demand and prices for natural rubber.‡ In 1975, our domestic consumption of synthetic rubber was over twice that of natural rubber (7.511 million tons vs. 3.699 million tons, respectively). Projections, however, show total rubber production increasing at the rate of nearly 5 million tons every five years to 1990, with only a modest decline in the market share of natural rubber (to about 28 percent from 33 percent in 1975).[120]

Projections made in 1977 for hevea prices, were 60¢ per lb (U.S.) in 1980§ and 78¢ in 1985.[114] The agricultural costs of guayule are expected to be in the 29 to 38¢ per lb range in the 1980s, and processing costs are expected to be 20.5 to 28.0¢ per lb.¶ Capital requirements for a medium-sized processing facility are estimated at $30 to $40 million. Taking into account the expected value of by-products, Glymph arrived at a total net guayule rubber cost ranging from 29 to 54¢ per lb in 1985 vs. his projected hevea price of 54 to 63¢ per lb for the same year.[123]

Gopher Plant (Gopherwood)   The gopher plant (*Euphorbia lathyrus*) has been touted as a potential "gasoline tree"—a source of substitute crude oil.[124] A grant has been awarded to the University of Arizona to search for hydrocarbon-producing plants that can grow on arid lands and provide petrochemical feedstocks. The plant studied initially was the gopher plant, a 4-ft desert shrub

---

*Brazil, the original source of hevea, is not presently a factor (1977 share less than 1 percent of world's supply), because South American leaf blight prevents plantation culture. Goodyear is now claiming success in chemical attack on blight in their Belem holdings. If the South American disease should spread to Southeast Asia plantations, it could have a tremendous impact on the future of hevea and guayule.[120]

†Wax from leaves, bagasse, "cork," resins, and volatile aromatic oils.

‡Demand is expected to increase at a rate of 5.9 percent while the production rate grows by 5.7 percent over the next few years.[114] A large factor in demand growth is increasing sales of steel-belted radial tires, which require 40 percent natural rubber, twice as much as required in regular tires.[121] Large aircraft and earth-moving equipment tires are virtually all natural rubber.[122] Much depends on economic conditions in the years ahead.

§Prices in 1977 were about 40¢ per lb.

¶For median cost estimate of about 58¢ per lb.

which produces a latex or hydrocarbon emulsion at the rate of about 10 bbl coe per acre.[125] There are 2000 to 3000 different species of *Euphorbia*. Among them are *E. tirucalli* (African milk bush), a Brazilian hedge shrub, and *E. lacetea*, which grows to a height of 10 to 15 ft and produces a flow of latex when tapped. Both require several years to reach maturity, whereas the gopher plant, an annual, grows to 4 ft in seven months. It yields about 8 weight percent of recoverable hydrocarbon.

This "oil" is largely an isoprenoid* similar to natural rubber but of considerably lower molecular weight (about 20,000 vs. 500,000 to 2 million for hevea). Other extractables are sterols, glycerides, and diterpenes.[124] The polyisoprenes and steroids comprise up to 80 percent of the dry weight of the latex, which contains 30 to 90 percent water, depending on the species and time in the growth cycle when harvested.[115] Proper breeding could possibly result in doubling the yield.

The gopher plant is a hardy shrub that survives on little water, grows rapidly, and propagates continuously by coppicing. The latex lacks the tars and other such impurities associated with crude oil.[126] Its greatest promise is as a feedstock for fluid hydrocarbon fuel.

Jojoba This plant (pronounced ho-ho'-ba) (*Simmondsia chinensis*) is yet another drought-resistant species, whose olive-sized seeds contain as much as 60 weight percent of a light yellow, odorless, liquid wax. This wax comprises unsaturated fatty acids esterified with unsaturated fatty alcohols, and is almost identical to the oil of sperm whales, an endangered species.† Sperm oil is an important component of industrial and automotive lubricating oils, because of its unique properties.[128] Jojoba oil can be hydrogenated to a hard, white, crystalline wax, much like carnauba. The seed residue is high in protein (about 32 percent) and can therefore be used as an animal feed.[128]

Jojoba can be grown on the semiarid lands of California and Arizona, with limited irrigation required only during its first year. It also has a high tolerance for alkaline soils and the salinity of irrigation water. At the present state of the art, oil yields should be at least 0.9 tpa-yr (2 tonnes per ha-yr) after five years, although some experimental plantings have produced 10 times that.[128] The plant grows as high as 15 ft with as little as 4 in. of annual rainfall.[129]

Leucaena Leucaena (loo-see'-nah) is a leafy, mimosa-related evergreen legume native to Mexico. It is characterized by an amazing versatility. It is capable of yielding wood, paper fiber, food, feed, fuel, fertilizer, and wax.[130] It grows in size from bushy shrubs to tall trees, in adverse climates and on marginal and steeply sloping land. Its nitrogen-rich foliage when harvested makes a good fertilizer for corn. Wood yields from leucaena are among the highest achieved for any species, and the stumps coppice rapidly. Its long tap root makes it quite drought-resistant. The plants fix their own nitrogen from air in the soil, thus

---

*About half of it being cyclic isoprenes.

†Meadow foam is another plant source of spermlike oil.[127] The sale of sperm oil in the United States has been banned since the early 1970s.

reducing fertilizer requirements. The high-protein leaves from the shrub variety are palatable and nourishing for ruminants (but contain an amino acid that can cause goiter in some animals). The treelike varieties, which grow up to 60 ft in six to eight years, are a potential source of paper fiber and fiberboard.[131]

Kenaf Kenaf or ambary (*Hibiscus cannabinus*) is a tough, stringy tropical plant (native to Southeast Asia), which is now being grown experimentally near Yuma, Arizona, in a joint project of the Department of Agriculture (DOA) and the American Newspaper Publishers' Association (ANPA), as a possible source of paper and substitute flax fiber.[132] The primary purpose is to develop an agricultural cash crop that can provide a hedge against expected long-term pulpwood shortages. Newsprint has risen in price from $175 per ton in 1973 to $320 per ton in 1978.[133]

The plant grows to 12- to 16-ft heights in four months and is then ready for harvesting.* It is being looked at as a possible crop for central New Jersey.[132] Experimental plantings have produced an average of 7 tpa-yr (vs. 3¼ tpa-yr for softwoods and as little as ¾ tpa-yr for hardwoods),[133] with little labor required. It has been touted as one of the easiest of all crops to grow.[135]

The thin stalks are cut down much more easily than trees are. The harvested stalks can be thermomechanically pulped and prepared for the papermill at about half the cost of wood pulp. Water does not drain from kenaf as readily as from wood pulp, however; hence the papermaking machines might have to be slowed down. The resulting newsprint has a yellowish cast, but the opacity and bursting strength are nearly comparable to wood-based newsprint.[133] Kenaf fiber also has the potential of supplanting flax, which is becoming increasingly scarce, for use in certain types of fine paper.†

The 60 percent nonfiber components of kenaf constitute refuse for potential conversion to fuels or feedstocks. Also, any massive new entry into the pulp and paper raw material picture could have an effect on the overall balance of crop and forest lands and thereby affect the availability of biomass for conversion. A less energy-intensive paper fiber could also release energy for other uses. The overall productivity of kenaf is sufficiently high to suggest it as a specific crop for fuel and feedstock biomass.

Poinsettia, quack grass, milkweed, and sow thistle are all possible sources of hydrocarbons for rubber and plastics.[136] Even cauliflower and cabbage have been claimed as potential fluid hydrocarbon fuel sources.[137]

Prospects

At present it appears that energy farming is not a promising answer to our long-term fuel and feedstock problems, except as a minor supplementary resource. Technically, crops can be grown in fairly large quantities for the specific purpose of providing these necessities for modern living, but food is a potent com-

---

*Generally one to two weeks after a killing frost. This is later than most crops.[132] About 40 percent of the product is fiber.[134]

†Loss of markets for flax products (linseed oil and linen) has reduced flax availability.

petitor for the soil, water, and phosphorus essential to biomass production, and food production must take first priority.

Phosphorus is an essential plant nutrient and, unlike nitrogen and potassium, it is both nonrenewable and in limited supply. The estimated lifetime of this resource is shorter than that of coal.[106] It is important that nitrogen and phosphorus in wastes from biomass conversion processes by recycled to the soil. Thermal conversion processes produce an ash rich in many nutrients that can be applied to the land, but some of the phosphorus in ash may be in the form of such low-solubility compounds that it cannot benefit plant growth. Phosphorus lost by runoff and percolation into the soil is lost on the ocean floor forever—or at least until the next geological uplift.

The cropland requirements for any significant contribution to our energy needs would be large, and fertile soils would not be available for this use. Furthermore, many energy crops would be costly to harvest, separate, and transport to conversion sites. In addition, the preferred growing regions for energy crops would be largely in the semiarid areas of the lower 48 states, far removed from most of the heavily populated industrial areas where the products are most needed.

AQUACULTURE*

Water covers about 70 percent of the earth's surface, but the overall productivity of plant life on land exceeds that of water plants by almost 2:1.[100] Cultivation of water plants has nonetheless been proposed as a means of producing biomass for conversion to fuels and feedstocks, since on an equal-area basis, aquaculture should be able to produce three to four times as much biomass as land does.†

In their natural state, the large water areas of the earth, as a whole, are far less productive due to the low concentration of nutrients, the relative overall inefficiency of light gathering by some aquatic plants, the reflection of sunlight off the surface of the water, and the great depths the sunlight must penetrate. The vertical array of leaves in a land plant is a much more efficient energy-gathering arrangement than is the pigmentation of algae in a mixed water column. Although nutrients are present in large quantities in open bodies of water, they are concentrated at the bottom, where the light does not penetrate.

Photosynthesis in tropical surface waters of the oceans rids these waters of nutrients. Macromarine life consumes the algae and other surface microorganisms, thus removing the haze-producing matter and leaving the water clear. This life ultimately dies and is deposited on the ocean bottom, disintegrates, and is carried by deep ocean currents toward the equator, where upwelling zones bring it back near the surface to serve as nutrients for the next

---

*The term "mariculture" is also used, but it refers more specifically to "marine" or saltwater culture.

†Kelp, for example, yields about 40 tpa (9 tonnes per hectare) per year.[138]

life cycle. These upwelling zones are limited, covering only about 0.1 percent of the total area of the oceans. [139]

Macro Aquatic Plants

California Giant Kelp

Perhaps the single marine plant with the greatest potential as an energy resource is California giant kelp* (*Macrocystis pyrifera*). A joint effort by the American Gas Association (AGA), California Institute of Technology, General Electric, Institute of Gas Technology (IGT), DOA, DOE, and Global Marine Development, Inc., is under way at a site 5 mi off Corona Del Mar, California, to assess the potential of kelp mariculture. It has been estimated that 140 million acres of water area (a square 470 miles on a side) could supply our total national needs for pipeline-grade gas. [140] Rather than using nutrient-poor, shallow coastal waters,† where the roots of native kelp attach themselves to the rocky ocean floor, the intent of this undertaking is to employ "farms" in deep-water areas (1000 to 1200 ft), but with the plants growing 60 to 80 ft below the surface on moored structures.‡ These artificial "seabottoms" will be tethered, floating, 108-ft-diameter metal structures, superficially like an array of inverted umbrella ribs. The total area of the initial test farm will be 1/4 acre. Nutrients will be provided by artificial upwelling of nutrient-rich waters from the seabed.§ The plants will be "mowed" and gathered four to six times per year for processing into pipeline-quality gas. The kelp grows rapidly (2 ft per day, and up to a length of 200 ft)[142] and has excellent regenerative capacity after harvesting. The yield of PLG (about 1000 Btu per scf) is expected to be approximately 3 to 4 scf per lb of dry kelp. [143]

Aquatic Weeds

Quite a different proposition is the large mass of aquatic weeds fouling our various inland and estuarine waterways. To date, the emphasis on aquatic weeds has focused on controlling their profligate growth by chemical, biological, or mechanical means, rather than on cultivating, harvesting, and converting them for fuels and feedstocks. The weeds, in most parts of the world, are winning this battle, and the benefits to mankind accrued to date are virtually nonexistent. A limited number of these weeds have been considered seriously as possible contributors as a biomass resource, mainly the free-floating weeds that have proven to be such pests in many of our southern inland waterways.

---

*A species of brown algae.

†Present annual productivity of kelp beds in these shallow coastal waters is about 4 tpa (9 tonnes per hectare). [139]

‡Growth rates at greater depths are reduced by the lower light intensity.

§Because deep ocean water is also CO_2-rich, one estimate suggests that this upwelling would release to the atmosphere as much CO_2 as would be evolved in the burning of an energy-equivalent (vs. kelp) amount of coal or oil. [141]

Water Hyacinth   Of the possible freshwater energy plants, water hyacinth (*Eichhornia crassipes*), a free-floating native of South America, is perhaps the most promising, and at the same time the most pernicious. It has been called "one of the environmental disasters of the century."[144] It has become an aquatic weed in many of the world's waterways, particularly in tropical climates. It grows so rapidly that it clogs watercourses and riverine fisheries. It also harbors bilharzia organisms and malaria-bearing mosquitoes. And it is exceedingly tenacious. Unknown in the Upper Nile in 1958, it has since about 1960 formed floating mats in adjacent waters thick enough to support a person, and it has now invaded the lower Nile.

Water hyacinth collected in Venezuela and brought to the United States by Japanese visitors in 1884 escaped into our waters, where it has since constituted a menace of the first order. It is a free-floating weed with very fine roots, and leaves measuring 4 to 6 in. (10 to 15 cm) across. It reproduces vegetatively. In warm, nutrient-rich waters, this plant can double in number every 8 to 10 days.[145] Its many seeds sink to the bottom of the water, where they may remain viable upward of 15 years. During active growth periods, leaves and roots annually build up a greater than 12-in. (30-cm) deposit of debris.[146] One of the major ecological impacts of water hyacinth is the loss of water through this plant via evapotranspiration. About 6 acre-ft of water can be lost by this means over a six-month period.* The annual financial loss in 1956 attributable to water hyacinth totaled about $43 million in Florida, Alabama, Mississippi, and Louisiana.[146] The loss has surely risen even higher since then.

Chemical control of the weed is so fraught with ecological hazards that mechanical harvesting was introduced in 1937. After 10 years of valiant but losing effort, the herbicidal plant hormone 2,4-D (2,4-dichlorophenoxyacetic acid) was resorted to. Nonetheless, by 1975, about 180,000 acres (400,000 hectares) of Louisiana waters were infested with the pest. So much effort has been devoted to trying to eradicate it that relatively little thought has been expended on learning to exploit its feed and fuel potential.[144]

Studies by the National Aeronautics and Space Administration (NASA) have shown that when grown in warm, sewage-enriched water, water hyacinth can produce biomass at the rate of nearly 8 tpa-day (17.8 tonnes per ha-day). This translates to a dry biomass production rate of about 95 tpa-yr (212 tonnes per ha-yr). With a retention time of two weeks, a 0.22-acre (0.5-hectare) lagoon of the plant could remove excess nitrogen and phosphorus from the sewage of 1000 people. Water hyacinth also displays a remarkable ability to take up toxic heavy metals, for example, cadmium and nickel, and phenols and other trace organic compounds.[144]

The high crude protein (17 to 22 percent) and fiber (15 to 18 percent) contents of the plant suggest its use as a feed supplement. An acre of cultivated

---

*Approximately three and a half times the evaporation rate from the same area of open water. In India's dry climate, the evapotranspiration rate may be nearly eight times that of an open water surface.[146]

water hyacinth can also yield sufficient dry biomass* each day to produce a fuel value of 7 to 14 million Btu (60 to 80 percent methane, with two-thirds the heating value of natural gas). When this biomass is fermented to methane, the sludge residue is an excellent fertilizer, because it retains most of the nitrogen and virtually all the phosphorus and other minerals.

Duckweeds One of the major drawbacks of water hyacinth as a biomass source in a temperate climate is the fact that it becomes dormant in cold weather. In this case, however, it could be combined with *duckweed* (e.g., *Lemna gibba*), which is more tolerant of the cold. Duckweeds are the smallest and simplest of flowering plants, having single fronds about the size of a pinhead.[146] They have about 10 to 12 times the productivity of soybeans. Even higher yields should be possible through selective breeding of the wild species. With natural strains, the growing rate in temperate climates is high over an eight-month growing season. During the rest of the year, the growth rate is only 20 to 25 percent that of the warmer period, and the protein content is down to about 15 percent. Relatively still fresh water is required with duckweeds. One or more strains of the *Lemnaceae* species can be found in climatic regions ranging between (but not including) waterless deserts and tundra. It is even found in highly polluted waters.

Harvesting and processing constitute a major difference between duckweeds and water hyacinth, which is a large plant requiring harvesting with heavy machinery and then chopping for easy handling. Hyacinth is also bulky (about 80 percent of the density of *Lemna gibba* when fresh), although *L. gibba*, in turn, is only two-thirds the density of other duckweeds, owing to its distinctive ventrally inflated fronds.[147]

Another characteristic of duckweeds that bears on their adaptability as an energy or feedstock source is their low solids content as harvested. *Wolffia arrhiza* (khai-nam or "eggs-of-the-water"), a native Burmese plant, for example, averages only 4 percent dry combustible matter.[148] This means that energy-intensive techniques must be used to dry it for pyrolysis, or the dilute suspension must be restricted to fermentation processing.

Water fern (*Salvinia auriculata* Aublet.) and *water lettuce* (*Pistia stratiotes*) are two other free-floating weeds with potential for conversion to fuels or feedstocks.[146]

Submersed weeds are particularly troublesome because they do not lend themselves to control by herbicides and are not easy to harvest mechanically.

Hydrilla This is a submersed weed that has already infested nearly 700,000 acres (280,000 hectares) of Florida's fresh water, has spread to Georgia, Alabama, and Mississippi, and now threatens lakes of the Tennessee Valley. Hydrilla is a hardy plant that spreads rapidly and becomes virtually impossible to eradicate with time. Its bad features suggest that, like water hyacinth and

---

*About 0.16 to 0.33 tpa-day (0.36 to 0.73 tonnes per ha-da).

duckweeds, it could be a suitable candidate as a biomass source of fuels and feedstocks.

Eurasian Watermilfoil This is yet another submersed aquatic weed, which has already taken hold in the Tennessee Valley waterways and is fouling boat propellers, infesting water supply systems, and creating breeding areas for mosquitoes. Applying herbicides and varying the water level has at least prevented this weed from getting out of hand.

Emersed weeds,* phreatophytes,† and floating-island weeds,‡ are other forms of aquatic weeds, but few of these types have been considered as potential sources of biomass for fuel and feedstocks.[146]§

It is quite obvious from the discussion above that aquatic weeds have considerable potential for scavenging pollutants from wastewater and providing animal and possibly human food and fertilizer from the sludge and/or biomass for fuel and feedstocks. Unlike land plants, aquatic plants would make no demands on croplands and would not contribute to soil erosion. Availability of water is essential, but aquaculture can use wastewater (with its nutrients) to good advantage and return it to the water cycle in an improved condition. Since they scavenge toxic metals and chemicals from wastewater, these weeds must be evaluated thoroughly before use of their products in food and animal feeds.

Microalgae

The higher aquatic plants have only about 2 percent efficiency in converting solar energy to combustible material. Micro- or unicellular algae, for example blue-green varieties of *chlorella* and *scenedesmus,* have been touted for their higher conversion efficiency when nourished with municipal sewage in tanks or ponds exposed to sunlight. Through a symbiotic relationship between bacteria and algae, oxygen and carbon dioxide are exchanged and the oxygen generated contributes to the bacterial oxidation of organic wastes. The resultant algal biomass is separated by centrifugation or other means, with the solids generally being fed to a digester for anaerobic conversion to methane. The nutrient liquor can serve as a fertilizer for the cultivation of subsequent algal growth.[149]

Algal ponds are of two types: *facultative* and *high-rate.* The former are 1 to 3 m deep, and mixing is solely the result of wind and possibly recirculation of effluent. As a result, the bottom is anoxic and fermentation of the settled sludge occurs. A detention time of weeks or months is required. The *high-rate* type of pond is more efficient in that it is only 30 to 50 cm deep and is mechanically

---

*Having roots beneath water, and stems and leaves above.

†Woody plants, perennial grasses, and broad-leaved plants growing at the water's edge or with roots extending into the capillary zone overlying the water table.

‡Exist as masses of free-floating dead aquatic vegetation ("sudds") supporting increasingly larger types of vegetation.

§Notable exceptions are cottonwood (*Populus* spp.), a phreatophyte, and papyrus (*Cyperus papyrus*), a floating-island weed.

stirred to prevent settling. Detention time in the summer months is only two to four days.

The dilute algal suspension has to be concentrated by at least two orders of magnitude. Since microalgae are generally less than 20 μm in diameter, filtration, even using microstrainers, is not practical, except with either filamentous (e.g., planktonic) or colonial algae. Centrifugation is a capital- and energy-intensive process. Chemical flocculation uses large quantities of chemicals and produces a chemical- and algae-laden sludge that is difficult to dewater and dispose of.[150] Separation or concentration of the algal-pond effluent poses one of the more difficult technological hurdles if micrological biomass is to become a factor in our energy future.

Another problem is attaining and maintaining control of the algal species present in the ponds. Often one or two species will comprise over 95 percent of the algae in a sewage pond. The tertiary treatment ponds, which are only lightly loaded with organic matter, will contain more varieties of algae than the primary and secondary treatment ponds. The changing conditions make it difficult to control the algal composition.[150]

Including energy inputs and conversion losses, the potential *net annual energy productivity* (NEP) of methane from microalgae should be about 200 million Btu per acre-yr. Growth rates of algae are generally considered to fall in the range 30 to 50 tpa-yr, but one investigator claims that yields up to 210 tpa-yr have been achieved under closely controlled conditions. The difference lies largely in the technologies of growing and collecting the algae.[151]

CONVERSION PROCESSES

Waste recovery systems serve several important purposes. They:

1   Reduce consumption of virgin resources.
2   Reduce the amount of residue to be disposed of in landfills.
3   Reduce air and water pollution.
4   Permit reductions in energy contents of resulting industrial products.

In order for recycling of refuse to function effectively, waste must be highly concentrated, be continuously (or at least, predictably) accessible in large quantity, and be efficiently collectible. Furthermore, the bulk of the waste must be combustible or be convertible by mechanical, chemical, thermal, or biological means to a form suitable for fuel or for use as an industrial resource. The collection of municipal solid waste is highly institutionalized, but because of its dispersion, waste from only 70 percent of our population is available in sufficient volume and concentration for consideration as feed for large electric power generating stations.[7] This could amount to as much as 25 percent of the total dry, combustible waste generated in the United States, which might, in turn, lead to production of 3 percent of our energy budget.[152]

A generalized scheme for recovery is shown in Fig. 7.3. Figure 7.4 illustrates a typical MSW dry reclamation process, of which there are many variations. Wet reclamation processes are also available (see reference 154, for example), to permit recovery of fibers, ferrous metals, aluminum, and glass. Such systems basically call for wet pulping of unsorted MSW, followed by magnetic separation of ferrous metals and liquid cycloning to separate the fibers from the glass/aluminum stream. A glass recovery unit then separates out the latter two products and diverts the nonrecoverable inorganic waste to a landfill stream. The combustible component(s) of MSW can be used as feed for virtually any of the thermal conversion processes (Table 7.15), to produce fuels or chemical feedstocks.

One of the major obstacles to commercializing any energy- or feedstock-oriented use of MSW lies in the problem of ensuring a continuous supply of refuse for the 20-year design lifetime of a normal conversion facility. Plant financing would be very problematical at best without such guarantees. The owner must expect to invest $50 million or more for such a facility, depending on the technology and the capacity involved, and interruption of the flow of refuse would severely affect the economics of the operation. Surge bins could provide only limited relief in a prolonged emergency.

Obstacles to adoption of waste usage as fuel by utilities and industry have been noted[156]: lack of reliability data based on commercial-scale operations; concern over compliance with current and future environmental regulations;

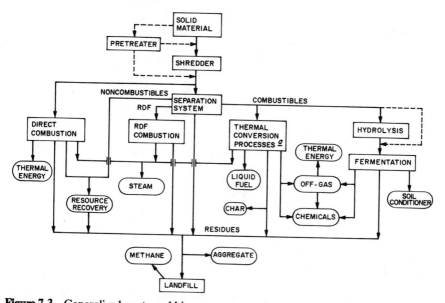

Figure 7.3 Generalized waste and biomass energy and product recovery system. (Adapted from reference 153.)

^{a}Pyrolysis; gasification; liquefaction.

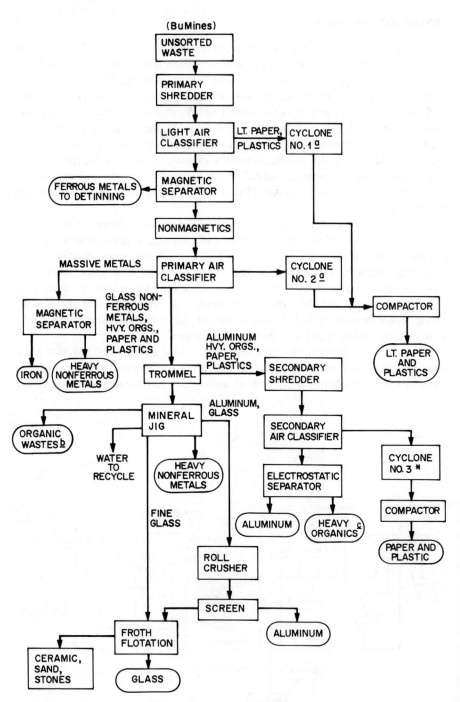

Figure 7.4 Municipal solid waste (MSW) recovery system (BuMines). (Adapted from reference 14.) *a*, Includes 93 percent of paper, 70 percent of plastics, 92 percent of fabrics, 46 percent of cardboard, 15 percent of leather/rubber, 14 percent of wood, and negligible amount of putrescibles; *b*, most yard wastes and putrescibles; *c*, most heavy-gage plastics, wood, leather, rubber.

Table 7.15 Biomass Conversion Technologies

Aerobic fermentation
Anaerobic fermentation
Biophotolysis
Chemical hydrolysis
Enzyme hydrolysis
Hydrogenation
Incineration
Partial oxidation
Pyrolysis
Separation
Steam reforming
Other

Source: Reference 155. Reprinted with permission of the copyright owner, The American Chemical Society.

limited economic incentives; siting problems; lack of coordination among various federal, state, and local regulatory agencies; and possibly most important, lack of information on the effect of refuse-derived fuel (RDF) on power plant maintenance and operation.

Explosion hazards of shredding dry waste prior to separation have been observed.[157,158] Two actual cases involved explosives discarded in municipal solid waste, although chance mixing of reactive ingredients in the waste could also present a hazard. Dust explosions can also be a cause for concern during dry shredding. One authority claims that processing of heterogeneous mixtures of municipal solid waste can lead to "one explosion . . . for every 20,000 tons of garbage shredded."[158]

Mechanical Beneficiation

Figure 7.4 illustrates a basically *mechanical beneficiation* technique whereby input components are reduced to manageable dimensions and then separated into various streams for recovery.

Air Classification

This procedure applied to MSW drops out a large quantity of putrescibles and yard waste, allowing 10 to 25 percent of the potential fuel to be carried out in the stream for landfill.[159] Heavy organics (wood, leather, rubber, heavy-gage plastics) collect in the middling stream, from which they can be sorted for recovery or for landfill, depending on the specific process. The fuels from such processes are relatively high in ash and moisture and are therefore normally co-fired with coal, at the rate of 10 to 15 percent RDF to 90 to 85 percent coal.[159] Residues requiring disposal in a landfill amount to about 13 weight percent of the incoming refuse.

Wet Pulping

This process typically starts with a 3:97 organic-to-water mix and yields a consistent 50:50 organic-to-water product having a heating value of about 4100 Btu per lb (vs. 4500 to 8000 Btu per lb for untreated or air-classified solid waste).[15]* This low heating value is perhaps more than offset by the efficient recovery of other products, the reduced explosion hazard, the reduced wear on the shredder, and the low fly ash content. The process also provides a relatively homogeneous fuel component which can be co-fired with coal or be burned alone in a bark-burning furnace. Some of the recycled fibers are reclaimed and used to manufacture roofing felts and other crude, fiber-based products. About 10 weight percent solids are landfilled, and wastewater is treated as sewage.

Biochemical Conversion

The formation of marsh gas was probably first noted by Benjamin Franklin in 1774. In a letter to his friend Joseph Priestly, he mentioned release of the gas from swamps, but it was not until the mid-1860s that Louis Pasteur determined that the gas was a product of biochemical reactions. In 1875, Popoff demonstrated the formation of methane from cellulose by living organisms, and in 1904 the first U.S. patents appeared for the formation of methane by anaerobic digestion. The first workable biogasification system made its appearance in India in 1939.[31]

Anaerobic Digestion

The appropriate pathway for *biochemical conversion* depends on the objective. The primary goal of *anaerobic digestion* is the formation of methane. This process has been touted as the possible source of up to 10 percent of our U.S. natural gas needs.[30]

Knowledge of the precise chemistry of digestion is rather sketchy, but the output of the digester has essentially three streams: *product gas* (CH_4, CO_2, H_2, H_2S), *recyclable supernatant liquid* (water plus dissolved and suspended organic compounds and bacteria), and *nondigestible sludge* (high solids with reduced organic fraction and bacteria) which must be disposed of by incineration or landfill.

Basically, with the aid of acid-producing bacteria, proteins, fats, and carbohydrates in the feed are solubilized by exocellular enzymatic hydrolysis to low-molecular-weight residues (sugars, fatty acids, amino acids), which are then taken up by the bacterial cell prior to further metabolic digestion. The next

---

*Wet-pulped fuels release about 50 to 55 percent of the incoming energy in useful form vs. 60 to 6 percent for dry mechanical systems.

step is production of volatile organic fatty acids (e.g., acetic and propionic acids),* and other low-molecular-weight acids (e.g., formic), together with lower alcohols, aldehydes, and so forth. These acids become substrates for methane-producing bacteria which yield methane and carbon dioxide.[160]

The basic chemistry of digestion can be suitably represented by the following reactions[30]:

Conversion of Amino Acid (e.g., leucine)

$$(CH_3)_2CHCH_2CH(NH_2)COOH + 3H_2O \rightarrow$$

Leucine

$$CH_3COCH_2COOH + CH_3COOH + NH_3 + 6H \tag{5}$$

Acetoacetic acid 1   Acetic acid 2

$$CH_3COCH_2COOH + H_2O \rightarrow 2CH_3COOH \tag{6}$$

1   2

$$CH_3COOH \rightarrow CO_2 + CH_4 \tag{7}$$

2

Conversion of Fatty Acid (e.g., octanoic acid)

$$2CH_3(CH_2)_6COOH + 12H_2O \rightarrow 8CH_4 + 8CO_2 + 24H \tag{8}$$

Octanoic acid 3

$$24H + 3CO_2 \rightarrow 3CH_4 + 6H_2O \tag{9}$$

the net reaction being

$$2CH_3(CH_2)_6COOH + 6H_2O \rightarrow 11CH_4 + 5CO_2 \tag{10}$$

3

Conversion of Alcohols and Other

$$2CH_3CH_2OH + 2H_2O \rightarrow 2CH_3COOH + 8H \tag{11}$$

Ethanol   2

*Responsible for more than 85 percent of methane formed.[33]

$$4CH_3OH \rightarrow 3CH_4 + CO_2 + 2H_2O \qquad (12)$$

Methanol

$$4HCOOH \rightarrow CH_4 + 3CO_2 + 2H_2O \qquad (13)$$

Formic
acid

Sewage is currently the primary feed for anaerobic digestion, although the process is also recommended for conversion of animal manures.[33]* Digesters or tanks for conversion of sewage are sized to contain 2 to 6 ft³ per person and are sealed from the atmosphere. The sewage is fed at 2 to 8 percent solids, with about 4 percent being considered optimal. Slug feeding† is suitable, but continuous feeding of finely divided sewage leads to more stable operation. A five-day residence time at 118°F (48°C) may convert about 43 percent of the input organic carbon to methane, with a ratio of 60:40 CH_4 to CO_2 in the product gas. One ton of dry organic solids yields nearly 13,000 scf methane using a 4000-ft³ digester.[30]

Organic compounds in the wash-out‡ and carbon dioxide represent carbon in undesirable forms. The latter dilutes the methane and lowers the heating value of the product gas, although the gas can be upgraded by scrubbing and drying operations.

Since the predominant organic component of typical solid waste is cellulose, the basic bioconversion reaction is

$$C_6H_{10}O_5 + H_2O \rightarrow 3CO_2 + 3CH_4 \qquad (14)$$

in which situation 1 lb of convertible waste should produce 6.65 scf of methane, together with an equal volume of carbon dioxide.[4] The ratio of methane to carbon dioxide increases with the fat and protein content of the input; hence the product gas from a mixed organic feed is usually 60 to 70 percent methane.[160]

Large-scale adoption of anaerobic digestion to produce SNG could be beneficial from a number of viewpoints[161]:

1 The purified gas could be introduced directly into existing gas pipelines for transport and distribution.

2 There would be no harmful combustion residues from its use.

3 Nuisance materials (wastes), otherwise requiring costly disposal, could be used as feed for the process.

---

*One ton of wet animal manure will produce about 390 lb of product gas (3088 scf of methane) by anaerobic digestion.[33]

†Periodic feeding of coarse, high-solids input.

‡Undigestible residue from treated sewage.

4 Thermal pollution and the greenhouse effect (CO_2 emission) from processing and burning fossil fuels would be partially offset.

5 Existing technology could be used.

But there are some potential drawbacks:

1 The digester sludge may be even more difficult to handle and dispose of than the wastes used for feeds.

2 In the case of agricultural feeds, marginal lands could be expected to yield marginal biomass crops.

3 Large-scale irrigation and heavy fertilization would be required for much of our available land, to produce adequate yields of biomass feeds.

Nonetheless, it would appear that bioconversion of refuse and biomass could make a useful, if small, contribution to our energy budget.

There are three major factors that have considerable bearing on the cost of providing fuel and feedstocks by anaerobic digestion: capital cost of the digesters, sludge treatment and disposal, and maintenance of digester stability. The digester capital cost is high* because of the retention times required. The disposal costs arise, in part, from a large residue comprising about half of the organic input and all the inorganic. A means of maintaining a balanced population of microorganisms is essential to digester stability, but such a system has not yet been achieved.[160]

Enzymatic Fermentation

Many forms of biomass have the cellulose in a matrix of lignin and hemicelluloses. The cellulose can be split into glucose by chemical hydrolysis, but with some difficulty because of the spatial and linkage constraints. Enzymes provide a valuable tool to enhance this splitting. Four major steps are involved in the industrial fermentation to alcohol by *enzymatic hydrolysis:* (1) fermentation specific to the formation of the enzyme cellulase, (2) cellulase recovery, (3) enzymatic hydrolysis of the biomass to glucose, and (4) glucose fermentation to produce ethanol. The cellulase can be derived from cellulytic fungi or yeast.[106] The enzymatic hydrolysis process is favored over chemical hydrolysis because of the ease with which the reaction occurs under rather mild conditions [about 50 hr at 121°F (50°C) and pH 5].[160]

Fermentation of glucose can be specific for the formation of ethyl alcohol, acetone, butanol, or other chemicals. Fermentation to alcohol yields ethanol and carbon dioxide in approximately equal weight ratios. The particular organism employed will depend on the specific feed. At a constant fermentation temperature of 68 to 86°F (20 to 30°C), yields, in the form of 10 to 20 percent

---

*Estimated at $1610 to $1910 capital per 1000 scf daily output.[33]

ethanol solutions, are typically in excess of 90 percent of theoretical, based o
fermentable sugars.[160]

The cost of conversion has become highly dependent on the cost of the ca
bohydrate feed and the markets competing for it. At present, about 80 percer
of the ethanol produced in the United States is derived from ethylene, in turn
petroleum or natural gas derivative. For fermentation alcohol to be competitiv
with this alcohol, the price of natural gas–based ethylene would have to be a
least 18¢ per lb, assuming $1.30-per-bushel corn is used as the fermentatio
feed.[159]

A mixture of butyl and isopropyl alcohols can be formed from carbohy
drates, by a specific organism, with a yield of 30 to 33 percent of a mix of th
alcohols, based on the substrate sugars. At 86 to 99°F (30 to 37°C), the fermer
tation is accomplished in 30 to 40 hr. The normal product composition falls i
the range 53 to 65 percent n-butanol, 19 to 44 percent isopropanol, 1 to 24 per
cent acetone, and zero to 3 percent ethanol.

Strains of yet another organism produce a mixture of butanol and acetone
at a maximum yield of about 30 percent, based on fermentable carbohydrates
Carbon dioxide and hydrogen (2:1) are by-products of the fermentation. Th
product mix is variable, depending on the way in which the process i
engineered. Such solvents are presently largely based on chemosynthesis c
petroleum derivatives.[159]

In a system involving bioconversion of waste energy crops—be they agricu
tural, silvicultural, or aquacultural—the process must also be thoughtful
assessed for its impact on the ecology as well as its effect on costs. The land o
water surface area commitment, the transport of materials, evapotranspiration
harvesting, and sludge disposal are all factors of concern. The sludge o
semisolid residue from the process offers some potential as a soil conditione
but even here some careful consideration must be given to the environmenta
aspects, since potential pathogens—viruses, bacteria, and so forth—in suc
sludge, as well as concentrates of toxic metals and organic compounds may re
main hazardous.

Aerobic Fermentation (Composting)

A Swiss process has been developed for *wet composting* or *aerobic thermophili
fermentation* of sludge to "pasteurize" the sludge prior to land disposal.[16
Aeration in the reactor produces intense microbial activity, causes rapi
metabolism (over a few hours), and creates enough heat to eliminate th
pathogens and kill parasitic ova and seeds while preserving the fertilizer value
of the product. This process does not, however, eliminate toxic metals o
organic chemicals (such as pesticides or herbicides). Alternatively, the sludg
may be dewatered to at least 20 percent solids and composted to form a mulc
aerobically for use as a soil conditioner for nonagricultural land.[163] Again, thi
process does not control potentially harmful materials. Sludge contaminate
with organic toxicants can probably be decontaminated, however, by exposur

to gamma rays from radioactive cesium (Sandia Labs) or to a high-energy electron beam (MIT).[164]

Ocean dumping of sludge by New York City must, by Environmental Protection Agency edict, cease by the end of 1981. By that time, a composting operation should be functioning, but it will probably last for only about seven years because of the limited land area available. Pyrolysis of the sludge is not in the immediate offing because of stringent air pollution regulations.[165] The biotreatment of sludge is not pertinent directly to the energy and feedstock aspects of the material, but thermal conversion treatments are. These will be discussed below.

Thermal Conversion

Direct Combustion (Incineration)

Combustion of raw MSW is a common (and generally wasteful) procedure for preparing waste for landfill disposal. Until relatively recently, it had not usually been tied directly to a heat-recovery system. About 30 million tpy is disposed of in this manner. Energy recovery generally requires shredding and mixing to achieve some degree of homogeneity in the feed. In direct combustion, most solid wastes and biomass are low in sulfur and removal is not required. Chlorides from poly(vinyl chloride) and other components in MSW do not pose a problem at combustion temperatures below 2000°F. Ferrous metals are normally removed by magnetic separation, and the balance of the residue is landfilled. Electrostatic precipitators are required, however, to remove the large amounts of particulates formed. The ash content is high. Mechanical beneficiation is one step toward a cleaner, more uniform feed.

A *circulating fluidized-bed combustion* (CFB) incinerator is claimed to be cheaper to build and operate than a conventional incinerator. It is more compact, accepts a wide range of fuels (coal through sewage sludge), and does not create air pollution.[166] Sand is heated to 1470°F (800°C) by blowing burning gas through it. Shredded, classified refuse and air are fed in. The air creates the circulation and ensures complete combustion. The facility can be used as a conventional incinerator or as a steam boiler.

Wet combustion is a relatively new development[167] whereby sewage sludge, wet garbage, and so forth can be injected into a closed pressure tank at 300°F (150°C) together with air or oxygen. It ignites readily with high efficiency. The heating value of the input is said to be the same as if the waste were dry. Capital requirements are claimed to be modest. This process might prove to be a valuable tool in disposing of sewage sludge.

A significant fraction of our sludge is currently incinerated, because this method of disposal is reported to be only half as costly as land application.[168] Still, one of the major costs of conventional combustion is dewatering the sludge to a point where it can be incinerated. Higher water contents require that more fossil fuel be co-fired with the sludge. A 40 percent solid sludge has a heating value of only about 500 Btu per lb, but 250 million Btu per day* can be recovered from an incinerator system servicing a population of 100,000. Perhaps the significant factor, compared to land application, is the fact that less than 0.03 percent of the DDT and 2,4,5-T and only about 6 percent of the PCBs in the sludge will survive incineration and be returned to the environment. These toxicants remain in sludge applied directly to the soil.[168]

Pyrolysis and Gasification

Fluid and solid fuels can be recovered from waste and biomass by *pyrolysis*— destructive distillation in the absence of air. Pyrolysis offers several advantages over direct combustion. Metals, glass, and some plastics often cause operating problems during incineration. The metals and glass can be removed from pyrolyzer feed, but it is not essential, and plastics make a beneficial contribution to the final product. The energy content of this product exceeds the energy consumed in the process. The volume reduction by pyrolysis is equivalent to incineration—75 percent—or higher if glass and metal are reclaimed. The residue is sterile and inert and can be landfilled safely as long as toxic metals are absent. Air and water pollution should be minimal. The pyrolyzer char is transportable and storable, and at the same time should be a marketable substitute for some activated carbons.[169]

The *flash pyrolysis* process (Occidental Research Corp.) emphasizes recovery of materials (fuels and feedstocks). It is carried out on finely shredded, air-classified MSW in less than 1 sec at about 900°F (490°C) and produces an oil-like liquid—*pyrolysis oil*.[170] The 7 percent char residue is landfilled.[2] The yield per ton of MSW is approximately 36 gal of pyrolysis oil, with a heating value equivalent to 27 gal of No. 6 fuel oil.[171] The net energy yield is about 29 percent.[172] The low heating value of the pyrolytic oil is largely because of its low carbon content (57.5 vs. 85.7 percent) and high oxygen content (over 33 percent).[172] Figures 7.5 and 7.6 show typical separation and flash pyrolysis schemes, respectively.

The pyrolysis oil formed by the process has an empirical formula of approximately $C_5H_8O_2$. Since the process begins with cellulose ($C_6H_{10}O_5$) as feed, the initial two steps in flash pyrolysis can be assumed to be the loss of carbon dioxide and water. The oil is not miscible with No. 6 fuel oil, but the two can be blended to form a relatively stable dispersion which is something of an improvement over each of the two, in that it is much less corrosive than pyrolysis oil[†] and has a much lower sulfur content than No. 6 fuel oil. Pyrolysis oil can be solvent refined into two fractions (40 percent solids and 60 percent water-soluble oil), both of which are very complex.[173]

Wood oil, which has a higher oxygen content, is more aromatic than cellulose-derived pyrolysis oil, because of the lignin ($C_{10}H_{11}O_2$) in the feed. Its

---

*The energy equivalent of about 43 bbl of crude oil.
†Corrosivity is similar to acetic acid; the pH is 2.2 to 3.0, owing to formic, acetic, glycolic, malic, and acrylic acids.

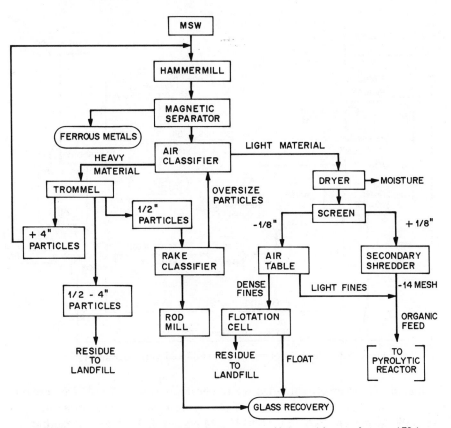

Figure 7.5   Separation for flash pyrolysis system. (Adapted from reference 172.)

fuel value is also no more than 80 percent as high, because of its oxygen content. One barrel of oil, equivalent to No. 6 bunker fuel, can be produced from about 900 lb of wood chips by a catalytic thermal process being developed at Lawrence Berkeley Laboratory in conjunction with Rust Engineering Co. and DOE. The crude can be manufactured for about $26 per bbl.[174]

Another pyrolysis system (Monsanto's *Landgard*) emphasizes disposal and claims to result in a 94 percent volume reduction.[169] It introduces a shredded combustible waste into a rotary kiln, with supplementary No. 2 fuel oil, at a temperature of 1200°F (650°C). The product gases* are combined with air and combusted at 1400°F (760°C) in an afterburner, from which the heat is directed to waste heat boilers.[170,171]

Purox (Union Carbide Corp.) is a *hybrid pyrolytic-incineration* process that produces a medium-heating-value gas (about 370 Btu per scf) at temperatures up to 3000°F. The off-gas is suitable for conversion to methane, methanol, or

*Heating value about 120 Btu per scf.

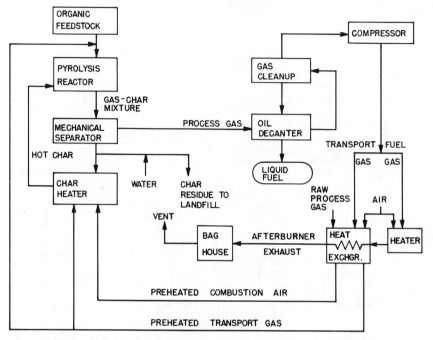

Figure 7.6 Flash pyrolysis system. (Adapted from reference 172.)

ammonia. The coarsely shredded waste decomposes to synthesis gas, organic liquids, and char. The latter is oxidized by oxygen. The gas composition is typically: 50 percent CO, 30 percent H_2, 14 percent CO_2, 3 percent methane, 2 percent mixed C_2's, and 1 percent mixed N_2 and argon. A plant with an input capacity of 350 tpd could produce about 2.62 billion Btu of this gas daily.[171] Volume reduction is about 95 to 98 percent.[169] This is perhaps the most advanced of the pyrolysis processes currently under development. Its efficiency is equal to dry mechanical beneficiation. About 8 percent of the available energy is required to run the plant and equipment, 17 percent is lost in chemical conversion, and another 10 to 15 percent loss is associated with combustion of the fuel and supplying the oxygen, for an overall efficiency of about 65 percent.[175]

At Georgia Tech, a mobile *slagging pyrolyzer* has been developed to convert forest or agwastes to solid fuel. A wet feed of 200 tpd can produce a free-flowing char oil fuel (40 percent char and 60 percent oil) similar to powdered coal and having a heating value of over 12,000 Btu per lb. A temperature in the range of 800 to 1400°F (425 to 760°C) is used. This system is being commercialized by Tech-Air Corp. for pyrolyzing wood waste at 1100°F (540°C). The products have the following approximate heating values: char, 11,000 to 13,500 Btu per lb; oil, 10,000 to 13,000 Btu per lb; gas, 200 Btu per scf.[176]

The *Torrax* process (Carborundum Environmental Systems) obtains the 3000°F heat required for *gasification* from air oxidation of the pyrolytic char. A fuel gas (140 Btu per scf) can be drawn off or the 160 to 170 Btu raw gas can be diverted directly to a furnace with a waste heat boiler.[171]

Molten salt pyrolysis (Atomic International) is yet another conversion process. The waste is fed into a molten sodium carbonate bath at 1700 to 1850°F (920 to 1015°C), where it is immediately and completely pyrolyzed. The off-gases pass through the melt, which absorbs the acid gases (H_2S and HCl) together with the ash. The waste and char are completely pyrolyzed and gasified without formation of tars, liquids, or significant quantities of NO_x, and the products are sterile and odor-free. The process works well with organic waste, rubber, and wood and has also been demonstrated on nitropropane and waste X-ray film. A typical off-gas composition range from these wastes is: H_2, 9 to 21 percent; CO, 8 to 20 percent; CO_2, 4 to 20 percent; CH_4, 2 to 5 percent; and C_2's, 0.2 to 1.2 percent. The higher heating value will be in the range 55 to 180 Btu per scf.[177]

SUMMARY

Very large quantities of radiant energy from the sun are intercepted by the earth on a continuing basis. Unfortunately, the photosynthetic efficiency of biological solar energy fixation has been assessed at only 2 ± 1.5 percent, on the basis of a review of 135 references.[178] A further serious shortcoming of biosynthesis is its dependency on phosphorus, a nonrenewable resource in limited supply. It is estimated that known reserves of phosphate rock will be mined out in no more than 400 to 500 years without introducing a biomass energy option. Mineral-containing ash from thermal conversion processes can be recycled to the soil, but there is no guarantee that the minerals will be in a biochemically available, soluble form. Even if this could be ensured, the economics of transporting such a lean form of nutrient back to the farm or tree plantation from a remote conversion facility could become an overriding economic factor.[106]

Yet another deterrent to large-scale energy farming is the ongoing competition with a growing population for arable land for food culture. Soil erosion is depleting the amount of our topsoil at an alarming rate, and urbanization and highway systems are reducing the available acreage of cropland. Biomass yields have been high in recent years, but there has to be a time when we reach a limit to the growth rate of agricultural productivity, and that time cannot be too far in the future.

If the United States is to continue to be the breadbasket for the U.S.S.R. and the less-developed countries of the world, we must relegate energy farming to marginal croplands so that it will not compete with food and feed cultivation. This generally means using relatively dry, infertile land with a topography perhaps not too suited to large-scale cultivation and harvesting. It must also be realized that at this stage, land sown in corn or grain for food and feed will produce about twice the financial return of the same land devoted to energy farming of trees. Large areas of land would have to be under cultivation to provide a reliable level of raw material output for an economically viable energy conver sion facility. Cultivation and harvesting costs could prove to be high in terms o energy, labor, and machinery. All in all, it must be expected that contribution: to our energy budget from such sources will be modest into the foreseeabl future.[179]

There is some concern over the possible ecological consequences o aquaculture. Carried out on a significant scale, upwelling of nutrients fo mariculture could reduce the surface temperature of the ocean (in the are under cultivation) sufficiently to affect the local climate, hence local rainfal and yields of cash farm products. Similarly, aquaculture in large inland im poundments could raise the local relative humidity and moderate loca temperatures.[106]

It is obvious that there are a number of bioengineering options for th generation of fuels. Biomass can be produced by bioconversion of solar energy in the course of supplying food, feed, and fibers. The organic waste can then be combusted directly for its thermal energy or be pyrolyzed to storable and transportable fuels and chemical feedstocks. Alternatively, the cultivation o biomass can be undertaken for the sole purpose of providing directly a com bustible mass, generally for generation of electric power. These are the options covered in this chapter.

Other possibilities are formation of hydrogen fuel by biophotolysis or by electrolysis—with the photoelectric effect providing the energy for the latter conversion. These options have been omitted from specific discussion in this book because the emphasis here is on hydrocarbons, although hydrogen may be an essential ingredient of hydrocarbons and processes for their synthesis.

One basic principle apparently overlooked in some publications on biomass energy projections is the fact that you cannot butcher the same hog twice. A writer will often discuss the amount of MSW being collected today as the basis for estimates of the availability of MSW for landfill gas tomorrow, forgetting that in the following pages he or she may also extol the virtues of RDF extract able from that waste before landfilling. This RDF is the major part of that organic component of the waste which is the source of landfill gas. Once it is ex tracted for RDF there can *be* no landfill gas. Other papers tell of the amount of forest and field residues that can be collected for conversion to fuel and feedstocks while ignoring the fact that this litter, allowed to remain and decom pose on the land, provides nutrients and soil conditioners that maintain high productivity of subsequent crops. Without them, yields are bound to decrease over the years. Keeping these points in mind, some of the rosy projections for future biomass energy may require some shading.

Estimating the economics of energy farming is fruitless without a prior determination of the crop and the conversion process to be used. Without doubt, introduction of a significant new fuel or feedstock source would have a measurable impact on regional economics and employment, environmental regulations, pricing of competing fuels, and international trade. The question is whether we shall be able to realize development of a source large enough to cause more than a ripple effect.

The annual per capita rate of generating wastes is approaching 3.5 lb for domestic solids and 1.5 lb for commercial and industrial solids, with a 2 to 4 percent increase per annum, or a doubling in approximately each generation.[180] This waste "commodity," for the most part, has a zero or negative value. Over the long haul in our affluent society, the cost of labor has outpaced the value of such commodities, thus making it cheaper to mine virgin resources for them than to separate them from scrap. Technology has made its contribution to this situation. For example, adoption of the basic oxygen process for steel has decreased the demand and price for scrap, thus leading to the abandonment of junk cars on the streets and over the countryside.

According to 1977 data, we payed over $25 billion for foreign oil* and over $6 billion for MSW disposal (collection and landfill). We are short of energy, yet we fail to take steps that would enable us to reduce both our import and our waste disposal bills.[181] Subsidies may be called for to stimulate recycling, not only to reduce consumption of virgin resources, but to reduce the cost of waste handling and disposal.†

While the per capita weight of solid waste is rising at 2 to 4 percent per annum, volume is rising even more rapidly because of increased fluffiness due to preprocessing and prepackaging of increasing amounts of our food. Volume is the important factor with landfill; however, this sort of volume increase largely should yield to compaction at the disposal site.

To dispose of all its solid wastes in sanitary landfills, New York City would require about 400 acres of land per year within economic hauling distance. Marshlands are readily accessible by barge, but they are now off limits for waste disposal since they are looked upon as vital links in our food chain.[180]

The upper few feet of a landfill can decompose aerobically (compost) to a humuslike material in a few months, and the lower portion, deposited below the water table, may form a similar material anaerobically. A market projection for compost, however, shows that demand is unlikely to amount to more than 1 percent of what can be produced.[180]

There are those who believe that renewable resources have a near-term potential of providing approximately 8 Q of our energy budget per annum (or the equivalent of about 1.4 billion bbl of oil at a 1977–1978 cost of $18 billion).‡ U.S. energy budget estimates for the year 2000 range from a low of 106 Q to a high of 225 Q,[182] relatively little of which is likely to be provided by renewable resources, despite the implied economic incentive. At present, the energy con-

---

*Estimate for 1978, about $23 billion.

†About three-fourths of the total waste handling cost goes into collection. Approximately a tenth as much is required to dispose of the waste in a sanitary landfill.[180]

‡Based on an average landed cost in the United States of $12.85 per bbl. At the second half 1979 OPEC price of crude, the value of this 8 Q becomes about $28 billion. With the maximum prices of Libyan crude announced in January 1980, 8 Q equates to $48.6 billion.

tribution of renewable resources is about 1.8 Q per year in the United States, and it is growing at a rate of over 5 percent annually, which would suggest a doubling time of less than 14 years.[31]

Fuels or feedstocks expected from waste conversion include: shredded and/or pelletized solids (e.g., RDF or Eco-Fuel II), heavy liquids (e.g., pyrolytic oil), light liquids (e.g., methanol), low-Btu (100 to 200 Btu per scf) gas, medium-Btu (300 to 400 Btu per scf) gas, and methane.[183] The bulk of these products undoubtedly would be used in the generation of steam for electric power, heating, or industrial processing. Relatively little is expected to find its way into chemical feedstocks over the near term.

The major problems to be faced in converting waste or biomass to fuel and feedstocks are: (1) acquiring a large enough supply of feed to provide a steady flow and to permit economical conversion, (2) producing a marketable fuel or feedstock for industrial use, and (3) doing so in a conversion facility capable of meeting environmental regulations while maintaining favorable costs.[183]

REFERENCES

1   C. G. Golueke and P. H. McGauhey, "Waste Materials," in J. M. Hollander and M. K. Simmons, Eds., *Annual Review of Energy*, Vol. 1, Annual Reviews, Palo Alto, CA, 1976, pp. 257–277.

2   J. H. Skinner, "Demonstration of Systems for the Recovery of Materials and Energy from Solid Wastes," in H. Alter and E. Horowitz, Eds., *Resource Utilization and Recovery*, ASTM Spec. Tech. Publ. 592, Amer. Soc. Test. Mater., Philadelphia, 1975, pp. 53–63.

3   *NCRR Bull.* **5**(1), 2–10 (1974).

4   D. L. Wise et al., "Fuel Gas Production from Solid Waste," in C. R. Wilke, Ed., *Cellulose as a Chemical and Energy Resource*, Interscience, New York, 1975, pp. 285–302.

5   *News Rep.* (NAS) **26**(4), 5, 16 (Mar. 1976).

6   Inst. Res. Land Water Resour. (1975); reported in reference 1, p. 262.

7   *Resource Recovery and Source Reduction*, 2nd Rep. to Congress, No. SW-122, EPA, Washington, DC, 1974.

8   D. A. Tillman, *Chemtech* **7**(10), 612–615 (Oct. 1977).

9   "Recycling Responds," Natl. Assoc. Recycling Ind., New York, 1978, p. 4.

10 "Energy Conservation Through Recycling," Natl. Assoc. Recycling Ind., New York, undated.

11 M. J. Mighdoll and P. D. Weisse, *Harvard Bus. Rev.* **54**(5), 143–151 (Sept.–Oct. 1976).

12 J. J. Fritz and J. Szekely, *Chemtech* **7**(4), 207 (Apr. 1977).

13 W. J. Landman and W. J. Darmstadt, Proc. 1st Int. Conf. Convers. Refuse Energy, Montreux, Switz., Nov. 3–5, 1975, pp. 589–595.

14 B. W. Haynes et al., "Metals in the Combustible Fraction of Municipal Solid Waste," *Rep. Invest. 8244*, DOI, Washington, DC, 1977, p. 4.

15 N. Rueth, *Mech. Eng.* **99**(12), 24–29 (Dec. 1977).

16 Reference 14, p. 15.

17 W. R. Niessen et al., *Systems Study of Air Pollution from Municipal Incineration*, Vol. 1, DOC, PB 192378, Natl. Tech. Inf. Serv., Springfield, VA, 1970.

18 A. J. Buonicore et al., *Trans. Amer. Nucl. Soc.* **23** (Suppl. No. 1), 31–34 (1976).

19 M. E. Henstock, presented at CHEMRAWN I—World Conf. Future Sources Organic Raw Mater., IUPAC, Toronto, July 10-13, 1978.

20 M. B. Bever, *Technol. Rev.* **79**(4), 23-31 (Feb. 1977).

21 R. C. Donovan et al., *Bell Labs Rec.* **55**(8), 215-219 (Sept. 1977).

22 J. A. Mock, *Mater. Eng.* **87**(2), 3, 8-12 (Feb. 1978).

23 B. Miller, *Plast. World* **34**(1), 22-28 (Jan. 19, 1976).

24 R. A. Kruppa, *Plast. Technol.* **23**(5), 63, 64 (May 1977).

25 D. S. Hill, *Kiwanis Mag.* (Nov.-Dec. 1977), pp. 16-18, 44-47.

26 (a) L. L. Anderson (1972); (b) A Poole (about 1975); (c) *Combustion* (Feb. 1977); (d) R. D. Smith; (e) H. W. Schultz et al. (Jan. 1976); reported by B. Sternlicht, *Mech. Eng.* **100**(8), 30-41 (Aug. 1978).

27 "Recycling Provides Resources for America's Future," Natl. Assoc. Recycling Ind., New York, undated.

28 R. A. Livingston, *Trans. Amer. Nucl. Soc.* **23** (Suppl. No. 1), 28-30 (1976).

29 V. E. Smay, *Popul. Sci.* **209**(3), 97, 164 (Sept. 1976).

30 "Technology for the Conversion of Solar Energy Fuel to Gas," Natl. Sci. Found.-Univ. Pennsylvania, PB-271142, Natl. Tech. Info. Serv., Springfield, VA, Jan. 31, 1973.

31 D. A. Tillman, "Uncounted Energy: The Present Contribution of Renewable Resources," in D. A. Tillman et al., Eds., *Fuels and Energy from Renewable Resources,* Academic, New York, 1977, pp. 87-104.

32 J. R. Greco, "Energy Recovery from Municipal Wastes," pp. 289-312, in reference 31.

33 F. T. Varani and J. J. Burford, Jr., "The Conversion of Feedlot Wastes into Pipeline Gas," in L. L. Anderson and D. A. Tillman, Eds., *Fuels from Waste,* Academic, New York, 1977, pp. 87-104.

34 J. Herman, *Courier-News* (Bridgewater, NJ) **93**(216), B-1 (Feb. 10, 1977).

35 R. Stone, *Civ. Eng.—ASCE* **48**(1), 51-53 (Jan. 1978).

36 *Chem. Eng. News* **56**(35), 25 (Aug. 28, 1978).

37 R. Todd, *Phila. Inquirer* **301**(29), 8-A (July 29, 1979).

38 J. Pacey; reported in reference 32.

39 G. R. Stephens and G. H. Heichel, "Agricultural and Forest Products as Sources of Cellulose," pp. 27-42, in reference 4.

40 G. M. Woodwell, *Sci. Amer.* **223**(3), 64-74 (Sept. 1970).

41 P. R. Ehrlich et al., *Ecoscience: Population, Resources, Environment,* Freeman, San Francisco, 1977, p. 71.

42 D. N. Moss; reported in *Chem. Eng. News* **55**(9), 22, 23 (Feb. 28, 1977).

43 H. L. Penman, *Sci. Amer.* **223**(3), 99-108 (Sept. 1970).

44 T. F. Ledig and D. I. H. Linzer, *Chemtech* **8**(1), 18-27 (Jan. 1978).

45 R. M. Thomson, *Dimensions* (NBS) **59**(1), 12-15 (Jan. 1975).

46 G. C. Szego and C. C. Kemp, *Chemtech* **3**(5), 275-284 (May 1973).

47 B. Wolverton and R. C. McDonald: reported by F. von Hippel and R. H. Williams, *Bull. At. Sci.* **33**(8), 12-15, 56-60 (Oct. 1977).

48 *Renewable Resources for Industrial Materials,* Comm. on Renewable Resources for Ind. Mater. (CORRIM), Natl. Acad. Sci.-Natl. Res. Counc., Washington, DC, 1976, pp. 51-56.

49 J. F. Saeman; reported in reference 42.

50 Reference 48, pp. 5, 6.

51 O. C. Doering; reported in reference 42.

52 Reference 48, p. 7

53 Reference 48, pp. 6-8, 14, 15.

54 Reference 48, pp. 64-74.

55 *Chem. Eng. News* **54**(27), 16 (June 28, 1976).

56 P. F. Hahn, *Energy User News* **2**(51), 22, 23 (Dec. 26, 1977).

57 D. Pimentel et al., *BioScience* **28**(6), 376-382 (June 1978).

58 Reference 48, pp. 39, 40.

59 J. R. Gosz et al., *Sci. Amer.* **238**(3), 93-102 (Mar. 1975).

60 G. E. Likens et al., *Science* **199**(4328), 492-496 (Feb. 3, 1978).

61 *The Outlook for Timber in the United States,* For. Resour. Rep. No. 20, DOA, Madison, WI, Oct. 1973.

62 *Forbes* **122**(2), 63 (July 24, 1978).

63 Reference 48, pp. 57, 58

64 Reference 48, p. 63.

65 Reference 48, pp. 78, 79.

66 Reference 48, pp. 80-82.

67 Reference 48, pp. 217, 218.

68 Reference 48, p. 237.

69 *Chem. Eng. News* **55**(37), 46-48 (Sept. 12, 1977).

70 Reference 48, p. 223.

71 Reference 48, p. 235.

72 T. R. Miles, "Logistics of Energy Resources and Residues," pp. 225-248, in reference 31.

73 Reference 48, p. 219.

74 R. E. Inman, "Summary Statement on the Substrate," pp. 1-7, in reference 4.

75 L. L. Anderson, "A Wealth of Waste; a Shortage of Energy," pp. 1-16, in reference 33.

76 J. A. Alich, Jr., et al., "Feasibility of Utilizing Crop, Forestry, and Manure Residues to Produce Energy," pp. 213-224, in reference 31.

77 G. H. Heichel, *Amer. Sci.* **64**(1), 64-72 (Jan.-Feb. 1976).

78 E. Cook, *Man, Energy, Society*, Freeman, San Francisco, 1976, p. 327.

79 G. H. Heichel, "Energetics of Producing Agricultural Sources of Cellulose," pp. 43-47, in reference 4.

80 A. B. Makhijani and A. J. Lichtenberg, *Environment* **14**, 10 (1972).

81 A. E. Humphrey, "Economical Factors in the Assessment of Various Cellulosic Substances as Chemical and Energy Resources," pp. 49-65, in reference 4.

82 M. J. Povich; reported in *Chem. Eng. News* **55**(36), 27, 28 (Sept. 5, 1977).

83 E. E. Robertson, *Bioconversion: Fuels from Biomass*, Franklin Inst., Philadelphia, 1977, p. 44.

84 *Civ. Eng.—ASCE* **47**(4), 26 (Apr. 1977).

85 *U.S. News World Rep.* **80**(20), 77 (May 17, 1976).

86 R. Clarke, *New Sci.* **69**(990), 512, 513 (Mar. 4, 1976).

87 *Chem. Eng. News* **56**(50), 6, 7 (Dec. 11, 1978).

88 *Ibid.* **49**(33), 43 (Aug. 16, 1971).

89 G. A. Mills; reported in *ibid*.

90 H. W. Anderson et al., CHEMRAWN I Conf., poster session, IUPAC, Toronto, July 10-13, 1978.

91 H. W. Anderson, private communication.

92 W. G. Pollard, *Amer. Sci.* **64**(5), 509-513 (Sept.-Oct. 1976).

93 D. Pimentel et al., *Science* **194**(4261), 149–155 (Oct. 8, 1876).

94 Bennett (1939); reported in reference 93.

95 U.S. Natl. Resourc. Bd. (1935); reported in reference 93.

96 L. J. Carter, *Science* **196**(4288), 409–411 (Apr. 22, 1977).

97 Counc. Agric. Sci. Technol. (CAST); reported in *ibid*.

98 W. Lockeretz, *Amer. Sci.* **66**(5), 560–569 (Sept.-Oct. 1978).

99 GAO (1977); reported in reference 96.

100 J. A. Bassham, *Covered Energy Farms for Solar Energy Conversion*, LBL 4844, DOE, Washington, DC, undated.

101 J. A. Bassham, *Science* **197**(4304), 630–638 (Aug. 12, 1977).

102 L. G. Nickell, *Chem. Eng. News* **56**(41), 18–34 (Oct. 9, 1978).

103 DOA (1974); reported by R. A. Brink et al., *Science* **197**(4304), 625–630 (Aug. 12, 1977).

104 Conf. Bd. (1977); reported by L. G. Hauser, *Public Util. Fortn.* **102**(7), 22–30 (Sept. 28, 1978).

105 M. J. Povich, *Chemtech* **6**(7), 434–439 (June 1976).

106 H. R. Bungay and R. F. Ward, "Fuels and Chemicals from Crops," pp. 105–120, in reference 33.

107 M. Calvin, "Hydrocarbons via Photosynthesis," presented at Amer. Chem. Soc., Div. Rubber Chem., San Francisco, Oct. 5–7, 1976.

108 V. P. Gutschick, *BioScience* **28**(9), 571–575 (Sept. 1978).

109 P. A. Waldheim, "Ammonia," in D. M. Considine, Ed., *Chemical and Process Technology Encyclopedia*, McGraw-Hill, New York, 1974, pp. 107–114.

110 S. H. Wittwer, *BioScience* **28**(9), 555 (Sept. 1978).

111 M. Calvin, *Chemtech* **7**(6), 352–363 (June 1977).

112 *Chem. Eng. News* **55**(37), 46–48 (Sept. 12, 1977).

113 E. S. Lipinsky, *Science* **199**(4329), 644–651 (Feb. 10, 1978).

114 *Guayule: An Alternative Source of Natural Rubber*, PB 264170, Panel on Guayule, Natl. Acad. Sci., Washington, DC, May 1977.

115 P. E. Nielsen et al., *Science* **198**(4320), 942–944 (Dec. 2, 1977).

116 E. V. Anderson, *Chem. Eng. News* **56**(35), 10, 11 (Aug. 28, 1978).

117 G. S. Schatz, *News Rep.* (NAS) **27**(6), 1, 4 (June 1977).

118 *Chem. Eng. News* **55**(21), 22, 23 (May 23, 1977).

119 H. Yokoyama et al., *Science* **197**(4308), 1076, 1077 (Sept. 9, 1977).

120 D. Griffin, *Fortune* **97**(8), 78–81 (Apr. 24, 1978).

121 E. Stiles, *Courier-News* (Bridgewater, NJ) (Sept. 12, 1978), p. A-13.

122 *Sciences* **17**(4), 4, 5 (July–Aug. 1977).

123 E. M. Glymph; reported in reference 121.

124 M. Calvin, *Chem. Eng. News* **56**(12), 30–36 (Mar. 20, 1978).

125 *Chem. Eng. News* **56**(44), 21 (Oct. 30, 1978).

126 *Technol. Rev.* **79**(2), 17 (Dec. 1976).

127 *Jojoba: Feasibility for Cultivation on Indian Reservations in the Sonoran Desert Region*, Natl. Res. Counc., Washington, DC, 1977.

128 T. H. Maugh II, *Science* **196**(4295), 1189, 1190 (June 10, 1977).

129 *Bus. Week* (No. 2504), 96F, 96I (Oct. 10, 1977).

130 *Sciences* **17**(8), 5 (Dec. 1977).

131 *Mech. Eng.* **100**(1), 55 (Jan. 1978).

132 D. Pothier, *Phila. Inquirer* **295**(130), 18D, 24D (Nov. 7, 1976).

133 *Bus. Week* (No. 2554), 94B (Oct. 2, 1978).

134 *Ocean City* (NJ) *Sentinel-Ledger* **96**(32), Sec. 1, 11 (Aug. 27, 1976).

135 W. Kammann; reported in *Phila. Inquirer* (Sept. 4, 1979), p. 8-B.

136 *Elizabeth* (NJ) *Daily J.* **200**(200), 2 (Aug. 23, 1978).

137 *New Sci.* **74**(1050), 276 (May 5, 1977).

138 M. Calvin, *Amer. Sci.* **64**(3), 270–278 (May–June 1976).

139 D. F. Othmer, *Mech. Eng.* **99**(11), 30–35 (Nov. 1977).

140 *Mach. Des.* **49**(24), 20–27 (Oct. 20, 1977).

141 C. E. Wise, *ibid.* **49**(12), 22–28 (May 28, 1977).

142 A. L. Hammond, *Science* **197**(4305), 745, 746 (Aug. 19, 1977).

143 D. Utroska, *Ind. Res.* **19**(10), 25, 26 (Oct. 1977).

144 *News Rep.* (NAS) **26**(14), 2, 3 (Dec. 1976).

145 B. Wolverton and R. C. McDonald, *New Sci.* **71**(1013), 318–320 (Aug. 12, 1976).

146 L. G. Holm et al., *Science* **166**(3906), 699–709 (Nov. 7, 1969).

147 W. S. Hillman and D. D. Culley, Jr., *Amer. Sci.* **66**(4), 442–451 (July–Aug. 1978).

148 K. Bhanthumnavin and M. G. McGarry, *Nature* (*Lond.*) **232**(5311), 495 (Aug. 13, 1971).

149 B. Brinkworth, *Chem. Br.* **11**(9), 311–316 (Sept. 1975).

150 J. R. Benemann et al., *Nature* (*Lond.*) **268**(5615), 19–23 (July 7, 1977).

151 J. G. Brown; reported in *Chem. Eng. News* **55**(37), 50 (Sept. 12, 1977).

152 "Research Needs Report: Energy Conversion Research," Amer. Soc. Mech. Eng., New York (1977); reported by G. P. Cooper, *Mech. Eng.* **99**(11), 22–28 (Nov. 1977).

153 R. H. Montgomery, *The Solar Decision Book,* Dow Corning, Midland, MI, June 1978, p. 4-11.

154 "Hydroposal/Fibreclaim," Black Clawson Fibreclaim, Inc., New York, undated.

155 *Chem. Eng. News* **54**(8), 24–26 (Feb. 23, 1976).

156 H. J. Young; reported by A. Morrison, *Mech. Eng.* **100**(7), 69–71 (July 1978).

157 A. R. Nollet and E. T. Sherwin, in *ibid.*

158 J. M. Kehoe; reported in *Bus. Week* (No. 2543), 86D (July 17, 1978).

159 D. A. Tillman, "Energy from Wastes: An Overview of Present Technologies and Programs," pp. 17–39, in reference 33.

160 J. D. Keenan, *Energy Convers.* **16**(3), 95–103 (1977).

161 A. H. Brown, *Chemtech* **5**(7), 434–437 (July 1975).

162 *Mech. Eng.* **100**(8), 67 (Aug. 1978).

163 K. L. Wasserman, *Civ. Eng.—ASCE* **48**(2), 60–65 (Feb. 1978).

164 *Phys. Today* **29**(12), 20 (Dec. 1976).

165 M. J. Bartos, Jr., *Civ. Eng.—ASCE* **48**(11), 80–84 (Nov. 1978).

166 *New Sci.* **75**(1064), 357 (Aug. 11, 1977).

167 D. F. Othmer, U.S. pat. 4,017,421 (Apr. 12, 1977).

168 *Environ. Sci. Technol.* **10**(12), 1080–1082 (Nov. 1976).

169 J. E. Liebeskind, *Chemtech* **3**, 537–542 (Sept. 1973).

170 M. LaBreque, *Popul. Sci.* **210**(6), 95–98, 166, 167 (June 1977).

171 E. M. Wilson and H. M. Freeman, *Environ. Sci. Technol.* **10**(5), 430–435 (May 1976).

172 S. J. Levy, "The Conversion of Municipal Solid Waste to a Liquid Fuel by Pyrolysis," 1st Int. Conf. Convers. Refuse Energy, Montreux, Switz., Nov. 3–5, 1975, pp. 226–231.

173 K. W. Pober and H. F. Bauer, "The Nature of Pyrolytic Oil from Municipal Solid Waste," pp. 73-85, in reference 33.

174 S. Ergun; reported in *Chem. Eng. News* **57**(25), 6 (June 18, 1979).

175 D. A. Tillman, *Environ. Sci. Technol.* **9**(5), 418-422 (May 1975).

176 F. Shafizadeh, "Fuels from Wood Waste," pp. 141-159, in reference 33.

177 S. J. Yosim and K. M. Barclay, "Production of Low-Btu Gas from Wastes, Using Molten Salts," pp. 41-56, in reference 33.

178 D. O. Hall; reported by L. A. Pilato and W. T. Reichle, *Chemtech* **8**(11), 643 (Nov. 1978).

179 J. B. Grantham, "Anticipated Competition for Available Wood Fuels in the United States," pp. 55-91, in reference 31.

180 M. M. Feldman, *Trans. N.Y. Acad. Sci.,* Ser. II, **31**(6), 648-655 (June 1969).

181 Reference 33, Preface, p. xiii.

182 C. W. Mottley, "How Much Energy Do We Really Need?" pp. 1-22, in reference 31.

183 D. A. Tillman, "Nontechnical Issues in the Production of Fuels from Waste," pp. 211-226, in reference 33.

8 BASIC ORGANIC CHEMICALS AND INTERMEDIATES

... the positions of oil and natural gas raw materials will not be seriously threatened in the years immediately ahead.

H. GRÜNEWALD

It is essential at this point to clarify the matter of terminology associated with this chapter. The term *petrochemical* has been adopted to describe a chemical derived wholly or in part from crude oil or natural gas. It may result, in its final form, from a mere separation or extraction from the raw material, without undergoing alterations in its chemical composition. *Botanochemical,* is a more recent term, used to imply a material, perhaps physically or chemically indistinguishable from a petrochemical, but derived from biomass sources. An *intermediate* is a chemical formed at a stage between one or more raw materials and the commerical petrochemical or botanochemical product.

The synthetic chemical industry had its beginnings in 1856, based on tar produced in coke ovens. The polycyclic aromatic hydrocarbons (PAHs), extracted by distillation, yield high-purity products used in paint vehicles, adhesives, roofing and impregnating materials, and for carbon black feedstocks and solvents.[1] Typical compositions of light oils coproduced in coke production are shown in Table 8.1. Petrochemicals appeared in the laboratory in about 1920, but the industry did not develop until approximately 1940.[2] Since that time its expansion has been phenomenal.*

In 1978, oil's $12 to $15 per bbl value escalated to 5 times that when converted into polymers, 10 times as much in the form of fibers, and 100 times the

*The current value of the *petrochemical* industry worldwide has been estimated at $300 billion by Mathis (Exxon),[2] and Shapiro (duPont)[3] has credited the world's *chemical* industry as a whole with contributing $300 billion (U.S.) to gross world product, along with 5 million jobs.

Table 8.1 Typical Compositions of Light Oil (Benzol) from Coal[a]

Component	Benzol Composition, wt %	
	Semifinished	Press-Refined
Benzene	61.7	72.4
Toluene	16.9	16.6
Xylenes and ethyl benzene	7.5	7.2
Trimethyl benzene, methylethyl benzene, styrene, dicyclopenta-diene, and methyl styrenes	4.2	—
Nonaromatics	3.9	1.5
Indane, indene, coumarone	3.2	—
Indane	—	1.0
Naphthalene, methyl naphthalene	1.5	—
Thiophene and derivatives	0.5	—
n-Butyl benzene	0.2	—
Other alkyl benzenes	—	0.8
Benzonitrile	0.2	—
Tetralin	—	0.2

Source: Reference 1.
[a]Boiling below 366°F (180°C).

initial value when converted to highly specialized chemicals and products (herbicides, plant growth regulators, X-ray films, etc).[3] There is every reason to believe that oil and natural gas will remain the major feedstocks for the chemical industry as long as they continue to be the least costly alternatives. "The United States has alternatives—politically, technologically, and physically,"[4] but there is no reason for optimism that these alternatives will be able to meet the economic criteria in the foreseeable future. Ultimately, we must face up to the necessity of converting to a postpetroleum world economy. This is not going to come about, however, until such time as the economics become attractive to the chemical industry—whether brought about by still higher Organization of Petroleum Exporting Countries (OPEC) prices, deregulation of domestic oil and gas prices, tax incentives, government participation in technology development, easing of environmental regulations, and so forth, either individually or collectively. For example, a recent Battelle Columbus Laboratories' study for USDA showed that wood-based oleoresin chemicals would be about three times as expensive as the same compounds derived from petroleum.[5]

We can look to the future in terms of three basic time frames[3]:

1 *Near-term* (to late 1980s), during which the supply of gas and oil will be adequate to meet demand worldwide for all uses.

2 *Middle-term* (1990–2000), when we must look to energy alternatives—coal liquefaction, oil shale, tar sands—but crude oil and natural gas should still be available* for synthesis of petrochemicals.

3 *Long-term* (beyond 2000), we must learn to adopt fusion and biomass and other solar-based options. In this period, the chemical industry should be drawing more heavily on coal as a primary hydrocarbon resource.

Petroleum and natural gas are expected to be the primary hydrocarbon feeds for the chemical industry long after these fossil fuels become unavailable for the electric power and transportation industries, because only about 3 percent of their output is consumed as feedstocks by the chemical industry, and conversion technology for this essential industry, based on other resources, lags use of alternative resources by other industries.[6] Tables 8.2 and 8.3 show output quantities and values of some major U.S. gas-based chemicals, by way of suggesting the magnitude of the task facing the chemical industry. There is no proven, practical, short-term technology for the volume manufacture of the products shown in Tables 8.2 and 8.3 from feedstocks other than natural gas,[3] and the lead time required to do so in the future is probably two to three decades.[6]

This chapter will address the subjects of feedstocks (current and potential) and intermediates for the formation of synthetic hydrocarbons derived from fossil and biomass sources. The actual syntheses and the facts and figures involving the resulting synthetics will be taken up in Chapter 9.

COMMODITY ORGANIC CHEMICALS

Virtually all commodity organic chemicals have their origins in fossil fuels—generally, petroleum or natural gas. Table 8.4 shows U.S. production

Table 8.2   U.S. Production of Major Gas-Based Chemicals, 1975–1977

Product ·	Quantity, 10^6 Tons	
	1976	1977
Ammonia (anhydrous)	16.7	17.4
Urea (primary solution)	4.1	4.5
Methanol (synthetic)	3.1	3.2
Formaldehyde (37%)	2.7	3.0

Source: Reference 7. Reprinted with permission of the copyright owner, The American Chemical Society.

*"Exhaustion" of a natural resource is not a realistic condition; "increasing scarcity and price" are. In other words, the "availability" is more economic than physical.

Table 8.3 Value of Major Gas-Based Chemicals, United States, 1976

| Product | 10^6 Dollars[a] | |
	Total Production Value	Merchant Value
Ammonia	1800	750
Urea	425	335
Methanol	385	155
Formaldehyde	240	78

Source: Reference 8. Reprinted with permission of the copyright owner, The American Chemical Society.

[a]Data not precise; scaled from graph.

Table 8.4 U.S. Production of Major Organic Chemicals

| Chemical | 10^6 lb[a] (% change) | | | |
	1976	1977	1978[b]	1979[b]
Ethylene	22,475	25,172 (+12%)	27,400 (+9%)	28,500 (+4%)
Propylene	10,030	13,328 (+33%)	14,700 (+10%)	15,600 (+6%)
Styrene	6,301	6,867 (+9%)	7,000 (+2%)	7,350 (+5%)
Methanol	6,242	6,453 (+3%)	6,400 (−1%)	6,500 (+2%)
Formaldehyde	5,449	6,046 (+11%)	6,350 (+5%)	6,400 (+1%)
Ethylene glycol	3,335	3,675 (+10%)	4,200 (+14%)	4,400 (+5%)
Butadiene	3,507	3,259 (−7%)	3,600 (+10%)	3,700 (+3%)
p-Xylene	2,911	3,172 (+9%)	3,500 (+10%)	3,700 (+6%)
Cumene	2,716	2,644 (−3%)	3,300 (+25%)	3,450 (+5%)
Phenol	2,121	2,338 (+10%)	2,700 (+15%)	2,825 (+5%)
Acetic acid	2,463	2,570 (+4%)	2,575 (0%)	2,600 (+1%)
Cyclohexane	2,187	2,265 (+4%)	2,325 (+3%)	2,425 (+4%)
Acetone	1,869	2,219 (+19%)	2,200 (0%)	2,250 (+2%)
Acrylonitrile	1,518	1,646 (+8%)	1,950 (+18%)	2,050 (+5%)
o-Xylene	854	984 (+15%)	1,075 (+9%)	1,140 (+6%)
Phthalic anhydride	902	926 (+3%)	990 (+7%)	1,030 (+4%)
Benzene, 10^6 gal	1,425	1,457 (+2%)	1,425 (−2%)	1,500 (+5%)
Toluene, 10^6 gal	999	1,018 (+2%)	1,150 (+13%)	1,200 (+4%)

Source: Reference 9. Reprinted with permission of the copyright owner, The American Chemical Society.

[a]Except as otherwise noted.
[b]*Chem. Eng. News* estimates.

figures for these materials. In turn, these chemicals are converted to the various organic products—plastics, fibers, antifreeze agents, paint vehicles, and so forth—which occupy such important places in our daily lives. In general, 1978 sales of industrial chemicals were estimated to be about 10 percent higher than in 1977, and 1979 sales were expected to be about 5 percent higher yet.[9] The 1977 sales of the U.S. chemical industry amounted to $113 billion, with industrial sales accounting for $55.5 billion. These chemical products contributed a net $5.4 billion to our trade balance in 1977,[7] thus offsetting, in some small measure, part of our trade deficit attributable to importation of crude oil.

The following year (1978) proved to be one of modest production growth over 1977 for most organics, with synthetics increasing about 6 percent overall. Chemical producers expected, however, to experience some continued slowing of this growth to 2 to 3 percent in 1979, with the Conference Board and the American Statistical Association (ASA) both projecting a growth in real gross national product of no more than about 2.4 percent.[9]

Typically, production of synthetic organics is a large consumer of energy—both as feedstocks and as process energy. A limited sampling of synthetic products shows from one-fourth to more than one-third of the energy is required for processing.[10] Overall, the chemical industry accounts for, at most, 8 percent of the U.S. energy budget, with projections to 1990 showing about 10.7 percent for chemicals, nearly doubling the energy consumption growth rate of the United States as a whole.[11] Table 8.5 indicates the sources of energy for the chemical industry, with projections to 1990. Natural gas is shown to be on the wane as a source of energy for the industry, but only in terms of percentage. Absolute consumption of natural gas by the industry is projected to increase during this period, as is the case with all other energy sources.[11] Half of the energy consumption by the chemical industry for 1971 was credited to the produc-

Table 8.5   Sources of Energy for Chemical Industry

	Quads (10^{15} Btu)								Annual Growth Rate, % (1971–
	1971		1980		1985		1990		
Source	Q	Percent	Q	Percent	Q	Percent	Q	Percent	1990)
Petroleum[a]	1.56	31.0	3.80	50.2	4.88	50.1	6.27	51.5	7.6
Natural gas	1.98	39.4	1.55	20.5	2.06	21.1	2.31	19.0	0.8
Coal	0.45	8.9	0.65	8.6	0.79	8.1	0.98	8.1	4.2
Purchased power	1.04	20.7	1.57	20.7	2.01	20.6	2.61	21.4	5.0
Total	5.03	100	7.57	100	9.74	100	12.17	100	4.8

Source: Reference 11. Reprinted with permission of the copyright owner, The American Chemical Society.

[a]Includes "other."

tion of olefins alone (Table 8.6). Later figures would probably show an even higher percentage for olefins, since polyolefin plastics are now among the faster growing segments of this industry (Table 8.7), and are expected to continue to lead the way, at least through 1985 (Table 8.8).

ALTERNATIVE FEEDSTOCKS

The impact of petrochemicals on our life-style and on our energy future should be sufficiently visible to encourage us collectively to accelerate the search for energy raw material alternatives. Unfortunately, the man-in-the-street is not yet convinced that there is a looming energy crisis, and private industry alone appears to be incapable of providing the capital necessary to get on with the job.

Table 8.6 Energy Consumption by Industrial Chemical Sector, 1971

Chemical Class	Quads	Percent
Olefins	2.0	49
Ammonia and acids	1.3	32
Cyclic intermediates	0.3	7
Others	0.5	12
Total	4.1	100

Source: Reference 11. Reprinted with permission of the copyright owner, The American Chemical Society.

Table 8.7 1977 Shipments of Chemicals—Growth by Category (Estimated)

Chemical Class	Annual Growth, %
Synthetic rubber	+22
Plastics materials and resins	+15
Soaps and detergents	+14
Organic chemicals	+14
Industrial organic chemicals[a]	+13
Nitrogen fertilizer	+10
Industrial gases	+8
Phosphatic products	+5
Chlor-alkalies	+4

Source: Reference 12. Reprinted with permission of the copyright owner, The American Chemical Society.

[a] $19.1 billion, 1976; $21.6 billion, 1977.

Table 8.8 Projected Average Annual Growth of Chemical Shipments through 1985

Chemical Class	Estimated Average Annual Growth Rate, %
Plastics materials and resins	9.9
Industrial organic chemicals	8.8
Industrial gases	7.4
Synthetic rubber	5.0
Industrial inorganic chemicals	4.5
Chlor-alkalies	4.2

Source: Reference 12. Reprinted with permission of the copyright owner, The American Chemical Society.

According to a report prepared for OTA by the American Institute of Chemical Engineers (AIChE), today's feedstocks for the chemical industry will undoubtedly remain the feedstocks of the future, but they will have their origins in different raw materials.[13] The first major shift will be toward the increased use of coal as a raw material.

Coal

There are two basic optional routes for coal or other fossil fuel conversions to useful fluids: either break the compound(s) down to smaller (fluid) units having a higher $H:C$ ratio (plus residual char) or build up the desired fluid product(s) from synthesis gas obtained from the raw material. Buildup can also be accomplished directly, by high-temperature/pressure catalytic hydrogenation of powdered coal suspended in oil.[14] Solids separation costs can become a significant factor in the breakdown method. In coal liquefaction, this cost may run from in excess of $2 to nearly $4 per bbl.[15]

Fluid fuels are important because they are the most readily adapted to storage, transport, and control in use, and the technology for their conversion to chemicals is well developed. Because of its domestic abundance and accessibility, coal is the prime prospect for an alternative raw material source for chemicals. Figures 8.1 and 8.2 give an indication of products available from the treatment of coal liquids and gases, respectively. Technologically, the hydrogen, benzene, light oil (benzol), ethylene, and so forth, can enter the same product streams as these materials derived from natural gas or crude oil. Figure 8.3 shows one process for co-producing methane (pipeline gas) and methanol (CH_3OH) from coal. The economics is another matter, but escalating natural gas and oil prices are making the coal option increasingly attractive. Derivation of intermediates from coal by carbonization is indicated by Fig. 8.4. The range of products was shown in Fig. 8.1. Table 8.9 points out the liquid fractions ob-

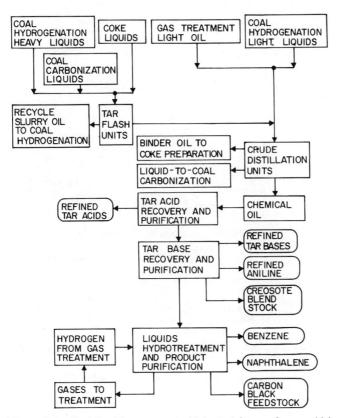

Figure 8.1 Coal liquids treatment. (Adapted from reference 16.)

tained from distillation of coking plant tars. The high-temperature tar from coal carbonization contains a dozen or so simple aromatics in quantities ranging from 1 to 10 percent of each, based on the crude tar. Various phenols and other compounds are also produced. Coal tar pitch comprises about half of the tar. Its composition is very complex, containing literally thousands of chemical compounds, many still unidentified.[1]

Table 8.10 shows typical composition ranges of coke oven gas, which is primarily hydrogen and methane. Before these gases can be used in chemical manufacture, sulfur- and nitrogen-containing impurities must be removed.

Figure 8.5 makes a comparison of the derivation of chemical intermediates from natural gas and from petroleum. It is immediately apparent that there is considerable overlapping of products, but in the case of coal feedstock, production of aromatics is emphasized, whereas aliphatic derivatives are more common from natural gas and crude oil.

Ideally, alternative chemical feedstocks should be low in sulfur and have a low carbon-to-hydrogen weight ratio.* Sulfur poisons gasification catalysts and

*Approaching that of methane.

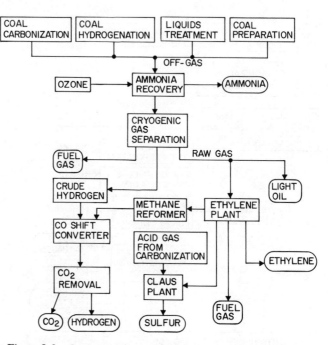

Figure 8.2 Coal gas treatment. (Adapted from reference 16.)

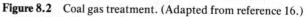
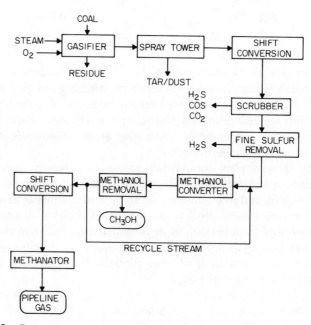

Figure 8.3 Co-product methane/methanol process. (Adapted from reference 17.)

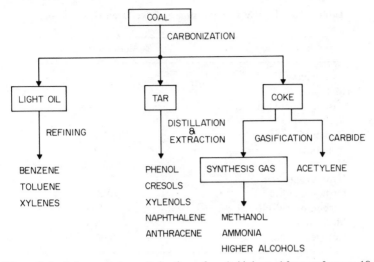

Figure 8.4   Intermediates—derivatives of coal. (Adapted from reference 18.)

Table 8.9   Liquid Fractions from Distillation of Coking Plant Tars

	Distillation Range	
Fraction	°F	°C
Light oil	Up to 366	Up to 180
Medium oil	366–446	180–230
Heavy oil	446–518	230–270
Anthracene oil	Beyond 518	Beyond 270
Pitch	Residue	

Source: Reference 1

is a notorious atmospheric pollutant. High-carbon liquid feedstocks form coke deposits that plug gasifiers and hinder catalyst performance, At the same time, they require addition of larger quantities of expensive hydrogen.[19]

The conversion of coal via syn gas is becoming increasingly important for our energy future. Figure 8.6 shows the general scheme, and Fig. 8.7 describes the *Mobil* process in more detail. This is probably the most advanced, but not yet commercialized coal conversion technology developed to date. Methanol produced from syn gas in a Lurgi gasifier could be converted to gasoline via the Mobil process, at an overall thermal efficiency of about 65.5 percent.[22] The only commercialized competitive technology is the Fischer-Tropsch process used in South Africa's SASOL II technology.

The latter process, however, in terms of gasoline output, is quite inefficient, because a broad spectrum of hydrocarbons and oxygenated hydrocarbons is

Table 8.10 Typical Composition of Coke-Oven Gas

Component	Volume Percent
Hydrogen	55–64
Methane	25–27
Carbon monoxide	4.6–5.8
C_nH_m	2–4
Carbon dioxide	1.5–2.5
Ammonia	~1.1
Hydrogen sulfide	0.3–3.0
Hydrocyanic acid	0.1–0.25
Carbon disulfide	0.016
Carbon oxysulfide	0.009
Mercaptans	0.003
Nitrogen oxides	0.0001

Source: Reference 1.

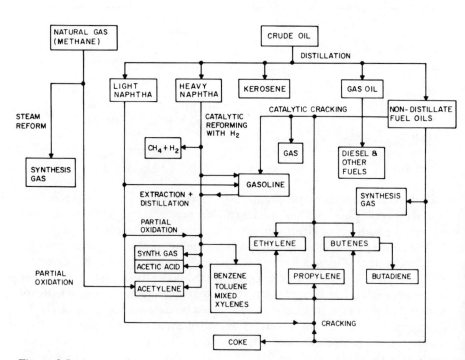

Figure 8.5 Intermediates—derivatives of natural gas and crude oil. (Adapted from reference 18.)

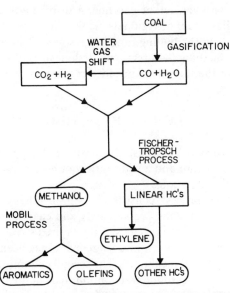

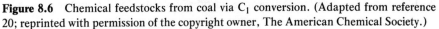

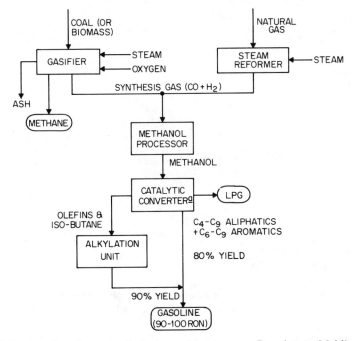

Figure 8.6 Chemical feedstocks from coal via C_1 conversion. (Adapted from reference 20; reprinted with permission of the copyright owner, The American Chemical Society.)

Figure 8.7 Gasoline from fossil fuels or biomass. *a*, Proprietary Mobil process. [Adapted from reference 21; reprinted with permission from S. L. Meisel et al, *Chemtech* **6**(2), 86–89 (Feb. 1976).]

produced, together with low-octane gasoline. A distinct advantage of the Mobil process is that no diesel fuel or residuals result. The bulk of the hydrocarbons produced are C_4-C_{10}, in the boiling range of gasoline. This would appear to be the ideal use for methanol—direct conversion to sulfur-, nitrogen-, and oxygenate-free motor fuels, with an unleaded research octane number (RON) of 90 to 100[21]:

$$xCH_3OH \rightarrow (CH_2)_x + xH_2O \qquad (1)$$

Using a family of related catalysts, the basic Mobil process could also be employed to convert ethanol and a wide variety of heteroorganic compounds to gasoline and other hydrocarbons.[22]

The essence of the process consists of using synthetic zeolite catalysts,* designed specifically to have the proper cavity and channel dimensions to provide selective penetrability by molecules of the desired size. The rate of conversion of methanol to hydrocarbons, in a single pass, has been determined to be 99+ percent (Table 8.11, first column). The second column shows the composition of a conventional C_{5+} gasoline fraction, but part of the butane fraction can be blended in to provide the proper gasoline volatility and increase the yield

Table 8.11 Conversion of Methanol to Hydrocarbons by Mobil Process[a]

Fraction Produced	Yield Based on (CH_2) Input, wt %			
	I[b]	II[c]	III[d]	IV[e]
$C_1{}^f + C_2{}^f$	1.3	—	—	—
$C_3{}^f$	6.6	—	—	—
$C_4{}^f$	15.8	—	3.7	3.5
C_5	19.3	35.3	35.3	35.5
C_6	16.4			
C_7-C_{9+}	13.2	13.2	13.2	22.8
A_6-$A_{9+}{}^g$	27.4	27.4	27.4	27.4
Total	100.0	75.9	79.6	89.2

Source: Reference 21. Reprinted with permission from S. L. Meisel et al., *Chemtech* **6**(2), 86–89 (Feb. 1976). Copyright by the American Chemical Society.
[a]Data not precise; scaled from graph.
[b]Single pass, methanol input.
[c]C_{5+}gasoline fraction.
[d]Formulated gasoline, with volatility adjusted by addition of *n*-butane.
[e]C_3-C_4 gases, olefins, and isobutane alkylated to additional gasoline. Balance is LPG.
[f]Gases.
[g]Aromatics.

*Crystalline aluminosilicates.

(third column). Finally, C_3 and C_4 gases, olefins, and isobutane can be alkylated to produce additional gasoline (nearly 90 percent overall yield), with most of the balance being LPG (fourth column). The beauty of the process is that the conventional natural gas- or coal–syn gas–methanol sequence can be carried through to high-octane gasoline by the addition of one relatively simple step. Biomass is another potential feedstock, of course.

The big question remaining is one of economics. The capital cost of the Mobil technology is expected to add less than 15 percent to the investment cost for the preparatory raw material-to-methanol part of the overall conversion. Keeping in mind that from the standpoint of energy input, nearly 2.5 gal of methanol is required to produce 1 gal of gasoline, and the Mobil process adds about 5¢ per gal to the gasoline cost, there is some question as to whether this technology can compete with petroleum refining at the present time. The product cost is estimated at 50 to 100 percent more than conventional gasoline. Furthermore, the sheer plant capacity required to meet only 1 percent of our domestic gasoline demand in the form of methanol* would cost billions of dollars.[21]

This raises the question of whether methanol should be converted at all or be used directly, since it is a cleaner burning fuel than the converted product. Despite current economics, there are those who feel confident that methanol will occupy an important place in our energy budget by the late 1980s, as both industrial and commercial fuel. They expect it to be a larger factor than either LNG or SNG, since it is easily transported and stored, and burns with little pollution.[23]

A technically feasible coal "refinery" for separating or converting solid coal into useful fluid fuels and intermediates has been proposed. The process basically comprises hot extraction of prepared coal with hydrogen-saturated solvent, and separation into liquid extract, gas, and sludge streams. The sludge is carbonized to produce BTX and cresols as by-products, with the char being gasified and fed to a Fischer-Tropsch synthesis unit. Solvent is extracted and recycled from the liquid stream and the residual coal extract is hydrogenated at 750 to 850°F and 3000 psi. Gas streams are combined, cleaned up, and separated to provide fuel gas, butane, propane, and SNG, with ammonia and free sulfur as by-products. Products of the liquid hydrogenation (hydrocracked and refined) and the Fischer-Tropsch synthesis yield fuel oil, gasoline, diesel fuel, and jet fuel.[24]

It has been estimated that by 1990, coal conversion could provide about 2 billion bpy of syncrude,† "with half the effort devoted to landing a man on the moon." The projection is based on the technology of the *H-Coal* process discussed in Chapter 5. The capital requirements would probably be about $30 billion—$20 billion for the conversion plants and $10 billion for opening the

---

*Equivalent to current domestic methanol capacity.

†Reference 25 quotes units of bpd, rather than bpy, but this is obviously a typographical error, since OPEC exports in 1976 totaled only 29 *million* (not *billion*) bpd, and the capital costs per unit of capacity quoted require bpy units to be credible.

new mines required.* These capital costs could be spread over 10 years and would represent less than 2 percent of the gross national product.[25]

Such projections seem to ignore the political and economic realities of such an undertaking—the problems of acquiring capital from the private sector with no promise of a sure market for a basically higher-priced product, and the long lead times and uncertainties of obtaining all environmental clearances for both mines and conversion facilities. There are also possible problems of water supply, and provision of transport for coal to the conversion facilities, but these problems relate largely to the more remote western coals, which must surely come into play in the long haul.

One expert has gone so far as to state that by 1985, ammonia and methanol could be manufactured from advanced gasification of coal as cheaply as from natural gas.[26] Any large-scale exploitation of coal resources, however, must deal with the societal factors, the questions of carbon dioxide and other pollution of the atmosphere, disturbance of the earth, use of considerable quantities of water, and erection of plants requiring large amounts of structural materials likely to become scarce if all the needed energy facilities are built.[27]

The major underlying factor in the adoption of alternative resources, whether for fuels or for chemical feedstocks, is economics. This is perhaps best illustrated by Table 8.12, which indicates the interaction of oil prices with the potential use of coal and biomass for chemical feedstocks. The actual numbers may now be outdated because of changing economic and technological conditions, but the principle still holds. The price of oil at any time will determine the role that coal (and biomass) can play in our energy and feedstock picture. Because of the long lead times required for the development of new resources and conversion technologies, we cannot afford to wait until that day, down the line, when oil prices finally make a given alternative option economic. The economic scale may have to be tipped to favor accelerated development of technology and facilities for coal and biomass conversion if we are to be properly

Table 8.12 Effect of Oil Price on Alternative Feeds

Oil Price per bbl (1974 dollars)	Scenario
$10	Coal economic only for heat and power
20	Coal economic for producing ammonia and methanol
25	Coal economic for producing ethylene and gasoline[a]
30	Biomass economic for producing ethylene and gasoline[a]

Source: Reference 20. Reprinted with permission of the copyright owner, The American Chemical Society.

[a]Assumes gasoline price $2 to $3 per gal (before taxes).

*Nearly 700 million tpy of coal would be required to feed such an undertaking.

prepared to make the switch to alternative raw materials. This "tipping" of the scales could take the form of federal subsidies, guaranteed price supports or markets, relaxed environmental regulations, tax relief, and so forth. Whatever the form of economic encouragement, we must begin to phase in capacity for the conversion of alternative resources, so that the long development lead times do not put our backs to the wall and leave us without viable, last-minute alternatives.

Production of substitute natural gas (SNG) from coal is one route toward an alternative to natural gas and crude oil as chemical feedstocks; however, the projected cost of SNG is in excess of $5 per 1000 scf. Coal, at $20 per ton, would account for about $0.80, and labor, management, distribution costs, return on equity, and reserve for maintenance and replacement would make up the balance. The low efficiency of the process (approximately 60 percent) is a contributing factor to the high product cost.[28]

The only thoroughly tested, dependable coal conversion processes now available are variants of the old *Lurgi* process, but they cannot compete economically with natural gas or crude oil as sources of fluid fuels and feedstocks, largely because of high capital costs. In 1977, OTA estimated it would cost $120 billion to establish a domestic coal-synthetics industry,* with the liquid fuel product costing $25 to $40 per boe. Since the official price of OPEC oil reached $32 per bbl by the third quarter of 1980, this option seems to be potentially a good bargain for the near future.[29] Another source's estimates of relative liquid fuel prices from various sources are shown in Table 8.13.

Technologically, the *Hygas* process of IGT appears to be the most advanced gasification process, but it has not yet been proven in on a full commercial scale. Estimates suggest that the cost of SNG produced from coal by the Hygas process could run as high as $4 per million btu[31]—more than half again the cost of domestic natural gas in mid-1980. The cost of a large-scale gasification plant

Table 8.13 Relative Costs of Liquid Fuels from Various Sources (Estimated, Based on Equal Energy)

Source	Gasoline	Diesel Fuel	Methanol
Petroleum	1.0	0.8	n.a.
Natural gas	n.a.	n.a.	2.3
Coal			
via syncrude	1.9	1.7	n.a.
via syngas	3.8	3.4	2.8
Oil shale	1.6	1.2	n.a.
Biomass	n.a.	n.a.	3.5

Source: Reference 30.

*Based on 80 commercial conversion facilities at $1.5 billion per plant.

(600,000 scf per hr) would be so high that the price of the product would be about the same as its heat equivalent in fuel oil.[32] As for syncrude, a substantial part of the production cost comes from the addition of hydrogen to increase the H:C ratio.[33]* Another major cost factor is the corrosiveness of carbon steel at the high temperatures required. This calls for the use of costly corrosion-resistant steels, such as 300 series austenitic CRES. If the process requires temperatures in excess of 800°F (427°C), clad or weld overlays must be used to prevent precipitation of carbides along the grain boundaries. Hydrogenator wall thicknesses in excess of 12 in. are not unusual, because of the high temperatures and pressures.[34]

All these factors add to the cost of producing synthetic fluids from coal. Capital costs and thermal inefficiencies are the real culprits. There is little hope that modification of current commercial conversion processes, for example, SASOL II, will have significant impact on conversion economics. What is most needed is really new technology, but this would require billions of dollars and in excess of 10 years for development to a commercial stage.[35]

According to the Bureau of Mines (BuMines), the United States has eight geographical areas with adequate coal and water to support multiplant conversion complexes capable of producing 1 billion scf per day of SNG. Three are in Illinois and one each is in Illinois–Kentucky, New Mexico, North Dakota, Montana–Wyoming, and Ohio–West Virginia–Pennsylvania.[36]

Oil Shale

Oil shale is another prime potential source of chemicals. If we could recover *all* the shale oil in the Green River Formation alone, we would have available to us (by one reckoning) an estimated 3 to 7 trillion barrels, or 3000 to 7000 times the amount of domestic crude oil pumped out of the ground since the Civil War. Approximately 500 billion bbl could be extracted from a limited stratum of the Piceance Creek Basin, which contains at least 25 gpt. The annual yield from this alone would be 125 times the 1973 domestic production of crude.[37] Unfortunately, all the oil cannot be extracted because of economic, ecological, and technological considerations.

Table 8.14 suggests that shale oil is on an equal footing with coal as a possible alternative source of liquid automotive fuels. Shale oil differs from petroleum in that it has a higher nitrogen and oxygen content. Furthermore, its carbon-to-hydrogen weight ratio is intermediate between those of petroleum and coal tar, but lying closer to the ratio of petroleum. Aromaticity of the product oil and composition of the retort gas can be altered somewhat by controlling retorting conditions.[37]

Shale oil is not directly substitutable for crude oil, but it can be integrated easily with other feedstocks in a conventional refinery. This calls for first reducing the nitrogen content. Initially, dedicated shale oil refineries probably would

---

*The energy cost of adding the required hydrogen for syncrude can be equivalent to about one-fourth of the energy available in the final product.[32]

Table 8.14 Alternative Sources of Potential Automotive Fuels

Fuel	Relative Prospects[a]				
	Crude Oil	Coal	Oil Shale	Biomass	Nuclear
Liquid fuels					
Gasoline	A	B	B	—	—
Diesel fuel	A	B	B	—	—
Methanol	—	B—	—	C—	—
Ethanol	—	—	—	C—	—
Gaseous fuels					
Hydrogen	—	D	—	—	D
Methane	—	E+	—	E+	—
Ammonia	—	—	—	—	E—
Propane	E+	E+	E+	—	—
Acetylene	—	E—	—	—	—

Source: Reference 30.

[a]On a scale of A = excellent to E = very poor.

not be used. Hence considerable transportation costs would be incurred in getting the oil to existing refineries located near petrochemical complexes concentrated along the Gulf coast. Shale oil could, however, be prerefined at the mine, at a premium cost of about $2.50 per bbl. If refining capacity situated in the Rocky Mountains were to be used, transportation costs of products to major markets would be the problem.[38]

Shale oil also suffers in comparison with conventional crude oil in that its yield of gasoline is low. On balance, the prime market for shale oil would appear to be boiler fuels for the Great Lakes region, while chemical feedstocks appear to be the most unlikely outlet for the foreseeable future.[38]

The U.S.S.R. has made production of petrochemicals a prime goal of their oil shale industry. About half[39] to two-thirds[40] of their shale output* is used directly as power plant fuel† and the balance (approximately 9 million tpy of raw shale)‡ is used for production of fuel gas, liquid hydrocarbons, and chemicals. Half of this latter shale is fed to continuous chamber ovens for high-temperature pyrolysis [1830 to 2060°F (800 to 900°C)] directed at high gas yields. Some crude naphtha is also produced and cracked to gas, with some

*The major deposits are kukersite shales of algal origin, deposited in the Baltic Basin (Estonian S.S.R. and the Leningrad area). This shale is approximately 35 weight percent organic[39] (with a range of 20 to 60 weight percent).[40]

†Because of the high ash content of this shale (60 to 75 percent), fouling and corrosion of heat-transfer surfaces is a serious problem that may ultimately lead to retorting the shale and using the ash-free fluid hydrocarbons as boiler fuels.

‡This amount of shale could yield about 38,000 bpd syncrude if their effort were oriented in that direction.

other liquids and high-temperature tars also being formed. These liquids are then processed further to oil products and petrochemicals. Gasoline, diesel fuels, and other conventional fractions are not produced from shale for economic reasons. Even gas production from shales has a dim economic future in the U.S.S.R., because of competition from natural gas now coming on stream in large quantities.[39]

The emphasis in the U.S.S.R. oil shale industry is on multipurpose plants—power generation and chemicals. In the northwestern part of the country, close to major oil shale mines, chemicals compete successfully with petrochemical production from crude oil. Phenol homologs (both water-soluble forms in wastewater and low- and high-boiling alkyl fractions from retorting) have received particular attention. Alkyl derivatives of resorcinol occur in the aqueous fractions, and alkyl derivatives of phenol and naphthol are predominant in the high-boiling oil fractions. All are valuable intermediates for the manufacture of surfactants, polymers, coatings, adhesives, and tannins.[40]

The crude gasoline (low-temperature distillate) from chamber-oven retorting contains as much as 70 percent unsaturated aromatic and aliphatic hydrocarbons. Fractionation and pyrolysis of this material produces high yields of C_6-C_9 aromatics (benzene, toluene, their methyl and ethyl homologs, styrene and homologs, indene, etc.) The kukersite kerogen alone is also usable as a chemical feedstock. Raw shale oil can be converted to 70 to 90 percent kerogen concentrates, which are then oxidized to mixtures of dibasic acids for use in the synthesis of polyurethane foams, plasticizers, plant growth stimulators, and other organics.[40]

These descriptions are merely indicative of the potential of a specific type of kerogen as chemical feedstocks. The Russians are considerably more advanced in this field than we are, but with an equal expenditure of time and effort, there is no reason why similar results could not be obtained with our own particular types of oil shales. Product mixes and yields would undoubtedly be different, but the general chemistry should be similar.

Figure 8.8 shows a schematic of the *Hytort* (IGT) process developed to extract hydrocarbons from eastern Devonian shales. Economics is the real key to the use of shale oil by the chemical industry. This is tied, in large measure, to the energy balance of the conversion of oil shale to fluids, a highly controversial subject at this stage of development. Table 8.15 shows one breakdown of the energy balance. Production of fluids alone would appear to be about 62 percent energy efficient if the energy values of the other products and wastes are ignored. The other major economic factor, of course, is capital expense. As was seen in Chapter 6, the oil shale industry is a highly capital-intensive industry. Table 6.6 highlighted some of the economic aspects of shale oil production.[43]

Tar Sands

Tar sands are not a prime consideration in our energy or petrochemical future because of our limited domestic supplies of this resource. But since we must be

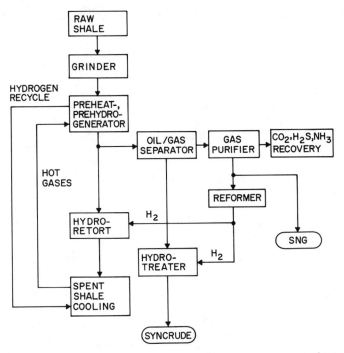

Figure 8.8 Retorting eastern Devonian shales: Hytort process (IGT). [Adapted from reference 41; reprinted with permission from *Environ. Sci. Technol.* **12**(9), 1001 (Sept. 1978); copyright by the American Chemical Society.]

prepared to use whatever energy resource is available, in whatever quantity, Fig. 8.9 is presented to show how bitumen from tar sands could be upgraded to a syncrude for further conversion to chemicals or for direct use as a fuel.

Wastes and Biomass

Ideally, a raw material for fuel or chemical feedstocks should be available close to the marketplace, be readily renewable on a relatively short cycle, be available in sufficient concentration and quantity to provide steady feed for a conversion plant without the need for large storage facilities, be relatively independent of the seasons and the whims of nature, have a low cost, and have little or no adverse impact on the environment. The factors of renewability and environmental impact alone suggest that coal, oil shale, and tar sands are less than satisfactory feedstocks for hydrocarbon fluids, except on an interim basis.

Wastes and biomass appear to meet the greatest number of these needs. The question is one of practicality. There is considerable pressure to recycle materials and to reduce overall consumption, in the interests of conserving energy required to produce them. If such a conservation campaign proves suc-

Table 8.15 Energy Balance—Conversion of Oil Shale to Gasoline and Distillate Fuel (50,000-bpd Plant)

	10^6 Btu/Day
Input	
Oil shale (85,780 tpd; 2640 Btu/lb)	452,918
Electricity (92,100 kWh/day; 3413 Btu/kWh)	314
Natural gas (3,610,000 scfd; 1000 Btu/scf)	3,610
Total Input	456,842
Output	
Motor gasoline (25,193 bpd; 5.4 × 10^6 Btu/bbl)	136,042
Jet fuel (4471 bpd; 5.4 × 10^6 Btu/bbl)	24,143
Distillate fuel (20,336 bpd; 5.6 × 10^6 Btu/bbl)	113,882
n-Butane (1613 bpd; 4.325 × 10^6 Btu/bbl)	6,976
Decant oil (191 bpd; 6 × 10^6 Btu/bbl)	1,146
Coke (932 tpd; 14,000 Btu/lb)	26,096
Sulfur (47 tpd; 3983 Btu/lb)	374
Ammonia (150 tpd; 9675 Btu/lb)	2,902
Spent shale (65,713 tpd; 305 Btu/lb)	40,085
Cooling water and air	64,852
Waste heat	40,344
Total Output	456,852

Source: Reference 42.

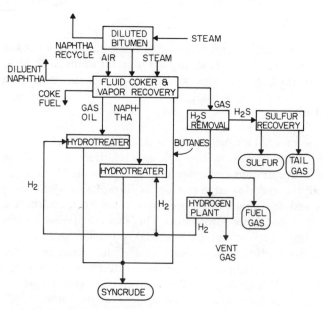

Figure 8.9 Upgrading tar sands bitumen. (Adapted from reference 44.)

cessful, it will mean a reduction in the waste available for conversion. Furthermore, terrestrial biomass competes with food for land, water, fertilizer, and labor. It is also highly dependent on weather, predators, and disease. Both wastes and biomass are more thinly distributed than fossil fuels, hence collection and transportation costs are important considerations.

Overall, there are serious questions as to their real-life future as raw materials for fuels and chemical feedstocks. Some "experts" claim that these resources are ready to be exploited on a commercial scale. Others state just as positively that the future prospects of wastes and biomass leave them "underwhelmed."

Wastes

Figure 7.3 indicated quite clearly the major options for dealing with waste and biomass conversion and the main products available. The most promising options for fluid fuels and chemical feedstocks are landfill gas, methane from anaerobic digestion, glucose from enzymatic conversion, and fluid hydrocarbons from thermal conversion. Anaerobic digestion of MSW is said to produce 6 to 9 scf of gas per pound of combustible solids,[45] but if the quantity of such solids is reduced by extraction of newspapers and plastics for recycling, garden debris for composting, and so forth, the yield of gas from MSW is bound to suffer. This is a rather insecure base from which to operate a major domestic energy and chemical resource recovery program. It is probably more reasonable to think of such wastes as supplemental fuels for power generation to stretch supplies of more conventional resources. The options along this line have been discussed in Chapter 7.

Most other potential waste resources are not gathered on a routine basis in such a manner that they can be delivered readily to a resource recovery facility. Lack of institutionalized collection is due, in part, to the wide dispersion of these wastes, making transport a major element of cost.

Biomass

The utilization of biomass for conversion to fluid fuels and chemicals is perhaps a more promising and stable prospect if the land requirements of such an option allow us to sustain food, feed, and fiber production. Klass[46] estimated the land requirements for various levels of SNG production from biomass, based on then current (1973) annual demand for natural gas (22 tcf). The calculation of land requirements for predicted 1978 levels of natural gas production (19.6 tcf),[47] on the same basis, are shown in Table 8.16. Table 8.17 indicates the distribution of land and water area in the lower 48 states. Only about 3 percent of our area is idle farmland and therefore presumably available for energy farming of field crops. With crops yielding 10 tpa-yr, less than 15 percent of our gas demand could be met by producing SNG from biomass, assuming *all* this idle farmland were so employed. At 25 tpa-yr yields, about 35 percent of our demand could be met. With yields of 50 tpa-yr, 62 percent of our natural gas requirements could be satisfied by biomass conversion.

Table 8.16 Land Requirements for Various Levels of SNG Production from Biomass

Level of Production, % of 1978 Production[a]	Productivity, tpa-yr		
	10	25	50
	Approximate Land Area Required, mi^2		
2	12,900	5,400	3,000
10	64,200	26,800	15,100
50	322,100	134,100	75,300
100	644,200	268,210	150,600

Source: Reference 46. Reprinted with permission from D. L. Klass, *Chemtech* 4(3), 161–168 (March 1974). Copyright by the American Chemical Society.
[a]Estimated (IGT) at 19.6 tcf.

Table 8.17 Division of U.S. Area by Type, Lower 48 States

Type	Square Miles	Percent of Total
Water	58,000	1.9
Idle farmland	94,000	3.1
Deserts	100,000	3.3
Active farmland	470,000	15.6
Forests	990,000	32.9
Other	1,300,000	43.2
Total	3,012,000	100.0

Source: Reference 48.

The assumptions made, however, are untenable for several reasons: some of this idle farmland would not be available for such use; some would occur in small, inaccessible plots too inefficient to cultivate and harvest economically; the idle farmland is most certainly less fertile than the average land under cultivation, hence it is doubtful if the yields would come up to expectations; and although demand growth of natural gas is declining, additional farmland will probably have to be brought under cultivation to keep up with the world's growing demand for food.

The annual yield of most field crops is higher than forest yields, but short-rotation forest species are receiving much attention and do show promise of offering high yields (see Chapter 7). Aquaculture generally provides the highest productivity of all, but natural water areas suitable for cultivation of water plants or algae are not readily available in large units. The water areas needed would undoubtedly have to be constructed at high capital cost. Technical differences also exist between conversion of fossil and nonfossil resources. Biomass has a high water content, requiring the expenditure of considerable energy to separate it from carbon-containing components. Plants extract essential inorganic nutrients from the soil, and these must be returned to the soil when water and carbon are removed. In some cases new nutrients must be used, since the thermal conditions of processing the biomass may leave the inorganics in an insoluble, hence biologically inaccessible form. Finally, disposal of the residues from thermal processing may pose an environmental problem.

Herein lie some of the problems or potential problems standing in the way of large-scale commercialization of biomass for energy or chemical production. This is not to say, however, that a judicious approach to biomass utilization could not pay off in a significant, if small, way. In the near-term future, such utilization is most likely to be based on forest resources.

Wood Derivation of chemicals from wood is illustrated schematically in Fig. 8.10. Figure 8.11 shows the use of ethanol as an intermediate for a wide range of chemicals.

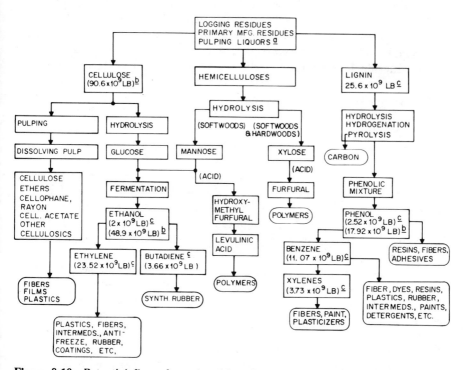

Figure 8.10 Potential flow of wood residues for chemical uses. *a*, Plus urban, agricultural, and animal residues; *b*, quantity needed for 1974 production of benzene, butadiene, ethanol, ethylene, phenol, and xylenes (cellulose requirements may include hemicelluloses yielding mannose); *c*, 1974 production (ethanol and phenol derived from ethylene and benzene; xylenes not derived from benzene). (Adapted from reference 49.)

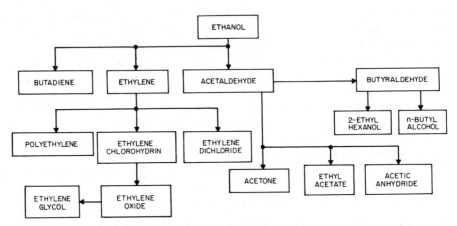

Figure 8.11 Chemical derivatives of ethanol. (Adapted from reference 18.)

Naval stores (turpentine, rosin, tall oil, and other oleoresins) are common products exclusive to wood feedstocks. The demand for these products is in a decline (since 1973), in part because rosin sizes for paper are being supplanted increasingly by the more efficient maleic anhydride- and fumaric acid–modified rosins.[50] Inasmuch as oleoresins are hydrocarbons, there is no technical reason why they cannot be used to produce petrochemicals. The major long-term deterrent is economics. As noted earlier in this chapter, hydrocarbon products so derived are much more expensive than identical petroleum-derived products, at 1979 crude oil prices.

Recent research has shown that application of paraquat herbicide to pine trees can increase the yield of oleoresin by two to five times that of untreated trees. From a practical standpoint, it is probably not reasonable to expect such treatment to be applied effectively to all stands of pine trees; hence the lower figure is probably a more sensible estimate of the amount of increase in yield that can be expected overall. Although the ratios of α- and β-pinenes and camphene may be altered, turpentine yields can also be increased at the same time, with no decline in product quality.[50]

The energy content of turpentine is slightly more than gasoline (130,000 to 150,000 Btu per gal vs. 120,000 to 130,000 Btu per gal for gasoline). But even with paraquat-enhanced production of turpentine (to 50 million gpy), available for conversion to a motor fuel, this would still only equate, from an energy standpoint, to about 0.1 percent of our annual gasoline demand. When the oleoresin is extracted or separated by pulping,* of course, the lignocellulose still remains available for conversion.[50] Lignin must be separated from the rest of the feed, since it is generally biologically inert and hampers processing and may even prove inhibitive to fermentation. Lignin can, however, be processed into phenol, and hence into other aromatics, but the technology is not as ad-

*For optimum oleoresin yields, extraction must take place promptly after harvesting the wood.

vanced as that of cellulose conversion. The economics of lignocellulose conversion could possibly enable oleoresin conversion to break even vs. petroleum. Pine has not been considered the lignocellulosic feedstock of choice because it cannot be coppiced as some other species can, but the growth-rate increase brought about by paraquat treatment could alter the economics quite markedly in favor of pine.[51]

Table 8.18 indicates the products of wood pyrolysis at two different temperatures, and Fig. 8.12 is a schematic for producing hydrocarbons or ammonia from wood by pyrolysis. It has been estimated (1975) that the cost of syncrude from wood via pyrolysis would be about $12.35 per bbl, but this does not seem to account fully for the cost of collecting and transporting the wood from widely disperse locations, nor does it appear to give sufficient weight to the high cost of the Fischer-Tropsch synthesis. Syncrude from coal via gasification and Fischer-Tropsch synthesis would call for only about 42 percent of the mass of feed compared with wood required to produce the same amount of syncrude.

Table 8.18 Products of Wood Pyrolysis[a]

	Pyrolysis Temperature	
	1490°F (810°C)	860°F (460°C)
Components	Composition, wt %	
Wood gas	89.5	15.7
Tar	8.0	13.5
Methanol	—	3.0
Acetone	—	1.0
Acetic acid	—	1.2
Water	—	31.6
Ash	2.5	—
Charcoal	—	34.0
Wood gas	Composition, vol %	
CO_2	21.2	28.7
CO	27.7	27.4
CH_4	16.9	37.4
C_2H_4 and higher	8.7	—
H_2	25.5	5.0
Heat of combustion (net), Btu/scf	441	483
Heat of combustion (total), Btu	13.4×10^9	2.22×10^9

Source: Reference 52.

[a]Based on 1000 oven-dried tons of bark (50% moisture content).

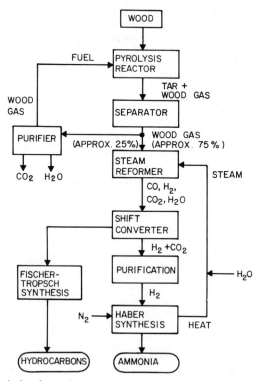

Figure 8.12 Pyrolysis of wood to form hydrocarbons and ammonia. (Adapted from reference 52.)

Basically, at the present stage of technology, the major restraints to producing liquid hydrocarbon fuels from wood are: high capital costs (production and refining), dispersion of raw material (collection and transportation costs), limited availability of wood, long cycle time, and poor overall thermal efficiency of conversion.[52]

A major intermediate from wood and wood waste would undoubtedly be methanol, commonly produced commercially from natural gas, although coal was noted as a potential alternative feedstock. Figure 8.13 is a schematic of a typical process for manufacturing methanol from wood waste. Unfortunately, wood conversion is not as efficient as natural gas, based on yield vs. energy input (Table 8.19). The feedstock requirements for a 100-million-gpy plant are shown in Table 8.20.

A major option for liquid transportation fuels would be the production of methanol from biomass to extend gasoline supplies. In 1977, our domestic gasoline demand was about 110 billion gpy. To extend this fuel by 10 percent would require 11 billion gpy of methanol, or better than 900 percent of available capacity. All but about 1 percent of methanol is now derived from fossil fluids.[53]

Current technology for methanol production comprises catalytic steam

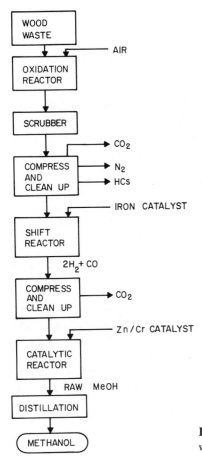

Figure 8.13 Synthesis of methanol from wood waste. (Adapted from reference 53.)

reforming of methane from natural gas to syn gas and then to methanol. Any carbonaceous feedstock could be used—fossil, waste, or biomass—but the conversion would be more energy-intensive, the raw material would be more costly to transport, and the yield would be lower than using natural gas. Nonetheless, at least eight different technically feasible processes have been developed for gasifying carbonaceous feeds. Figure 8.14 shows a general scheme for the derivation of many chemicals from synthesis gas.

The derivation of methanol from natural gas was noted in Chapter 3. Hydrocarbons and alcohols, alone or in combination, can be formed from synthesis gases. The particular products formed are dependent on the specificity characteristics of the catalyst, for example[55]*:

$$CO \;+\; H_2 \to HCHO \tag{2}$$

*Not all these reactions have been proven commercially.

$$CO + 2H_2 \rightarrow CH_3OH \tag{3}$$

$$CO + 3H_2 \rightarrow CH_4 + H_2O \tag{4}$$

$$nCO + 2nH_2 \rightarrow CH_3OH + \text{higher alcohols} + H_2O \tag{5}$$

$$\rightarrow CH_4 + \text{higher straight-chain HCs} + \text{olefins} + CH_3OH +$$
$$\text{higher alcohols} + \text{waxes} + \text{oxygenated HCs} + H_2O \tag{6}$$

$$\rightarrow \text{long, straight-chain paraffins} + H_2O \tag{7}$$

$$\rightarrow \text{long, straight-chain alcohols} + H_2O \tag{8}$$

$$\rightarrow HOCH_2{=}CH_2OH + CH_3OH + \text{other products} + H_2O \tag{9}$$

$$CO_2 + 3H_2 \rightarrow CH_3OH + H_2O \tag{10}$$

Table 8.21 shows some of the methanol production processes that have been investigated. Thermal efficiencies range from 50 to 63 percent. Much controversy has developed over the prospects for replacing gasoline with methanol for automotive fuel, at least in part. It appears likely that 5 to 20 percent blends of methanol in gasoline may be able to improve general performance over gasoline used alone. There are, however, some potential pitfalls. Compatibility of water with methanol and the limited solubility of methanol in gasoline could lead to separation, with spotty performance and corrosion of tanks, fuel lines, car-

Table 8.19 Fuel-to-Methanol Conversion Efficiencies

	Natural Gas	Coal	Wood Waste
Heating value (input)	1000 Btu/scf	8600 Btu/lb[a]	9000 Btu/lb[b]
Process efficiency, %	91.0	84.6	50.8
Plant efficiency, %	61.3	59.0	38.0
Capital costs			
($ million)[c]			
50 Mgpy capacity	23.1	74.4[d]	64.0
200 Mgpy capacity	61.0	178.0[d]	169.0

Source: Reference 53.
[a]NM coal, 19% ash.
[b]Douglas fir, 25% bark.
[c]Including 25% contingency; no escalation.
[d]Higher cost due to pressurized system, increased steam requirements, and desulfurization.

Table 8.20 Feedstock Inputs for Methanol Plants (Plant Capacity, 100 million gpy)

Feedstock	Daily Input Requirement
Natural gas	32.6×10^6 scf
Coal	2760 odt
Wood	3000 odt

Source: Reference 53.

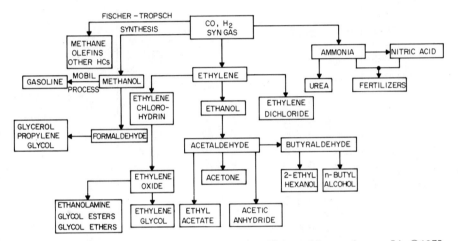

Figure 8.14 Derivation of chemicals from syn gas. (Adapted from reference 54; ©1975; reprinted by permission of the copyright owner, John Wiley & Sons, Inc.)

buretors, and so forth. Methyl-Fuel,* with its blend of higher-alcohol "impurities" (Fig. 8.15), could prove to have sufficient compatibility with gasoline (a complex blend of C_4–C_{10} hydrocarbons) to eliminate part of the problem.[17] Water miscibility problems are another matter and could prove difficult to overcome, without drastic redesign of internal combustion engines and of storage and transport facilities.

The effect of methanol on the environment has come under discussion. In the internal combustion engine, its use is probably more benign than gasoline because of lower NO_x, CO, and hydrocarbon emissions. Aldehyde emissions, however, are higher with methanol. In case of massive spillage, heavy-metal-free methanol can be toxic to various forms of life in estuarine ecosystems at levels of 1 percent or more. With heavy-metal contaminants, levels below 1 per-

---

*Controlled proprietary blend of methanol with C_2–C_4 alcohols (Vulcan-Cincinnati Co.).

Table 8.21 Methanol Production Using Different Feedstocks

Input[a]		Abstract of Process	Thermal Efficiency, %	Environmental Aspects
1.	Coal Oxygen	Gasify to CO + H_2; convert catalytically at 40–50 atm/390–570°F (200–300°C)	50	S-removal problem, similar to coal gasification
	2 tons (8000 Btu/lb)[b] 1 ton			
2.	SNG	Reform gas to syn gas; compress at 40–50 atm and convert to CH_3OH over Cu catalyst at 390–570°F (200–300°C)	63	Minimal pollution problems
	30.2×10^6 Btu (feed + fuel)			
	Power 75 kWh			
	CO_2 7300 scf			
	Feed water 144,016 lb[c]			
	Cooling water 13,200 gal			
3.	Naphtha[d]	Steam-convert naphtha to syn gas, then convert catalytically to CH_3OH at 390–570°F (200–300°C)	59	Must desulfurize S-containing feedstocks
	1148 lb			
	Fuel 9.7×10^6 Btu			
	Power 58 kWh			
	Feedwater 1600 lb			
	Cooling water 12,700 gal			
4.	Bunker-C oil 2020 lb	Same as above	50	S-removal problems similar to coal gasification
	Power 130 kWh			
	Feedwater 1680 lb			
	Cooling water 19,800 gal			

Source: Reference 17.

[a]For production of 1 ton of CH_3OH plus small amount higher hydrocarbons

[b]Corrects error in source.

[c]Assumed. Units omitted from source.

[d]Derived from coal or oil shale.

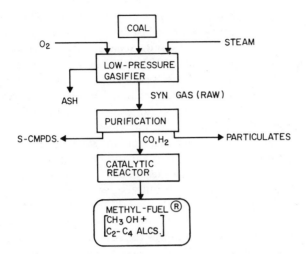

Figure 8.15 Methyl-Fuel from coal. [Adapted from reference 56; reprinted with permission from *Environ. Sci. Technol.* 7(11), 1002, 1003 (Nov. 1973); copyright by the American Chemical Society.]

cent are toxic. The overall effect of methanol on the environment is likely to be minimal, however, because of its miscibility, volatility, and biodegradability.[57] Quite apart from the technical aspects of methanol usage, it has only half the energy content of gasoline, so transport and storage would entail about double the volume of fuel.

Ethanol (ethyl alcohol, C_2H_5OH) is yet another potential biomass-derived fluid fuel or gasoline extender. *Gasohol* (90 percent unleaded gasoline plus 10 percent agriculturally derived ethanol) has been evaluated as an automotive fuel.* Those opposed to this use for ethanol claim that its manufacture from biomass is too energy-intensive and will worsen our need for foreign oil rather than diminishing it, while decreasing domestic crops available for the export trade. *Gasahol†* is an undefined blend of gasoline and alcohol (methanol or higher), the alcohol being derived from any renewable source. Methanol would most likely be formed by gasification of wood or wood refuse. Ethanol could be produced by fermentation and distillation of various grains or by hydrolysis of cellulosic biomass to sugar followed by fermentation. Similar processes could be used to manufacture higher alcohols or blends, any of which might be used to extend gasoline.[58]

The key issues inevitably will be economics and energy efficiency. For the United States, neither *gasohol* nor *gasahol* is attractive economically as long as less costly fossil-derived fuels are available. Furthermore, fermentation of

---

*Actually, the ethanol used in a large Nebraska road test was wood-derived; the ethanol for a later demonstration in Washington, D.C., was produced by a corn sweetener plant; and some for use in Illinois came from cheese whey.[58]

†Note difference in spelling.

grains does not appear to help from an energy standpoint (Fig. 8.16), since up to 44 percent of the input energy is lost in the process, according to Cray (Midwest Solvents). On the other hand, Scheller (head of the department of chemical engineering at University of Nebraska) shows a positive energy gain of 27,700 Btu per gal of alcohol. Scheller's ethanol production costs are 43 to 78.5¢ per gal vs. 98¢ for the lowest estimate by Cray. The Scheller figures project to a standoff in pump price for gasohol vs. unleaded regular gasoline,[58] but this is subject to change with escalating crude oil prices.

Kendrick (an agricultural economist from the same institution) believes that the net energy loss of a grain alcohol plant will be 122,000 to 145,000 Btu per gal or about 15 to 18 percent. He also takes issue with many of Scheller's assumptions. Experts from the state universities of Iowa, North Dakota, and South Dakota also counter Scheller's claims.[58] Weisz and Marshall[58a] claim that at the current state of our agriculture and technology, each gallon of ethyl alcohol produced from biomass requires the consumption of 2 to 3 gal of high-grade fuel equivalent from petroleum or natural gas. The only way to achieve a positive net productivity is to eliminate (or reduce sharply) the input of high-grade fossil fuels to the alcohol process facility. Following current U.S. agricultural practices, only about 0.35 to 0.60 gal of *new* high-grade fuel results from the consumption of 1 gal of gross (visible) fuel. This means the net per gallon cost of new fuel ranges from $2.60 to $5.00, or several times the cost of coal-derived synfuels. Most experts question the wisdom of the gasohol option based on economics and energetics, and some other opponents of gasohol also cite possible distribution and operational problems, as well as questionable impacts on the agricultural community. Possibly the most sensible response to the controversy is that of API, which favors continuing research and development of *all* potential domestic energy sources, including alcohol, with the final decision being controlled by the marketplace, not by bureaucratic edict.[58]

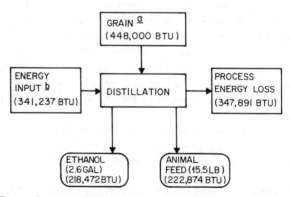

Figure 8.16 Energy balance of fermentation alcohol. *a*, One bushel; *b*, 4 percent is electrical. (Adapted from reference 58; reprinted with permission of the copyright owner, The American Chemical Society.)

One major deterrent to the adoption of a national gasohol fuel policy is that even a 5 percent blend would require the construction of 86 new plants, each with a 75.8-million-gpy alcohol capacity, by 1990,* at a cost of about $10.9 billion.[58] A study by the Department of Agriculture (DOA) indicates that use of a 10 percent blend would call for a federal subsidy of 10.4¢ per gal of gasohol. The study further shows that each Btu invested in growing and processing corn would return only 0.5 to 0.8 Btu as ethanol; acreage would have to be diverted from soybeans and wheat, causing a sharp rise in food and feed grains and a drop in livestock production; and net farm income would enjoy only a modest increase.[59]

Despite these and similar findings, there is much pressure by politically oriented organizations to launch federally subsidized gasohol programs. Ultimately, fermentation alcohol may be able to replace natural gas–derived synthetic ethanol from ethylene, but there is little to suggest that ethanol from grain will become a serious contender as a transportation fuel, until such time as the economics become more favorable and the energy efficiency of conversion is greatly improved.

Sucrochemistry has been touted recently as a potentially valuable source of intermediates for organic chemicals (Fig. 8.17). The potential range of chemicals produced is wide, but few of the products involved are commodity chemicals, except for some of those produced by fermentation (Fig. 8.18 and Table 8.22).

SUMMARY

Some of the many potential alternatives to fossil fuels as energy and feedstock sources have been discussed above. All fall short of being ideal answers to our needs, but, at the same time, we must pursue each option to such an extent as to

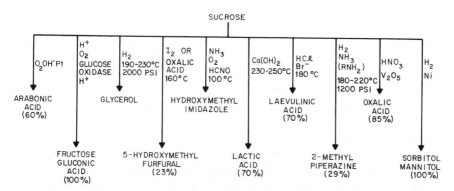

Figure 8.17 Potential products of sucrochemistry. (Adapted from reference 60.)

*Based on raw sugar juice feedstock.[58]

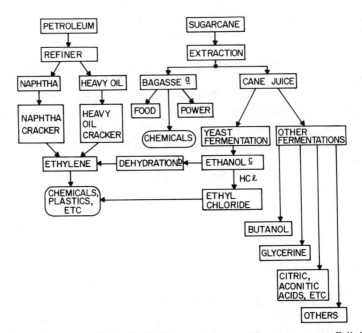

Figure 8.18 Production of chemicals from petroleum and sugarcane. *a*, Cellulose and lignin; *b*, to about 50 percent aqueous; *c*, about 10 percent aqueous. [Adapted from reference 61; reprinted with permission from M. Calvin, *Chemtech* **7**(6), 352–363 (June 1977); copyright by the American Chemical Society.]

Table 8.22 Domestic Production Levels of Major Industrial Intermediates (Derivable from Fermentation), 1978

Product	Production, 10^9 lb	Per Capita Production, lb
Acetic acid	2.79	12.6
Acetone	2.12	9.6
Ethyl alcohol (synthetic)	1.27	5.7
n-Butyl alcohol	0.65	2.9

Source: Reference 62.

plumb all its potential and recognize all its shortcomings. We may not, in the near-term future, be able to derive more than 5 percent or so of our energy or feedstock needs from any single new source, but each such increment is bound to have a favorable impact on our consumption of declining, nonrenewable resources. The combined effect of the various options cited could be enough to relieve some of the pressures now being felt.

Conceptually, wood products alone can be the source of over 95 percent of the organic chemicals, polymers (resins, plastics, and elastomers), and noncellulosic fibers used today, with little or no sacrifice in properties.[63]

The problem is acquiring a given raw material in sufficient quantity to provide significant relief for our fossil energy budget. Such a substitution is not commercially acceptable if the resulting conversion products are not sufficiently competitive in the marketplace. At present, such resource alternatives cannot compete with crude oil and natural gas, when the cost of collection and conversion and of plant investment are taken into consideration. As the prices of fluid fossil fuels rise, however, there should come a time when coal, oil shale, refuse, or biomass begin to become competitive. Unfortunately, we cannot afford to wait for that day to arrive before we initiate large-scale development of the alternatives. If we should fail to act soon enough, we could find ourselves faced with a delay of 5 to 10 years before we could bring new facilities on stream, and longer yet before significant production volumes could be produced.

Since the business community tends to invest its money in ventures with some promise of a near-term payout, there is a limit to the interest of industry in supporting new energy options until such time as governmental regulations are stabilized and the alternative products are proven to be marketable at a profit. Our ability to develop production of synthetic fuels to a commercial scale at a tolerable cost may well hinge on industry's skill in acquiring necessary federal seed money without the exquisite torture of ever-present, stultifying, and uncertain federal regulation. Although it seems doubtful if biomass can become a major source of energy and feedstocks, it could well be a first step in changing from dependence on the capricious OPEC nations to increasing dependence on the capricious elements and on society's ability to make an enduring peace with our environment.

REFERENCES

1   G. Kölling, *Pure Appl. Chem.* **49**(10), 1475–1482 (1977).

2   J. F. Mathis, CHEMRAWN I—World Conf. Future Sources Organic Raw Mater., IUPAC, Toronto (July 10–13, 1978); reported by H. Egan, *Chem. Ind.* (No. 15), 542, 543 (Aug. 5, 1978).

3   I. S. Shapiro, *Science* **202**(4365), 287–289 (Oct. 20, 1978).

4   A. H. Abdel-Rahman, presented at CHEMRAWN I—World Conf. Future Sources Organic Raw Mater., IUPAC, Toronto (July 10–13, 1978).

5   *Chem. Eng. News* **55**(34), 17 (Aug. 22, 1977).

6   I. S. Shapiro, in reference 4.

7   *Chem. Eng. News* **56**(24), 45–75 (June 12, 1978).

8   B. F. Greek and W. F. Fallwell, *ibid.* **55**(5), 8–12 (Jan. 31, 1977).

9   BuCensus, Int. Trade Comm., Soc. Plast. Ind., and C & EN; reported in D. M. Kiefer, *ibid.* **56**(51), 21–26 (Dec. 18, 1978).

10 E. T. Hayes, *Science* **191**(4227), 661–665 (Feb. 20, 1976).

11 Pace Co.; reported in *Chem. Eng. News* **54**(37), 12 (Sept. 6, 1976).

12 "U.S. Industrial Outlook, 1977," Off. Bus. Res. Anal., U.S. GPO, Washington, DC, 1977; reported in *Chem. Eng. News* **55**(3), 6 (Jan. 17, 1977).

13 AIChE; reported in *Chem. Eng. News* **56**(48), 23 (Nov. 27, 1978).

14 D. F. Othmer, *Mech. Eng.* **99**(11), 29-35 (Nov. 1977).

15 K. Migut; reported in *Chem. Eng. News* **55**(37), 60 (Sept. 12, 1977).

16 K. A. Schowalter and N. S. Boodman, "Carbonization/Hydrogenation Method for Producing Metallurgical Coke," in D. M. Considine, Ed., *Energy Technology Handbook*, McGraw-Hill, New York, 1977, p. *1*-252.

17 A. Landman, "Methanol as an Automotive Fuel," PB 270 401, Natl. Tech. Inf. Serv., Springfield, VA, April 1977.

18 A. W. Taylor, *Chem. Ind.* (No. 1), 13-22 (Jan. 1, 1977).

19 F. C. Schora, "Substitute Natural Gas (SNG)," in D. N. Lapedes, Ed., *McGraw-Hill Encyclopedia of Energy*, McGraw-Hill, New York, 1976, pp. 652-655.

20 D. S. Davies, *Chem. Eng. News* **56**(10), 22-27 (Mar. 6, 1978).

21 S. L. Meisel et al., *Chemtech* **6**(2), 86-89 (Feb. 1976).

22 *Chem. Eng. News* **56**(5), 26, 28 (Jan. 30, 1978).

23 Roger Williams (Tech. & Econ. Serv.) and Chas. A. Stokes (Consulting Grp.); reported in *Chem. Eng. News* **55**(42), 8 (Oct. 17, 1977).

24 N. P. Cochran, *Sci. Amer.* **234**(5), 24-29 (May, 1976).

25 C. Gulledge; reported in *Environ. Sci. Technol.* **12**(9), 999 (Sept. 1978).

26 L. H. Weiss, presented at CHEMRAWN I—World Conf. Future Sources Organic Raw Mater., IUPAC, Toronto (July 10-13, 1978).

27 S. M. Dix, *"Energy: A Critical Decision for the United States Economy,"* Energy Education Publ., Grand Rapids, MI, 1977, p. 118.

28 R. Hughes, *Chemtech* **8**(10), 581 (Oct. 1978).

29 *Bus. Week* (No. 2508), 84D, 84F (Nov. 7, 1977).

30 General Motors Res. Labs advertisement; *Chem. Eng. News* **56**(42), 16 (Oct. 16, 1978).

31 R. Culbertson; reported by H.-J. Peters, *Energy User News* **2**(39), 11 (Oct. 3, 1977).

32 F. Calhoun, *ibid.* **2**(47), 23 (Nov. 28, 1977).

33 F. J. Weinberg, *Technol. Rev.* **79**(1), 20 (Oct.-Nov. 1976).

34 J. B. O'Hara et al., *Met. Prog.* **110**(6), 33-38 (Nov. 1976).

35 P. E. Rousseau; reported in *Energy User News* **2**(43), 21 (Oct. 31, 1977).

36 *Chem. Eng. News* **55**(31), 14 (Aug. 1, 1977).

37 M. T. Atwood, *Chemtech* **3,** 617-621 (Oct. 1973).

38 G. E. Ogden; reported in *Chem. Eng. News* **56**(40), 19 (Oct. 2, 1978).

39 G. U. Dineen, "Retorting Technology of Oil Shale," in T. F. Yen and G. V. Chilingarian, Eds., *Oil Shale*, Elsevier, New York, 1976, pp. 180-198.

40 C. H. Prien, "Survey of Oil-Shale Research in the Last Three Decades," in *ibid.*, pp. 235-267.

41 *Environ. Sci. Technol.* **12**(9), 1001 (Sept. 1978).

42 H. R. Linden, "The Proper Role of Synthetic Fuels," presented before Subcomm. on Energy Res. Water Resour., Comm. Interior Insular Affairs, U.S. Senate, Mar. 3, 1975; reported in *Energy Facts II*, Sci. Policy Res. Div., Libr. Congr., Serial H, U.S. GPO, Washington, DC, 1975, p. 404.

43 C. L. Wilson, Dir., *Energy: Global Prospects 1985-2000*, WAES, McGraw-Hill, New York, 1977, p. 220.

44 Hon. B. Dickie and M. Carrigy, "Fuel from Tar Sands," in D. M. Considine, Ed., *Energy Technology Handbook*, McGraw-Hill, New York, 1977, pp. *3*-159 to *3*-167.

45   S. A. Klein (1970); reported by C. G. Golueke and P. H. McGauhey, "Waste Materials," in J. M. Hollander and M. K. Simmons, Eds., *Annual Review of Energy*, Vol. 1, Annual Reviews, Palo Alto, CA, 1976, pp. 257–277.

46   D. L. Klass, *Chemtech* **4**(3), 161–168 (Mar. 1974).

47   J. D. Parent and H. R. Linden, "IGT Predicts Moderate Increase in U.S. Energy Consumption in 1978," Inst. Gas Technol., Chicago, Dec. 20, 1977.

48   *Statistical Abstracts of the United States*, 93rd ed., DOC, Washington, DC, 1972.

49   *Renewable Resources for Industrial Materials*, CORRIM, Natl. Acad. Sci., Washington, DC, 1976, p. 203.

50   T. J. Collier, "Feasibility of Petrochemical Substitution by Oleoresin," Lightwood Res. Coord. Counc. Proc., Annu. Meet., Atlantic Beach, FL (Jan. 18–19, 1977), U.S. GPO, Washington, DC, 1977, pp. 172–187.

51   H. R. Bungay and R. W. Ward, pp. 188–191, in reference 50.

52   Pulp Paper Res. Inst. Can. (May 1975); reported by L. Gardner, "Production of a Hydrocarbon-Type Synthetic Fuel from Wood," NRC No. 15638, Natl. Res. Counc. (Can.), Ottawa, Sept. 1976, p. 2.

53   A. E. Hokanson and R. M. Rowell, "Methanol from Wood Waste: A Technical and Economic Study," DOA, Forest Serv., Gen. Tech. Rep. FPL 12, June 1977.

54   V. H. Edwards, "Potential Useful Products from Cellulosic Materials," in C. R. Wilde, Ed., *Cellulose as a Chemical and Energy Resource*, Interscience, New York, 1975, p. 335.

55   A. B. Stiles, *AIChE J.* **23**(3), 362–375 (May 1977).

56   *Environ. Sci. Technol.* **7**(11), 1002, 1003 (Nov. 1973).

57   R. K. Pefley et al., "Characterization and Research Investigation of Methanol and Methyl Fuel: Final Report," PB 271 889, Natl. Tech. Inf. Serv., Springfield, VA, Aug. 1977.

58   E. V. Anderson, *Chem. Eng. News* **56**(31), 8–12, 15 (July 31, 1978).

58a P. B. Weisz and J. F. Marshall, *Science* **206**(4414), 24–29 (Oct. 5, 1979).

59   "Gasohol from Grain—The Economic Issues," DOA, Washington, DC, 1978.

60   A. Vlitos, *Chem. Br.* **13**(9), 340–345 (Sept. 1977).

61   M. Calvin, *Chemtech* **7**(6), 352–363 (June 1977).

62   Int. Trade Commis.; reported by A. L. Compere and W. L. Griffith, "Bioconversion of Wastes to Higher-Valued Organic Chemicals," in T. N. Veziroğiu, Ed., Forum Proc., *Energy Conservation*, Clean Energy. Res. Inst., Univ. Miami, Coral Gables, FL, Dec. 1–3, 1975, pp. 181–198.

63   L. G. Stockman; reported in *Bus. Week* (No. 2545), 68B (July 31, 1978).

9 SYNTHETIC POLYMER PRODUCTS AND PRECURSORS

In the future, raw materials, such as oil and gas, must be utilized for better and more essential purposes ... [than for fuel].

SHAH MOHAMMAD REZA PAHLAVI

There appears to be no firm agreement among the experts as to just how much of the petroleum budget (crude oil plus natural gas and associated liquids) is used by the petrochemical industry. Numbers mentioned range from about 3 percent of the world total for feedstocks alone,[1] or 4 percent of all hydrocarbons consumed in the United States for feedstocks,[2] to "never ... more than 8 percent ... in the U.S." for feedstocks and fuel.[3] Evaluation of nine different published sources suggests that 6 percent would be a reasonable approximation based on usage for both feedstocks and fuels. Despite this relatively modest demand made on our energy budget, petrochemicals constitute 70 to 80 percent of all our chemical output.[4]

Even so, one might well ask whether we should not return to natural materials and set petroleum-based synthetics aside. Commoner, for example, decries the use of synthetic fibers in place of natural fibers (cotton and wool) on the basis of energy content, pollution, biodegradability, and waste disposal.[5] On the other hand, he has remarkably little to say about the demands that would have to be made on our arable land to replace synthetics with cotton or wool fibers, and how we would then be pressed to continue as the world's breadbasket. Both cotton and wool cultivation are notorious for their denudation of the soil. This is a fact that cannot be overlooked. It has been suggested that the Garden of Eden may well have been reduced to desert by overgrazing of sheep and goats,[6] and it is a well-documented fact that cotton cultivation results in high loss of topsoil, thus guaranteeing decreasing yields with time.[7]

A closer inspection of Commoner's charges is in order. Cotton lint is indeed less energy-intensive than polyester fibers by a ratio of about 1:3.5—for electricity, fuel, fertilizer, and chemicals* (for cotton) vs. oil recovery through polymerization and spinning (for polyester). In manufacture of the cloth, cotton suffers seven times the processing loss of polyester. The total energy for manufacturing a shirt finds the ratio closing to 1:1.25 for cotton vs. the common 65:35 polyester/cotton blend shirting.† The real difference comes in the energy for maintenance (washing, drying, and ironing) of cotton vs. the blend and in the fact that three cotton shirts must be provided to give the same number of wear cycles as two polyester/cotton blend shirts. The bottom-line energy content ratio over the lifetime of the shirts is 1.88:1 cotton vs. polyester/cotton blend. Furthermore, the amount of energy from nonrenewable resources used in the manufacture and maintenance of cotton is 20 times the amount of renewable energy required in the production of the raw cotton fiber. Details of this analysis are given in reference 7.

Perhaps of even greater importance to society is the fact that a return to cotton would entail competition for arable land with the cultivation of food, feed, and possibly biomass for energy. The land requirements would be large and would have to be restricted to warm climates and to soil suitable for cotton cultivation.

It has been calculated that, in 1974, it would have been necessary to plant nearly 5 million more acres of cropland in cotton (or better than a 35 percent increase) to replace the polyester fiber then being used. (This assumes that the added marginal cropland be suitable for cotton cultivation and be as productive, on the average, as the current acreage planted to this crop.) This addition would be about four times the new acreage brought under cultivation each year for *all* crops, or twice the amount of arable land withdrawn each year for highways, urbanization, and special uses.[7] There is little hope that we could afford to dedicate such a large amount of arable land to replacing synthetic fibers with cotton. The foregoing assessment ignores altogether the recently identified health hazards associated with working in cotton mills.‡

Our best compromise, from the standpoint of both energy conservation and ecology, would appear to be to continue making use of synthetic fibers, while minimizing their impact on the environment. This means using the best possible pollution control procedures during synthesis and processing, and pursuing viable means for rendering the synthetics biodegradable or, alternatively, of assuring their pollution-free reclamation.

Many other everyday products also have been shown to have lower energy contents when designed and fabricated of synthetics (Table 9.1). On an average, for 13 major plastics, energy as feedstock represents 38 percent of their total energy content—ranging between 28 percent for nylon 6 to 67 percent for low-density polyethylene (LDPE) (Table 9.2). Most good-quality products do

---

*But excluding human and solar energy.
†Cotton is also a denser fiber, requiring 0.68 oz of fabric per shirt vs. 0.59 oz for the blend.[7]
‡Byssinosis, an ailment somewhat similar to pneumoconiosis ("black lung"), is believed to be caused by dust carried into the mills in the cotton bales.[8]

Table 9.1 Energy Content of Some Commercial Products

	Total Energy, boe[a]	
Product	Traditional Materials	Synthetic Materials
Bottles, 1-liter	Glass 1680/M[b]	PVC 710/M[b]
Containers	Tinned steel 550/M	PVC 590/M
	Aluminum 1035/M	
Drainage pipe, 4-in.	Cast iron 145/km	PVC 30/km
	Clay 35/km	
Fertilizer sacks	Paper 5110/M	LDPE 3430/M
Shirts	Cotton 7/k[c,d]	Nylon 15/k[c,d]

Source: Reference 9.

[a] Bbl oil equivalent.
[b] Million.
[c] Thousand.
[d] Does not consider maintenance energy costs during lifetime of product.

not benefit from simple-minded, direct substitution of plastic for a common material (wood, metal, glass) on a same-shape, same-thickness basis. They generally call for a total redesign in the plastic to take full advantage of the material's particular attributes. Nonetheless, it is an accepted fact that such materials generally are bought by weight and used by volume. The fuel-energy content of the least energy-intensive metal (steel) is higher than the total energy content of an equal volume of the most energy-intensive major plastic (acetal) (Table 9.2). Quite apart from the frequent cost and energy advantages of substituting plastics for other materials, in recent years synthetics have also been more stable against inflation than other commodities, as shown in Table 9.3. The merits, in specific applications, of using plastics to replace more conventional materials must be judged on the basis of the design of the specific product and the properties and economics of the candidate materials.

PETROCHEMICALS

With few exceptions, petrochemicals currently suffer from underutiliztion of capacity. The fourth quarter 1978 operating rate overall was 74 percent, and a continuing overcapacity is forecast for the next 10 years.[4] Prior capacity expansions were for the purpose of providing for growth. With the slowing of the economy, and petrochemicals in particular, the coming years will see capacity expansion for the sake of improving production efficiency and adding sufficient flexibility to accommodate heavier feeds. This period will be accompanied by ups and downs in capacity as new plants come on stream and old plants are

Table 9.2 Energy Content of Metals vs. Plastics

| | 10^6 Btu/in.$^{3\,a}$ | | |
Material	Feedstock	Fuel	Total
Copper[b]	—	8.7	8.7
Magnesium[b]	—	8.2	8.2
Yellow brass[b]	—	7.8	7.8
Aluminum[b]	—	7.7	7.7
Zinc[b]	—	5.9	5.9
Steel[b]	—	5.4	5.4
Acetal	1.6	3.7	5.3
Polyphenylene oxide	1.4	2.8	4.2
Nylon 6	1.1	2.9	4.0
Nylon 6:6	1.5	2.2	3.7
Polyester	1.3	2.0	3.3
Polycarbonate	1.1	2.1	3.2
Acrylic	1.1	1.9	3.0
ABS	0.7	1.5	2.2
Poly(vinyl chloride)	0.6	1.5	2.1
Polystyrene	0.8	1.2	2.0
HDPE	0.6	0.8	1.4
Polypropylene	0.5	0.8	1.3
LDPE	0.6	0.3	0.9

Source: Reference 10.

[a]Data not precise; scaled from graph.

[b]Assumes recycling of 46% aluminum, 30% copper, 20% magnesium, and 5% zinc.

Table 9.3 Change in Commodity Prices between December 1975 and October 1978

Commodity	Price Change, %
Lumber	+63.0
Aluminum	+35.0
Iron and steel	+26.3
Paper	+19.8
Plastics	+7.6

Source: Plastics World Magazine (Oct. 1978). Reference 11.

retired. Plants designed to employ heavier feeds (naphtha and gas oils) are three to four times as costly as plants suited to natural gas liquid (NGL) feeds, since the facility must be capable of extracting sulfur at the front end of the process, and it must be able to handle the large amounts of by-products (propylene, butadiene, etc.) that result from the use of heavy feeds. In the case of "high-severity cracking"* of crude oil directly to olefins, the high temperatures and pressures required call for more costly manufacturing facilities.

The level of natural gas feed is expected to hold fairly steady through 1990, with virtually all the production growth resulting from the use of heavier feeds. There is some indication that producers may delay converting to new, expensive plant facilities in hopes that deregulation of interstate gas prices† might stimulate discoveries that could extend the lives of their old ethylene crackers. A limited number of these cracking units are able to tolerate as much as 40 percent heavier feedstocks mixed with the light feeds.[2]

Conceptually, production of basic organic chemicals begins with the catalytic reforming of naphtha to benzene–toluene–xylene (BTX), and with the thermal cracking of naphtha,‡ refinery gas, or gas oil to ethylene, propylene, and butylenes (C_4 fraction). There is little likelihood that U.S. producers will follow the lead of the European and Japanese petrochemical industries and jump into the use of naphtha feeds for olefins. Our own needs call for reforming most naphtha catalytically§ to the aromatics required as octane boosters for our gasolines, as well as for other feedstocks. Rather, for the near-term future, our heavy feed will consist of gas oil (diesel or heating fuel), the next higher distillation fraction of petroleum, which will be thermally cracked to olefins.¶ At the present state of technology, both gas oil and naphtha yield excess quantities of unsaturated C_4's.[12] This could lead to a market surplus of butylenes, which would affect petrochemical economics.

The ultimate goal, of course, is to achieve commercialization of direct cracking of crude to olefins, a development that is well under way, with the Carbide-Kureha-Chiyoda prototype advanced cracking reactor (ACR) (Fig. 9.1) becoming operational in 1979. The ethylene yield is about 30 percent, with significant amounts of acetylene, butadiene, and other products that bring yields of high-value chemical products to 60 to 70 percent.[13] The capital cost is about 4 percent higher than a conventional crude oil cracker with the same ethylene nameplate capacity, but the process should be more competitive with higher-priced crude. The primary advantage of direct cracking of crude oil is that the petrochemical manufacturer can purchase raw material directly from the producer rather than feedstocks from the refinery.[14] Dow Chemical and Exxon also have direct cracking processes under development. The Dow process, based on partial combustion cracking, yields 65 to 70 percent olefins and requires less

---

*Union Carbide Corp. development.
†Under the Natural Gas Policy Act of 1978.
‡A mixture of C_4–C_9 straight-chain and cyclic aliphatic hydrocarbons.
§Usually "platforming" or reforming with the use of platinum catalysts.
¶Gas oil does not produce as clean a reaction as naphtha.

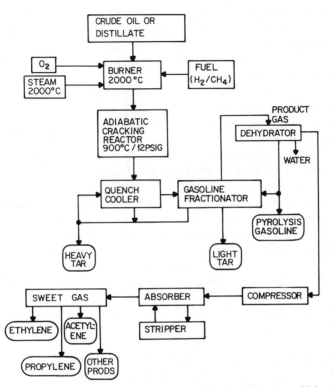

Figure 9.1 Flame-cracking process of Carbide-Kureha-Chiyoda. (Adapted from reference 13; reprinted with permission of the copyright owner, The American Chemical Society.)

capital than naphtha crackers. Dow expects to commercialize the process in the mid-1980s.[15]

In catalytic reforming, straight-chain hydrocarbons [e.g., hexane, CH_3-$(CH_2)_4CH_3$] and cycloaliphatics [e.g., methylcyclopentane, $CH_3CH(CH_2)_4$] are isomerized to cyclohexane, which is then dehydrogenated to benzene. Dehydrogenation has been called the single reaction most important to the chemical industry, since it is essential to the production of six of our seven most basic commodity organics.* The most valuable aromatic is p-xylene, a precursor of terephthalic acid used in the manufacture of polyester fibers.[12] Since BTX is a by-product of conventional refineries, the two largest manufacturers of petrochemicals in 1977 were oil companies: Shell† $6.3 billion, and Exxon $4.2 billion in chemical revenues. Other oil companies also dominate the field, for example, Occidental, Standard Oil of Indiana, Phillips Petroleum, Atlantic

---

*Ethylene, propylenes, butylenes, benzene, toluene, and xylenes. The seventh commodity organic is methane.

†Includes Royal Dutch Shell and Shell Oil Company.

Richfield, Gulf, and Mobil, each with from $1.6 to $1.2 billion in chemical revenues.[4] It is likely that smaller oil companies will encounter difficulty competing in the building of new, efficient, heavy-feed reactors, because of the high capital cost. They may opt for gambling that they will continue to obtain sufficient gas to operate their old facilities.

The end result of all this corporate maneuvering means a great deal to our economic future. Excessive price increases (due to costly feeds, inefficient plants, excess unmarketable by-products, high capital cost, etc.), could affect our ability to compete on world markets. Petrochemicals have been a mainstay of our world trade in recent years. In 1974, the U.S. petrochemical industry enjoyed a positive trade balance of $3.6 billion. But this advantage was due almost solely to our artificial feedstock price, which is still controlled below the cost of European and Japanese feedstocks.* Our petrochemical trade balance could vanish altogether unless we can improve process efficiency or reduce the cost of feeds. Producers have advocated diverting gas from use as fuel to almost exclusive use as chemical feedstocks. Continuance of a healthy petrochemical industry is essential to the country's economy. One job in the industry is said to create five to eight more jobs downstream.[2] In 1975, nearly $130 billion in finished products were produced from Texas-manufactured petrochemicals alone. This was the equivalent of 13 percent of our gross national product and provided employment for 3 million persons. As fuel, the equivalent feed would have been worth considerably less.[16]

The problems facing the industry can be illustrated by two rather typical examples. Construction of 1.3 billion-lb of gas-based capacity in Texas since 1973 cost one firm $51 million. An equivalent new facility based on naphtha feed would now cost over $400 million. Another facility finds itself paying six times as much for only 90 percent of the energy purchased five years ago, and the product yield is about the same. In Texas, which is the focus of our petrochemical industry, energy costs have soared by a factor of 12 since the early 1970s.[2]

Over the period 1967–1976, the chemical industry's total capacity grew at an average annual rate of 6 percent, but capital investment growth rate was only about 1.4 percent. The difference reflects the construction of larger facilities, which operate more efficiently and cost less to build per unit output. Individual plants built in 1978 have capacities 200 percent greater than corresponding facilities built in 1967. Maximum practical limits of capacity are expected to be reached for ethylene, styrene, and ammonia plants during the 1980s.[17] The expected domestic capital needs for major petrochemicals are shown in Table 9.4.

Fears have been expressed that the Organization of Petroleum Exporting Countries (OPEC) nations will build up a petrochemical industry and dominate international markets through the advantage of possessing virtually limitless quantities of low-cost raw materials.[4] These fears overlook the facts that these countries lack the technical expertise and support technicians, and it is far more expensive to construct plants in the Middle East or north Africa and to

---

*In an energy policy speech on April 5, 1979, President Carter proposed a phased decontrol of crude oil prices, to be completed in September 1981. This is now in effect.

Table 9.4 Expected Capital Spending by U.S. Chemical Industry for Major Petrochemicals, Fourth Quarter 1977 to 1988

Products	Billion 1977 Dollars	Number of New Units
Ethylene and derivatives	20.6	240
LDPE alone	(4.09)	(63)
Propylene and derivatives	4.4	70
Polypropylene alone	(2.4)	n.a.
Benzene and derivatives	8.5	233
Total	33.5	543

Source: Reference 17. Reprinted with permission of the copyright owner, The American Chemical Society.

transport the resulting chemical products to U.S. and other industrial markets.[18] Furthermore, they must build roads, housing, and the vast infrastructure necessary to make the industry function. These factors would add greatly to the cost of their products and effectively narrow the price gap between their products and ours.[4] It has been said that "nine-tenths of what's been announced for the Middle East will never happen."[19] Offsetting this is the fact that Libya did manufacture methanol and dump it on the market in the second quarter of 1978, thereby dropping the European price about 20 percent.[4]

The OPEC nations' best bet appears to be to manufacture selected petrochemicals* directly from the natural gas now being flared or vented to the atmosphere, and market them to developing nations or other OPEC countries. Refined fuel products could be their best outlets for crude oil.[18]

In 1975, our major petrochemical problem was one of insufficient refinery and extraction capacity, *not* a raw materials shortage. To this, five years later, we might add regulatory and economic uncertainties as unsettling elements in the future of the industry. The technological problems are relatively easy to solve compared to those imposed by people. Without the capital investment, which can be ensured only through a sense of assurance as to what the nature of the gamble is, we are likely to find ourselves with antiquated facilities, unable to compete in the marketplace with imports. From 1967 to 1974 (through the Arab oil embargo), the capital component of cost in the price of ethylene, our major plastics precursor, rose by as much as 600 percent[20] and was still climbing at an alarming rate, with the added specter of preparing for heavier feeds hanging over the petrochemical scene.

POLYMERS

The five main branches of the polymer industry (in 1972), were: plastics and resins, market volume $4.5 billion (32 percent of the petrochemical market);

*For example, ammonia, methanol, and ethylene.

manmade fibers, $4.0 billion (28 percent); paints and allied products, $3.4 billion (24 percent); synthetic elastomers, $1.3 billion (9 percent); and adhesives and sealants, $0.9 billion (7 percent). The total market volume of the industry was $14.2 billion.[21]

Plastics and Resins

In the following pages of this chapter, these materials are discussed essentially in the order indicated above, with the exception that fibers will be dealt with under the sections in which their plastics counterparts are considered. This is done to simplify discussion of the feedstocks and synthesis of polymer-related products—the main thrust of this chapter.

Polyethylene

The feedstock picture is changing, with increasing emphasis on heavier feeds as the supply picture for natural gas worsens. By 1990, it is expected that only about 27 percent of the ethylene output, the major derivative monomer of natural gas, will come from natural gas liquids, as opposed to 72 percent in 1976 (Table 9.5). Table 9.6 shows the fuel and feedstock situations in 1974 for the petrochemical industry as a whole. The change to heavier feedstocks poses some complex problems, as noted earlier in the chapter. Heavy-feedstock olefin plants are more flexible, both in selection of feedstock and in product mix. This multiplicity of products, however, could be a dubious advantage, since one or more of the co-products could encounter overproduction or strong sales resistance and leave the producer with no market and no storage facilities for an important fraction of its output. Furthermore, new plant facilities might be needed to store and to consume these co-products. Besides the higher capital costs required for employing heavier feeds, larger quantities of the heavier feedstocks are required to produce a given amount of ethylene. Propane requires nearly twice as much, naphtha about 2.5 times, and heavy gas oil over three times as much feed to yield the same quantity of ethylene as ethane feed

Table 9.5 Ethylene Feedstock Sources, 1976–1990

	Billion lb		
Source	1976	1980	1990
Natural gas liquids	15.6 (72%)	17.9 (59%)	14.6 (27%)
Naphtha	3.3 (15%)	5.9 (19%)	18.4 (35%)
Gas oil	2.7 (13%)	6.6 (22%)	20.3 (38%)
Total	21.6 (100%)	30.4 (100%)	53.3 (100%)

Source: Reference 22. Reprinted with permission of the copyright owner, The American Chemical Society.

Table 9.6 Petrochemical Industry Consumption of Natural Gas, 1974

	Units	Natural Gas	NGL[a] from NG[b] Processing Plants	Other Sources	Total
Fuel (4.6% of NG output)	(10^{12} Btu)	1031 (53%)[c]	—	912 (47%)	1943 (100%)
Feedstocks (2.6% of NG output)					
Natural gas	(10^9 Btu)	550 (25.0%)	—	—	550 (25.0%)
Ethane	(10^6 bbl)	—	118 (13.8%)	7 (0.7%)	125 (14.5%)
Propane[d]	(10^6 bbl)	—	60 (9.3%)	25 (3.7%)	85 (13.0%)
Butane[e]	(10^6 bbl)	—	49 (8.7%)	6 (1.3%)	55 (10.0%)
Butane–propane mix[f]	(10^6 bbl)	—	6 (1.0%)	4 (0.5%)	10 (1.5%)
Heavy liquids	(10^6 bbl)	—	—	150 (36.0%)	150 (36.0%)
Feedstock totals		550 (25.0%)	233 (32.8%)	192 (42.2%)	975 (100.0%)

Source: Reference 23. Reprinted with permission of the copyright owner, The American Chemical Society.

[a]Natural gas liquids, including ethane.
[b]Natural gas.
[c]Percentages of Btus.
[d]Includes propylene.
[e]Includes butylene.
[f]Includes isobutane.

produces. With increasingly heavier feeds, more propylene, C_4's, and heavier products are formed. For example, heavy gas oil yields about 60 percent as much propylene as ethylene, whereas, with ethane feed, propylene is only 2 to 3 percent of the ethylene yield.[24] Furthermore, the heavier feeds are more costly than natural gas, a condition that is only partially offset by the added market value of the co-products.

The United States is presently the major supplier of ethylene, but this picture is expected to change as competition increases from developing nations (Table 9.7). The major derivative of ethylene, by far, is polyethylene (Table 9.8). Other derivatives of ethylene are also important in the manufacture of polymers. Thermoplastics, and polyethylene in particular, dominate the polymer field (Table 9.9), and this dominance is expected to continue through at least 1990 (Table 9.10). The major end use of ethylene is in fabricated plastics (Table 9.11)—LDPE, poly(vinyl chloride) (PVC), high-density polyethylene (HDPE), and polystyrene. The scope of ethylene's role in polymer synthesis is

Table 9.7 Sources of Ethylene, 1976 and 1990

	Percent	
Source	1976	1990 (est.)
United States	36	32
Western Europe	34	26
Japan	13	11
Other western hemisphere	4	8
Eastern Europe	8	11
Asia/Pacific	4	6
Middle East/Africa	1	6

Source: Reference 25. Reprinted with permission of the copyright owner, The American Chemical Society.

Table 9.8 Major Derivatives of Ethylene

Derivative	Percentage
Polyethylene	45
Ethylene oxide/glycol	20
Vinyl chloride monomer	15
Styrene monomer	10
Other	10

Source: Reference 26. Reprinted with permission of the copyright owner, The American Chemical Society.

Table 9.9 U.S. Plastics Consumption (Major Resins),1976–1977

Plastic	Million lb		Annual Growth Rate, %
	1976	1977	
Thermoplastics			
LDPE	5,777	6,480	12
PVC/copolymers	4,716	5,248	11
HDPE	3,133	3,572	14
Polystyrene	3,151	3,510	11
Polypropylene	2,542	2,750	8
ABS	926	1,032	11
Acrylic	498	529	6
Nylon	221	243	10
Total	20,964	23,364	11
Thermosets			
Polyurethane foams	1,599	1,755	10
FRP[a]	1,400	1,544	10
Phenolic	1,312	1,407	7
Urea; melamine	992	1,133	14
Unsaturated polyester	961	1,052	9
Epoxy	248	276	11
Total	6,512	7,167	10

Source: Reference 27.

[a]Includes reinforcements.

best shown by Fig. 9.2. Domestically, ethylene is derived presently from the cracking of ethane (and some propane) extracted from natural gas. This is the source of about three-fourths of our ethylene. The balance is obtained from the steam cracking of liquefied refinery gas (LRG) (15 percent) and naphtha (10 percent) distilled from crude oil.[29]

Straight-run naphtha, the initial distillate from crude, is fed to steam crackers for olefins and other petrochemicals, and to catalytic reformers and crackers for high-octane gasoline components. The uncertainties of the gasoline situation—demand, grade, and the future of lead alkyl antiknock additives—have curtailed commitments to refinery expansions. Reformers are therefore being run hard to produce sufficient aromatic octane builders for current gasoline needs, thus placing added demands on limited naphtha supplies.

A new, high-efficiency process (Unipol) for polymerizing ethylene to LDPE is under development by Union Carbide. It is a gas-phase process in which ethylene and a catalyst are reacted in a fluid-bed reactor at 100 to 300 psi and temperatures below 100°C. The product is in granular form.[30] Compared with conventional technology, it requires only half the capital and one-fourth the

Table 9.10 Projected Consumption Growth Rate
of Precursors and Polymers, 1977–1990

Precursors/Polymers	Average Annual Consumption Growth Rate, %
Ethylene	10.3
LDPE	8.6
VCM	7.6
PVC	8.3
Ethylene oxide	8.2
Ethylene glycol	8.8
HDPE	8.1
Propylene oxide	7.6
Acrylonitrile	7.3
Isopropyl alcohol	6.2
ABS	5.7
DMTA/PTA[a]	7.2
Styrene	5.2
PET[b]	8.8
Polystyrene	6.9
Cumene	7.9
SBR	3.3
Cyclohexane	4.7
PA[c]	6.8
Caprolactam	6.9
TDI[d]	7.8

Source: Reference 17. Reprinted with permission
of the copyright owner, The American Chemical
Society.

[a]Dimethylene terephthalic acid/purified terephthalic acid.
[b]Polyethylene terephthalic acid.
[c]Polyamide.
[d]Toluene diisocyanate.

operating energy.* The current LDPE polymerization process is carried out at
up to 50,000 psi pressure† and 660°F (350°C) temperature, by free-radical initiation, with reaction taking place in solution. The polyethylene is recovered
from the ethylene by solvent stripping and is then prepared for fabrication by

---

*Operating energy would represent only 8.5 percent of the total manufacturing cost, or 40[31] or 50[32]
percent when the ethylene raw material is excluded. The process energy required in LDPE
manufacture is largely electricity for operating high-pressure pumps, and electricity (per Btu) is
approximately three times as expensive as gas.[32]
†Approximately the pressure developed in an artillery piece firing a projectile 20 mi.[31]

Table 9.11 Major End Uses of Ethylene

Use	Percent
Fabricated plastics	65
Antifreeze	10
Fibers	5
Solvents	5
Other	15

Source: Reference 26. Reprinted with permission of the copyright owner, The American Chemical Society.

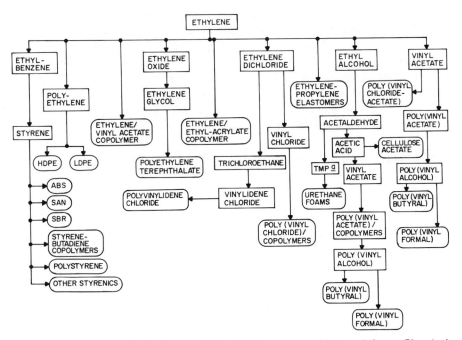

Figure 9.2 Ethylene derivatives. *a*, Trimethylol propane. (Adapted from *Chemical Origins and Markets*, SRI International, reference 28.)

extruding, pelletizing, and drying. Conversion per pass now stands at about 30 to 35 percent.[30]

The new technology is based on a unique and as yet undisclosed family of proprietary catalysts. Resin properties can reportedly be controlled precisely by varying catalyst composition rather than by altering equipment or operating parameters, which requires sacrificing production efficiency. Dual-purpose facilities could produce either LDPE or HDPE, as desired. The product from the reactor is in the form of ready-to-use granules (similar to soap powder) rather than the usual pellets. It is estimated that the new process could yield a price advantage of 6 to 12 cents per lb, but it may be necessary to modify bulk-handling equipment.[33]

At the time the Unipol process was first disclosed to the public (November 1977), Exxon Chemical USA indicated some concern over whether the product properties would allow it to compete with conventional LDPE,[31] but in April 1979, Exxon became the first licensee to commit itself to use the process in a new facility scheduled for startup late in 1981.[34]

The amount of natural gas flared annually (1973) in the Persian Gulf[35] amounted to nearly 12 percent of the total U.S. natural gas produced,[36] or nearly enough to provide all U.S. feedstock gas (3 tcf) for the year.[37]

Both natural gas producers and users have long espoused the cause of price decontrol of natural gas as their only salvation, seeing this as a means of providing the incentive needed to increase discoveries. Since about 80 percent of the natural gas fuel for the petrochemical industry comes from intrastate supplies not under federal control, only 20 percent of the fuel requirements should be affected by federal decontrol. Even if the decontrol measures provided by the Natural Gas Policy Act of 1978 were to stimulate drilling,* it is generally felt that resulting new discoveries will be only marginally effective in extending the lifetime of natural gas supplies. As always, there are dissenters—those who claim that there is no actual or imminent natural gas crisis. Until such time as really major new finds may be made, however, we cannot afford to continue consuming natural gas at present rates and merely assume that somehow the situation will take care of itself.

Polypropylene

Meaningful data on production, capacity, and operating rates for basic petrochemicals, such as propylene (Table 9.12), are hard to come by because of a basic lack of agreement between ITC and the petrochemical industry. The situation is made worse by the fact that only about 40 percent of the propylene is sold on the open market, while the balance is converted to derivatives by the producers (Table 9.13). Further projections are made difficult by unstable business conditions. For producers, the 1978 performance of propylene was disappointing, and near-term projections were expected to prove no better. A major problem is that of capacity growth outstripping demand and resulting in a softening of prices. Steam crackers currently are shifted toward light-end feeds—ethane and propane—in order to use feeds that are selling at low prices. Nonetheless, continuing growth in propylene capacity is resulting in marginal profitability for producers.

Production of propylene from co-product streams of steam-cracking ethylene plants is overshadowed by the volume recovered from catalytically cracked refinery gas streams. About three-fourths of the refinery propylene goes

*More than 48,000 new wells scheduled for completion in 1978 vs. 27,600 in 1973.[38]

Table 9.12 Production and Capacity Use Data for Propylene, 1975-1979[a]

	10^9 lb		Capacity Use, %
Year	Production	Capacity[b]	
1975	7.6	n.a.	n.a.
1976	11.3	16.4	69
1977	13.1	17.4	75
1978	14.3	18.4	78
1979[c]	15.0	20.9	72

Source: References 26 (1977–1979 data), 39 (1975 data), 40 (1976 data).
Reprinted with permission of the copyright owner, The American Chemical
Society.
[a]Data not precise; scaled from graph.
[b]Nameplate capacity for extraction and dehydrogenation for first quarter.
[c]Estimated.

Table 9.13 Major Derivatives of Propylene

	Percent		
Derivative	1976	1977	1978
Polymers	23	25	25
Acrylonitrile	16	15	15
Isopropyl alcohol	14	12	10
Propylene oxide	13	13	10
Oligomers	n.a.	12	10
Cumene	⎱ 10	10	n.a.
Phenol	⎰	n.a.	n.a.
Other	24	13	30

Source: References 26 (1978 data), 39 (1976 data), and 40 (1977 data).
Reprinted with permission of the copyright owner, The American Chemical
Society.

into gasoline and other fuels, frequently blended with propane, but mostly as
propylene alkylates. These alkylates require addition of lead alkyls to improve
antiknock properties, and with lead being phased out of gasolines, the value of
these alkylates is now low. This could result in a surplus of propylene supply
even without capacity continuing to increase.[26]

Chemical-grade propylene (92 to 94 percent pure) is easier to make than
polymer grade (99+ percent pure), hence is more attractive to the refiners,

leading to a 2¢ per lb spread in prices between the two grades.* Unfortunately for producers, the increase in propylene markets due to soft pricing is not expected to be enough to overcome marginal profits. Even potential increases in derivatives sales are not expected to have a beneficial impact.[26] A few years ago it was felt that polypropylene consumption would grow at a rate of 14 percent.[41] Today, it looks as though polypropylene consumption will exceed the average 5 percent growth expected for propylene (Table 9.14), but other uses (e.g., acrylonitrile) will be well below average.

Figure 9.3 is a flowchart of polypropylene production options, but polypropylene is only one end use of propylene, as can be seen in Fig. 9.4, which illustrates the derivation of numerous polymers from this monomer. The major outlets for propylene monomer (about two-thirds of the market) are fabricated plastics and fibers, with the 1978 ratio of the two being a little better than 3:1. From 1976 to 1978, the solvent share of the market dropped from 20 to 10 percent. Miscellaneous chemical uses accounted for about one-fourth of the market, up from 10 percent in 1976.[26,39,40] Nearly a third (30 percent) of the polypropylene used ends up in fiber products. Car and truck parts and packaging account for 15 percent each. Toys and housewares and appliance parts use 5 percent each, and the remaining 30 percent ends up in a variety of products.[44]

The long-term prospects for polypropylene are probably enhanced by the fact that it is a relatively new polymer line for which the manufacturing and fabricating technologies are not fully matured. Thus there is a greater opportunity for major technological breakthroughs than in the case of LDPE, polystyrene, and other older polymers.

Table 9.14 Sales and Use of Polypropylene, 1973–2000

Year	Sales and Use	
	10^9 lb	% Annual Change
1973	2.2	n.a.
1974	2.2	0
1975	1.9	−14
1976	2.5	+32
1977	2.7	+8
1978 (est.)	2.9	+7
1979 (est.)	3.2	+10
1980 (est.)	3.5	+9
2000 (est.)	27.5	—

Source: Reference 42.

*About twice the normal spread and over 2.5 times the spread when propylene supplies are tight.

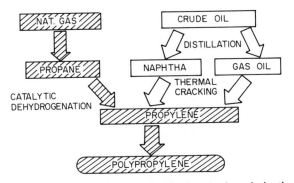

Figure 9.3   Simplified flowchart of polypropylene derivation.

Poly(vinyl chloride)

A flowchart for the manufacture of poly(vinyl chloride) (PVC) is shown in Fig. 9.5. It can be seen that vinyl chloride monomer (VCM) is the polymer precursor. A few years ago, VCM was described as the "most troubled large-volume U.S. chemical,"[46] because of its political, economic, and medical problems. The monomer (17 plants) and polymer (41 plants) producers were being forced to conform to stringent Occupational Safety and Health Administration (OSHA) regulations on VCM emissions into the atmosphere and into the water. The producers expected to have to spend $198 million on capital improvements to control air emissions and $83 million to limit water emissions. In addition, they estimated $70 million and $17 million, respectively, in annual expenses to operate the equipment.[47]

Fortunately, producers of both monomer and polymers introduced viable technology which resulted in only minor capacity losses, soon offset by innovative refinements. The price increases deemed necessary to cover environmental compliance never materialized. Return on investment for emission control equipment has been difficult to attain, however, because of low product prices brought about, in part, by overcapacity. From 1976 to 1978, annual production of vinyl chloride monomer rose from 5.67 to 6.20 billion lb, but capacity rose even faster, causing a decline in operating rate from 85 to 79 percent.[48]

Despite low operating rates, capacity additions are continuing in units of 1 billion lb or more.[49] At the low point in VCM's fortunes, accompanying the uncertainties of the impact of OSHA regulations on future supplies and markets, it appeared that VCM capacity might become the bottleneck, but this situation also never developed.

Chlorine   There are two major intermediates for vinyl chloride monomer—ethylene and chlorine (Fig. 9.5). The former has already been discussed. Chlorine is a co-product with caustic soda (sodium hydroxide) of electrolytic cells. Traditionally, chlor-alkali production has been carried out at operating rates of 90+

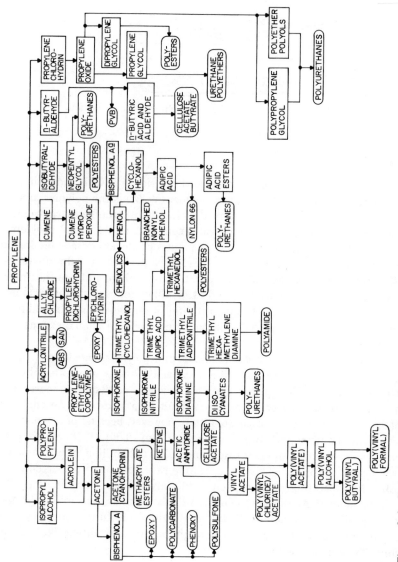

Figure 9.4 Propylene derivatives. *a*. See derivatives of propylene via isopropyl alcohol and acrolein. (Adapted from *Chemical Origins and Markets*, SRI International, reference 43.)

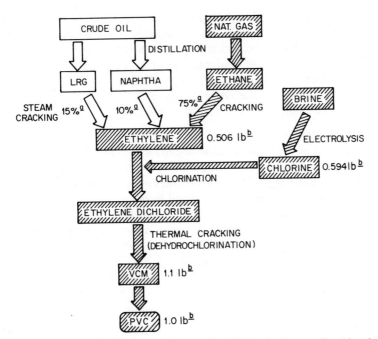

Figure 9.5 Simplified flowchart of poly(vinyl chloride) derivation. *a*, Cracking fractions (from reference 29); *b*, feed and output data (from reference 45).

percent of nameplate capacity but, in recent years rates have been no higher than 82 percent, because of a critical imbalance in domestic supply and demand.* Production of chlorine rose from 9.3 to 11.0 (estimated) million tons between 1975 and 1978. Capacity increases were somewhat slower, resulting in operating rates rising from about 74 percent in 1975 to an estimated 80 percent in 1978, after peaking at about 82 percent in 1977.[49,50]

Chlorine has suffered from slow demand growth due to the loss of certain markets—chlorofluorocarbon aerosol propellants and trichloroethylene cleaning fluids. Co-product caustic, primarily a neutralizing agent, disappears rapidly in aqueous effluents or recycled streams. Chlorine, on the other hand, is a highly persistent chemical material, a fact that has deterred its approval by environmentalists for use in insecticides, propellants, and solvents.[50]

The chlor-alkalies (chlorine, caustic soda, and soda ash) are all energy-intensive, both in manufacture and in transport. Energy costs comprise about 60 percent of total production costs.[50] DuPont and Diamond Shamrock jointly have developed a new membrane technology, based on DuPont's Nafion membrane material (copolymer of tetrafluoroethylene and a perfluorosulfonylethoxy ether reinforced with Teflon mesh) used in Diamond Shamrock's chlor-alkali cells. Output from these membrane cells, worldwide, amounted to 500 to 700

*Exports of this highly corrosive commodity are virtually nonexistent.

tpd in the first quarter of 1978. The product caustic solution from these cells is more concentrated than solutions from conventional diaphragm cells and are basically salt-free. Steam requirements for concentration of the caustic are therefore much reduced*; hence overall energy consumption is less in the membrane process than in the diaphragm process. The membrane process also lacks the asbestos diaphragm of the latter process and the mercury of the mercury-cell process, which can lead to pollution problems.[51]

Vinyl Chloride Monomer   Virtually 100 percent of the VCM output is used in the manufacture of PVC homo- and copolymers. Table 9.15 shows the consumption patterns of PVC. A different breakdown is shown in Table 9.16, by more specific product type. PVC plastic takes about 20 percent of the chlorine output.[44] The monomer ranks twenty-third in our domestic commodity chemicals, in terms of volume, while PVC and copolymers rank second in volume among our plastics, accounting for a larger volume than the entire thermosetting resin category.[52] Following the 1973 oil embargo, the sales and use of PVC did not approach preembargo levels (4.7 billion lb) until 1976, but from then until 1980, the annual growth rate averaged nearly 13 percent.[42] The U.S. manufacturing technology for PVC consumes about 20 percent more energy than do Western Europe technologies, according to a NATO study (Table 9.17). The main difference lies in our use of ethane as the feedstock rather than naphtha, as well as higher polymerization energy inputs in the United States. But there is

Table 9.15   Poly(Vinyl Chloride) and Copolymer Consumption, 1976–1977

	Billion lb	
Market	1976	1977
Calendering (flooring, textile coating, film, sheet)	0.706 (15%)	0.747 (14%)
Coating (flooring, textile and paper coating, protective coatings, adhesives)	0.481 (10%)	0.485 (9%)
Extrusion (film, sheet, pipe, conduit, siding, wire, cable)	2.390 (51%)	2.825 (54%)
Molding (bottles, housewares, records, pipe fittings)	0.523 (11%)	0.569 (11%)
Paste processes	0.154 (3%)	0.170 (3%)
Export	0.273 (6%)	0.243 (5%)
Other	0.190 (4%)	0.209 (4%)
Total	4.717 (100%)	5.248 (100%)

Source: Reference 27.

*Evaporation required (pounds of water per ton of 50% caustic soda) 1570 (membrane cell) vs. 5730 (diaphragm cell).[51]

Table 9.16   Major Uses of Poly(Vinyl Chloride)

Use	Percent
Pipe and fittings	35
Films and sheet	15
Flooring materials	10
Wire and cable insulation	5
Automotive parts	5
Adhesives and coatings	5
Other	25

Source: Reference 44. Reprinted with permission of the copyright owner, The American Chemical Society.

Table 9.17   Energy Consumption in Domestic Manufacture of PVC

Process	Percent
Natural gas production	0.3
Ethane extraction from natural gas	2.2
Ethylene production by ethane cracking	4.9
Sodium chloride mining	2.7
Chlorine production from NaCl	19.1
VCM production from ethylene and chlorine	17.8
Polymerization of VCM to PVC	17.0
Subtotal	64.0
Feedstock	36.0
Total	100.0

Source: Reference 53. Reprinted with permission of the copyright owner, The American Chemical Society.

a developing trend toward the use of heavier feeds in the United States, and this could bring the technologies closer together.

Phthalic Anhydride   Consideration of PVC applications suggest that no story about this polymer would be complete without a discussion of plasticizers, since they constitute a sizable component of over half the PVC products.

Table 9.18 shows the domestic consumption of plasticizers by application. A tabulation of sales of PVC between 1968 and 1975 showed that in the mid-1970s the market share of flexible PVC had leveled off at about 54 percent vs. 46 percent for rigid compound.[55] Flexible compound commonly contains 25 to 35 percent plasticizer, the most important types being phthalates (Table 9.19), which account for about 68 percent of domestic plasticizer sales. Dioctylphthalate (DOP) has become the workhorse of the industry,* but since the mid-1970s,

*Dioctyl phthalate comprises about one-third of the phthalate diesters.[56]

Table 9.18   Domestic Consumption of Plasticizers in PVC, 1976–1978

	10^3 Tons		
Application	1976	1977	1978 (est.)
Film and sheet	125	127	127
Molding and extrusion	90	101	116
Wire and cable coating	85	99	103
Textile and paper coating	75	77	76
Flooring	81	84	70
Other	91	96	105
Total	547	584	597

Source: Reference 54.

Table 9.19   Domestic Consumption of Plasticizers by Type, 1976–1978

	10^3 Tons		
Plasticizer	1976	1977	1978 (est.)
Phthalates (exc. linear)	331	336	346
Linear phthalates	130	142	154
Epoxy	53	54	57
Adipates	27	28	29
Polyesters	23	24	24
Trimellitates	11	12	14
Azelates	5	5	5
Others	102	103	105
Total	682	704	734

Source: Reference 54.

producers failed to add DOP capacity, with the result that the supply became tight and prices rose.[55] A major bottleneck for this and other monomeric plasticizers has been oxoalcohols.[57] Phthalic anhydride is another intermediate that is crucial to DOP supply.[55]

Figure 9.6 is a flowchart of the derivation of phthalic anhydride. The greatest growth in phthalic anhydride capacity is occurring with *o*-xylene feed (835 million lb or 68 percent in 1977 to 1165 million lb or 75 percent in 1980). Capacity of coal tar naphthalene (295 million bbl) and petroleum naphthalene (90 million lb) plants has remained stable over this period, resulting in reductions in the market share for each.[58] In the years between 1965 and 1974, petronaphthalene fell from 44 to 36 percent of the total naphthalene production

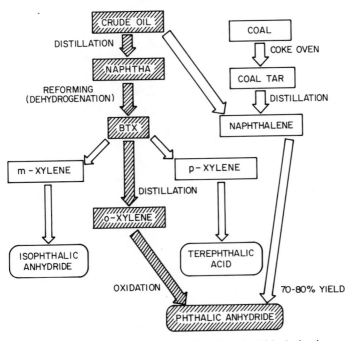

Figure 9.6 Simplified flowchart of phthalic anhydride derivation.

for economic reasons. Over the same period, total naphthalene production decreased from 810 to 590 million lb per yr.[59] In the early 1970s, between 65 and 70 percent of the phthalic anhydride was derived from o-xylene, but even so, over 70 percent of the naphthalene output found its way into phthalic anhydride in the mid-1970s.[59]

Nearly 100 percent of the o-xylene produced goes into phthalic anhydride. Almost half (48 percent) of the 1977 phthalic anhydride output went into plasticizers, and polyester resins accounted for about one-fourth (24 percent). Alkyd resins used up 21 percent, with the remaining 7 percent going into miscellaneous applications.[60] Capacity is not a limiting factor in the case of either p- or o-xylene, because demand for both has eased off and operating rates are low. Still more xylenes can be pulled from the gasoline pool, at a price, if needed.[39] Table 9.20 shows production trends for major phthalate plasticizers and intermediates. During the period 1970 to 1977, the average annual production growth rate has slowed to about 5 percent, compared to an average AGR in excess of 11 percent for 1962 to 1967.

Over the past decade virtually all phthalic anhydride capacity additions have been designed for o-xylene feeds. Any new capacity based on naphthalene feed has been by firms having captive naphthalene sources.[67]

Exports of o-xylene are high, running 30 to 35 percent of the 1 billion lb per yr output. The reason is that hydrocarbon raw material costs have been attrac-

Table 9.20 Production Trends for Major Phthalate Plasticizers and Intermediates,[a] 1970–1977

Material	Production, 10^6 lb (annual growth rate, %)							
	1977	1976	1975	1974	1973	1972	1971	1970
Intermediate								
o-Xylene	943 (+10)	854 (+22)	703 (−33)	1056 (−1)	1068 (+28)	832 (+6)	785 (−2)	799
Phthalic anhydride	932 (+3)	902 (+28)	702 (−28)	977 (−4)	1023 (+10)	933 (+18)	794 (+8)	734
Plasticizer								
Dioctyl phthalate	395.6 (+26)	314.0 (+0.5)	312.5 (−21)	395.2 (−8)	429.5 (+8)	467.3 (+7)	437.3 (+0.4)	435.5
Diisodecyl phthalate	156.6 (+9)	143.1 (+35)	105.7 (−28)	146.7 (−14)	170.7 (+11)	153.3 (+18)	135.7	—

Source: References 61 (1975 to 1977 data), 62 (1974 data), 63 (1973 data), 64 (1972 data), 65 (1971 data), and 66 (1970 data). Reprinted with permission of the copyright owner, The American Chemical Society.

[a] Data not precise; scaled from graph.

tive in the United States. Our raw material prices are now rising, however, so exports may decline.[26]

Figure 9.7 shows a general scheme of how xylenes and derivatives fit into the plastics picture.

Polystyrene and Copolymers

The derivation of polystyrene is shown in a simplified flowchart (Fig. 9.8). Polystyrene is one of the five major commodity resins, but many styrenics are modified to improve impact resistance, processing characteristics, or other properties. Among the more important modified styrenics is ABS (acrylonitrile-butadiene-styrene) terpolymer. Flowcharts for acrylonitrile ($CH_2:CHCN$) (Fig. 9.9) and butadiene (Fig. 9.10) and the stages leading to the terpolymer (Fig. 9.11) are also shown.

By far the major route for the production of polystyrene is dehydrogenation of ethylbenzene (Fig. 9.8). Other routes have not yet proven economically attractive, although two new technologies offer some promise.*

Aromatics, in general, were in an undersupply–overdemand situation in the first half of 1979, but with prospects of easing in the second half of 1979, despite the fact that the gasoline market for aromatics was expected to peak in the fourth quarter of 1979. Domestic aromatics are pegged close to world prices,

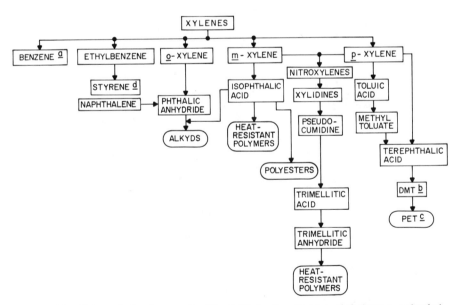

Figure 9.7 Xylene derivatives. *a*, See Fig. 9.13; *b*, dimethyl terephthalate; *c*, polyethylene terephthalate; *d*, see Fig. 9.8. (Adapted from *Chemical Origins and Markets*, SRI International, reference 68.)

*Mobil-Badger process[69] and POSM (propylene oxide styrene monomer by-product) process.[70]

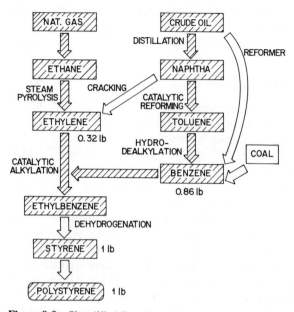

Figure 9.8 Simplified flowchart of polystyrene derivation.

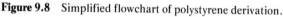
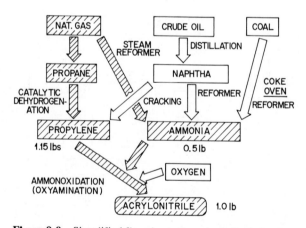

Figure 9.9 Simplified flowchart of acrylonitrile derivation.

but olefins are abundant in the United States at prices about half those in Western Europe and Japan.[71] Styrenic plastics should face increased competition from polyolefins because of the price differential.

Toluene This is a key aromatic petrochemical, as can be seen from an examination of Figs. 9.12 and 9.13. Figure 9.8 indicates that toluene is a major precursor of benzene, which, in turn, is an intermediate for many polymers, including the styrenics. Table 9.21 shows projections of the sources of toluene.

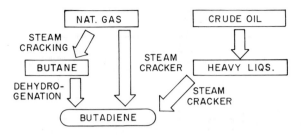

Figure 9.10 Simplified flowchart of butadiene derivation.

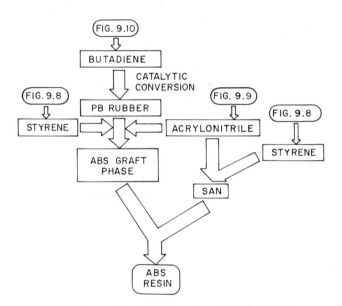

Figure 9.11 Simplified flowchart of ABS derivation.

Catalytic reforming at the refinery is, by far, the largest source, accounting for about two-thirds of the toluene in 1978, but expected to give ground over the following seven years to increasing amounts from olefin steam crackers. Coal is not expected to be a real factor over this time frame.[73] Table 9.22 lists the end uses of toluene, exclusive of motor fuels. The major outlet by far is benzene. Solvent uses of the aromatics are on a decline, partly because of concern over air pollution, but use in gasoline is expected to show an annual growth rate of 12 percent through 1981.[71]

Benzene   Adequate domestic production capacity exists for benzene, an essential intermediate for styrene, but quality feedstocks are lacking. The nameplate capacity is based on the assumption that naphthas with relatively high contents of naphthenes and other precursors capable of being reformed into aromatics will be employed. Naphthas currently available, however, have low aromatic

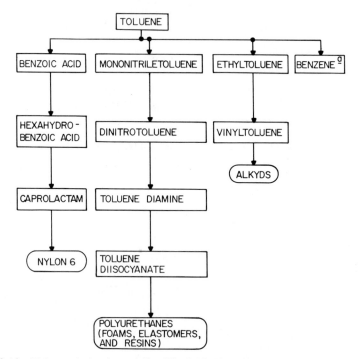

Figure 9.12   Toluene derivatives. *a*, See Fig. 9.13. (Adapted from *Chemical Origins and Markets*, SRI International, reference 68.)

potential. Because of this decline in reformer feed quality, benzene produced from refinery reformate will decrease and olefin steam crackers will provide increasing amounts of aromatics[71] (Table 9.23).

Ethylbenzene is by far the major derivative of benzene, accounting for half the output. Cumene and cyclohexane production use 15 percent of the output each. Aniline consumes about 5 percent, and the remaining 15 percent is divided among maleic anhydride and chlorobenzene and other products.[26] Table 9.24 shows the projected demand pattern through 1981. Polystyrene consumes one-fourth of the benzene output, with an additional 10 percent going into other styrene resins, and 5 percent into styrene–butadiene rubber. Twenty percent ends up in nylons, and the remaining 40 percent is consumed in a wide variety of products.[26]

A large portion of the benzene produced goes into styrenic resins. The growth of sales for these materials appears to be destined for a healthy future to at least the year 2000 (Table 9.25). The demand by polymer type is shown in Table 9.26. Packaging and containers are by far the largest outlets (together 35 percent) for polystyrene. Toys and recreational equipment, housewares, appliance parts (e.g., cabinets), and disposable food containers and utensils each accounts for about 10 percent of the polystyrene output. The remaining 25 per-

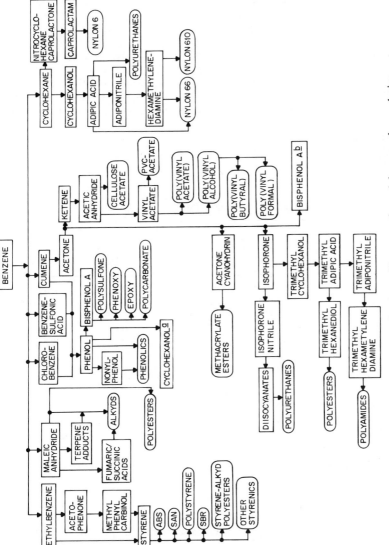

Figure 9.13 Benzene derivatives. *a*, See derivatives of benzene via cyclohexane; *b*, see derivatives of benzene via chlorobenzene, benzene sulfonic acid, or cumene and phenol. (Adapted from *Chemical Origins and Markets*, SRI International, reference 72.)

Table 9.21   Sources of Toluene, 1978–1985

Source	Million gal			
	1978	1980	1982	1985
Catalytic reforming	930 (70%)	930 (66%)	880 (61%)	700 (46%)
Pyrolysis gasoline[a]	225 (17%)	320 (23%)	420 (29%)	675 (44%)
Styrene by-product	89 (6%)	90 (6%)	90 (6%)	100 (7%)
Coal	35 (3%)	45 (3%)	45 (3%)	50 (3%)
Subtotal	1270	1385	1435	1525
Imports	55 (4%)	20 (1%)	0 —	0 —
Total	1325 (100%)	1405 (99%)[b]	1435 (99%)[b]	1525 (100%)

Source: Reference 73. Reprinted with permission of the copyright owner, The American Chemical Society.

[a]From olefin steam crackers.
[b]Not exactly 100% due to rounding errors.

Table 9.22   Nongasoline Uses of Toluene, 1978–1985

Use	Million gal			
	1978	1980	1982	1985
Benzene	485 (62%)	460 (61%)	460 (63%)	455 (62%)
Solvents	110 (14%)	100 (13%)	90 (12%)	80 (11%)
TDI	55 (7%)	60 (8%)	65 (9%)	85 (12%)
Benzyl chloride	17 (2%)	18 (2%)	20 (3%)	20 (3%)
Benzoic acid	15 (2%)	17 (2%)	20 (3%)	20 (3%)
Phenol	8 (1%)	10 (1%)	15 (2%)	15 (2%)
TNT/DNT	10 (1%)	10 (1%)	10 (1%)	10 (1%)
Others (including exports	80 (10%)	75 (10%)	55 (8%)	50 (7%)
Total	780 (99%)[a]	750 (98%)[a]	735 (101%)[a]	735 (101%)[a]

Source: Reference 73. Reprinted with permission of the copyright owner, The American Chemical Society.

[a]Not exactly 100%, due to rounding errors.

cent goes into many diverse products.[44] Toy usage is on the decline and insulation uses are increasing.[42] Many of the end uses are for disposable or short-lived items. One of the problems ahead for styrenic plastics is the fact that they are the highest priced of the commodity resins and are encountering competitive pressures from polypropylenes. The more durable types (ABS and SAN) are in a favorable supply situation, in that their high-value-added aspect should permit them to compete for feedstocks.

Table 9.23 Benzene Sources, 1974–1985

	Production, % of total					
Source	1974	1975	1977	1978	1981 (est.)	1985 (est.)
Refinery reformate	52	n.a.	45	44	38	40
Olefin-derived[a]	12	13	15	21	29	33
Toluene hydrodealkylation	29	n.a.	21–22	n.a.	n.a.	15
Coke ovens[b]	7	n.a.	18–19	n.a.	n.a.	12

Source: References 39 (1975 and 1977 data), 71 (1978 and 1981 data), and 74 (1974 and 1985 data). Reprinted with permission of the copyright owner, The American Chemical Society.
[a] Pyrolysis gasoline from olefin steam cracker.
[b] Including imports.

Table 9.24 Benzene Demand, 1978–1981

	Billion lb			
Product	1978	1979	1980	1981
Styrene	6.2	5.9	6.4	7.0
Cumene–phenol	2.2	2.4	2.4	2.6
Cyclohexane	2.0	2.0	2.2	2.2
Other	2.0	1.8	2.0	2.0
Total	12.3	12.1	13.0	13.9

Source: Reference 71. Reprinted with permission of the copyright owner, The American Chemical Society.

Phenolics

Phenolics are the largest volume thermosetting resin, except for polyurethane foams, but demand is in a decline, largely because of stiff competition from new engineering thermoplastics—for example, polyesters and bulk molding compounds (BMCs). In 1975, admittedly a bad year for the plastics industry, consumption of phenolics declined 40 percent from the preceding year. The rate of decline was twice that of the plastics industry as a whole.[76] Usage of phenolic resins as bonding agents and adhesives follows closely the demands of the housing construction industry, which has had its ups and downs. Figure 9.14 is a flowchart for the derivation of phenolic resins.

Phenol The precursors are phenol (via benzene–cumene–cumene hydroperoxide) and formaldehyde (via methanol). About 93 percent (3365 million lb) of the total 1978 plant capacity of 3600 million lb was for the cumene peroxidation

Table 9.25 Sales and Use of Styrene-Based Resins, 1973–2000

Year	Sales and Use (Annual Change)	
	10^9 lb	Percent
1973	4.5	n.a.
1974	4.1	−9
1975	3.2	−22
1976	4.1	+28
1977	4.8	+17
1978 (est.)	5.6	+17
1979 (est.)	6.0	+7
1980 (est.)	6.5	+8
2000 (est.)	27.0	—

Source: Reference 42.

Table 9.26 U.S. Styrene Demand, 1975–1980

Use	Million lb			Annual Change, %
	1975	1976	1980	1976–1980
Polystyrene	2310	2700	3393	5.9
Expanded polystyrene	350	450	710	12.0
ABS	347	500	750	9.5
SAN	78	92	125	9.5
Polyester	270	375	520	8.5
SBR	513	600	650	2.0
Other copolymers	410	566	720	6.2
U.S. total	4278	5283	6868	6.7
Exports	575	900	685	
Total	4853	6183	7553	

Source: Reference 75. Reprinted with permission of the copyright owner, The American Chemical Society.

process. Benzene sulfonation represent a distant second process (150 million lb or 4.2 percent), with toluene oxidation and natural phenol from coal tar and petroleum being very minor sources.[77] The operating rate for phenol production steadily declined from 79 to 73 percent from 1977 to 1979 in the face of an accompanying increase in capacity from 3.0 to 3.6 billion tons.[78] Table 9.27 lists derivatives of phenol, and the end uses are given in Table 9.28. About three-fourths of the phenol output ends up in phenolic and engineering plastics.

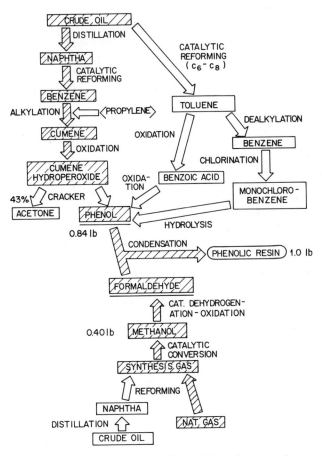

Figure 9.14 Simplified flowchart of phenolic derivation.

Formaldehyde This chemical is formed by oxidation of methanol. It is the material that is reacted with phenol to form phenolic resins, as well as with urea to form urea–formaldehyde resins. The latter are widely used as binders for particleboard, which is replacing large quantities of plywood. These resins are used primarily in the construction industry. Merchant sales of formaldehyde take only about one-third of the output. The balance goes into captive uses, since the low price and the large volume of solvent (water and/or alcohol) make it uneconomical to ship over long distances.[79] Formaldehyde (37 percent solution) and phenol (synthetic), the two major components of phenolic resin synthesis, ranked twenty-first and thirty-seventh among U.S. chemicals in 1977, with outputs of 6.08 and 2.38 billion lb, respectively. These outputs represent steady increases over 1971 figures, except for the recession years resulting from the 1973 oil embargo.[61-66] The major derivatives of formaldehyde are polymers and resin intermediates: urea–formaldehyde resins, 30 percent; phenol–formaldehyde

Table 9.27   Derivatives of Phenol, 1977

Use	Percent of Total
Phenolic resin	46
Bisphenol A	18
Caprolactam	16
Methylated phenol	4
Plasticizers	3
Adipic acid	2
Nonyl phenol	2
Salicyclic acid	1.5
Dodecyl phenol	1
2,4-D	1
Pentachlorophenol	0.5
Other alkyl phenols	2.5
Other chlorophenols	1.5
Refining	1

Source: Reference 77. Reprinted with the permission of the copyright owner, The American Chemical Society.

Table 9.28   Phenol End Uses

| | Million lb | | Annual Growth |
Use	1977	1986	Rate, %
Phenolic resins	1170	1810	5.0
Engineering plastics	472	1278	11.7
Textiles	430	640	4.5
Chemicals	187	217	1.7
Total	2259	3945	6.4

Source: Reference 77. Reprinted with permission of the copyright owner, The American Chemical Society.

resins, 20 percent; acetals, 10 percent; and pentaerythritol, 5 percent. Many other derivatives account for the remaining 35 percent of formaldehyde output.[79] The major end uses of formaldehyde are: in adhesives and binders, 55 percent; engineering plastics, 10 percent; and various chemicals, 35 percent.[79]

Polyesters

The major polyester resins are of two basic types: saturated (polyethylene terephthalate) and unsaturated (often spoken of as styrenated alkyds). The former constitutes polyester for fibers and films.* The latter find their way largely into fiberglass-reinforced plastics (FRP).

Polyethylene Terephthalate (PET) Through the mid-1970s fiber polyester resins have retained about 78 percent of the total polyester resin market, with unsaturated polyesters accounting for the remaining 22 percent, as the total output increased from 3829 to 4703 million lb.[61] Polyester film takes only about one-tenth the amount of PET that is used for fibers.[80]† These resins are saturated derivatives of *p*-xylene, as is the PBT resin. The *p*-xylene is catalytically oxidized to terephthalic acid (TPA), which is then esterified with methanol to dimethylterephthalate (DMT). Ethylene glycol is added to produce an ester interchange to the PET.[81] Only a very small amount of *p*-xylene finds its way into other polyester resins, and there are no significant markets for this isomer outside the plastics and fiber fields.

In the chemical products area, synthetic fibers have a reputation for standing out above most other products for erratic market behavior. There are two essential intermediates for the synthesis of polyester, the largest volume synthetic fiber: DMT/PTA‡ and ethylene glycol. Because of the eccentric behavior of the market for polyester fibers, it has proven virtually impossible for producers to maintain a high operating rate (Table 9.29). Operating rates for the fiber plants are far better than for the intermediates. (The latter have been only

Table 9.29 Production and Capacity for Key Intermediates for Polyethylene Terephthalate,[a] 1976–1978

Chemical	Units	1976	1977	1978
DMT/PTA[b]				
Production	Billion lb	3.96	4.40	4.81
Capacity	Billion lb	5.17	6.24	6.24
Operating rate[c]	Percent	77	70	77
Ethylene glycol				
Production	Billion lb	3.47	3.58	3.76
Capacity	Billion lb	5.17	5.35	6.42
Operating rate	Percent	67	67	59

Source: Reference 82. Reprinted with permission of the copyright owner, The American Chemical Society.

[a]Data not precise; scaled from graph.
[b]Combined data.
[c]PTA operating rates are poorer than DMT.

*Polybutylene terephthalate (PBT) is a relatively low-volume (saturated) molding thermoplastic.[27]
†Polybutylene terephthalate is a small item—about 4.5 percent as much as unsaturated resin.[27]
‡Dimethyl terephthalate and purified terephthalic acid are lumped together, since they represent two related, commercial precursors.

about 70 percent over the past few years.) The reason producers continue to put up with this kind of performance is that in 1976, for example, these intermediates* accounted for 26 percent of merchant volume and 43 percent of merchant sales dollars, while amounting to only 16 percent of the total petrochemical production volume.[82]

The ethylene glycol used in the PET synthesis starts with the hydration of ethylene oxide. Half of the glycol output goes into PET, and about 40 percent is used for antifreeze.[82]

A new technology (acetoxylation or direct hydration of ethylene) was introduced commercially by Oxirane in 1978 (Fig. 9.15). The glycol is formed directly from ethylene without the intermediate formation of ethylene oxide.[83] The glycol yield is 90 percent, compared to about 70 percent for the best ethylene oxide/glycol operation. Capital costs are essentially the same for both technologies. With rising ethylene prices, the economic advantage of the acetoxylation process is bound to become even more apparent.[74] An alternative synthesis via toluene–benzoic acid–potassium benzoate–potassium terephthalate is also possible.[81]

During the fourth quarter of 1978, synthetic fiber business appeared to take a turn for the better, based on a shift in women's apparel to garments and accessories using more fabric, and to an increasingly favorable export market. Although polyesters are gradually taking an increasing share of the tire cord market, this part of the polyester fiber business uses only about 8 percent of the total output. Most of the balance goes into apparel (61 percent) and home furnishings (17 percent).[84]

Polyester tire cords were introduced in the mid-1950s, largely to overcome the "flat-spotting" tendency of nylon cords. By the mid-1970s, it had displaced nylon as the industry leader. In 1979, polyester cords (275 million lb) led nylon tire cords (235 million lb) by about 17 percent, and held nearly 36 percent of the total tire cord market (773 million lb), despite extremely rapid growth of steel

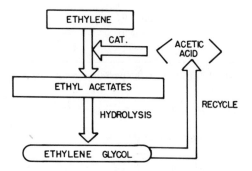

Figure 9.15 Oxirane acetoxylation process for ethylene glycol. (Adapted from reference 83; reprinted with permission of the copyright owner, The American Chemical Society.)

*Including cyclohexane, an intermediate for nylon.

(200 million lb) for radial tires. Meanwhile, rayon fibers (17 million lb) have all but disappeared from the tire market.[85]

Capacity for p-xylene has expanded far too rapidly for the demand being made for the chemical. The operating rate was 77 percent in 1978 and only about 68 percent in the first quarter of 1979. As a result of this overcapacity, the price and profitability are low and imports are negligible, despite an oversupply in Western Europe. The future of aromatics as octane improvers for gasoline probably holds the key to xylene prices.[39]

Unsaturated Polyesters The major use of unsaturated polyester (about 80 percent) is for fiber-reinforced plastics.* Polyester chemistry is complex and varied. In the following discussion, any comments regarding raw materials and resin derivations should be looked upon as typical, not specific. The flowchart for unsaturated polyester resin derivation (Fig. 9.16) is a case in point. The particular anhydrides, glycols, and cross-linking agent will differ depending on resin supplier, desired properties, and end use of the product.

The chemistry of unsaturated polyesters is such that these resins must compete with virtually every other plastic and many synthetic fibers and other petrochemicals for at least one of their principal feedstocks: benzene, propylene, ethylene, and xylenes. The composition of a typical general-purpose, unsaturated polyester is approximately 34 percent styrene and 14 percent maleic anhydride (both benzene derivatives),† 27 percent ethylene or propylene, and 25 percent isophthalic or phthalic anhydride (xylene derivatives). About 70 percent of our total domestic maleic anhydride output,[56] 30 percent of our isophthalic or phthalic anhydride,[87] and 6 percent of our styrene[56] go into unsaturated polyesters. The impact of polyester resins on maleic anhydride markets suggests that this is a key material that must be monitored carefully, particularly since it has occasionally been in a tight supply situation.

The largest C_8 fraction in mixed xylenes is the meta isomer (44 percent), used in making isophthalic anhydride. The other petroleum-based isomers of mixed C_8 aromatics are: ethylbenzene, 18 percent; o-xylene, 20 percent; and p-xylene, 18 percent.[88] Neither isophthalic nor phthalic anhydride should pose a supply problem unless PVC, the largest user of phthalates, should experience a period of unusually high demand.

Styrene monomer is added to polyesters as a cross-linking agent and reactive diluent. Other cross-linkers are used—diallyl phthalate (DAP) or triallyl cyanurate (TAC)—in limited situations, but usually only where high heat resistance or other specific properties are required. Styrene is the commercial cross-linker for general-purpose use because of its moderate cost, its wide range of curing characteristics, and its good balance of properties.

---

*Unless otherwise noted, fiber-reinforced plastics (FRP), as used here, will mean unsaturated polyester resin reinforced with glass fibers and generally extended with mineral fillers. Data supplied will be based on the "neat" (i.e., unreinforced, unfilled) resin unless specifically stated otherwise. About 20 percent of FRP resins are actually nonpolyester (usually epoxy).[86]
†Maleic anhydride can also be derived from butene.

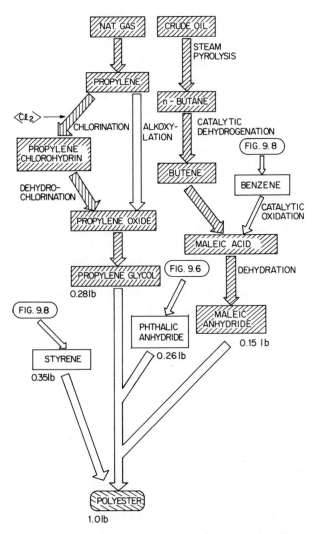

Figure 9.16   Simplified flowchart for unsaturated polyester derivation.

Unlike PET resin, which requires the use of ethylene glycol, unsaturated polyesters permit numerous options in their chemistry, one of which is the use of propylene glycol. This glycol is being used increasingly in the unsaturated resins, but often with some sacrifice in specific properties compared with ethylene oxide. Propylene oxide operating rate has been low in recent years, hovering around 70 percent, so supply should not be critical, even though polyurethane foams take about three times as much as do unsaturated polyesters.[48] One of the most demanding and fastest-growing applications of FRP is for corrosion-resistant uses—equipment used for production, transport, and storage of oil and gas; chemical processing equipment; water and sewage treatment facilities; and other such applications. These uses call for resins specially tailored for the environment. The principal resin types for such use are vinyl esters, isophthalate polyesters, and bisphenol A-fumarates. In extreme cases, furfuryl alcohols may be used. The general-purpose orthophthalic polyesters are usually inadequate, except in mildly corrosive environments.[89]

In the automotive industry, efforts to increase gasoline mileage have prompted various approaches to reducing car weight, not the least of which is increasing the use of plastics in place of metals. Fiber-reinforced thermoplastics, particularly polypropylene and polybutylene terephthalate, have become the major resins for this market. In 1976, land transportation applications took over from marine uses as the largest market sector for FRP.[90]

The fibrous reinforcement is the component that makes FRPs the "premium-quality, premium-priced" products of the plastics industry. Glass fibers are by far the most widely used reinforcements, accounting for over 99.5 percent (358 thousand tons) of the fibers used in the industry (360 thousand tons). Aramid fibers are a distant second (938 tons) followed by carbon fibers (276 tons).[91] The latter would see more service but for their high cost. Nonetheless, carbon (or graphite)-reinforced FRP is being investigated as a possible replacement for metals in some critical automotive components.

Experts view the long-range glass fiber supply situation with some concern. In 1977, some fabricators were placed on allocation by their fiber suppliers, but the situation was expected to ease in 1979 as new capacity came on stream. Even so, this added capacity is not the sole answer, because restrictions on natural gas and other fuels have, in recent winters, reduced average operating rates of fiber producers.* Furthermore, increased demand for glass-fiber thermal insulation has created a competition for capital for manufacture of FRP-grade fibers. Perhaps the most worrisome factor is the low profitability of fiberglass manufacture—high capital costs of added capacity and high operating costs (+170 percent energy costs and +80 percent other production costs between 1973 and 1977).[92]

It has long been felt that unlike asbestos fibers, fiberglass posed no health hazards other than irritation of skin, eyes, and upper respiratory tracts, some fibrotic lung changes, and an increased mortality due to nonmalignant respiratory disease.[93] Now the possibility has been raised that lung calcification and pneumoconiosis, a lung disease similar to asbestosis,† may result from inhalation of glass-fiber dust.[94] Further testing is needed to assess the hazard. In an effort to reduce the amounts of reinforcing fibers required, many fillers (extenders) and property improvers have been evaluated for addition to FRP. The major commercial additives are: carbonates‡, 80.5 percent; clays, 7.5 percent; starches and cellulosics 4.3 percent; wood flour, 3.5 percent; and silicas, 2.9

---

*To less than 80 percent, in 1977, for some producers.
†A nonmalignant scarring of the lungs caused by mineral fibers.
‡Calcium carbonate, chalk, limestone, and so forth.

percent.* The balance is made up of nonfibrous glass (bubbles, spheres, etc.), perlite, shell flour, cork, and others (each less than 1 percent of the total). In 1977, a total of 1335 $\times$ 10^3 tons was used in FRP. This represents an increase of 10 percent over 1976, but ratios of the various additives were unchanged.[95] These additives not only reduce material costs and fiber loading, but they often also improve specific properties.

The major end uses of polyester resins are shown in Table 9.30 and specific market sectors for thermoset and thermoplastic FRPs are shown in Table 9.31.

Epoxies

An epoxy group (also called epoxide, ethoxyline, or oxirane) is a three-membered ring comprising two joined carbon atoms attached to a common oxygen atom. An epoxy resin is an intermediate molecule containing more than two epoxy groups, which make it possible to thermoset (cure) them into a cross-linked, three-dimensional network. The epoxy rings are opened and reacted by either basic or acidic materials (catalysts, hardeners, curing agents). The most common epoxies (DGEBA)† are those manufactured from bisphenol A and epichlorohydrin, as shown in Fig. 9.17. More on the chemistry of epoxies can be found in reference 98. Because of their unique chemistry, epoxies are very ver-

Table 9.30   Polyester Resin Market, 1973–1987

Form	Million lb[a]				Annual Growth Rate, % 1977–1987
	1973	1975	1977	1987	
Unsaturated polyester for reinforced plastics	800	600	800	1917	9.1
Thermoset surface coatings	10	9	12	20	5.0
Cultured marble and other unsaturated thermosets	205	110	210	310	4.0
Thermoplastic PET, PBT	10	23	38	322	24.0
Strapping[b]	—	2	6	72	28.0
Film	210	218	273	578	9.2
Elastomers	1	3	5	16	12.3
Export	16	9	10	11	1.0
Total	1252	974	1354	3246	9.2
Fibers	2851	2995	3400	5033	4.0

Source: Reference 96. Reprinted with permission of the copyright owner, The American Chemical Society.

[a] Neat resin.

[b] Mostly scrap.

---

*Novaculite, sand, synthetic silicas, quartz, and so forth.

†Diglycidyl ether of bisphenol A.

Table 9.31 Markets for Reinforced Plastics, 1977

Market Sector	Percent	
	Thermosets	Thermoplastics
Transport	23	51
Marine	25	—
Construction	17	2
Anticorrosion	12	—
Electrical	7	19
Appliances	—	21
Consumer goods	—	6
Other	16	1
Total	100	100
	(1.61 billion lb)	(190 million lb)

Source: Reference 97. Reprinted with permission of the copyright owner, The American Chemical Society.

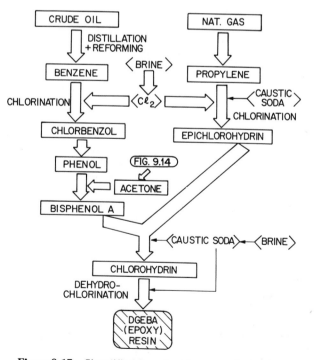

Figure 9.17 Simplified flowchart for epoxy derivation.

satile and have a wide variety of ends uses—adhesives, coatings, encapsulants, mortars, molding compounds, and FRP resins.

Epoxies represent a small volume but are important members of the growing family of resins. Demand for epoxies was in a stage of rapid growth when the oil embargo struck the entire business community, but epoxies have since made a good recovery, showing a growth rate of over 10 percent between 1977 and 1978 (Table 9.32).

Those companies basic in bisphenol A and epichlorohydrin (Shell Chemical and Dow Chemical)* have found themselves in a favored position as suppliers of epoxies. At various times in the recent past, producers dependent on merchant supplies of raw materials or precursors have been forced to reduce output and allocate supplies because of short-term shortages of intermediates or ancillary chemicals (or electric power, in one local situation).

Assuming adequate supplies of raw materials and intermediates, variations in demand pose little problem, because epoxies are based on "pots-and-pans" chemistry and capacity can be diverted pretty much at will to or from the manufacture of several other resins. Furthermore, the types of equipment required are not particularly capital-intensive.

The energy pinch has emphasized the necessity for conservation. This suggests the desirability of employing high-performance, less-energy-intensive materials such as epoxies. This is showing up in a number of potential automotive applications for epoxy–carbon fiber composites,† probably the *crème de la crème* of high-performance plastics.

Table 9.32   Epoxy Consumption, 1976–1977

| | 10^3 Tons | |
Market Sector	1976	1977
Protective coatings	55.7 (44.5%)	61.5 (44.5%)
Reinforced uses[a]	18.1 (14.5%)	21.8 (15.8%)
Tooling, casting, molding	10.6 (8.5%)	10.3 (7.5%)
Bonding and adhesives	9.4 (7.5%)	9.2 (6.7%)
Aggregates (paving, flooring)	4.9 (3.9%)	6.1 (4.4%)
Other	13.8 (11.0%)	15.4 (11.1%)
Export	12.7 (10.1%)	13.8 (10.0%)
Total	125.2 (100%)	138.1 (100%)

Source: Reference 27.

[a] Does not include reinforcements.

*In 1973, these two producers together provided over half of the total industry output of epoxies.
†For example, driveshafts, springs, and frames.

Polyurethanes

Urethane chemistry is even more complex than polyester chemistry and depends on a larger variety of raw materials: benzene, propylene, ethylene, chlorine, propylene oxide, toluene, adipic acid (for polyester polyols), castor oil, aniline, and others. The urethane prepolymer is formed from *methylene diphenyliso-cyanate* (MDI) or *toluene diisocyanate* (TDI), and *polyether* or *polyester polyols*. The MDI is derived from benzene by nitration to nitrobenzene, reaction with ammonia to aniline, and condensation with formaldehyde plus phosgenation. The TDI is derived from toluene by nitration to dinitrotoluene followed by reduction and phosgenation. Polyether polyols are formed from aliphatic chlorhydrin by dehydrochlorination, followed by oxyalkylation of the resulting aliphatic oxides with propylene or dipropylene glycol. Polyester polyols are derived from carboxylic acid esterified with glycol or triol. The urethane prepolymer is then reacted with castor oil or its derivatives and diisocyanate to form the polyurethane.[99,100] A shortage, in any one of these key stages could upset the resin supply situation.

Three different isocyanates play major roles in urethane technology: TDI (toluene diisocyanate), PMPPI (polymethylene polyphenylene isocyanate), and MDI (methylene diphenylisocyanate). The latter two are aniline-based and were, indeed, its major derivatives (Table 9.33) back in 1974. For such a "mature" commodity chemical, the production growth rate of aniline has been remarkably high, owing largely to the demands of polyurethane producers. Through 1982, at least, isocyanate is expected to maintain an 8 percent annual growth rate, which means an increase to 1.65 million tons from 1.10 million tons in 1977.[102] To meet demand, aniline operating rates must be high, a feat made difficult by the "dirty" chemistry involved. There has been little incentive for investing in new capacity because of a 50 percent operating rate as recently as 1971 to 1972.[101] Worldwide AGR for TDI production is expected to average 5

Table 9.33 Aniline End Uses, 1974

End Uses	Consumption 10^6 lb	Percent of Total
Isocyanates	215	40
Rubber chemicals	190	35
Dyes/intermediates	30	6
Hydroquinone	30	6
Drugs	20	4
Miscellaneous	50	9
Total	535	100

Source: Reference 101. Reprinted with permission of the copyright owner, The American Chemical Society.

to 7 percent for the period 1977–1982. The annual growth rate for polymer and for pure MDI is expected to be 10 to 12 percent, with MDI overtaking TDI production in the mid-1980s.[103] The major outlet for TDI (about 85 percent) is for flexible foams (bedding, furniture, and transportation vehicles), while MDI is widely used as thermal insulation. Table 9.34 shows a breakdown of polyurethane consumption by applications and market share. Urethane foam is becoming a significant factor in home insulation, because of the energy situation, with these uses growing at about twice the rate for plastics in general. Early in the 1970s, urethane foam insulation was involved in some spectacular fires, but this problem apparently has been solved with the development of fire-retardant foams conforming to the Uniform Building Code (UBC).[104] The thermoplastic urethane is used for fabric coating, injection molding (e.g., RIM—reaction injection molding*), wire coating, and adhesives.

Nylon

The chemical key to nylon is cyclohexane (Fig. 9.18). Table 9.35 shows the dependency of the various grades of nylon on this and other chemicals.

Table 9.34 Polyurethane Consumption, 1976–1977

	Billion lb		Percent	
			Market	Annual Growth
Form/Use	1976	1977	Share	Rate
Flexible foam				
Furniture and bedding	0.521	0.589	28	+13.1
Transportation (seating)	0.351	0.382	18	+8.8
Rug underlayment	0.113	0.151	7	+33.6
Other[a]	0.215	0.278	13	+29.3
Total—flexible	1.2	1.4	67	+14.3
Rigid foam				
Building construction	0.179	0.241	11	+34.6
Appliances	0.074	0.085	4	+14.9
Other[a]	0.129	0.146	7	+13.2
Total—rigid	0.382	0.472	22	+23.6
Total foam	1.6	1.9	90	+18.8
Millable gums, castable				
resins, thermoplastics	0.203	0.225	11	+10.8
Total polyurethanes	1.8	2.1		+16.7

Source: Reference 42.
[a]By difference.

*A low-energy process, claimed to require less than half the process energy to produce equivalent or better products.

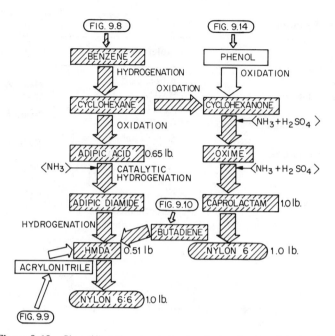

Figure 9.18   Simplified flowchart of nylon 6:6 and nylon 6 derivations.

Cyclohexane   Beginning in the late 1960s, the production of cyclohexane, the critical component in most nylon manufacture, became unprofitable, and several producers withdrew from the business or at least shut down inefficient units. The major producers of cyclohexane now are largely basic in petroleum refining, since the benzene feed is extracted from the aromatic stream in the reforming process. Most cyclohexane (about 85 percent) is made by hydrogenating benzene, although small amounts are also obtained fom natural gas liquids (NGL). The hydrogen, too, is a co-product of the refinery reformers. At present, benzene is in adequate supply. Because of cyclohexane's close ties to "the vagaries of oil refinery economics," it could easily become a major bottleneck in nylon production.

Adipic Acid   Approximately 55 percent of the cyclohexane output goes into the manufacture of adipic acid, a key nylon intermediate. About 95 percent of this acid is derived from cyclohexane. The balance is made from phenol.[106] Ninety percent of the adipic acid output is used in the manufacture of nylons. This intermediate was in tight supply until about the mid-1970s, at which time new capacity came on stream.

Hexamethylene Diamine (HMDA)   This is yet another key nylon intermediate, which has four possible synthetic routes, two of which (hexanediol and adipic acid) are dependent on cyclohexane. Actually, over half of the commercial HMDA comes from butadiene. Some also comes from acryloni-

Table 9.35 Nylon Precursors and Products

Feedstocks	Intermediates	Product
Cyclohexane	Adipic acid ⎱	Nylon 6:6[a]
Butadiene/cyclohexane	HMDA ⎰	
Oleic acid	Azelaic acid ⎱	Nylon 6:9
Butadiene/cyclohexane	HMDA ⎰	
Castor oil	Sebacic acid ⎱	Nylon 6:10
Butadiene/cyclohexane	HMDA ⎰	
Coconut oil	Dodecanoic acid ⎱	Nylon 6:12
Butadiene/cyclohexane	HMDA ⎰	
Butadiene/cyclohexane	Caprolactam	Nylon 6[a]
Castor oil	Undecanoic acid	Nylon 11
Coconut oil	Lauryllactam	Nylon 12

Source: Reference 105.
[a]Used in fibers.

trile.[106,107] Approximately 10 percent of the butadiene production goes into HMDA.[26]

Caprolactam   This is another nylon intermediate made from cyclohexane—via cyclohexanone (or from phenol). It appears to have adequate production capacity. About 90 percent of the output goes into nylon 6.

Indications are that adipic acid capacity is the more crucial item in the production of nylon 6:6, but cyclohexane, one step further back in the chain, is unquestionably the controlling factor in most nylon production.

Butadiene   The projected ample availability of butadiene, arising from the use of heavier petrochemical feedstocks, suggested other production options for nylon intermediates. Actually, however, U.S. output of butadiene, in 1979, was running about the same as 1978, in the face of a 2 percent higher 1979 demand. Furthermore, imports from Europe, our primary foreign source, were expected to decline because of increasing overseas demand in 1979.[108]

Technically, butadiene can be used to synthesize adipic acid by catalytic carbonylation or to make mixed cyclohexene (65 percent)/cyclohexane (35 percent) by hydrogenation. Cyclohexene can then be oxidized to adipic acid in higher yields than the 75 percent yields obtained from cyclohexane. The butadiene process, however, is economically unattractive at this time, and will remain so until such time as the price gap between benzene and butadiene closes somewhat.[107]

Consumption of nylon plastics was up 9.9 percent in 1977 over the preceding year, with injection molding accounting for nearly 2.5 times as much as extrusion.[27] The annual growth rate of nylon fiber production in 1978 (over 1977) was a slightly more modest 8.7 percent. Tire cord takes a larger percentage of the total nylon fiber output (11 percent) than of the polyester fiber (8 percent), but the total volume of nylon fiber is lower (2.5 billion lb vs. 3.8 billion lb for polyester). Acrylic fiber (polymerized acrylonitrile) is a poor third among synthetic fibers (730 million lb in 1978), and it is the most vulnerable to market fluctuations, because it has no significant industrial outlets.[84]

Other Plastics

The major commercial plastics have been discussed above. There are many other so-called engineering and specialty plastics which, considered individually, are produced in relatively small quantities but which are important to various industries. Altogether, these 20 or so plastics had a total production of 3.34 billion lb in 1977 and a total sales and captive use value of $1.47 billion. Nonetheless, it would be beyond the intended scope of this book to enter into detailed discussions of these individual products, except for polyesters and nylons. These were discussed above because the fiber uses of these materials are of sufficient volume to have a significant impact on the petrochemical industry and on feedstocks. It does seem pertinent, however, to add to the discussion a particular specialty plastic that offers a lesson in practical economic issues. Similar issues may be raised with the development of other new technologies.

Methyl Methacrylate (MMA) This is currently produced via the acetone-acetone cyanohydrin-hydroxybutyrate esters route. Oxirane International, with a captive source of *t*-butyl alcohol,* has developed a C_4-oxidation process via isobutylene-*t*-butanol-methacrolein-methacrylic acid.[109]

Since current MMA production is the largest end use of acetone, the new technology could have far-reaching effects on this and other petrochemicals. Phenol, co-produced with acetone from cumene (isopropyl benzene) (Fig. 9.14) and isopropyl alcohol, an intermediate between propylene feedstock and the acetone product (Fig. 9.4), could end up in serious economic straits. Reduced demand for acetone could force up prices of co-product phenol, since the price of the latter is subsidized somewhat by its co-product. This could result in price increases for phenol derivatives, leading to redirection of priorities, alteration of processes, redesign of products, and other disruptions. Isopropyl alcohol plants could be forced to close because of reduced demand,[110] placing supply pressures on other isopropyl alcohol derivatives.

These are real threats to the established order of petrochemical economics—threats that cannot be appraised fully at this writing. Other new technologies, developed or developing, could have similar impacts on the industry, if commercialized on a large scale.

Paints and Allied Products

This topic includes paints, varnishes, lacquers, printing inks, and associated products—materials frequently covered under the blanket heading "coatings." The coatings industry has been characterized as parasitic, in that it has traditionally "piggybacked" on plastics technology. Paint vehicles or binders are

*By-product from propylene oxide plants.

often little more than plastic resins dissolved or dispersed in a volatile organic solvent or water, or a solid resin ground to a powder. Other ingredients are added to control flow, shelf life, curing speed, toughness, or to provide color or other specific properties.

A coating is not employed in bulk, hence the coatings business does not enjoy the volume of the plastics industry. The value of the paint applied rarely exceeds 1 percent of the value of the substrate product, and more often is less than 0.1 percent of its value. The value of the service rendered by the coating is immense, however, because it not only enhances the aesthetics of the product, but protects it from the elements as well.[111,112]

Only the larger producers carry on research programs in coatings technology. Most paint companies count on resin and other suppliers to provide product innovation. In the industry as a whole, only about 2 percent of sales revenues is spent on research. In times of recession, the suppliers could withdraw their technical service and leave the formulator on his own. To offset this possibility, some smaller paint companies have banded together to form cooperative research organizations.

Since the paint business, by nature, is highly competitive, the profit margin is low, but this is no real cause for alarm, because capital investment in this industry is also very low. If epoxy resin manufacture can be characterized as "pots-and-pans" chemistry, most paint manufacture would have to rate as a "bucket-and-stick" operation.

Because of the high shipping weight of liquid paints, they are normally supplied within a narrow geographical range. This accounts for the fact that there are 1500 to 1800 paint companies in the United States, only a handful of which are national producers, and none accounts for more than 10 percent of the total market. Approximately 1 percent of the domestic paint producers account for 40 to 45 percent of total paint sales. Sherwin-Williams is the largest U.S. producer, with nearly 10 percent of the market. DuPont is the number 2 producer and is first in industrial finishes, but paint accounts for only a small percentage of duPont's total sales.[112] Industrial finishes for automobiles are supplied by a handful of paint companies, and some automobile manufacturers, notably Ford and Chrysler, make paints for their own internal use. Ten or 15 years ago, 7 gal of paint was required to coat a typical automobile. More recently, using high-performance paints in reduced thicknesses, 5 gal has become adequate, and 4 gal may become a reality.

The paint industry serves two separate markets: *trade sales* ("shelf goods") sold over the counter for consumer use, and *industrial finishes* ("chemical coatings") for manufacturers' use in the factory. Table 9.36 shows a history of the market division between these two categories from 1970 to 1977. The market is now essentially equally divided between the two, but in the early 1960s, trade sales had a larger market share.

Alkyd resins are the workhorse for industrial coatings, and in the early to mid-1970s had a volume over twice that of its nearest competitor (Table 9.37). Alkyds are essentially polyesters of a polyhydric alcohol (e.g., glycerol or pentaerythritol) and a polybasic acid (typically phthalic anhydride) modified with

Table 9.36 Domestic Coatings Output, 1970-1977

Year	Million gal		
	Trade Sales	Industrial Finishes	Total (Annual Growth Rate, %)
1977	428 (50.7%)	416 (49.3%)	844 (−9.1)
1976	474 (51.0%)	455 (49.0%)	929 (+4.4)
1975	452 (50.8%)	438 (49.2%)	890 (−9.3)
1974	475 (51.0%)	457 (49.0%)	932 (+3.9)
1973	424 (47.3%)	473 (52.7%)	897 (−3.2)
1972	452 (48.4%)	475 (51.6%)	927 (+6.1)
1971	431 (49.3%)	433 (50.7%)	874 (+5.8)
1970	427 (51.7%)	399 (48.3%)	826 (n.a.)

Source: References 61 (1975 to 1977 data), 62 (1974 data), 63 (1973 data), 64 (1972 data), and 113 (1970 and 1971 data). Reprinted with permission of the copyright owner, The American Chemical Society.

Table 9.37 Domestic Consumption of Paint Vehicles, 1971-1974

Vehicle	Average Annual Consumption	
	10^6 lb	Percent of Total
Alkyd	660	42.7
Acrylic	310	20.1
Poly(vinyl acetate)	215	13.9
Epoxy	95	6.2
Urethane	70	4.5
Aminoplast	70	4.5
Cellulosic	65	4.2
Styrene–butadiene	30	1.9
Phenolic	30	1.9
Total	1545	[a]

Source: Reference 114.

[a]Not exactly 100, due to rounding errors.

an oil (Table 9.38) or a fatty acid. As other, higher-performance resins phase in (e.g., acrylic, epoxy, polyurethane), alkyd's share of the market will probably decline, together with the phenolics and nitrocellulose.

Environmental Aspects

Concern over air pollution originally focused on conformance to Los Angeles County's Rule 66, but more recently on the federal Clean Air Act of 1970 and

Table 9.38   Average Annual Domestic Usage of Nonfossil Oils in Coatings, 1963–1973

Oil	Average Annual Usage	
	10^6 lb	Percent of Total
Linseed	297	34.8
Soybean	178	20.8
Tall	130	15.2
Fish	76	8.9
Castor	72	8.4
Secondary oils	40	4.7
Tung	30	3.6
Oiticica	6	0.8
Other primary oils	25	2.9
Total	854	a

Source: Reference 114.

aNot exactly 100, due to rounding errors.

subsequent amendments. These regulations have forced the reformulation of many coatings to omit smog-producing (i.e., photochemically reactive) solvents. Basically, five options are available to accomplish this objective: (1) exempt aliphatic solvents can be used, (2) water can be used to disperse the paint vehicle, (3) solvent content can be reduced by using higher solids solutions of lower-molecular-weight resin vehicles, (4) paints can be formulated as radiation-curing systems thinned with reactive diluents that copolymerize with the vehicle during cure, and (5) solventless systems of dry powders or reactive liquids can be used. Table 9.39 shows how these options are expected to affect markets for industrial finishes in 1982.

Poly(vinyl acetate) (PVAc) resins are the most widely used vehicles for latex coatings (67 percent). Acrylic resins make up 27 percent of such coatings. Styrene–butadiene is 6 percent, down from 16 percent in 1970, with the difference being made up largely by PVAc.[116]

Although water-base paints have grown rapidly in trade sales grades, penetration into the industrial-finishes market has been modest to date. As the technology of water-dispersible paints advances, however, it is expected that these formulations will become a dominant force in the industry. One reason for the delayed acceptance is the high cost of removing the water. Another is the marginal level of corrosion protection afforded metals, both during application and in service. Relatively poor adhesion to metals is yet another factor.[111] Table 9.40 indicates the solids–volatiles relationships in various types of typical coatings.

Although there is no officially designated classification of coatings on the basis of volume solids content, a practical listing would suggest the following:

Table 9.39 Market Shares of Industrial Finishes

	Percent	
	1975	1982 (est.)
Exempt solvent-thinned	15	31
Water-thinned	4	30
High-solids, solvent-thinned	2	15
Powder	3	5
Radiation-cured[a]	1	4
Conventional solvent-thinned	75	15

Source: Reference 115. Reprinted with permission from the *Journal of Coatings Technology* and the Federation of Societies for Coatings Technology, the copyright owner.

[a]Generally contain only reactive diluents that copolymerize during cure.

Table 9.40 Volume of Solvents in Various Categories of Paints

	Volume, %	Volume, gal[a]	
Category	Solids	Solvent	Paint (total)
High solids	80	0.2	1.0
Exempt solvents	30	1.86	2.67
Waterborne	30	0.37[b]	2.67
		1.49[c]	
Industrial lacquer	15	4.53	5.33

Source: Reference 115. Reprinted with permission from the *Journal of Coatings Technology* and the Federation of Societies for Coatings Technology, the copyright owner.

[a]Required to apply 0.8 gal of solids.
[b]Coupling solvent(s).
[c]Water.

high solids, 80 to 100 percent volume solids; intermediate solids, 60 to 80 percent; conventional solids, 30 to 45 percent; and low solids, 10 to 15 percent.[115]

Energy of Coatings

A great deal of development effort is being expended now to reduce energy consumption in coatings technology. Table 9.41 shows that in a metal-coating operation, incineration of volatile solvents to eliminate air pollution, metal pretreatment, and provision of makeup air for paint application and cure take over 85 percent of the energy in a typical case. Elimination of the incinerator

Table 9.41 Metal Coating Line Energy Consumption (Typical)

Process Step	Percent of Total
Metal pretreatment	20.7
Metal drying	9.1
Application makeup air[a]	9.1
Cure	2.3
Makeup air[a]	10.9
Hanger cleaning	3.3
Incinerator[b]	44.6

Source: Reference 115. Reprinted with permission from the *Journal of Coatings Technology* and The Federation of Societies for Coatings Technology, the copyright owner.

[a] Air input necessary to keep air in curing oven below explosive limit.

[b] Tabulation would be altered considerably by eliminating incineration or using energy from incinerated products as auxiliary process heat.

might be possible if exempt solvents (e.g., methyl ethyl ketone or other aliphatics) were used. Energy from the products incinerated can be used as auxiliary process heat to reduce energy consumption. Makeup air is necessary to keep the solvent-laden air below explosive limits.

Although the actual energy required for curing the paint film is a modest fraction of the overall process energy, more efficient cure systems have also been developed. In processing a fast-curing sandable wood sealer, infrared curing was found to require 1600 kJ per m^2. Radio-frequency curing reduced the energy requirement to 330 kJ, and ultraviolet curing and electron-beam curing called for only 64 and 30 kJ per m^2, respectively.[115] Electron-beam curing is by far the most energy-efficient, but it is competitive only for coating relatively flat areas in excess of 20 million ft^2 per yr, because of high capital costs.[117] Other references show less disparity in energy expenditures for infrared and ultraviolet curing, but the rank order remains the same. Radio-frequency curing has been used for a number of years to cure coatings on wood panels, but much of this market is now being taken over by ultraviolet curing. Infrared curing energy varies widely with the specific application, because of substrate material and thickness factors. Ultraviolet cures are now being widely used also with printing inks.[118]

A great deal of attention is being given to electrocoating and powder coating. The former is applicable only to electrically conductive substrates, because it is essentially similar to electroplating of metals. These coatings have found good acceptance as automotive primers, but they cannot be used for finish coats, since the primer forms an insulating layer over the substrate surface. Electrocoating does offer the ability to coat recesses, undercuts, sharp edges, small openings, and so forth, with a more uniform film than can be achieved with conventional finishes.

Powder Coating

Powder coating is a system that could well make big inroads in the automotive field, but for one major shortcoming. So-called metallic finishes comprise a large fraction of automotive finish coats, and powder coatings do not lend themselves to forming metallic finishes completely acceptable to this market. Powder coatings can be applied by several basic techniques: fluidized bed, electrostatics, air or flock spray, and cascade or waterfall coating.

The latter two methods bring the powder into contact with substrates heated above the sintering temperature of the resin powder. The fluidized-bed system does the same thing but calls for immersing the substrate in an air-suspended bed of powder. For thin coatings and for substrate materials that cannot tolerate extended preheating, an electrical charge differential can be maintained between the substrate and the powder, causing the powder to collect on the part. The amount of heat required to coalesce and/or cure the coating is much less than that required for preheating.[119]

Fluidized bed and electrostatics are the most popular application techniques. The former gives good thickness control in the range from 8 to 20 mils (0.20 to 0.50 mm), whereas the latter is best at 1 to 5 mils (0.03 to 0.13 mm). Thin-film (<3-mil) powder coatings are primarily decorative, while coatings from 3 to 60 mils serve utilitarian purposes.

Coating resins used can be either thermoplastic or thermosetting. Low-molecular-weight thermosetting resins are often used for thin decorative coatings, while high-molecular-weight resins (either thermosetting or thermoplastic) are generally employed for thick, functional coatings.[120] Table 9.42 shows powder coating consumption data, by resin type, for the mid-1970s. Epoxies and vinyls dominated the market.

Powder coating offers a number of advantages over conventional coatings: zero solvent consumption, complete compliance with EPA regulations, low energy demand, high build, and low air requirements, and they are also cost-competitive. On the debit side, they require completely different application equipment and technology from conventional coatings.

The coatings industry as a whole faces two major challenges: increasingly stringent environmental regulations and tightening energy and feedstock supplies. The former seems well on the road to satisfactory solution and the latter is being solved, in part, by the response to the environmental issue—reduced solvent and energy consumption. Aside from occasional spot shortages of specific solvents and compounding ingredients, supplies for coatings are in relatively good shape, since the volumes of petrochemicals involved in coatings are completely overshadowed by the plastics industry's demand for the same or similar materials. Furthermore, coatings have far greater formulating flexibility than most plastics, since essentially identical results can be obtained by using dif-

Table 9.42   Powder Coatings Consumption

Material	Consumption, 10^6 lb (Market Share, %)	
	1975	1976
Epoxy	22.0 (65.5)	25.4 (64.5)
Vinyl	7.1 (21.1)	8.4 (21.3)
Polyester, thermoset	1.5 (4.5)	1.9 (4.8)
Surlyn ionomer	1.4 (4.2)	1.8 (4.6)
Nylon	0.6 (1.8)	0.6 (1.5)
Cellulosics	0.4 (1.2)	0.5 (1.3)
Polyester, thermoplastic	0.4 (1.2)	0.5 (1.3)
Acrylic	0.2 (0.6)	0.3 (0.8)
Total	33.6	39.4

Source: Reference 121.

ferent combinations of oils, plasticizers, flow agents, fillers, and solvents. Modern-day paint formulators have tended to favor petroleum-based ingredients but in a pinch, they could return to natural materials with, perhaps, tolerable sacrifice in properties or costs.

This would possibly suggest a return to the cultivation of raw materials for paints, but it must be remembered that pressures are already being applied on the agricultural community for more food, fiber, energy, and feedstocks. Also, coatings based on synthetics are generally more uniform, have somewhat greater durability, and are easier to use. A few years ago, a professional house painter would have frowned on the use of synthetic latex house paint. Today, most are using the latex and even exterior latex trim enamels.

Synthetic Elastomers

A synthetic elastomer is merely a synthetic polymer with elastic or rubberlike properties. There are seven major commercial types, but just two—SBR (styrene–butadiene rubber) and polybutadiene—dominate the market, with slightly over 75 percent of the consumption (Table 9.43). Styrene–butadiene rubber is a polymer comprising about 3 parts of butadiene to 1 part of styrene. Polybutadiene production is only 25 to 30 percent of SBR production (Table 9.44), but their combined output makes butadiene the most important feedstock for elastomers. The supply tightness of this chemical was discussed in the section of this chapter dealing with nylons, and will not be repeated here.

The low operating rate (74 percent) for SBR plants seems incompatible with the fact that SBR supply is tight, but producers can only operate at about 90 percent of capacity under the best of conditions. The "practical" capacity is lower for several reasons: some of the major plants were built during World

Table 9.43 Consumption of Synthetic Elastomers, 1977–1979

Elastomer	10^3 Tons[a]		
	1977	1978	1979
SBR	1595	1565	1505
Butadiene	456	452	463
Ethylene–propylene	157	154	163
Butyl	154	143	140
Neoprene	130	132	132
Nitrile	79	83	85
Isoprene	76	84	77
Total	2647	2613	2565

Source: Reference 122. Reprinted with permission of the copyright owner, The American Chemical Society.

[a]Data not precise; scaled from graph and converted to English units.

Table 9.44 SBR and Polybutadiene Production, Capacity, and Operating Rate,[a] 1977–1979

Elastomer	Units	1977	1978	1979
SBR				
Production	10^3 tons	1520	1510	1465
Capacity	10^3 tons	2060	2045	1990
Operating rate	%	74	74	74
Polybutadiene				
Production	10^3 tons	400	420	425
Capacity	10^3 tons	420	465	460
Operating rate	%	95	90	92

Source: Reference 122. Reprinted with permission of the copyright owner, The American Chemical Society.

[a]Data not precise; scaled from graph and converted to English units.

War II and are in such condition that they cannot be pushed. Inefficient operation also results from users demanding a large number of elastomer grades, and the industry has long been plagued by crippling strikes. Each of these factors effectively reduces the real capacity to far less than the nameplate capacity (Table 9.45).

The dominant outlet for the rubber industry is automotive tires. Nearly 65 percent of the synthetic elastomers end up in tires and tire products. In the cases of SBR and polybutadiene, 65 and 91 percent, respectively, go into these products. As a result of this close tie to the automotive industry, a bad year for

Table 9.45 Domestic Nameplate Capacities for Synthetic Elastomers, 1979

Type	10^3 Tons	Number of Producers
SBR	1999	9
Polybutadiene	458	7
Butyl	246	2
Ethylene–propylene	221	5
Neoprene	217	2
Nitrile	122	5
Polyisoprene	66	2^a

Source: Reference 123. Reprinted with permission of the copyright owner, The American Chemical Society.

[a] One 63-tpy plant mothballed; not included in tabulation.

new-car sales usually means a poor year for the rubber industry, even though about 72 percent of tire shipments are for the replacement market. Furthermore, several other elastomers depend heavily on automotive mechanical goods (hoses, weatherstripping, gaskets, etc.) for their markets. Styrene–butadiene rubber is a versatile elastomer, being used also for: nonautomotive mechanical goods, 15 percent; automotive mechanical goods, 5 percent; latex, 10 percent; and miscellaneous uses, 5 percent. Polybutadiene rubber has only one significant outlet other than tires and tire products. That is for high-impact resin modifiers—9 percent of total sales.[122]

Radial tires are still growing in relation to bias-ply tires.* This means an erosion of some of the synthetic rubber market, since a bias-ply tire uses three times as much synthetic as natural rubber, while radials use the two in essentially equal amounts, on a rubber hydrocarbon basis. Some tire makers may take another look at polybutadiene because of its price advantage (43 to 44¢ per lb vs. 58 to 60¢ for natural rubber), and in light of the fact that it too is a low-hysteresis rubber, as needed for radial tires.

Ethylene-propylene co- and terpolymers collectively (EPDM) also have a piece of the tire markets, largely for white sidewalls. This is one of the newer (1960s) and faster growing† of the synthetics. Despite the high operating rate (92 to 93 percent) and demand growth, the low profit margin has discouraged investment in new capacity.

Polyisoprene is unique in that it is a synthetic "natural" rubber, but real natural rubber still holds a price advantage. If natural rubber prices were to rise above 60¢ per lb for a prolonged period of time, polyisoprene could be in the

*Up to about 47 percent penetration of the replacement market in 1979 vs. 42 percent in 1978.
†Probably an annual growth rate of 10 to 12 percent in the long term.

picture, but meanwhile, nearly half of the potential capacity is mothballed, waiting for that time.[122]

Yet another alternative to a petroleum-based "natural" elastomer is rubber from guayule shrubs. This homegrown product shows considerable promise as a substitute for natural rubber in radial tires. Guayule culture and processing were discussed in Chapter 8.

Thermoplastic olefinic (TPO) elastomers are one of the newest (introduced in 1973) and fastest-growing* polymer types. By 1980, nearly 55 percent of its projected demand of 44 million lb is expected to be for automotive applications (body filler panels, air deflectors, stone shields, bumper strips, interior horn shrouds, sound deadeners, etc.). Advantages offered by TPO include: the ability to be processed with plastics machinery, elimination of waste by scrap recycling, and conservation of process energy by eliminating curing.[124]

Carbon Black

A necessary ingredient of most rubber products (particularly tire compounds) is carbon black. Despite the low demand growth rate expected for this product (2 to 3 percent annual growth rate in the 1980s), 70 to 100 million lb of annual capacity must be added each year to keep up with this demand. Even so, new plant construction is not assured, because of the industry's declining return on investment and because of the Clean Air Act Amendment of 1977, which imposes some very demanding requirements on new plant construction. As a result, grass-roots carbon black plants will probably be set aside in favor of de-bottlenecking additions of capacity to existing plants. This places a practical limit on the potential capacity of the industry.[125]

The predicament of the producers can be attributed to two factors. As in the case of SBR, users demand a large number of carbon black grades (30 to 35 grades, with a number of varieties in each grade). Because of the turnaround time expended in changing output from one variety to another, high operating rates are unattainable.[125] The second factor is the energy-intensive nature of the industry. By 1978, feedstock costs alone had risen to more than two-thirds the total production costs,[126] and with the current mood of OPEC, the situation is worsening rapidly.†

A further complication is the fact that most carbon producers are merely divisions of large corporations, and hence are forced to compete with sister divisions for capital. Because of their poor economic bargaining position, a carbon black shortage appears to be certain.

Again, the tire industry is the controlling factor. This one industry consumes about 93 percent of the domestic carbon black, and these consumers must assume the responsibility for demanding the endless varieties of carbon black that so inhibit supply. The only alternative for the carbon black industry is to

---

*Demand (in 1977) 28.5 million lb, up from 1.5 million lb in 1973.
†Carbon black is produced commercially by the thermal or oxidative decomposition of hydrocarbons, principally natural gas.

reduce somehow the number of grades and varieties offered, so operating rates can be raised.[126] This, of course, would be only a short-term measure. The long-term solution demands the addition of new capacity.

Adhesives and Sealants

Adhesives and sealants date back to biblical days, when natural resins, waxes, and bitumens were used as caulks and adhesives. The modern adhesives industry, however, is an outgrowth of the plastics industry. Estimates of the domestic consumption ratio of natural-to-synthetic adhesive products vary considerably. One study indicates that the market share for synthetics rose from one-third to one-half between 1963–1965 and 1976.[127] Another study states that, at the end of 1976, synthetics accounted for three-fourths of the adhesives market.[128] There are two areas of total agreement, however: synthetics are taking an ever-increasing share of the total adhesives market, and the adhesives industry as a whole will continue to grow at an annual growth rate (consumption) of +4.6 percent to 1985[129] or +7 percent to 1990,[127] depending on the source, but in any case, at a rate higher than the domestic economy.

The reasons for the rapidly increasing interest in synthetic adhesives* are that, in general, they have certain inherent advantages over natural products: greater moisture resistance, increased durability, higher joint strength, better clarity, greater versatility, improved uniformity, and better working properties. Furthermore, in the case of "carbamides" (urea- and melamine-formaldehydes), studies in the U.S.S.R. have shown that 1 ton of resin takes the place of 2.5 tons of scarce animal glue and permits increasing productivity (based on personnel) by 50 percent.[130]

Feed for the adhesive industry's synthetic resin output is derived from only about 3 percent of the petrochemical industry's share of feedstocks,[131] and the product yield (in dollar volume) amounts to 6.6 percent of all polymers shipped.

Between 1965 and 1973, there were reported to be about 175 domestic manufacturers of polymers[132] supplying resins to an adhesives industry of 463 (in 1972) manufacturers.[133] There were over 5000 varieties of adhesives, and about 500 new formulations were being added each year.[134] In 1974 to 1975, there were 745 companies with 1114 plants,[134] and the top 50 accounted for less than one-third of the total sales.[135]

In the mid-1970s, only five adhesives manufacturers, all multinationals, had annual adhesives sales volumes in excess of $100 million. They were: H. B. Fuller Company, USM Company, National Starch and Chemical Company, 3M Company, and Henkel & Cie.[136] In 1977, the four largest U.S. producers of adhesives (National Starch, Borden, H. B. Fuller, and duPont), supplied only 19 percent of the domestic output.[131] At the same time, there were reported to be over 500 formulators, mostly small operators, specializing in specific markets. These accounted for 30 percent of the output (weight basis) and over

---

*Projected growth rate over 30 percent to 1985, partly at the expense of natural products and partly due to growth of the economy.[129]

30 percent of the dollar volume. About 70 percent of the producers are reported to be concentrated in the Midwest and Northeast.[131] Because of the fragmented nature of the industry, meaningful statistics are not easy to acquire, particularly in the area of captive applications. Captive uses (plywood, particleboard, paperboard, rug backing, thermal insulation binders, etc.) approximately equal the market consumption.[135]

The major U.S. adhesive resin is phenol–formaldehyde (including resorcinols), which accounts for nearly 50 percent of the total output. In the rest of the world, aminoaldehydes (urea– and melamine–aldehydes, known also as carbamides) dominate production.[132] Phenolics generally are more durable than the aminoaldehydes, particularly as wood adhesives. Phenolic adhesive resins take many forms and serve numerous functions in adhesive formulations. One-step, two-step, substituted heat-reactive and non-heat-reactive, novolac, and terpene-modified phenolic resins are added to various elastomer or resin adhesive bases to serve as tackifiers, adhesion promotors, or cross-linking agents.[137] The quantity added to any specific elastomer-base adhesive (particularly nitrile and neoprene) is significant, but consumption data are not available for these applications.

It is very difficult to put a finger on meaningful production volume data for these and other adhesive resins, because bulk plastic uses dominate their consumption in adhesives and sealants. Even the Bureau of the Census category, S.I.C. 2891 Adhesives and Sealants, is misleading for purposes of this chapter, inasmuch as S.I.C. 2891 includes natural-base materials, household iron cement, household porcelain cement, pipe thread and joint sealing compounds, and sealing wax, none of which is pertinent to a discussion of synthetic hydrocarbons. Table 9.46 cites the major adhesive and binder uses of key resins. The list fails to include many adhesive materials, for example hot melts, pressure sensitives, and general-purpose rubber cement, to name a few, nor does it take note of the poly(vinyl acetate) derivatives* used as modifiers for epoxy- and phenolic-base adhesives or used alone as hot melts. (See Fig. 9.19 for their derivations.) Consumption figures for 1978 show that the epoxies were the most widely used of the so-called "engineering" adhesives, with 55.8 percent of the market, followed by silicones (18.4 percent) and urethanes (8.8 percent). Other engineering adhesives were relatively small factors: anaerobics (2.7 percent), cyanoacrylates (1.2 percent), and modified acrylics (0.3 percent). The total market in 1978 was 59.1 million lb. In 1988, this figure is expected to be 131.5 million lb, with epoxies and silicones having essentially equal market shares (35.2 and 35.7 percent, respectively). Big gains are expected in the market shares of modified acrylics (to 11.4 percent), anaerobics (to 4.6 percent), and urethanes (to 11.9 percent).[138]

The major end uses for adhesives are in primary wood products (particleboard, plywood) (31 percent), packaging and converted paper products (19 percent), and construction (15 percent). Large growth is expected in hot melts,

---

*Poly(vinyl alcohol), poly(vinyl formal), and poly(vinyl butyral). The latter is the base of laminated safety glass.

Table 9.46 Major Resin Use in Adhesives, 1976 and 1977

	Million lb	
Resin	1976	1977
Phenolic[a]	706 (48.1%)	762 (47.1%)
Amino[b]	672 (45.8%)	758 (46.9%)
Poly(vinyl acetate)[c]	50 (3.4%)	56 (3.5%)
Epoxy	19 (1.3%)	18 (1.1%)
Poly(vinyl alcohol)[c]	18 (1.2%)	20 (1.2%)
Polyurethane (thermoplastic)	2 (0.1%)	3 (0.2%)

Source: Reference 27.

[a] Fiber-insulation binder, resin-bonded wood, coated/bonded abrasives, friction materials, foundry/shell moldings.
[b] Resin-bonded woods.
[c] Packaging usage only.

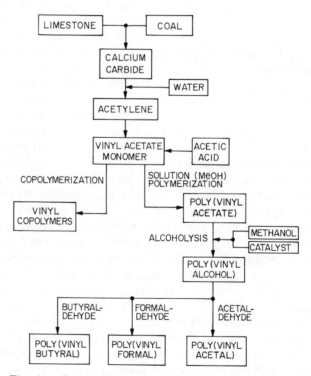

Figure 9.19 Flowchart for poly(vinyl acetate) and derivatives for use in adhesives. See also Fig. 9.2.

pressure sensitives, and water-base and high-performance (structural and heat-resistant) adhesives.[131]

There is every reason to believe that as advances continue to be made in the technology of applying and employing adhesives, use of these materials will continue to grow at a pace exceeding the growth of the economy as a whole. In 1965, the industrial fastener market (nuts, bolts, rivets, etc.) amounted to $1.5 billion and the adhesives and sealants market was about $0.56 billion, or 37 percent of that. By 1976, the fastener market had risen to $2.6 billion (an increase of 73 percent) and adhesives/sealants had grown to nearly $1.2 billion, or 45 percent of the fastener market, an increase of 114 percent over the same period.[128] This bodes well for the future of synthetic resin and elastomer adhesives and applies additional pressures on fossil feedstocks. These products, however, benefit from an even higher value-added advantage than do plastics. They are materials used in a thin film rather than as a bulky mass, hence much can be achieved with a relatively small volume of material. There should be, at worst, only spot shortages of specific compounding ingredients as long as fossil feedstocks are available for petrochemicals.

ALTERNATIVE RAW MATERIALS

It is generally felt that today's feedstocks will remain virtually unchanged in the foreseeable future, but the raw materials for these feedstocks probably will be different, with biomass most likely to become the ultimate universal source of organic chemicals. For the present, however, the economic incentive for making such a change is lacking.[139]

Tables 9.47 through 9.49 show the amounts of lignocellulose feedstocks that would be required to produce all major synthetic plastics, fibers, and elastomers at 1978 levels. Slightly over 68 million tons of lignocellulose would be needed to provide the feed for all but a few of these products. If lignocellulose output could be reserved exclusively for feedstock uses, there would be little problem. Many other essential demands, however, will continue to be made on these resources in the future. Moreover, the conversion technology based on fossil resources is already well established, whereas lignocellulose conversion technology, although conceptually feasible, would require further development to make it a commercial reality.

Perhaps as early as the 1980s, coal is likely to be the raw material of choice for production of synthesis gas, but probably not for other chemicals.[139] Approximately 10 percent of the world's organic chemicals were derived from coal in the early 1970s, and this quantity is likely to continue growing. Unfortunately, coal as an alternative source of feedstocks poses some technical problems. A key fact of life is that no two coals are identical—not from the same coal rank and often not even from a few feet apart in the same seam. The mix and quantity of conversion products obtained are bound to vary with the blend of input coals, with their reaction in the equipment, and with the process rate.[141]

Table 9.47 Lignocellulose Feedstock Requirements for U.S. Production of Synthetic Plastics, 1978

Resin	10^3 Tons	
	1978 Production	Lignocellulose Feedstock Required
Thermosets		
Epoxy	150	426 (L)[a]
Polyester	605	1,622 (L)
Urea	506	n.a.
Melamine	100	n.a.
Phenolic and other tar acid resins	810	2,316 (L)
Thermoplastics		
Polyethylene		
LDPE	3,555	14,200 (C)[b]
HDPE	2,100	8,400 (C)
Polypropylene and copolymers	1,535	6,140 (C)
Styrene and copolymers	2,540	7,549 (L)
Poly(vinyl chloride) and copolymers	2,860	4,983 (C)
Subtotals		11,913 (L)
		33,723 (C)
Total	14,761[c]	45,636 (LC)[d]

Source: Reference 140.

[a] Lignin-derived.

[b] Cellulose-derived.

[c] Less urea and melamine.

[d] Lignocellulose.

Even so, coal, as a raw material, might be used in a number of ways: (1) regional gasification to medium-Btu gas, then manufacture of SNG or synthesis gas for use as chemical feeds; (2) production of SNG for fuel with the excess capacity being applied to the production of synthesis gas for feedstocks; or (3) gasification of coal in dedicated chemical plants for ammonia or methanol. In any case, capital and operating costs would be high, because of the complexities of the process,[139] and overall efficiencies would be low.

A less plentiful source of feedstocks could be "mined" from our solid waste stream. Plastics along constitute about 3.4 percent of MSW by EPA estimates. In less than a year's time, packaging plastics, representing over 13 billion lb per yr, end up as scrap.[142] Conceptually, this scrap can be converted to feedstocks by pyrolysis at 1365°F (740°C) in a mixed bed under an inert atmosphere. The product stream can yield over 95 percent of the polymer input, in the form of aliphatic and aromatic hydrocarbons, largely reflecting the nature of the polymer feed.[143]

Table 9.48 Lignocellulose Requirements for U.S. Production of Synthetic Fibers, 1978

| Fibers | 10^3 Tons | |
	1978 Production	Lignocellulose Feedstock Required
Cellulosic		
Rayon	300	n.a.
Acetate	155	n.a.
Noncellulosic		
Nylon	1,275	3,645 (L)[a]
Acrylic	365	730 (C)[b]
Polyester	1,900	5,092 (L)
Olefin	345	1,380 (C)
Subtotals	455 (C)[b]	2,110 (C)
	3,885 (NC)[c]	8,737 (L)
Total	4,340	10,847

Source: Reference 140.

[a]Lignin-derived.
[b]Cellulose-derived.
[c]Noncellulosic.

Table 9.49 Lignocellulose Requirements for U.S. Production of Synthetic Rubber, 1978

| Synthetic Rubber | 10^3 Tons | |
	1978 Production	Lignocellulose Feedstock Required
Styrene–butadiene	1,515	5,347 (C)[a]
		1,800 (L)[b]
Butyl	170	1,001 (C)
Polybutadiene	420	2,474 (C)
Ethylene–propylene	190	1,119 (C)
Neoprene and others	955	n.a.
Subtotals		9,941 (C)
		1,800 (L)
Total		11,741

Source: Reference 140.

[a]Cellulose-derived.
[b]Lignin-derived.

The main problem lies in extracting this polymer feed from the waste stream. In-plant recovery is obviously the most efficient, with extraction downstream being almost prohibitively costly. Sorting and transportation costs are particularly high. To make most effective reuse of this waste stream, separation into specific plastics would be necessary, since many polymers are incompatible. This is technically but not economically feasible at this time, other than for a few specific captive situations. Reuse of the plastics can save up to 95 percent of their energy content, over merely burning the plastic as a fuel, but the current abundance of virgin plastics reduces the incentive to pursue their reuse.[142]

These and other raw materials options (oil shale, aquaculture, energy crops, etc.) do exist for chemical feedstocks, but they remain far removed from reality at this time. The most difficult hurdle to clear is probably economics. Until investors have some assurance that an option can compete in the marketplace, in the face of price-controlled competition, and until they have some inkling of the impact federal regulation will have on their proposed new options, they are going to be loathe to commit the necessary capital, as well as the research and development effort.

REFERENCES

1   I. S. Shapiro, *Science* **202**(4365), 287–290 (Oct. 20, 1978).
2   *Bus. Week* (No. 2492), 44–47, 50, 54 (July 18, 1977).
3   R. S. Wishart, *Science* **199**(4329), 614–618 (Feb. 10, 1978).
4   P. Gibson, *Forbes* **122**(7), 35–38 (Oct. 2, 1978).
5   B. Commoner, *The Closing Circle*, Knopf, New York, 1971, pp. 158–163.
6   J. E. Guillet, *Plast. Eng.* **30**(8), 48–56 (Aug. 1974).
7   T. L. van Winkle et al., *Amer. Sci.* **66**(3), 280–290 (May–June 1978).
8   R. P. Timmerman, reported in *Mod. Text.* **58**(5), 22, 24, 25 (May 1977).
9   A. W. Taylor, *Chem. Ind.* (No. 1), 13–22 (Jan. 1, 1977).
10 *Mod. Plast.* **54**(6), 16 (June 1977).
11 J. Neth, *Plast. World* **36**(10), 21, 22 (Oct. 1978).
12 H. Wittcoff, *Chemtech* **7**(12), 754–759 (Dec. 1977).
13 *Chem. Eng. News* **56**(13), 29, 30 (Mar. 27, 1978).
14 SRI International; reported in *ibid.* **57**(16), 36, 37 (Apr. 16, 1979).
15 *Chem. Eng. News* **55**(48), 6 (Nov. 28, 1977).
16 *Ibid.* **54**(13), 6 (Mar. 29, 1976).
17 *Ibid.* **55**(41), 19–21 (Oct. 10, 1977).
18 *Ibid.* **56**(43), 12 (Oct. 23, 1978).
19 S. H. Wamsley; reported in reference 4.
20 *Chem. Eng. News* **52**(37), 7 (Sept. 16, 1974).
21 BuCensus; reported by I. Skeist, *Adhesives Age* **19**(4), 37–41 (Apr. 1976).
22 Exxon Chem. USA; reported in *Chem. Eng. News* **55**(37), 18 (Sept. 12, 1977).
23 A. D. Little, Inc., and BuMines; reported by E. V. Anderson, *Chem. Eng. News* **53**(34), 12, 15, 16 (Aug. 25, 1975).

24 J. R. Lambrix et al. (1969); reported by J. E. Wallace, "Ethylene," in D. M. Considine, Ed., *Chemical and Process Technology Encyclopedia*, McGraw-Hill, New York, 1974, pp. 429–437.

25 ChemSystems; reported in *Chem. Eng. News* **56**(43), 12 (Oct. 23, 1978).

26 B. F. Greek and W. F. Fallwell, *Chem. Eng. News* **56**(47), 12–17 (Nov. 20, 1978).

27 *Mod. Plast.* **55**(1), 41–53 (Jan. 1978).

28 *Chemical Origins and Markets*, SRI International; reported in *Facts and Figures of the Plastics Industry*, 1976 ed., Soc. Plast. Ind., New York, Sept. 1976, p. 17.

29 P. Collinswood, "Ethylene in Europe," AIChE, 68th Annu. Meet., Houston, Mar. 1971.

30 *Chem. Eng. News* **55**(49), 21, 22 (Dec. 5, 1977).

31 *Bus. Week* (No. 2512), 44L (Dec. 5, 1977).

32 M. J. Millenson, *Energy User News* **2**(47), 1, 10 (Nov. 28, 1977).

33 *Plast. World* **35**(13), 12, 13 (Dec. 1977).

34 *Chem. Eng. News* **57**(16), 7 (Apr. 16, 1979).

35 "On the Trail of New Fuels—Alternative Fuels for Motor Vehicles," Bundesminist. Forsch. Technol., Bonn, 1974.

36 A. J. Dahl, *Public Util. Fortn.* **98**(9), 24–32 (Oct. 21, 1976).

37 H. Enzer et al., "Energy Perspectives"; reported in *Energy Facts II*, Sci. Policy Res. Div., Libr. Congress, Washington, DC, Aug. 1975, p. 53.

38 J. Berry, *Forbes* **122**(5), 32, 33 (Sept. 4, 1978).

39 B. F. Greek and W. F. Fallwell, *Chem. Eng. News* **54**(47), 11–18 (Nov. 15, 1976).

40 *Ibid.* **55**(45), 12 (Nov. 7, 1977).

41 B. F. Greek, *Chem. Eng. News* **54**(7), 12–14 (Feb. 16, 1976).

42 *Plast. Eng.* **XXXIV**(9), 16–23 (Sept. 1978).

43 Reference 28, p. 18.

44 B. F. Greek and W. F. Fallwell, *Chem. Eng. News* **56**(36), 10–16 (Sept. 4, 1978).

45 R. R. MacBride, *Mod. Plast.* **51**(5), 44–46 (May 1974).

46 B. F. Greek, *Chem. Eng. News* **53**(32), 8–10 (Aug. 11, 1975).

47 *Chem. Eng. News* **53**(51), 10 (Dec. 22, 1975).

48 B. F. Greek, *ibid.* **56**(30), 10–13 (July 24, 1978).

49 B. F. Greek and W. F. Fallwell, *Chem. Eng. News* **55**(8), 12, 13 (Feb. 21, 1977).

50 *Ibid.* **56**(6), 8, 9 (Feb. 6, 1978).

51 *Chem. Eng. News* **56**(12), 20–22 (Mar. 20, 1978).

52 W. J. Storck, *ibid.* **56**(18), 31–41 (May 1, 1978).

53 NATO; reported in *Chem. Eng. News* **55**(43), 17, 18 (Oct. 24, 1977).

54 *Mod. Plast.* **55**(9), 60 (Sept. 1978).

55 *Ibid.* **52**(4), 44, 45 (Apr. 1975).

56 E. V. Anderson, *Chem. Eng. News* **53**(26), 10, 11 (June 30, 1975).

57 *Plast. World* **37**(3), 68, 69 (Mar. 1979).

58 E. V. Anderson, *Chem. Eng. News* **55**(44), 8, 9 (Oct. 31, 1977).

59 U.S. Tariff Comm.; reported in *Chem. Eng. News* **52**(27), 14, 16 (July 8, 1974).

60 Monsanto; reported in *Chem. Eng. News* **56**(50), 12, 13 (Dec. 11, 1978).

61 *Chem. Eng. News* **56**(24), 48–50 (June 12, 1978).

62 *Ibid.* **55**(23), 42–44 (June 6, 1977).

63 *Ibid.* **54**(24), 36–38 (June 7, 1976).

64 *Ibid.* **53**(22), 32–34 (June 2, 1975).

65 *Ibid.* **52**(22), 24–27 (June 3, 1974).

66 *Ibid.* **51**(23), 12–14 (June 4, 1973).

67 *Ibid.* **55**(45), 21 (Nov. 7, 1977).

68 Reference 28, p. 22.

69 N. Rosato, *Plast. World* **34**(1), 12 (Jan. 19, 1976).

70 F. Knight; reported by R. R. MacBride, *Mod. Plast.* **52**(5), 16, 18 (May 1975).

71 Reference 34, pp. 11–12.

72 Reference 28, p. 20.

73 Shell Chem. Co.; reported in reference 60, pp. 12–13.

74 W. C. Bowman, *Chem. Eng. News* **53**(13), 10 (Mar. 31, 1975).

75 Foster Grant; reported in *Chem. Eng. News* **55**(14), 11 (Apr. 4, 1977).

76 *Mod. Plast.* **53**(1), 38–51 (Jan. 1976).

77 Monsanto; reported in *Chem. Eng. News* **56**(39), 15, 16 (Sept. 25, 1978).

78 B. F. Greek and W. F. Fallwell, *Chem. Eng. News* **56**(44), 10, 12, 14 (Oct. 30, 1978).

79 B. F. Greek and W. F. Fallwell, *Chem. Eng. News* **55**(5), 8, 12 (Jan. 31, 1977).

80 B. F. Greek, *Chem. Eng. News* **53**(39), 11–13 (Sept. 29, 1975).

81 B. Golding, *Polymers and Resins,* Van Nostrand, Princeton, NJ, 1959, pp. 283–289.

82 B. F. Greek and W. F. Fallwell, *Chem. Eng. News* **55**(49), 10–14 (Dec. 5, 1977).

83 *Ibid.* **54**(18), 8, 9 (Apr. 26, 1976).

84 W. J. Storck and W. F. Fallwell, *Chem. Eng. News* **56**(49), 8–12 (Dec. 4, 1978).

85 E. V. Anderson, *Chem. Eng. News* **57**(7), 9, 10 (Apr. 23, 1979).

86 W. F. Fallwell, *Chem. Eng. News* **54**(5), 8–10 (Feb. 2, 1976).

87 A. S. Wood, *Mod. Plast.* **52**(1), 59, 60 (Jan. 1975).

88 M. J. Sterba, "Petrochemical Complex," in D. M. Considine, Ed., *Chemical and Process Technology Encyclopedia,* McGraw-Hill, New York, 1974, p. 846.

89 F. W. Tortolano, *Plast. World* **33**(10), 51–56 (Oct. 20, 1975).

90 *Chem. Eng. News* **57**(6), 12 (Feb. 5, 1979).

91 *Mod. Plast.* **54**(7), 50, 52 (July 1977).

92 *Ibid.* **54**(11), 50, 51 (Nov. 1977).

93 *Plast. Technol.* **24**(3), 42 (Mar. 1978).

94 *Ibid.* **25**(2), 131 (Feb. 1979).

95 *Mod. Plast.* **54**(7), 56–58 (July 1977).

96 Business Communications Co.; reported in *Chem. Eng. News* **56**(5), 11 (Jan. 30, 1978).

97 Reinforced Plast./Composites Inst.; reported in *Chem. Eng. News* **56**(7), 11, 12 (Feb. 13, 1978).

98 H. Lee and K. Neville, *Handbook of Epoxy Resins,* McGraw-Hill, New York, 1967, pp. 2-3 to 2-5.

99 J. H. Saunders and K. C. Frisch, *Polyurethanes Chemistry and Technology,* Pt. 1, Interscience, New York, 1962, pp. 21, 34, 47, 50.

100 A. S. Wood, *Mod. Plast.* **51**(12), 60–62 (Dec. 1974).

101 *Chem. Eng. News* **52**(36), 7, 8 (Sept. 9, 1974).

102 *Ibid.* **56**(37), 7 (Sept. 11, 1978).

103 Int. Isocyanate Inst., in reference 102.

104 *Mod. Plast.* **54**(10), 44–48 (Oct. 1977).

105 A. S. Wood, *Mod. Plast.* **52**(2), 52, 53 (Feb. 1975).

106 B. F. Greek and W. F. Fallwell, *Chem. Eng. News* **54**(42), 11, 12 (Oct. 11, 1976).

107 *Chem. Eng. News* **53**(50), 20, 21 (Dec. 15, 1975).

108 *Ibid.* **57**(17), 8 (Apr. 23, 1978).

109 *Chemtech* **8**(12), 745 (Dec. 1978).

110 *Chem. Eng. News* **56**(35), 8 (Aug. 28, 1978).

111 D. M. Kiefer, *ibid.* **47**(53), 31-48, 53-57 (Dec. 22, 1969).

112 S. Lauren, *Preprints,* ACS, Div. of OCPL **36**(1), 198-201 (1976); 171st Meet., New York, Apr. 5-9, 1976.

113 DOC; reported in *Chem. Eng. News* **51**(23), 12-14 (June 4, 1973)—1970 and 1971 data.

114 L. H. Princen, *J. Coatings Tech.* **49**(635), 88-93 (Dec. 1977).

115 R. N. Price, *J. Coatings Tech.* **50**(641), 33-40 (June 1978).

116 W. Brushwell, *Coatings Update,* American Paint Journal Co., St. Louis, MO, 1974, p. iii.

117 A. N. Wright, Inst. Elec. Eng. Meet., Dec. 8, 1975.

118 M. A. Parrish, *J. Oil Colour Chem. Assoc.* **60**(12), 474-478 (1977).

119 W. L. Toth, *Insul./Circuits* **21**(13), 37-40 (Dec. 1975).

120 J. A. Mock, *Mater. Eng.* **89**(1), 50-53 (Jan. 1979).

121 C. J. Schroeder, *Mod. Plast.* **52**(12), 34-37 (Dec. 1975).

122 E. V. Anderson, *Chem. Eng. News* **57**(10), 8-11 (Mar. 5, 1979).

123 Int. Inst. Synth. Rubber Producers; reported in reference 122.

124 J. E. Marsland, ACS Rubber Div. Meet., Boston (1978); reported in *Chem. Eng. News* **56**(43), 9 (Oct. 23, 1978).

125 *Chem. Eng. News* **56**(43), 9 (Oct. 23, 1978).

126 E. V. Anderson, *ibid.* **56**(1), 10 (Jan. 2, 1978).

127 Predicasts, Inc.; reported in *Adhes. Age* **21**(10), 38 (Oct. 1978).

128 Forst & Sullivan, Inc.; reported in *ibid.* **20**(1), 40 (Jan. 1977).

129 Delphi Marketing Services, Inc.; reported in *ibid.* **20**(12), 51 (Dec. 1977).

130 V. G. Raevsky, *Adhes. Age* **19**(3), 23-29 (Mar. 1976).

131 J. B. Toogood, *ibid.* **19**(1), 37-39 (Jan. 1976).

132 V. G. Raevsky, *ibid.* **19**(2), 17-26 (Feb. 1976).

133 *1972 Census of Manufacturers,* DOC, Washington, DC, 1972, 28H-13.

134 *Adhesives Red Book,* 7th ed., Palmerton, New York, 1974-75.

135 D. Kuespert, *Adhes. Age* **18**(1), 26-29 (Jan. 1975).

136 H. S. Holappa; reported in *ibid.* **19**(1), 24-29 (Jan. 1976).

137 B. P. Barth, "Phenolic Resin Adhesives," in I. Skeist, Ed., *Handbook of Adhesives,* 2nd ed., Van Nostrand Reinhold, New York, 1977, pp. 382-416.

138 Springborn Labs; reported in *Adhes. Age* **22**(3), 43 (Mar. 1979).

139 AIChE study; reported in *Chem. Eng. News* **56**(48), 23 (Nov. 27, 1978).

140 Adapted from "Renewable Resources for Industrial Materials," CORRIM, Natl. Acad. Sci., Washington, DC, 1976, p. 202; updated from SPI and ITC production data for 1978, reported in *Chem. Eng. News* **57**(19), 25 (May 7, 1979).

141 H. C. Messman, *Chemtech* **5**(10), 618-621 (Oct. 1975).

142 R. H. Wehrenberg II, *Mater. Eng.* **89**(3), 34-39 (Mar. 1979).

143 W. Kaminsky; reported in *Chem. Eng. News* **53**(30), 17 (July 28, 1975).

10 THE ROAD AHEAD

[the energy crisis]. . .festers like a cancer, sapping away the basic strength of our nation.

PRESIDENT JIMMY CARTER

Of all the necessities of life, even in the most primitive of societies, there are none more essential than food and potable water. Human beings have survived in some climates without shelter, heat, or clothing, but never without food and drink. The world's population is growing at a rate that threatens to outreach food supplies. To make matters worse, modern agriculture has become highly dependent on fossil fuels—for energy and for fertilizers—hence the people problem further compounds the energy problem. To magnify this problem still more, energy distribution, geographically and politically, does not coincide with energy need. Most inherently energy-rich areas are very thinly populated, and some of the most densely populated areas are virtually without energy resources other than marginal quantities of wood and manure.

Furthermore, consumption of fossil fuels inevitably leads to "fouling of the nest." Debris from extraction, transport, and combustion of these fuels despoils the landscape and poisons the soil and the waterways with a growing accumulation of hazardous wastes. Gaseous products of combustion—oxides of carbon, nitrogen, and sulfur, in particular—poison the air we breathe and threaten our health and the long-term stability of our climate. There is really little to commend such fuels, except that they are available (albeit in questionable and poorly distributed quantities) and we have the technology to exploit them. This exploitation has been so extensive that we now find ourselves facing an energy crisis, our second in less than a decade—both orchestrated by OPEC.

In considering our crisis, there are several factors one should keep in mind[1]: (1) Earth is a finite, closed system in a material sense, but it is open to the flow of energy. (2) It is an interlinked system in which whatever we do may affect our neighbors. (3) There is no free lunch; we "pay" for our energy in one way or another. (4) In the same sense, when we use material resources, we ultimately disperse them to a point where they can no longer be reclaimed economically. (5) Destructive processes are more rapid than constructive ones; we are exhausting, in little more than a century, the fossil fuels that nature required millions of years to form. (6) Natural homeostatic systems are finite; we cannot abuse our environment indefinitely without causing irreversible harm to it.

THE ALTERNATIVES

Perhaps energy "dilemma" is a more meaningful term for the energy crisis. Energy dilemma suggests a predicament whereby we are faced with equally unsatisfactory alternatives. Certainly, we are confronted with a *practical* shortage of fluid fossil fuels—but it's a "local" shortage and it is controlled by the price we are able to pay and the willingness of the political entities owning the supplies to divest themselves of enough to meet our demands. Ultimately, there will be a *real* physical shortage, when we can no longer afford to extract the desired fluids from the earth or our technology is no longer able to respond to our insatiable appetite for them.

One of the unsatisfactory alternatives to this situation entails substitution of synfuels—synthetic crude oils (SCO) and gases (SNG)—produced from coal, oil shale, or tar sands. We have already seen in Chapters 4 through 6 the unsatisfactory environmental and social consequences of wide dependency on these energy sources. No thinking person can look forward, with any pleasure, to the time when these raw materials, with their inherent problems of water consumption and pollution, health hazards in both mining and combustion, transport, manpower, waste disposal, and others, will be called upon to replace dwindling supplies of more conventional fossil fluids.

Yet another unsatisfactory energy alternative is conversion of biomass to forms usable as fuels or feedstocks. On the surface, this appears to be a benign, ideal solution to our energy dilemma. The raw material is infinitely renewable, with the aid of the sun, and the conversion can be relatively free of environmental consequences.

What, then, of the negative side to biomass energy? Foremost on the negative side of biomass energy is the necessity of competing with food, feed, and fiber for arable land. As human beings have squandered their fossil fuel heritage, they have also been profligate in their desecration of the soil. Through poor cultivation practices, people have allowed the topsoil to be washed down streams and out to sea at a much faster rate than it can be replenished. This is happening in the face of a still-growing population and continuing removal of arable land for urbanization, parking lots, and superhighways. Perhaps just as serious is the fact that cultivation of biomass, on a scale sufficient to meet a significant portion of our energy needs, would require large amounts of energy-rich fertilizers but more specifically, large amounts of phosphorus, an essential element in plant physiology. It is one of our least plentiful, essential, nonrenewable resources. Once it has been applied to the soil, much of it ultimately is washed out to sea or otherwise dispersed in such a manner that it is lost to our use until the next geological uplift. Some, of course, can be recovered in the process of conversion to botanochemicals or during combustion of the biomass. The chemical form of phosphorus (and other minerals) recovered from thermal processes can be such, however, that recovery in a usable state is not practical.

It would appear that the solution to our energy dilemma, particularly with regard to providing fuel for personal transportation, has not yet been conceived. This need poses some difficult problems. To meet the demands of current internal combustion engines, the fuel must be liquid; have high energy content; be noncorrosive; be readily transportable, storable, and transferable; and be non-polluting, within limits of current and planned environmental guidelines. Furthermore, it must be available in huge, continuing quantities at a price the general public can afford to pay. This fuel does not now exist, and there is some question of whether it will exist in the lifetimes of those now living.

In the minds of most thinking people, there is urgent need for us to take major steps today to conserve all possible fluid fossil fuels, to reduce the drain on our nonrenewable resources, to minimize our dependency on an unstable cartel, to buy time to decide where we go from here, to develop the policies and technologies to solve our dilemma, to reduce the cost of our energy, and to shore up our national security. This course is most important for us, since we have become increasingly dependent on insecure foreign supplies of oil, and discoveries of new domestic natural gas and crude oil are trailing extraction rates. The need for conservation is further highlighted by recent curtailments of oil shipments from Iran, and by production cutbacks and continuing price increases by our other Organization of Petroleum Exporting Countries (OPEC) suppliers. Even our Western allies have been excoriating us for our failure to reduce consumption to something approaching their own per capita levels.

It has been estimated that 40 percent of the difference in energy-to-GDP ratios between the United States and Western Europe is due to "compositional and structural features." Americans historically favor single-family dwellings, have a more disperse population, and cover a much larger area, making transport of both people and goods unavoidably more energy costly. The remaining 60 percent difference arises largely from relative energy efficiencies of motor transport and manufacturing processes. The effects of these differences have been magnified by the high fuel taxes abroad and the regulated fuel prices in the United States; hence the domestic practices and technology are perhaps less wasteful than they appear on the surface.[2]

Industry has been the one bright spot in our energy conservation picture.[3] Out of concern for the future stability of their energy supplies and the necessity for offsetting rising fuel prices, insofar as possible, industry has been finding and implementing means of recycling waste heat, insulating process facilities, and otherwise accomplishing their production objectives by less-energy-costly methods.

Nearly 40 percent of all our crude oil—domestic and foreign—goes into the production of gasoline.[4] But the recent, initial decrease in our overall demand for petroleum has resulted largely from industry's efforts.* Conservation of

---

*The annual growth rate in 1977 was 5 percent. In 1978, it was down to 2 percent, despite increasing demands for its major product, gasoline.[3]

energy in our transportation sector did not keep up with industry, despite mandated higher mileage performance and increasing gasoline prices.* One problem arose from the fact that only passenger cars and light trucks came under this regulation. Furthermore, about 20 million drivers and 24 million vehicles had been added to the domestic scene between 1973 and mid-1979. But a big factor is that it takes 5 to 10 percent more crude oil to produce unleaded gasoline than it does leaded gasoline, and by mid-1979, unleaded gasoline represented about 40 percent of all gasoline sold in the United States.[4] Department of Energy estimates made in mid-1979 were that by 1982, demand would peak and start to decline, as the old (pre-1975) gas-guzzlers headed for the junkyards. This estimate presumably also took into consideration the fact that essentially all vehicles on the nation's roads then would be using unleaded gas, thus requiring an across-the-board 5 to 10 percent increase in crude feeds (above the levels required for leaded gas).

In mid-1980, weekly crude oil and petroleum products import data of the American Petroleum Institute (API) showed a decline of about 23 percent over the same period in 1979. This was a consequence of conservation of both gasoline and heating oil because of high prices and the recession.

DECLINE FROM POWER

Perhaps the best way to gain some insight into the thinking of our chief protagonists in the energy arena is to listen to one of their representatives. Ali M. Jaidah, secretary-general of OPEC in 1978, noted that prior to the 1973 embargo, the Seven Sisters were in full control of production and pricing of OPEC oil. As a result, the United States, as home base for five of the companies, was in a position to receive favorable treatment in the acquisition of oil. The events of 1973–1974 brought us face to face with the fact that a shift was occurring in international power. OPEC did three things to show this: they allocated their oil, they fixed prices without the consent of the international oil companies, and they started in motion an action that resulted in ownership of Mideast oil reverting to them. Our responses [President Ford's 1975 reaffirmation of Nixon's Project Independence and President Carter's proposed National Energy Plan (NEP), submitted to Congress in April 1977] disappointed OPEC.[6]

With our high energy consumption rate per capita, we have much potential for conservation,† yet neither plan pressured industry enough for conservation, for fear it might suppress economic growth and employment. In the area of alternative-source development, the United States has 34 percent of the world's recoverable reserves of coal—a potential source of syncrude. The proposed National Energy Plan, however, suggested that this was not too likely to become a

---

*Gasoline demand in the first quarter of 1978 was up about 1.5 percent over the same period in 1977, and first quarter 1979 demand was up 4.3 percent over the same period in 1978,[4] or nearly 80 percent over the 1979 annual rate estimated in the first quarter of 1978.[5] December 1979 consumption was, however, down from December 1978.

†Domestic potential for conservation is estimated at 30 to 40 percent of energy budget, with no lowering of standard of living.[7] Global potential for conservation is about 46 percent.[6]

factor, because of environmental, legal, and technical constraints. Similar problems also limited the prospects for expanded use of nuclear electric power generation. The most serious flaw of the NEP, in the eyes of OPEC, was the fact that the policy was pointedly directed at encouraging development of non-OPEC sources of crude (Alaska, OCS, etc.), in an attempt to weaken OPEC's position.

Yet another element of the NEP that proved unpalatable to OPEC was the interference "with the natural forces of supply and demand in the oil market," by price controls and by our intent to amass large, strategic stockpiles of crude oil and petroleum products.*

Another factor that annoyed OPEC, was our determination to maintain our independence of action in the international community. Our foreign trade was only 10 percent of our gross national product in 1976 vs. more than 25 percent for West Germany in the same year. Some other nations import much higher percentages of their energy with no serious harm to their economies, in the eyes of OPEC.[6] The OPEC nations are becoming increasingly dependent on the industrialized nations for goods and services needed for their planned development objectives.† Our inability to stave off inflation has caused OPEC to suffer large financial losses, since the U.S. dollar is the currency in which oil is traded,‡ and in which OPEC holds many international investments.

Basically, OPEC seems to be saying we are not the power we once were and should accept the fact gracefully and assume a less preeminent position in world affairs. At the same time, we should get our own economy and energy situation in order before we "export" our inflation to the Mideast. They are suggesting that we must also learn to place our trust in OPEC and the rest of the world by increasing our international trade, particularly in the energy field. In the eyes of OPEC, our traditional devotion to independence (or "arrogance," depending on the viewpoint) must give way to increasing economic interdependence, in keeping with other nations, as exemplified by the Communist bloc, OPEC, the European Economic Community (EEC), and so forth. This could prove to be a bitter pill for us to swallow.

Representative James Wright has said: "Unless we reduce our reliance on oil, it is the beginning of the end of the U.S. as a major power."[8] Our national policies have accomplished remarkably little toward restoring any degree of our preembargo independence. Seemingly, we have done little or nothing concrete to prepare ourselves for a renewed embargo or for the possible closing of the Strait of Hormuz, that strategic passageway from the Persian Gulf to the Indian Ocean. The same oil barons who curtailed our supplies in 1973–1974 now pro-

---

*The oil industry accepts a six-week reserve as adequate for commercial operations. In general, policies of industrialized nations called for twice this, but NEP was aimed at a domestic crude oil stockpile of 1 billion bbl, or about a four-month allotment of our imported oil.

†About 75 percent from the OECD sector, with the United States supplying 18 percent.

‡As this book was going to press, it was rumored that OPEC would soon be abandoning the dollar and resorting to a "basket of currencies" to price their oil, an action that would lead to even higher OPEC oil prices for the United States.

vide 80 percent of our oil imports, a figure that is now about 45 percent of our total oil supply. Approximately 20 percent of our oil imports come from Saudi Arabia alone, a land of about 4 million people and the only petroleum source large enough to take up the slack if another supplier should fail to meet its commitment, for whatever reason. In the first quarter of 1979, their wells were producing about 10 million bpd. They need revenues from only half that amount to provide their necessary income. With their oil price up to $28 per bbl (mid-1980), they can sell half as much and still receive their necessary cash.[9] We would have no immediate recourse, and no hope of *beginning* to recover from the situation for another 5 to 10 years, the time required to start bringing shale and coal syncrudes on stream in commercial quantities.[10]

Perhaps the most distasteful aspect of the current energy situation is that we find ourselves in a position where we are pressured to erode our independence and alter our historical conduct, merely to maintain communications with those nations controlling available foreign oil supplies. Their position should be understandable, if not acceptable to us. With our past wealth in both power and natural resources, we have been in a position to call the tune. What most of us fail to realize is that we are being ousted from the driver's seat. In the foreseeable future, we could find ourselves on the verge of becoming a has-been world power, as we empty our coffers to buy overpriced foreign oil to assuage our gluttonous energy appetite.

There is no alternative. We must allocate our fluid fossil fuels to essential, high-priority uses, and we must reduce their consumption. They should be replaced, insofar as possible, by more plentiful energy resources over which we have some control. This may mean altering our life-style, but it does not mean destroying it. It merely involves eliminating unessential or wasteful consumption of our scarce fluid fossil fuels.

WHAT SHORTAGES?

The man-in-the-street is still not fully convinced there is a fossil fuel crisis; hence he sees no pressing need to concern himself with conservation, except to save money. Recent polls have indicated that he also distrusts the government and big business—particularly the energy industry. Lacking the full support of their constituencies, our legislators lack the motivation needed to draft effective legislation that threatens to cost more or to otherwise inconvenience the voter. The energy industry needs high profits to invest in expanded exploration and the development of new technologies. Labor is inclined to see any form of conservation as a threat to workers, and the economist views it as a harbinger of a shrinking economy.* With so many vested interests adding to the confusion, it

*A National Academy of Sciences (NAS) energy conservation study, directed by J. Gibbons (OTA), has concluded that the United States could double its economic output by 2010 while reducing energy consumption 25 percent through the use of proven energy conservation practices. Industry did, in fact, reduce its energy consumption about 10 percent (or 1 Q) between 1973 and 1978 while *increasing* its output of goods and services by 12 percent, according to this study.

is no wonder that a matter touted as "the moral equivalent of war" has failed to stir the nation to action.

To most people, "shortage" implies "today"—at the gas station or in the supermarket. They do not look ahead to the thermostat next winter or the gas pump next summer. There is no "tomorrow" until they stand at its threshhold. They cannot be convinced that the way they order their life-styles today is going to have a severe impact on generations that follow. They forget that the fossil fuels they use unnecessarily today cannot be used by future generations. The result is continuing waste of limited and crucial resources. There are still only limited incentives to switch to energy-efficient automobiles, appliances, or space conditioners, or to energy-saving life-styles. The only real incentive—higher energy prices—is just beginning to have an effect. There appears to be an assumption abroad that if we *really* run out, some new technology will be waiting in the wings to take over from petroleum.

Some reports of energy experts have noted that there is enough crude oil in the world to last us for several hundred years. This is probably a true statement and, since it is reassuring, it is the statement that makes headlines and molds opinions of the average citizen. What it fails to say is that the poor accessibility of the oil and the limitations of our present technology will prevent us from extracting most of it at a cost the consumer can afford to pay. In other words, there is no *real physical* shortage of fluid fossil fuels, at this time, but there is a *practical* shortage that we cannot avoid for the foreseeable future—and there is no "miracle" solution to this dilemma.

There is no question that our earth and everything in it are finite. It holds a limited amount of accessible minerals and fossilized hydrocarbons. It has enough arable land to support a limited amount of both vegetation and animals of all species. It has a limited amount of fresh water—and not always where it is most needed. The question is: How much can we continue to expand in terms of food and fiber output, mineral production, and economic strength?

Common sense tells us that we can obtain our nonrenewable resources from only a very limited fraction of the earth, even if we are willing and able to pay astronomical prices for them. We cannot yet plumb the deep-ocean trenches, the polar caps, the mountain ranges, or many deserts, for minerals and fossil fuels, or if we have the technology to do so, economics and common sense tell us we shouldn't. Even accessible land masses yield to only limited economic recovery of these necessary materials, possibly 30 percent for secondary recovery of crude oil and 50 percent for deep-mined coal. Many pockets in accessible regions have to be left untouched because they cannot be exploited economically or simply because they underlie cities.

All the world's most erudite scholars, scientists, engineers, and economists cannot tell us how much specific fossil fuels or minerals exist in our planet. It is known that earth comprises a relatively self-sufficient ball of matter 7900 mi in diameter, with about 70 percent of its surface covered by water. Working only with such facts, it has been calculated that if earth were, in effect, a ball of crude oil of the same diameter and we were to start using it today at 1977 consumption levels, plus the 5 percent annual demand growth rate experienced as recently as 1977,[3] this hypothetical earth would cease to exist in 500 years[11]—about the same length of time ahead as the time past when Columbus made his momentous voyage. But that future day of reckoning becomes even more real when we consider that our earth is not *all* oil. Indeed, it is largely mineral and water, with a sprinkling of living and fossil vegetation and animal matter. The fossil matter is, in fact, a miniscule fraction of our earth. Even if we were to reduce our annual oil consumption growth rate to a 2 or 3 percent average, that still smaller amount of *accessible* and *extractable* fossil matter would be consumed in *much less* than the five centuries called for in our limiting example.

Whatever the time frame, there *is* a real energy crisis somewhere ahead. As earth's present caretakers, it is our privilege and our duty to deal with the situation in as intelligent and thoughtful a manner as we can muster. We must discipline ourselves to husband earth's limited nonrenewable resources—whether they be chromium, phosphorus, hydrocarbon fuels, or whatever—as if they were priceless family heirlooms. We must give thoughtful consideration to the "why" and the "how much" of our demands on these materials before we commit ourselves irreversibly to their use. We must ask ourselves first whether we are committing these materials to the most essential and efficient use. Where this is not the case, we should try to substitute a material based on a renewable resource.

Early depletion of our resources merely hastens the day when our destinies must succumb to the embrace of unproven alliances. If our trust proves to be poorly placed, we could be leading future generations to the sacrificial altar.

We must be concerned not only for those living today, but also for the security of future generations. There is no reason to expect that we would receive more than token sympathy from many of our present allies abroad if we were to fall from economic and military power because of an energy shortage. Yet there are many political, business, economic, and military leaders who say we could become a "paper tiger" as a result of the energy issue.

POLICY

One of the most serious needs we face today is able leadership. Depending on one's own particular political convictions, few of us have not had reason to be concerned, at one time or other, with the slowness and the vacillation with which our energy policy has been forming.

One damaging effect of our current indecision on energy policy is that now some people are probably even more firmly convinced that there is no energy shortage, other than the one conjured up in smoke-filled boardrooms. The recent price controls on domestic oil and natural gas could only reinforce their belief that shortages were a figment. In the meantime, the business world—petroleum industry and financial sector alike—are left to wonder which way the political winds will blow tomorrow. Congress knows it will take five to seven or more years to develop most new energy-related technologies to a point where they are ready to begin producing on a small commercial scale. Yet we still lacked concrete plans for such developments, on any significant level, until June 1980, when the National Security Act of 1980 was enacted.

Perhaps as important as leadership is the need for stability of objectives and policies, so that we can achieve credibility and so that the business world can enter into energy commitments with some assurance that the ground rules will not be changed after costly decisions have been effected.

In truth, we cannot say that we have *no* domestic energy policy. We have had, for many years, a pervasive, long-standing policy based on price-controlled "cheap energy." Unfortunately, price controls have prevented market prices from signaling to the consumer that because a product such as natural gas was cheap, it was not abundant. The low price has made it difficult for the average citizen to be convinced that the lifetime of natural gas as a major fuel is winding down. Furthermore, the price differential between natural gas and fuel oil, on an energy basis, has long been unfair to consumers of the latter.[12] But this is now changing, with price controls being phased out.

Phase One of President Carter's National Energy Plan (NEP), proposed on April 20, 1977, was aimed at conservation and direct combustion of coal. Phase Two (NEP-2) was drafted through April 1978, when NEP-1 stalled in Congress. The plan, finally passed by Congress on October 15, 1978, consisted of five acts which form the nucleus of President Carter's National Energy Plan: The Public Utility Regulatory Policies Act of 1978 (concerning the cost effectiveness of various alternatives), the Powerplant and Industrial Fuel Use Act of 1978 (including conversion-to-coal measures), the National Energy Conservation Policy Act of 1978 (directed at conservation in homes), the Natural Gas Policy Act of 1978 (aimed eventually at decontrolling natural gas prices), and the Energy Tax Act of 1978 (covering various unrelated relief measures). Taken together, the Department of Energy has estimated that these measures will save 2.39 to 2.95 million bpd of crude oil, with natural gas price decontrol accounting for approximately 45 percent of the saving.[13]

Overall, our energy policy still fails to deal with the major defects of past policies—it is still founded on regulation amounting to paralysis. Before construction can begin on a coal mine, up to 100 permits must be obtained. In recent years, the lead time had doubled, from 5 to 10 years. The delays in bringing coal-fired power plants on line is now approaching the time required for nuclear plants—10 to 12 years. Regulatory conditions (together with funding problems) have slowed development of shale oil to a crawl. As recently as June 1980, Congress refused to enact a "fast-track" bill that would set up an Energy Mobilization Board, empowered to waive some federal, state, and local regulations which threatened to stall large energy projects. The major reasons for the failure of this bill were environmental and states' rights issues, the wisdom of which must await the test of time.

There is a considerable body of evidence to suggest that "the energy problem starts with overregulation."[14] Some regulations have been contradictory, counterproductive, capricious, and plain confusing. They have misguided consumers by failing to feed back warnings of trouble ahead, and they have inhibited commitments by producers to respond to needs of the energy situation.

The political aspects of the energy scene have broad ramifications. The oil embargo of 1973 to 1974 was brought to bear against us by the Islamic oil-producing nations because of our support of the Israelis in their earlier conflict. The action was only partially effective because OPEC was not united. In the early months of 1979, a number of situations began to come to a head: the successful Iran revolt, displacing an oppressive, non-Moslem, pro-American government with an ultraconservative Moslem government inimical to the United States; a threatening double-digit U.S. inflation; the worst balance of payments in our history; increasing U.S. dependency on OPEC oil; worsening U.S. diplomatic relations with Mexico, a potential oil competitor of OPEC; a peace treaty signing between Egypt and Israel, with the participation of the United States but with the opposition of the major oil-producing nations; and now an Iraq-Iran war that threatens the stability of the entire Persian Gulf.

Now we find that control of the world's lifeblood—oil—has passed to the hands of "a small, noncompetitive and irresponsible group ... the OPEC cartel, who are at the moment wholly beyond our control."[15] Levy, a respected petroleum consultant, adds; "OPEC is not a normal cartel, and a confrontation is unlikely to work."[16] They can afford to do without oil revenues from us for some months, but we can afford to do without their oil (8 million bpd) for only a very short time. "We will never have any influence on OPEC until we develop a positive energy policy and a new generation of energy technology to replace oil,"[17] according to Sawhill, president of NYU and a former FEA administrator.

Because of the 1979 increases in OPEC oil prices, the world's purchasing power was expected to be drained of an additional $35 billion, only half of which would result in additional OPEC oil.[16] The ones affected most adversely were the world's billion or so subsistence-level people. The less-develped countries throughout the world already had built up a $210 billion cumulative indebtedness to pay for oil, with 1979 only in its first quarter.[3]

Five leading independent energy experts and economists* appear to agree on several critical aspects of the energy situation. The oil problem or crisis *is* real. We are entirely at the mercy of the OPEC cartel, and we have no alternative to fall back on at this time. Higher energy prices (i.e., decontrol) are necessary to convince people that we have a problem, encourage more efficient use of energy, and provide an incentive for increased exploration for new sources. Finally, we need to pull out all the stops to develop safe nuclear energy, low-sulfur coal, new sources of natural gas, and whatever other sources of energy offer any promise whatsoever.[18]

*M. Adelman (professor of economics, MIT), W. Levy (dean of petroleum consultants), J. Lichtblau (head of PIRF), A. Safer (Irving Trust economist), and J. Sawhill (president of NYU and former FEA administrator).

THE PRICE TAG

In the eyes of one concerned and respected observer, we are headed for a 26-Q energy shortfall by the year 2000.* To put this in proper perspective, the 1973 oil embargo produced a shortage of only *one* quad and precipitated our worst recession in over 40 years.[20]† There are three potential major options toward meeting this shortfall. We can begin immediately to build 300 to 400 new large nuclear power plants costing no less than $1 billion each, with transmission and distribution systems adding another 50 percent. Alternatively, we can develop 600 to 800 new, massive coal mines, and provide the necessary transportation systems (barges and waterways; pipelines; hundreds of thousands of hopper cars, locomotives to haul them, suitable rights-of-way; etc). Coal conversion plants and coal-fired power plants with all their pollution control equipment would also have to be provided. Coal, our most probable major source of energy in our midterm future, poses some very serious problems which are far from solution. Its development promises to be inhibited or disrupted by overregulation and labor unrest. Furthermore, major regulatory issues involve air pollution (particularly sulfur oxides), health and safety of extraction and combustion, and disturbance of the land by mining. These issues have been discussed in some detail in Chapter 4 and will not be treated further here. Both the nuclear and the coal routes pose unsatisfactory environmental, logistic, and societal problems. The only other alternative would appear to be intensive conservation efforts, with possible radical changes in our life-style.[19]

The energy dilemma has become more a socioeconomic problem than a technological one. Probably one of the major problems facing us is provision of enough capital to make sufficient energy available to meet our requirements. Annual capital needs for domestic energy, for the years 1976–1985, have been estimated, by consensus of a number of individual experts and agencies, at between $50 and $90 billion, with little consideration given to nonfossil resource development.[21] In another 1978 study, a total energy capital requirement of $397 billion for 1977 to 1982 was estimated. This would mean an average annual capital investment of $66 billion, but this study was based on an assumed 6 percent inflation rate.[22] With inflation reaching a double-digit level in 1979, it is difficult to estimate realistically what the capital needs will actually be, other than "high."

Contrary to popular belief, adoption of "soft energy" (i.e., solar-related—direct solar, winds, tides, etc) for electric power generation appears to offer no relief. One source has suggested that just to accommodate growth and replacement of retired electric power generating plants alone with "soft energy" generators would call for more capital than the nation creates.[23]

---

*Based on projections of a special National Academy of Sciences committee, and including conservation of 65 Q.[19]

†A quad is the energy equivalent of 7.5 billion gal of gasoline or enough to fuel 10 million automobiles for a year.[19]

The physical inputs to any economic system are labor, capital, and energy.[24] Because of our historical low energy costs, we have come to substitute energy for labor and capital.[23] This means that there have to be limits to our economic growth, since our nonrenewable energy forms are finite. But more important, it presages upsetting socioeconomic changes as we encounter higher energy costs and are forced to alter our strategy in dealing with these three major elements of our economic system.

Collectively, individuals control about a third of the entire U.S. energy budget. As individuals, our near-term handling of our energy dilemma could have a profound effect on the political, economic, and social structure of the world by the turn of the century.[21]

Conversion to unconventional sources (oil shale, biomass, MSW, and so forth) of fluid fuels offers no relief from the energy cost trend. These new synfuels probably will cost two or three times conventional fuels, since the former are more difficult to convert into useful forms. Since crude oil is now many times its cost when our economy growth pattern was established, it is likely that dependence on even-higher-priced synfuels would completely alter the picture, because of further escalation in the cost of energy. How we deal with this situation can have an overriding effect on our economic health.[25]

In the petrochemical industry, there is now considerable excess production capacity, and profits are marginal for many commodity chemicals. Producers are hesitating to invest in expansion of their own plants for these obvious reasons, and others as well. The cost of new plants and the interest rates are high, and the products must compete with those manufactured in older, depreciated facilities. Added to the cost of production facilities are the costs of meeting environmental and energy conservation regulations, and the rising costs of building sites. For these reasons, there is a growing trend toward mergers instead of new construction, with the result that available bank credit is being used up with little overall addition of productive capacity. In 1977 and 1978, acquisitions in the chemical industry took half the investment capital, compared with a fifth just three years earlier.[26]

In 1978, about $683 million was spent by the domestic chemical industry on pollution control. This was down from a total of about $765 million in 1976. Over half the outlay ($376 million) went for water pollution control and $260 million for air. The balance ($48 million) was for solid waste.[27]

All these drains on available capital are likely to lead to shortages of some key chemicals in the 1980s.[26]

FUTURE FEEDSTOCKS

We must expect that feedstocks for petrochemicals (plastics, fibers, elastomers, fertilizers, and other organic chemicals) will remain essentially unchanged for some time to come. It is reasonably assured that ethylene (as an example) will be made increasingly from heavy feeds (e.g., gas oil) rather than from natural gas, but both sources are nonrenewable fossil materials. The substitution of renewable resources, or even coal or shale oil, is clouded by the absence of economic incentives, at this time.

Coal is likely to provide increasing amounts of synthesis gas, in the 1980s, as crude oil and natural gas supplies become tighter and prices rise. But there is not too much hope in the near- to midterm that coal will become the feedstock for new families of chemical products now derived from oil or natural gas, except, possibly, for some aromatic chemicals. A major problem with coal as a feedstock, of course, is the inherent lack of uniformity in its composition and reactivity. Shale oil, with its relatively high concentrations of nitrogen and metals, which poison reaction catalysts, is even less likely than coal to become a significant chemical feedstock.

Economics is unquestionably the real key to alternative feedstocks. Conversion cost is only one consideration, but an important one. Environmental approvals, long lead times, costs of capital, and conversion efficiencies all play a role in the economics. The high cost of labor has also made it difficult to cultivate, harvest, collect, transport, and convert biomass, and to collect, transport, sort, and convert waste, in competition with extracting and converting virgin materials concentrated in natural deposits.

Even though wood products theoretically can be used to provide over 95 percent of the polymers, noncellulosic fibers, and organic chemicals used today, with essentially no sacrifice in properties, they must still prove that they can be competitive feedstocks in the marketplace. As in the case of any biomass source, one cannot ignore the competition of food, feed, and fibers for arable land, nutrients, and water. Under the pressures of soil depletion and dedication of arable farmland for other uses, biomass cultivation for energy and feedstocks would be forced into less fertile and less accessible lands, and lands that are more arid and have shorter growing seasons. This means that the productivities based on intensive cultivation of prime lands, which serve as the basis of all current biomass projections, are unlikely to be matched in any real-life commercial venture. Furthermore, because of the vagaries of nature, assurance of a steady, predictable flow of raw materials from biomass is uncertain at best.

Wastes have been touted as a reliable, predictable source of raw materials. Its collection is highly organized, its composition is reasonably stable, and it is concentrated around highly populated urban centers. To be of value for petrochemical feedstocks, however, some sorting would have to be carried out to concentrate those components that would provide the richest sources of the desired hydrocarbons. In addition, only the most populous metropolitan regions would be able to provide needed volumes of suitable feeds to support a petrochemical complex. Again, it becomes a matter of economics. The most likely outlet for the combustible component of municipal solid wastes is as a fuel to generate steam for electric power, heating, or industrial processing.

Regardless of the above-mentioned shortcomings of potential alternative petrochemical raw materials, at the present state of art, efforts should be continued to develop the technology further. At some time in the future, economics are likely to favor these materials. Even now reclamation of clean wastes at the source, during a manufacturing operation, can possibly compete with virgin feedstocks in restricted cases. Efficient reclamation and recycling of metals and glass from MSW may alter the economics sufficiently to make a reassessment of MSW practical. Harvesting of juvenile forest growths and coppicing may completely alter the economics of biomass production.

Although no possibility should be overlooked in the search for new renewable feedstocks, it should be remembered that, as high value-added products, many petrochemicals should be able to compete more successfully than fuels for high-priced fossil hydrocarbons.

THE CHALLENGE

We are perilously close to a real crisis on crude oil unless we are willing to bow to the will of OPEC in fashioning our economic and foreign policies. Natural gas discoveries are trailing production and consumption. We have coal and oil shale in great abundance, but the environmental and social costs of their use are monumental. We can ill-afford to abandon our role as the world's breadbasket for humanitarian reasons and for the economic reason of balance of payments. We are therefore in a poor position to take up the large-scale cultivation of biomass for energy production.

There is no cure-all in the offing for our oil shortage. This means that our crude oil should be reserved for essential uses only—basically, transport and home heating fuels and petrochemicals. Natural gas should be used primarily for petrochemicals, industrial processes dependent upon this form of fuel, and for home heating. Coal, shale oil, and biomass must be developed to the maximum degree compatible with environmental and socioeconomic considerations. Despite Three Mile Island, we must learn to live with nuclear power generation, because it offers the only practical means of supplanting oil- and gas-fired power plants that will have to be retired from service, and of providing growth in the power generation sector. Needless to say, more exacting procedures, engineering, construction, inspection, and waste disposal need to be employed to ensure safety.

It should be clear that we must develop every possible new technology for generating electricity and for producing fuels and feedstocks, even in small increments. There appears to be no single technology that can stave off our energy shortfall in the early part of the next century.

Our energy policy must include a number of essential elements. (1) We must reduce consumption of nonrenewable fluid fossil fuels to an absolute minimum, because there is virtually no likelihood of our discovering major new domestic fields. (2) Those fluid fossil fuels that are available should be allocated first to essential uses where there is no suitable substitute: petrochemicals, transportation, and home conditioning and water heating. (3) No new electric power plants should be constructed to use fluid fossil fuels, except for topping cycles or in locations where coal and nuclear power cannot be used safely. (4) The maximum possible fluid synfuels should be produced commercially from coal, oil shale, and biomass. (5) Efforts should be increased to solve extraction, social, and environmental problems inherent with coal and nuclear power, particularly pollution from coal combustion and nuclear waste disposal. (6) Solar energy technology should be developed and applied where economically viable.

It should be obvious that the key to a successful energy plan must be superb leadership. Without it we face a very bleak future.

REFERENCES

1   H. G. Cassidy, *Bull. At. Sci.* **33**(3), 31, 32 (Mar. 1977).

2   J. Darmstadter, *Energy User News* **3**(9), 20 (Feb. 27, 1978).

3   *Time* **113**(19), 70-79 (May 7, 1979).

4   *Ibid.* **113**(20), 60, 63 (May 14, 1979).

5   C. S. Nicandros, *Energy User News* **3**(13), 22 (Mar. 27, 1978).

6   Ali M. Jaidah, *Public Util. Fortn.* **102**(10), 66-69 (Nov. 9, 1978).

7   D. Yergin, "Conservation: The Key Energy Source," in R. Stobaugh and D. Yergin, Eds., *Energy Future,* Random House, New York, 1979, pp. 136-182.

8   Reported by J. Flanigan, *Forbes* **124**(2), 34, 35 (July 23, 1979).

9   R. Abramson, *Phila. Inquirer* **300**(147), 3-F (May 27, 1979).

10 H. Enzer et al.; reported in *Energy Facts II,* Sci. Policy Res. Div., Libr. Congr., Serial H, U.S. GPO, Washington, DC, 1975, p. 87.

11 *ESSO Mag. Suppl.,* 1973, Summer; reported by M. A. Parrish, *J. Oil Colour Chem. Assoc.* **60**(12), 474-478 (1977).

12 *Citibank Mon. Econ. Lett.* (Feb. 1977); reported by H. J. Kandiner, *Chemtech* **7**(4), 208, 209 (Apr. 1977).

13 *Public Util. Fortn.* **102**(10), 28-32 (Nov. 9, 1978).

14 O. P. Thomas, *Chem. Eng. News* **56**(45), 5 (Nov. 6, 1978).

15 M. Adelman; reported in *Time* **113**(17), 60, 61 (Apr. 23, 1979).

16 W. Levy; reported in *ibid.*

17 J. Sawhill; reported in *ibid.*

18 *Time* **113**(17), 60, 61 (Apr. 23, 1979).

19 A. M. Bueche, *The Chemist* **54**(4), 5-7, 18 (July 1977).

20 A. M. Bueche, *Chem. Eng. News* **55**(15), 5 (Apr. 11, 1977).

21 H. Bucknell, in J. E. Bailey, Ed., *Energy Systems: An Analysis for Engineers and Policy Makers,* Dekker, New York, 1978, pp. 27-38.

22 H. W. Krupp; reported in *Chem. Eng. News* **56**(39), 47, 48 (Sept. 25, 1978).

23 J. E. Bailey, pp. 3-9 in reference 21.

24 B. M. Hannon, pp. 11-26 in reference 21.

25 D. E. Gushee, *Chemtech* **8**(8), 474-477 (Aug. 1977).

26 J. G. Brookhuis; reported in *Chem. Eng. News* **57**(7), 6 (Feb. 12, 1979).

27 DOC; reported in *Chem. Eng. News* **56**(23), 4 (June 5, 1978).

Appendix A

ACRONYMS AND ABBREVIATIONS

A	Acre.
AAAS	American Association for the Advancement of Science.
AAR	Association of American Railroads.
ABS	Acrylonitrile–butadiene–styrene.
ACR	Advanced cracking reactor.
ACS	American Chemical Society.
ADL	Arthur D. Little, Inc.
AEC	Atomic Energy Commission.
AGA	American Gas Association.
AGR	Annual growth rate.
AIChE	American Institute of Chemical Engineers.
AMD	Acid mine drainage.
ANPA	American Newspaper Publishers' Association.
AOSTRA	Alberta Oil Sands Technology and Research Authority.
AP	Associated Press.
API	American Petroleum Institute.
AR	Agricultural refuse.
ARAMCO	Arabian American Oil Co.
ARCAN	Atlantic Richfield Canada, Ltd.
ARCO	Atlantic Richfield.
ASA	American Statistical Association.
ASCE	American Society of Civil Engineers.
ATM	Advanced technology mining.

BARR	Board on Agricultural and Renewable Resources.
BAU	"Business-as-usual" scenario; no special energy conservation measures, no policy changes.
bbl	Barrel (42 gal).
Bcf	Billion (10^9) cubic feet.
bdt	Bone-dry ton.
BMC	Bulk molding compound.
BOD	Biological oxygen demand.
boe	Barrels oil equivalent.
BOM	Bureau of Mines.
BP	British Petroleum.
bpcd	Barrels per calendar day.
bpd	Barrels per day.
bpd-oe	Barrels per day–oil equivalent.
bpt	Barrels per ton.
bpy	Barrels per year.
Btu	British thermal unit; the quantity of heat required to raise the temperature of 1 pound of water 1 degree Fahrenheit.
BTX	Benzene–toluene–xylene.
BuMines	Bureau of Mines.
CAST	Council on Agricultural Science and Technology.
CCPs	Coal conversion processes.
C-E	Combustion Engineering.
CEQ	Council on Environmental Quality.
CFB	Circulating fluidized bed.
cfd	Cubic feet per day.
CIA	Central Intelligence Agency.
coe	Crude oil equivalent.
COFCAW	Combination of forward combustion and waterflooding.
CORRIM	Committee on Renewable Resources for Industrial Materials.
COS	Oxysulfides of carbon.
CRES	Corrosion-resistant steel.
CRS	Congressional Research Service.
CS	"Conservation" scenario; calls for maximum practical reduction in energy consumption.
CSG	Consolidated Synthetic Gas.
DAP	Diallyl phthalate.
DCF	Discounted cash flow.
deNO$_x$	Exxon Thermal Denox process for scrubbing flue gas.

DGEBA	Diglycidyl ether of bisphenol A.
DMT	Dimethyl terephthalate.
DOA	Department of Agriculture.
DOC	Department of Commerce.
DOE	Department of Energy.
DOI	Department of Interior.
DOT	Department of Transportation.
dte	Dry ton equivalent.
dws	Dry wood solids.
dwt	Dead-weight tons (lifting capacity of ship in long tons).
EDS	Exxon Donor Solvent process.
EEC	European Economic Community.
EIS	Environmental impact statement.
EOR	Enhanced oil recovery.
EPA	Environmental Protection Agency.
EPDM	Elastomeric terpolymers of ethylene, propylene, and diene monomers.
EPRI	Electric Power Research Institute.
ERDA	Energy Research and Development Administration.
ETSI	Energy Transportation Systems, Inc.
FBC	Fluid-bed combustion.
FEA	Federal Energy Administration.
FERC	Federal Energy Regulatory Commission. Replaces Federal Power Commission (FPC).
FGD	Flue-gas desulfurization.
foe	Fuel oil equivalent.
"4-D's"	Drunkenness, depression, delinquency, and divorce. Symbols of decline in social values of boom towns.
FPC	Federal Power Commission. Replaced by Federal Energy Regulatory Commission (FERC).
FPL	Forest Products Laboratory (USDA).
FRP	Fiber-reinforced plastic. (Usually implies fiberglass-reinforced polyester.)
g	Gram.
GCOS	Great Canadian Oil Sands, Ltd.
GDP	Gross domestic product.
GNP	Gross national product.
gpd	Gallons per day.
GPP	Gross primary production.

gpt	Gallons per ton.
gpy	Gallons per year.
Gte	Giga (10^9) tonnes or 10^{15} grams.
GURC	Gulf Universities Research Consortium.
ha	Hectare.
HC	Hydrocarbon.
HDPE	High-density polyethylene.
HMDA	Hexamethylene diamine.
HTR	High-temperature (nuclear) reactor.
ICC	Interstate Commerce Commission.
IEA	International Energy Agency (Australia, Austria, Belgium, Canada, Denmark, Great Britain, Greece, Ireland, Italy, Japan, Luxembourg, The Netherlands, New Zealand, Norway, Spain, Sweden, Switzerland, Turkey, and United States.)
IGT	Institute of Gas Technology.
IMCO	UN-sponsored Inter-governmental Maritime Consultative Organization.
IOCC	Interstate Oil Compact Commission.
IRA	Industrial Reorganization Act.
ITC	International Trade Commission.
IUPAC	International Union of Pure and Applied Chemistry.
k	Kilo; thousand; 10^3. (*See also* M.)
kJ	Kilojoules.
kWh	Kilowatt-hours.
LAI	Leaf area index.
LBL	Lawrence Berkeley Laboratory.
LDCs	Less-developed countries.
LDPE	Low-density polyethylene.
LERC	Laramie Energy Research Center.
LLL	Lawrence Livermore Laboratory.
LNG	Liquefied natural gas.
LOOP	Louisiana Offshore Oil Port.
LPG	Liquefied petroleum gas (largely propane and butane).
LPM	Liquid-Phase Methanation process.
LPM/S	Liquid-Phase Methanation with Shift process.
LRG	Liquefied refinery gas.
lt	Long tons (2240 lb).
ltpcd	Long tons per calendar day.
ltpd	Long tons per day.

LV	Liquid volume.
m	Meter; milli- (thousandth).
M	Mega; million; 10^6. In articles on gas, it usually means thousand (10^3), but it will not be used in this way in this book, to avoid confusing the average reader.
m.a.f.	Mineral-ash free.
Mbpd	Million bbl per day.
m.c.	Moisture content.
MDI	Methylene diphenyl isocyanate.
MHF	Massive hydraulic fracturing.
MIT	Massachusetts Institute of Technology.
Mgpy	Million (10^6) gallons per year.
MM	Means million (10^6) in publications that use M to signify thousand (10^3). Will not be used in this book.
MMA	Methyl methacrylate.
mmf	Mineral-matter free.
MSW	Municipal solid waste (urban refuse).
Mt	Metric tons (or tonnes) in some publications; million tons (short). Abbreviation te for tonnes is used in this book.
Mte	Million tonnes (million metric tons).
MW$_e$	Megawatts electric.
n.a.	Not applicable; not available.
NAE	National Academy of Engineering.
NARI	National Association of Recycling Industries.
NAS	National Academy of Sciences.
NASA	National Aeronautics and Space Administration.
NATO	North Atlantic Treaty Organization.
NCA	National Coal Association.
NCB	National Coal Board (U.K.).
NCRR	National Center for Resource Recovery.
NEB	National Energy Board (Canada).
NEP	1. Net energy production. 2. National Energy Plan; NEP-1 proposed April 20, 1977; NEP-2 adopted Oct. 15, 1978; NEP-3 under study.
NG	Natural gas.
NGL	Natural gas liquids.
NIOC	National Iranian Oil Co.
nm	Nanometers; 10^{-9} meter.
NOAA	National Oceanic and Atmospheric Administration.

NO_x	Oxides of nitrogen. (Primarily NO and NO_2.)
NPC	National Petroleum Council.
NPP	Net primary production (annual).
NRC	National Research Council.
NSF	National Science Foundation.
NTIS	National Technical Information Service.
N-T-U	Nevada–Texas–Utah (oil shale extraction process).
NYSE	New York Stock Exchange.
OAPEC	Organization of Arab Petroleum Exporting Countries, founded in 1968. Membership includes Saudi Arabia*, Kuwait*, Libya*, Abu Dhabi, Algeria, Bahrein, Dubai, and Qatar. (*Charter members.)
OCR	Office of Coal Research.
OCS	Outer continental shelf.
O.D.	Oven dried.
odt	Oven-dry ton.
OECD	Organization for Economic Cooperation and Development. (Western industrial nations plus Japan, less France.)
OPEC	Organization of Petroleum Exporting Countries, founded in 1960. Present membership: Iran*, Iraq*, Saudi Arabia*, Venezuela*, Gabon, Algeria, Indonesia, Libya, Nigeria, Qatar, United Arab Emirates, and Ecuador. (*Charter members.)
ORNL	Oak Ridge National Laboratory.
OTA	Office of Technology Assessment.
PA	Polyamide.
PAH	Polycyclic aromatic hydrocarbon.
PAR	Photosynthetically active radiation (400 to 700 nm).
PBD	Polybutadiene.
PBT	Polybutylene terephthalate.
PCB	Polychlorinated biphenyl.
PEMEX	Petroleos Mexicanos, the national Mexican oil company.
PERC	Pittsburgh Energy Research Center (BuMines).
PET	Polyethylene terephthalate.
pH	Negative log of hydrogen ion concentration; measure of acidity.
PLG	Pipeline gas.
PMPPI	Polymethylene polyphenylene isocyanate.
POSM	Propylene oxide styrene monomer.
PP	Polypropylene.
ppb	Parts per billion (10^9).

ppm	Parts per million (10^6).
PS	Polystyrene.
PTA	Purified terephthalic acid.
PVA	Poly(vinyl alcohol).
PVAc	Poly(vinyl acetate).
PVB	Poly(vinyl butyral).
PVC	Poly(vinyl chloride).
PVF	Poly(vinyl formal).
q	Means quad (10^{15}), where Q is used to mean quintillion (10^{18}), but not in this book. (*See* Q.)
Q	Quad; a large unit of energy—10^{15} Btu; quadrillion. Sometimes means quintillion, but not in this book. (*See* q.)
RDF	Refuse-derived fuel.
R.H.	Relative humidity.
RI	Report of Investigation (DOI).
RMOGA	Rocky Mountain Oil and Gas Association.
ROI	Return on investment.
ROM	Run-of-mine coal.
RON	Research octane number.
SAC	Starved air combustion.
SAN	Styrene–acrylonitrile polymer.
SBM	Single-buoy mooring.
SBR	Styrene–butadiene rubber.
scf	Standard cubic foot, measured at 60°F (15.6°C) and a pressure 6 oz per in.^{2} above atmospheric.
scfd	Standard cubic feet per day.
SCO	Synthetic crude oil.
SNG	Substitute (or synthetic) natural gas.
SoCal	Standard Oil of California.
SOHIO	Standard Oil of Ohio.
SPG	Synthetic or substitute pipeline gas.
SRC	Solvent-refined coal.
SRC-II	Modified SRC process yielding fluid product.
SRI	Stanford Research Institute.
STA	Slurry Transport Association.
t	Short ton (2000 lb). (*See also* lt and te.)
TAC	Triallyl cyanurate.
TAPS	Trans-Alaska Pipeline System (oil), operated by Alyeska consortium.

tcf	Trillion cubic feet; 10^{12} ft^3. In natural gas, the approximate energy equivalent of 1 quad.
TDI	Toluene diisocyanate.
te	Metric ton (tonne); 1000 kilograms.
tpa	Tons per acre.
tpcd	Tons per calendar day.
tpd	Tons per day.
TPO	Thermoplastic olefinic elastomers.
tpy	Tons per year.
UAE	United Arab Emirates.
UBC	Uniform Building Code.
UCG	Underground coal gasification.
UMW	United Mine Workers.
UPI	United Press International.
USDA	United States Department of Agriculture.
USGS	United States Geological Survey.
VCM	Vinyl chloride monomer.
WAES	Workshop on Alternative Energy Strategies (MIT).
ZPG	Zero population growth.

Appendix B

GLOSSARY

Ablation Process of reducing mesh of tar sands ore prior to or during pulping.

Acetal Resin (Polyacetal) High-molecular-weight, linear polymer of formaldehyde (homopolymer), or of formaldehyde with trioxane (copolymer).

Acidize To add hydrochloric acid to limestone, dolomite, or other acid-soluble reservoir formation, to stimulate recovery of natural gas.

Acid Mine Drainage (AMD) Leakage of acid from mines and spoil banks into streams and aquifers.

Acid Rain Low pH rain resulting from SO_x emissions from combustion of fossil fuels.

Acrylic Fiber Polymerized acrylonitrile fiber.

Adhesive Substance capable of joining materials by surface attachment.

Advance Mining (*See* Longwall Mining.)

Aerobic Digestion Decomposition (liquefaction) of organic matter by microbes in the presence of oxygen.

Agricultural Refuse (AR) Wastes of crop, animal, and forest origin.

Agwaste Farm or agricultural waste, including both plant and animal origins.

Albedo Earth's surface reflectivity to solar radiation.

Alcan Pipeline Proposed pipeline system chosen for delivery of Alaskan natural gas to U.S. markets, and Mackenzie Delta gas to Canadian markets.

Algal Kerogen (*See* Kerogen, Algal.)

Algal Pond (*See* Facultative Algal Pond *or* High-Rate Algal Pond.)

Aliphatic Relating to organic compounds of carbon and hydrogen having a straight chain of carbon atoms.

Alkyd Resin Essentially polyester of a polyhydric alcohol (e.g., glycerol or pentaerythritol) and a polybasic acid (typically phthalic anhydride).

Alkyl   A monovalent organic radical of the form C_nH_{2n+1}.

Alpha-Cellulose   Lignin-free cellulosic material insoluble in cold 17.5 percent NaOH; consists essentially of long chains of glucose molecules with beta-oxygen linkages between the 1 and 4 carbon atoms.

Alyeska   Consortium constructing and operating TAPS (Trans-Alaska Pipeline System) oil pipeline from Prudhoe Bay to Valdez, Alaska.

Amino Acid   An organic compound containing one or more basic amino groups and acidic carboxyl groups; the basic building block of proteins.

Aminoaldehyde Resin (Carbamide)   Urea– or melamine–formaldehyde resin.

Anaerobic   In the absence of oxygen

Anaerobic Digestion   Decomposition of organic matter in the absence of air or oxygen not in chemical combination.

Anchor Bolt   Bolt driven into mine working roof to prevent cave-in.

Angiosperm   Member of the plant division Magnoliophyta (includes hardwood trees).

Anhydrous Solvent Extraction   Process involving mixing tar sands ore with recycled hydrocarbon solvent, draining to strip solvent from solids, and solvent recovery from bitumen product.

Annual Plant   A plant having a one-year growing cycle; must be replanted each year.

Anoxic   A condition of oxygen deprivation.

Anthracite   Hard coal; high-grade metamorphic coal with semi-metallic luster, high fixed-carbon and low volatiles content, and high density.

Antisolvent   A liquid that promotes the agglomeration of fine particulate matter in a coal solution.

Antrim Shale   Deep, thin Devonian shale deposits found in the lower peninsula of Michigan.

Appalachia   A geographical region including West Virginia, western Pennsylvania, western Virginia, eastern Kentucky, and Tennessee.

Apparent Reserves   (*See* Measured Reserves.)

Aquaculture (Aquiculture)   Cultivation of natural plant resources of water. (*See also* Mariculture.)

Aquifer   A permeable, water-bearing bed of stone, sand or earth.

Arab Oil Embargo   Reduction of Arab oil exports to Western nations during the fourth quarter of 1973 through the first half of 1974. The term "embargo" is a misnomer, since the supply was merely reduced (e.g., United States, 14 percent), rather than curtailed completely.

Area Strip Mining   Surface mining of large open area, as contrasted to, for example, contour mining.

Aromatic   Structure of organic compound characterized by the presence of at least one benzene ring.

Asbestosis   Nonmalignant scarring of the lungs caused by mineral fibers.

Ash   Largely mineral residue from coal combustion.

Asphaltenes   Polynuclear aromatics linked with alkyl chains.

Assimilation   Photosynthesis together with root absorption.

Associated Gas (Casinghead Gas)   Gas dissolved in or, more generally, accumulated above the crude oil pool in a trap.

Athabasca Tar Sands   Oil sands occurring in the Northeastern portion of Alberta, Canada.

Auger Mining   Technique employing large (5- to 8-ft-diameter) auger(s) to break coal and carry it into open.

Auxin   Plant hormone that causes enlargement of plant cells.

Bacterial Treatment   The use of sulfur-oxidizing bacteria (*Thiobacillus*) to effect separation of organic material from oil shale.

Bagasse   Remains of sugarcane after extracting juice.

Baltimore Canyon   Large geological OCS zone extending from Long Island to North Carolina.

Barrel   A unit of volume measure for oil; 42 gallons.

Basic Reclamation   Restoration of land by grading, revegetation, and drainage control following mining operations.

Bed (Seam; Stratum)   Layer of coal or other mineral in natural formation.

Behind-the-Pipe   Describes natural gas existing in pockets above producing reservoirs.

Beneficiation (Preparation; Cleaning)   Upgrading of coal by chemical and/or physical means.

Level A: Absence of preparation or upgrading of coal.

Level B: Coal preparation comprising breaking and removal of some trash.

Level C: Coarse preparation of coal by breaking, screening, and washing coarse ($+3/8$-in.) fraction. Fines are shipped dry.

Level D: Deliberate preparation of coal by breaking, screening, and washing. Fine fraction (minus-28 mesh) is either discarded or dewatered and shipped with clean coal.

Level E: Elaborate preparation of coal entails breaking and screening, washing, and thermal drying of fines.

Level F: Full preparation reduces size more than Level E and produces two or more product streams, usually a clean coal for steam generation and a middlings stream for noncombustion, industrial use.

Benzene The most important aromatic chemical; an unsaturated, resonant, C_6 ring with three double bonds and six hydrogen atoms.

Benzol (Light Oil)

1 Low-boiling distillate fraction of crude oil, below first lubricating oil fraction.

2 Sometimes benzol designation is applied to benzene.

Berm Coal bed exposed by stripping away overburden.

Binder

1 Resin used to bind components of plastic together to form a fabricated product.

2 (*See also* Paint Vehicle.)

Biochemical Conversion Modification of organic compounds to specific materials by action of living organisms.

Biodegradability Ability to be decomposed by microorganisms.

Biofixation (of Nitrogen) Formation of ammonia from nitrogen and water by biological means.

Biomass Total weight content of plant life over given area at a particular time; includes main stem, branches, twigs, leaves, fruit, bark, and roots.

Biosphere Life zone of earth; that part of the earth where life exists; between essentially 36,000 ft below sea level to at least 33,000 ft above sea level (but mostly between −650 and +20,000 ft).

Biosynthetic Conversion The cultivation of biomass specifically for fuels and feedstocks.

Biota Living and dead organic matter of a given region.

Bitumen

1 Dark or black, naturally occurring or pyrolytically formed material composed almost entirely of carbon and hydrogen, with very little oxygen, nitrogen, or sulfur.

2 Hydrocarbon component of tar sands.

Bituminous Coal Soft coal; dark brown to black coal high in carbonaceous matter, and containing 15 to 50 percent volatile matter.

Black Liquor Spent cooking liquor from alkali pulping; contains inorganic chemicals and dissolved organic components of pulpwood.

Black Lung (Pneumoconiosis) Chronic inflamation of the lungs caused by accumulation of inhaled black coal particulates.

Blind Borehole Technique A shaftless method of UCG pregasification involving formation of a single borehole with concentric pipes.

Board Foot Nominally, a piece of lumber 1 in. thick and 1 ft square. (Actual seasoned lumber will have a minimum cross section of slightly over 70 percent of nominal, because of shrinkage on seasoning and/or finishing.)

Bog   Naturally occurring deposit of peat.

Boghead Coal   A fossil product intermediate in heating value between high-volatility bituminous coal and high-grade oil shale.

Bole   Trunk or main stem of tree.

Bolewood   Wood in or from main stem of tree.

Boreal Zone   Biotic communities between Arctic and temperate zones.

Borehole Producer Technique   A shaft method of UCG pregasification calling for preparation of parallel underground galleries between which small boreholes are driven.

Botanochemical   Chemical product derived from biomass. (Term coined to suggest renewable-resource base for "petrochemicals.")

Bottled Gas   LPG or liquefied petroleum gas; a mixture of propane and butane stored under pressure in liquid form

Bottoms   Ships.

Brown Coal   A low-grade lignitic coal.

Brown Shale   Devonian shale.

Bunkers   Fuel for outgoing ships; not considered as exports.

Butylene   Any of three alkene hydrocarbon isomers having the formula C_4H_8.

Byssinosis   A form of pneumoconiosis resulting from inhaling dust from cotton bales.

Caking Coal   Coal (generally eastern bituminous) having a high sulfur content, which causes particles to clump together when heated.

Canopy Culture (Closed-Cycle Cultivation)   Cultivation of field crops under cover of a transparent plastic film tent.

Capping   (*See* Overburden.)

Captive Mine   Mine owned by and worked for owner-consumer (e.g., power company or steel mill).

Carbamide Resin   (*See* Aminoaldehyde Resin.)

Carbohydrate   A carbon–hydrogen–oxygen compound having the general formula $C_nH_{2n}O_n$.

Carbon Cycle   Cycle of carbon in biosphere, in which plants photosynthesize CO_2 to organic compounds that are, in turn, consumed by plants and animals, with the carbon being returned to the biosphere in the form of inorganic compounds by respiration and decay.

Carbon Efficiency   Hydrogen-to-carbon ratio.

Carbonification Formation of coal from plant material by the geologic processes of diagenesis and metamorphism; coalification.

Carbonization (*See* Pyrolysis.)

Carbonyl Group A radical (CO) in which atoms are joined by a double bond.

Carboxyl Group The COOH radical, in which one oxygen atom is joined to the carbon atom by a double bond. This radical determines the basicity of an organic acid.

Carcinogen Substance shown to cause cancer in human beings.
Presumptive carcinogen: Substance shown to cause cancer in one or more laboratory animals, but not yet in human beings.

Carried Interest One hundred percent of exploratory drilling cost.

Carrier Bed Permeable sandstone, limestone, or dolomite deposit which receives petroleum as it is expelled from compacting clay and mud.

Cartel An association of nations or businesses operating in concert to control the supply or price of a product.

Casinghead Gas (*See* Associated Gas.)

Catalytic Reforming (*See* Reforming.)

Cellulase Any of a number of extracellular enzymes (from fungi, bacteria, insects, and lower animals) that hydrolyze cellulose.

Cellulose Main polysaccharide (carbohydrate polymer) occurring in primary and secondary cell walls of living plants; our most abundant natural product.

Chamber Technique (Warehouse Technique) A shaft method of UCG pregasification, in which a large underground gallery is excavated and isolated by brickwork.

Char Carbon-rich residue of partial combustion.

Char-Oil Dry, free-flowing powder composed of a mixture of char and pyrolytic oil.

Chemical Coatings (*See* Industrial Finishes.)

Chemical Comminution Fracturing of raw coal along natural fault lines by treating with chemicals (ammonia; methanol).

Chemical Flooding Injection of special polymers, surfactants and/or alkali for enhancing recovery of oil from reservoirs.

Chlor-Alkalies Chlorine, caustic soda, and soda ash.

Cleanup Purification or removal of gaseous diluents and pollutants from raw gas stream to prevent catalyst poisoning and/or to upgrade heating value of product gas.

Clone (n.) A live stick 4 to 20 in. long, cut from a live plant, from which similar plants can be propagated by vegetative reproduction (as contrasted to reproduction from seed).

Closed-Cycle Cultivation (*See* Canopy Culture.)

Coal A complex organic polymeric product resulting from geologic decomposition of organic matter. Consists of aromatic clusters of fused rings held together by assorted hydrocarbon and heteroatom (O, N, S) linkages. Approximate composition is $CH_{0.8}O_{0.1}$.

Coal-Bound The inability to extract coal and transport it to the surface while removing interburden, because of physical (spacial) constraints. (Applies to deep, multiseam beds.)

Coal Tar Pitch Dark brown or black thermoplastic, amorphous residue from redistillation of coal tar; melting point about 150°F (66°C).

Coatings Paints, varnishes, lacquers, and other associated products, including printing inks.

COFCAW Combination of forward combustion and waterflood; process in which air is injected under sufficient pressure to fracture tar sand deposit and allow air and water drivers to force liquid bitumen toward production wells(s).

Cogeneration Combining electricity generation with production of steam or process heat, to reduce wasteful thermal emissions.

Coke A coherent, cellular, solid residue left from the dry distillation of coal, pitch, petroleum, or other hydrocarbons; comprises chiefly carbon, mineral, and volatile matter.

Coked Solids Particulate minerals (e.g., sands) encased in coke.

Coke-Oven Gas A gas produced during carbonization of coal to form coke.

Coker Still for separating volatiles from coked feed.

Cold-Water Extraction Process using a combination of cold water and kerosene to extract bitumen from tar sands.

Collection Well (*See* Production Well.)

Combined Cycle Process in which waste heat from large gas turbines is used to generate steam for conventional steam turbines.

Combustion, Wet (*See* Wet Combustion.)

Commercial Forest Forest land with minimum 20 ft^3 per acre annual yield, whether or not used to produce sawtimber or roundwood for commerce.

Composting Aerobic fermentation of organic matter. (*See also* Fermentation.)

Connate Water (or Air) Water (or air) trapped in the interstices of igneous rocks at the time of their formation.

Consist (Con'-sist, n.) Solid material suspended in slurry (e.g., pulverized coal in a water carrier).

Continental

Margin: The transition subsurface region between continent and ocean floor.

Rise: Subsurface region (average gradient 1:300) or continental margin lying at the foot of the continental slope.

Shelf: Shallow, subsurface region (average gradient 1:1000) of continental margin bordering continent.

Slope: Steep subsurface region (average gradient 1:40 to 1:6) of continental margin lying between continental shelf and continental rise.

Continuous Mining Machine   Machine for shearing and breaking coal from underground seam on a relatively steady basis.

Continuous Steam Stimulation (Continuous Steam Injection)   Tar sands in situ extraction process using steady, high-pressure steam injection as the driver.

Contour Mining   Mining of outcrop underlying a hill or mountain, by removing overburden around perimeter.

Conventional Reserves   Estimates of reserves based solely on tonnages or volumes. (*See* Standardized Reserves.)

Conversion

1 The transformation of higher-quality energy to work plus lower-quality energy.

2 The manufacture of fluid fuels from coal, peat, tar sands, oil shale, refuse, and so forth.

3 Switch from the use of oil- or gas-fired boilers to coal-fired boilers.

Conversion Efficiency   Percentage of chemical or latent-heat energy in coal that can be obtained by burning the conversion product.

Copolymer   A high-molecular-weight organic molecule formed by combining two or more dissimilar monomers.

Coppice (Stem Sprout) (v.)   To propagate new shoots from old stumps.

Coppice, Juvenile   (*See* Juvenile Coppice.)

Cotton Gin Trash   Leaf fragments, sticks, and other plant parts, and linty material (motes) from ginning of cotton.

Cotton Lint   Cotton fibers.

Countercurrent Combustion   (*See* Reverse Combustion.)

Countercurrent Internal-Combustion Process   Technique for extracting shale oil, which calls for upward motion of spent shale in retort, counter to downward advancing flame front; liquid product stream is drawn off ahead of flame front.

Crack (v.)   To break down or fracture an organic compound into smaller, simpler molecules.

Crib (v.)   To support the roof of a mine working.

Critical Temperature   Temperature above which a substance can exist only in a gaseous state, regardless of the pressure.

Crown   Top branches and upper stem of tree.

Crude Oil A fluid mixture of hydrocarbons that exists in natural underground reservoirs, less associated natural gas products.

Culm Fine fraction from screened coal; contains mixture of minerals and coal.

Cultivate To raise or foster the growth of plants.

Cultural Energy Fuel, fertilizer, and other forms of energy invested in crop. Does not include solar energy.

Curtailment (of Fuel) Restriction of deliveries.

Cutting (n.) Clipping of stemwood.

Cytokinin Plant hormone that stimulates cell division.

Datum Arbitrary benchmark water level for periodic monitoring of water depth on Great Lakes.

Dawsonite Mineral $NaAlCO_3(OH)_2$, associated with Piceance Creek Basin oil shale.

Dead-Weight Tons Total lifting capacity of ship in long tons (2240 lb).

Decarburization Removal of carbon from surface of ferrous alloy by heating in presence of reactive medium.

Decontrol To remove price or other restrictive regulations.

Deep Cleaning (*See* Beneficiation, Level F.)

Deep Mining (Underground Mining) Extraction of coal or minerals from a network of shafts and tunnels below ground level, down to depths of 1000 to 1500 ft for coal.

Dehydrogenation Removal of hydrogen atoms from an organic compound.

Demonstrated Reserves Combined measured and indicated reserves.

Denitrification Reduction of nitrates to molecular nitrogen (an anaerobic process).

Desulfurization Removal of sulfur before or after combustion.

Development Well Hole drilled to extend limits of known reservoir. (*See also* Exploratory Well.)

Devonian Shale (Brown Shale) Hydrogen-lean shales formed over a large portion of the eastern and midwestern United States.

Diamagnetic Having a magnetic permeability less than 1.

Digestion Decomposition of organic matter.
    Aerobic: (*See* Aerobic Digestion.)
    Anaerobic: (*See* Anaerobic Digestion.)
    Mesophilic: (*See* Mesophilic Digestion.)
    Thermophilic: (*See* Thermophilic Digestion.)

Direct Combustion (*See* Incineration.)

Direct Fluid Coking   Extraction of bitumen from tar sands by heating ore in a fluid-bed coker or still at 900°F.

Directional Drilling   (*See* Longwall Generation.)

Divestiture.

Horizontal: Division of major oil companies into independent organizations based on, for example, energy form (oil, natural gas, coal, and nuclear).

Vertical: Division of major oil companies into independent production, refining, transportation, and marketing organizations.

Dolomite   A white or colorless carbonate mineral, $CaMg(CO_3)_2$.

Double Bottoms   Double hulls providing buoyant outer compartments and separate interior cargo holds or tanks.

Doubling Time   The time required to achieve net doubling of a parameter (e.g., energy consumption, inventory of fissionable material in a breeder reactor, etc.). (*See also* Rule of 70.)

Down-Dip Mining   Working of a mine face sloping down away from the shaft, thus trapping drainage water in the stope. (*See also* Up-Dip Mining.)

Downstream   Farther removed from the source (e.g., downstream processes—refining, petrochemical production).

Drag Stream   Contaminated water left after separation of froth and sand from the hot-water process stream.

Drawdown   Net reduction of reserves through extraction at a rate exceeding new discoveries.

Dredge Mining   Removal of overburden by dredging under water to expose coal seam.

Drive   Indigenous or artificial energies providing the force required to expel petroleum from a reservoir.

Driver   Fluid substance injected under pressure into fossil fluid reservoir to force product to collection well(s).

Dry and Lean Gas   High-methane gas.

Dry Gas   (*See* Nonassociated Gas.)

Dry Hole (Duster)   Well that does not yield economically recoverable crude oil or natural gas; may contain water.

Dry Organic Solids (Volatile Solids)   Solid material volatilized in a furnace at 1200°F (650°C).

Duckweed Lemna gibba.   Free-floating aquatic weed. The smallest and simplest flowering plant, having single fronds about the size of a pinhead.

Dump   Uncovered land disposal site for solid and/or liquid waste.

Dumpground Gas   (*See* Landfill Gas.)

Dust Bowl   Large portions of Texas, New Mexico, Colorado, Oklahoma, and Kansas.

Duster   (*See* Dry Hole.)

Eastern Province (Coal)   Pennsylvania, West Virginia, eastern Kentucky, eastern Tennessee, and Alabama.

Eco-Fuel II   Pelletized solid fuel from MSW.

Ecology (Environmental Biology)   The science of interrelationships between living organisms and their surroundings.

Ecosystem   All living and nonliving things in some specified area, such as pond, field, or forest.

Effective Capacity   Nameplate capacity reduced to practical level to take into account maintenance downtime, holidays, unscheduled shutdowns, strikes, and so forth.

Efficiency   Ratio of number of calories stored to those absorbed.

Efficiency, Thermal   Ratio of heat converted to useful purposes, to total heat units available in fuel consumed.

Electrical Resistance Heating   Process in which alternating current is applied to tar sands deposit to raise temperature enough to extract bitumen with steam flush.

Electrochemical Corrosion (Electrolytic Corrosion)   Corrosion of metal associated with well, in contact with soil, due to electrical potential.

Electro-Linkage   Use of electrical potential to dry out and pyrolyze coal seams in situ, between injection and producer boreholes.

Electrostatic Precipitator   Device for removing dust or other finely divided particulates from a gas by charging the particles inductively and attracting them to highly charged collector plates.

Emersed Weeds   Plants having roots beneath water, and stems and leaves above.

Emulsion-Steam Drive   Surface-active agent and steam (or hydrocarbon diluent) are used as drivers to sweep emulsified bitumen from tar sands deposit.

Energy   The capacity for doing work.

Energy Content (of Material or Product)   Energy required for production (including feedstock).

Energy Farming (Fuel Farming)   Raising of dedicated crops for their cellulose.

Energy-Intensive   Aspect of a product or process requiring unusually large inputs of energy in the form of feedstocks and/or fuel.

Energy Plantation   A facility devised for the specific purpose of producing and storing fuels in the form of plants, by means of photosynthesis.

Enhanced Gas Recovery   Technology for extracting gas after exhaustion of natural drive.

Enhanced Oil Recovery   (*See* Tertiary Recovery.)

Entitlements   Payments made by refiner of domestic oil to refiner of foreign oil to compensate for unequal costs of crude oil.

Entrained Bed (Suspension Bed)   Reactor design in which solid and gaseous streams impinge and achieve complete reaction in short time at high temperature.

Environmental Death   Fatality due to heart or lung ailment from power plant or other emissions.

Enzymatic Fermentation   Process of splitting cellulose into glucose by enzymatic hydrolysis using cellulase.

Epoxide   (*See* Epoxy.)

Epoxy (Epoxide; Ethoxyline; Oxirane)

   1 A reactive group comprising an oxygen joined to each of two carbon atoms which are also joined.

   2 A resin in which compounds containing epoxy groups are reacted with amines, acid anhydrides, phenols, carboxylic acids, or other unsaturated compounds.

Ester Sulfates   Organic sulfur compounds characterized by carbon-oxygen–sulfur linkages.

Estuarine   Associated with a semienclosed body of water with a free connection to the sea, yet diluted with fresh water.

Ethoxyline   (*See* Epoxy.)

Ethylene   An aliphatic gas with the formula C_2H_4; a major petrochemical feedstock.

Eurasian Watermilfoil   A submersed aquatic weed which has infested waters as far north as the Tennessee Valley.

Exploratory Well   Hole drilled in previously unexplored region.

Explosive Fracture   Breakup of coal seams by use of shaped explosive charges.

Exponential Reserve Index   The number of years before the yearly demand exceeds the total supply.

Face   Surface of seam on which mining operations are performed.

Facultative Algal Pond   Deep (1 to 3 m) aquacultural impoundment lacking forced circulation.

Feedstock   Raw material fed to process equipment to become part of a new product.

Fermentation (Zymosis)   Enzymatic transformation of organic matter (especially carbohydrates), generally accompanied by the evolution of gas.

Fine Cleaning   (*See* Beneficiation, Level F.)

Fireflooding   (*See* Thermal In Situ Extraction.)

Fischer Assay   Method of determining hydrocarbon content of oil shale ore by retorting at 500°C (932°F) under specific conditions; yields unreliable results on hydrogen-lean Devonian shales, due to formation of trapped coke.

Fischer-Tropsch Process   A catalytic process for synthesizing hydrocarbons and oxygenated derivatives from synthesis gas.

Fixation   (*See* Biofixation.)

Fixed Bed (Moving Bed)   Reactor bed in which additive solid materials fall slowly through heated countercurrent gaseous stream.

Fixed Carbon   Carbon in compound formed by photosynthesis.

Flameless Vapor Explosion   Low-pressure explosion of LNG in contact with water.

Flare (v.)   To burn off waste gas.

Flash Pyrolysis   Rapid formation of pyrolysis oil from shredded and air-classified MSW at about 900°F (490°C).

Floating Island Weeds   Masses of free-floating dead aquatic vegetation (sudds) supporting larger vegetation.

Flue Gas   (*See* Stack Gas.)

Fluffing   Expansion of overburden and ore during mining and processing, due largely to inclusion of air.

Fluid   A form of matter in which molecules are mobile relative to each other; includes liquids and gases.

Fluidized Bed   Reactor bed kept in motion by injecting air through a porous base plate.

Fly Ash   Fine particulate, essentially noncombustible refuse, carried from furnace in a gas stream.

Forest Residue   Stumps, roots, limbs, crowns, foliage, and so forth.

Forward Combustion (Thermal In Situ Extraction).   In situ combustion in which deposit is ignited near air injection well and flame front is propagated forward toward the production or collection well(s). (*See also* Reverse Combustion.)

Fossil Fuel   Any combustible hydrocarbon deposit formed by geologic metamorphism of living matter.

Four Corners   Common point where New Mexico, Arizona, Utah, and Colorado meet.

Fourth World   (*See* World, Fourth.)

Framboidal Pyrite   A form containing microscopic aggregates of pyrite grains, often in spheroidal clusters.

Froth Flotation   Separation of pulverized coal and mineral contaminants by collection of coal in the froth.

Frozen Coal Cracker Single-roll crusher used primarily for breaking frozen coal unloaded from hopper cars.

Fuel A substance that can be burned to produce thermal energy.

Fuel Cycle Complete cycle of fuel from mining (recovery), extraction, and refining through burning (fission), recycling, and enrichment.

Fuel Farming (*See* Energy Farming.)

Fuel NO$_x$ Oxides of nitrogen formed by oxidation of nitrogen present in fossil fuels, during combustion.

Fuel, Refuse-Derived (*See* Refuse-Derived Fuel.)

Full Cleaning (*See* Beneficiation, Level F.)

Fusinite The micropetrological component of mineral charcoal or fusain, comprising carbonized woody tissue.

Future Gas Unconventional natural gas, substitute natural gas (SNG), and gas obtained by enhanced recovery technology.

Gasahol Undefined blend of gasoline and alcohol (methanol or higher), with the alcohol being derived from any renewable source. (*See also* Gasohol.)

Gas Cap Expansion Well drive force created by gas under pressure trapped above crude oil in reservoir.

Gas Combustion Process Oil shale extraction method in which low-heating-value product fuel is recycled as retort fuel to heat the oil shale.

Gas Condensate Well Natural gas well yielding large volumes of gas with droplets of light, volatile hydrocarbon liquids entrained in the gas stream.

Gasification
   1 Conversion of hydrocarbons to high-heating-value gas.
   2 A stage in UCG involving: input of gasification medium, ignition, maintenance of flame front, and process control.

Gas-Inerting System Means for flooding empty portions of compartment or tank with inert gas to prevent fires or explosions.

Gas, Landfill (*See* Landfill Gas.)

Gas, Marsh (*See* Marsh Gas.)

Gasohol An automotive fuel comprising 90 percent unleaded gasoline plus 10 percent agriculturally derived ethanol. (*See also* Gasahol.)

Gasoline A refined petroleum distillate having a boiling range of 80 to 400°F at atmospheric pressure.

"Gasoline Tree" (*See* Gopher Plant.)

Gas, Synthesis (*See* Synthesis Gas.)

Gas, "Unnatural" (*See* Landfill Gas.)

Geopressured (Geopressurized) Zone Very deep, high-temperature (>150°C) aquifers lying beneath Louisiana, much of the adjacent Gulf Coast states, northeastern Mexico, and offshore Gulf Coast areas, and containing natural gas in solution and under very high pressure.

Geothermal   Relating to the heat of the earth's interior (e.g., geothermal steam, geothermal gradient).

Giant Field   One containing a minimum of 500 million bbl (68 million tonnes) of recoverable crude oil or 3 tcf (86 billion m^3) of recoverable natural gas.

Gibberellin   Plant hormone that stimulates cell division, cell enlargement, or both.

Gin Trash   (*See* Cotton Gin Trash.)

Glowing Gas   Old interstate gas regulated at low price.

Glucose   An organic compound with the formula $C_6H_{12}O_6$; the most common sugar.

Glyphosine   A plant growth regulator which accelerates ripening of sugarcane.

Gob (Goaf)   Mineral waste removed from coal during beneficiation. (*See also* Slag.)

Gopher Plant (Gopherweed)   *Euphorbia latherus.* A hydrocarbon-producing shrub capable of growing in arid climates. Spoken of as the "gasoline tree."

Greenhouse Effect   The entrapment of heat from the sun by the earth's atmosphere.

Green River Basin   Oil shale deposit site in southwestern Wyoming; part of Green River Formation.

Green River Formation   World's richest deposit of oil shale, within 125-mi radius of juncture of Wyoming, Utah, and Colorado.

Gross Primary Production   Measure of total energy fixed by a community through process of photosynthesis.

Gross Production   Observed increase in biomass over unit time, plus correction loss (i.e., estimated amount of matter lost due to leaf and root shed).

Growing Stock (Standing Volume; Commercial Volume)   Wood in main stems (boles) of growing trees over 5 in. in diameter inside the bark at breast height, above a 1-ft stump, up to a top diameter of 4 in.

Guardian Process   A hot-water extraction process for tar sands, employing a proprietary chemical. Produces a semiliquid tar that can be cracked for cold pipelining.

Guayule   *Parthenium argentatum* Gray. A member of the sunflower family yielding a rubber latex equivalent to *Hevea brasiliensis,* or natural rubber. Grows in arid regions.

Gymnosperm   Common name for members of division Pinophyta; includes softwood trees.

Hardwood   Wood of an angiospermous tree; primarily deciduous trees.

Harvest (v.)

   1 To mine or extract peat from naturally occurring deposits or bogs.

   2 To reap or gather a crop.

Haul-Back Mining   Variation of contour mining in which a peripheral strip of overburden is removed sequentially and hauled back into the void left by removal of the coal or ore.

H-Coal Process   Direct hydrogenation, solvent-extraction process for producing feedstock and fuel-grade oil from coal.

Heavy Oil   (*See* Tar Sand.)

Hemicellulose   Simple or mixed polysaccharides of smaller dimensions than cellulose.

Heterocyclic   Cyclic organic compound in which ring structure contains more than one type atom.

Heterotroph   Organism obtaining nourishment from the ingestion and breakdown of organic matter.

Hevea Brasiliensis   Plant that provides natural rubber.

High-Rate Algal Pond   Shallow (30 to 50 cm) aquacultural impoundment with mechanical agitation.

High-Sulfur Coal   Coal containing over 3 percent sulfur by weight.

Holocellulose   The carbohydrate fraction of extractive-free wood.

Home Scrap   Residual material generated in primary production and returnable directly to the production process.

Homolog   One in a series of organic compounds differing from each other by a CH_2 group.

Hopper Car   Freight car with sloping, hinged bottom for discharging bulk cargo by gravity.

Horizontal Divestiture   (*See* Divestiture, Horizontal.)

Hormone, Plant   (*See* Plant Growth Regulator.)

Hot-Water Extraction   Bureau of Mines tar sands process similar to hot-water separation process of GCOS, but adding fuel oil solvent at conditioning stage to aid in extraction.

Hot-Water Separation   GCOS tar sands extraction process comprising three major steps: conditioning (mixing; pulping), separation (settling), and scavenging (flotation) in hot water.

Huff-and-Puff (Huff-'n'-puff)   (*See* Intermittent Steam Stimulation.)

Humic Acid   A complex organic acid found in the organic portion of soil.

Humic Kerogen   (*See* Kerogen, Humic.)

Humus The combined decomposition-resistant organic components of soil.

Hutch Sulfur-laden dust from dust-collector system of coal crusher.

Hydraulic Fracturing (*See* Hydrofracking.)

Hydraulic Mining Use of water jet to separate coal from spoil.

Hydrilla Submersed aquatic weed.

Hydrocarbon An organic compound comprising a combination of carbon and hydrogen atoms.

Hydrofining (Hydrorefining) A fixed-bed catalytic process to desulfurize and hydrogenate a raw hydrocarbon feedstock.

Hydrofracking (Hydrofracturing; Hydraulic Fracturing) Injection of high-pressure water into a seam to increase porosity.

Hydrogasification Reaction of coal and hydrogen to form methane and char.

Hydrogenation
1 Catalytic reaction of hydrogen with organic compounds (usually unsaturated) to form more stable compounds.
2 Deoxygenation or chemical reduction process for extraction of oxygen from cellulose.

Hydrologic Cycle (Water Cycle) Cycle of water through oceans, atmosphere, land, and back to the oceans.

Hydrorefining (*See* Hydrofining.)

Hydroretorting Process Hydrotreatment of kerogen in a retort.

Hydrotreatment Process Catalytic refinery process in which petroleum product is treated with hydrogen to remove impurities (e.g., sulfur, oxygen, nitrogen, or unsaturated hydrocarbons).

Hypothetical Resources Undiscovered deposits which may reasonably be expected to be found in known regions.

Hysteresis The ratio of energy absorbed to total energy input.

HYTORT Hydroretorting of oil shale to form SNG and syncrude.

Ichthyol Bituminous schist; a variant of oil shale.

Identified Resources Specific deposits whose existence and location are known.

Illite Sedimentary rock containing iron, aluminum, potassium, magnesium, and silicon with very little carbonate; mineral matrix of Devonian shales.

Incineration (Direct Combustion) Combustion of organic material to produce heat and ash.

Incineration, Run-of-Mine (*See* Run-of-Mine Incineration.)

Incremental Pricing Charging industrial and commercial users for subsidization or distributing costs for residential users of gas.

Index, Exponential Reserve   (*See* Exponential Reserve Index.)

Index, Leaf Area   (*See* Leaf Area Index.)

Index, Recycling   (*See* Recycling Index.)

Indicated Reserves   Deposits, the quantity and grade of which have been determined partly from specific measurements and partly from projections over moderate distances.

Industrial Finishes (Chemical Coatings)   Coatings marketed for manufacturers' use in the factory.

Inferred Reserves   Deposits, the quantity of which are estimated on the basis of geological knowledge.

Inhibitor   Plant hormone that retards physiological or biochemical processes.

In-Place Resource   (*See* Resource, In-Place.)

Insolation   Solar energy received; rate of delivery of direct solar energy per unit of horizontal area.

Institutional Factors   Factors such as federal regulations, featherbedding practices, union rules, and so forth, that affect production and delivery.

Interburden   Weak, thin rock layers separating multiple seams of coal in surface formations.

Interior Province (Coal)   Iowa, Kansas, Missouri, Illinois, Indiana, Oklahoma, Arkansas, and western Kentucky.

Intermediate   Compound formed between initial material or feedstock and final product.

Intermittent Steam Stimulation (Huff-and-Puff)   Process in which steam is injected into a tar sands deposit under high pressure (350°C and 2000 psi) for a period of four to six weeks ("huff"), and bitumen is then pumped out for up to six months ("puff"), after which the cycle is repeated; also applied to enhanced recovery of heavy crude oil following exhaustion of natural drive.

Interstate Gas   Gas produced in one state and distributed across state lines.

Intrastate Gas   Gas produced and distributed within a state.

Isomers   Two or more organic compounds with the same kind and number of atoms but having different molecular structures and properties.

Isoprene   A conjugated diolefinic liquid, C_5H_8.

Jack   Adjustable post emplaced to provide temporary roof support while the stope is worked.

Jet Fuel   Both naphtha- and kerosene-type fuels.

Jig Flotation   Mechanized scale-up of panning operation for separating coal or mineral from refuse.

Jojoba   *Simmondsia chinensis.* Source of spermlike oil. Can be grown in semi-arid regions.

Jones Act   Act that imposes restriction on interstate transport of goods by foreign carriers.

Juvenile Coppice (n.)   One- to two-year growth by stem sprouting.

Kelp, California Giant   *Macrocystis pyrifera,* a macroaquatic plant that grows off the California coast.

Kenaf (Ambary)   *Hibiscus cannabinus,* a tough, stringy tropical plant that produces high yields of paper and substitute flax fiber. Can be grown in the arid Southwest and in south-central New Jersey.

Kerogen   Organic component of oil shale; complex, cross-linked, high-molecular-weight combustible organic material occurring naturally with marlstone rock.

> Algal: Base of aliphatic hydrocarbon products formed from decomposition of algae.

> Humic: Base of aromatic hydrocarbon products; formed from decomposition of higher organic materials.

> Mixed: Combination of humic and algal kerogen which degrades to yield complex mixture of hydrocarbons.

Kiln, Rotary   (*See* Rotary Kiln.)

Kukersite   Algal-base Ordovician oil shale.

Landed Cost   Cost put ashore at port of entry.

Landfill ("Reverse Mine")   Area for disposal of solid waste or sludge.
> Sanitary: (*See* Sanitary Landfill.)
> Secured: (*See* Secured Landfill.)

Landfill Gas (Dumpground Gas; Unnatural Gas)   Combustible gas (nominally 50 to 55 percent methane to 45 to 40 percent CO_2) generated by anaerobic digestion of biodegradable matter in landfill.

Leaching   Process for extracting oil shale, in which very hot water is injected to leach nahcolite from mineral matrix, opening deposit for retorting kerogen in situ by steam injection.

Leaf Area Index   Ratio of the canopy leaf area to its projected surface area on the ground.

Leucaena   Mimosa-related, leafy evergreen plant which grows up to height of 60 ft in dry climates. Possible source of wood, paper fibers, food, feed, fuel, fertilizer, and wax.

Life Index   Current reserves divided by annual production rate.

Lifetime   Estimate of the length of time for which a resource will be economically recoverable in commercial quantities.

Lift (**v.**)   To pump (oil or gas) from wells.

Lighter **(v.)**   To transfer cargo from large ship to smaller vessel for trip to shore destination.

Light Oil   (*See* Benzol.)

Lignin   An indigestible component of the woody cell walls of plants that cements the cellulose together; empirical formula $C_{10}H_{11}O_2$.

Lignite   Coal of recent origin, intermediate between peat and bituminous coal.

Lignocellulose   Any of a combination of substances in woody plant cells consisting of cellulose and lignin.

Lime   Calcium hydroxide, $Ca(OH)_2$.

Limestone   Calcium carbonate, $CaCO_3$.

Lipoid (Lipid)   A fatlike substance.

Liquefaction   Conversion of solid or gaseous hydrocarbons to liquid derivative.

Liquefied Natural Gas (LNG)   Natural gas cooled to about $-160°C$ and compressed (600:1) for shipment and storage. Primarily methane.

Liquefied Petroleum Gas (LPG; Bottled Gas)   Consists primarily of compressed propanes and butanes recovered from natural gas and in the refining of petroleum.

Liquor, Black   (*See* Black Liquor.)

Litter, Forest   Branches and leaves on forest floor.

Live Stockpile   Accumulation of raw material or fuel used to feed process on day-to-day basis.

Lockhopper   Pressure-lock feeder for introducing solid material into high-pressure reactor.

Log Rule   (*See* Log Scale.)

Log Scale (Log Rule)   Measuring device for estimating volume of wood recoverable from log. May be used in conjunction with volume tables.

Long-Term   Beyond year 2000 A.D.

Longwall Generation (Directional Drilling; Preliminary Mining)   In situ gasification of coal in which combustion takes place between two parallel, horizontal tunnels driven in seam.

Longwall Mining (Advance Mining)   System whereby faces are worked advancing from shaft toward boundary, jacks are remotely emplaced, and roof is ultimately allowed to collapse behind miners as work progresses.

Lower 48   States of United States, excluding Alaska and Hawaii.

Low-Sulfur Coal   Coal containing 1 percent or less sulfur by weight.

Lurgi Process   Fixed-bed gasifier system for coal; the only proven gasification process deemed suitable at present to produce pipeline gas from coal.

Mahogany Zone   Subsurface zone containing richest oil shale ores of Piceance Creek Basin (110 ft thick, 25 to 30 gpt assay).

"Man-made" ore   Waste or refuse from domestic, commercial, or industrial sources.

Mariculture   Cultivation of marine or saltwater plants. (*See also* Aquaculture).

Marker Crude   A specific grade of crude oil used by a producing nation to serve as a standard for developing price schedules for all crudes.

Marketed Production (Gas)   Gas sold for pipeline distribution and LNG, following process losses, NGL and LPG production, repressuring of reservoirs for secondary recovery of oil, and so forth.

Marsh Gas   Combustible gas (chiefly methane) formed by decay of vegetation in stagnant water.

Massive Hydraulic Fracturing   Large-scale breakup of coal or other ore seams by hydrofracturing.

Measured Reserves (Apparent Reserves; Proved Reserves)   Deposits, the location, quantity, and quality of which are known from geologic evidence supported by engineering data.

Mechanical Beneficiation   Shredding refuse to manageable size, followed by air or other classification means, and finally separation into various product streams.

Medium-Sulfur Coal   Coal containing 1.1 to 3.0 percent sulfur by weight.

Mesophilic Digestion   Moderate-temperature, anaerobic digestion carried out at about 95°F (35°C). (*See also* Thermophilic Digestion.)

Metabolism   Physical and chemical processes involved in assimilation, catabolism, and provision of energy for use by an organism.

   C-3: Photosynthesis mechanism involving reaction of five-carbon compound with CO_2 to form six-carbon compound which then cleaves to form two stable three-carbon compounds.

   C-4: Photosynthesis mechanism involving formation of four-carbon compounds which are then decarboxylated to three-carbon compounds.

Metamorphism   Change of form, structure, or constitution of mineral or organic matter due to natural agencies, such as pressure and heat. (*See also* Carbonification.)

Methacoal   Slurry of pulverized coal in methanol.

Methanation

   1 Catalytic reaction of syn gas to form methane, with water as a by-product.

   2 Formation of methane from methanol and hydrogen.

Methane   The lightest compound in the paraffinic series of hydrocarbons (generalized formula C_nH_{2n+2}); composition CH_4; the major constituent of natural gas.

Methyl-Fuel   Controlled proprietary blend of methanol with C_2–C_4 alcohols (Vulcan-Cincinnati Co.)

Metric Ton **(te)**   1000 kg, also known as tonne.

Micellar Flooding (Surfactant Flooding)   Technique used for tertiary recovery by injecting alcohol(s), surfactant, and polymer into a reservoir.

Microalgae   Algae with diameters less than 20 μm.

Microbial Extraction   Tar sands process employing bacterial or fungal strains to break up bitumen and separate it from minerals.

Middle-Term   The period A.D. 1990 to 2000.

Middlings   Coal product intermediate between cleaned coal and tailings.

Milling   Harvesting peat by shredding it on the surface of the ground for drying.

Minirotation (of Forests)   One- to three-year harvest cycle.

Mixed Kerogen   (*See* Kerogen, Mixed.)

Modified In Situ Retorting (Stoke Mining)   Process in which about 20 percent of the oil shale deposit is mined out and extracted above ground, while balance is rubblized and extracted in situ.

Molten Salt Pyrolysis Process   Molten sodium carbonate is used as heat-transfer medium and absorbent for impurities in waste, feed, and off-gas, to form low-heating-value product gas.

Monopoly   A group of four or fewer companies accounting for 50 percent or more of an industry's business.

Mote   Linty material.

Mountain Province (Coal)   Idaho, Utah, Colorado, Arizona, and New Mexico.

Moving Bed   (*See* Fixed Bed.)

Multimineral Process   Method that extracts both minerals and kerogen from oil shale, thus reducing volume of spent shale by about half.

Multiseam Mining   Strip mining of multiple layers of coal separated by thin layers of interburden.

Muskeg   Semifloating mass of partially decayed vegetation and scrub trees (largely spruce, tamarack, and jack pine).

Nahcolite   Mineral sodium bicarbonate ($NaHCO_3$); major mineral component of Piceance Creek Basin oil shale.

Naphtha   Petroleum fraction between gasoline and kerosene.

Natural Coke   A rather rare, high-carbon, low-volatile, solid fossil fuel.

Natural Gas   A naturally occurring gaseous fossil fuel comprising basically 75 to 99 percent methane, plus small amounts of one or more of the following: ethane, propane, olefinic hydrocarbons, hydrogen sulfide, carbon monoxide or dioxide, hydrogen, helium, and/or water.

Natural Gas Liquids (NGL)   Condensate of propane, butanes, and pentanes recovered from natural gas.

Natural Gasoline   Light, volatile hydrocarbon liquids of higher molecular weight than butane, entrained in natural gas stream; used for blending with refinery gasoline.

Naval Stores   Turpentine, rosin, tall oil, and other oleoresins derived from wood.

Near-Term   Period to late 1980s.

Net Assimilation Rate (Photosynthetic Rate)   Weight added by photosynthesis per unit leaf area per unit time.

Net Energy Production (NEP)   Excess energy over amount required for maintenance of source plants; energy potentially available to consumers.

Net Photosynthesis   Photosynthesis less photorespiration of plant metabolism.

Net Primary Production (NPP)   Output of biomass for fuel and feedstocks less energy input.

Net Production   Production of cleaned coal (i.e., excluding mineral slag or culm).

New Gas   Gas brought into production after 1975.

New Oil   Oil discovered after 1973.

New Scrap (Prompt Industrial Scrap)   Waste generated in a manufacturing or fabricating process.

Nonassociated Gas (Dry Gas)   Gas found in traps not associated with crude oil. Contains less than 0.1 gal (condensed) natural gasoline vapor per 1000 scf.

Noncommercial Forest   Forest land yielding less than 20 ft^3 per acre-year.

Nonrenewable Resource   One of geological origin or, in a practical sense, a biological resource having an extremely long growing cycle (e.g., redwood trees).

Nonused Residues   Residues of biological growth not gathered for sale or material use (e.g., crop wastes burned in the fields and forest litter).

Normalized Reserves (Coal)   Quantities standardized on basis of heat content or sulfur content.

Northern Great Plains Province (Coal)   Montana, Wyoming, North Dakota, and South Dakota.

N-T-U Process   Batch-feed variation of countercurrent internal-combustion process of extracting shale oil.

No. 6 Residual Fuel Oil (Bunker Oil)   Heavy diesel fuel remaining after lighter oils are distilled off in the refinery.

Offshore   Subsurface region extending beyond tide level.

Oil, Char   (*See* Char-Oil.)

Oil, Pyrolysis (Pyrolytic Oil)   (*See* Pyrolytic Oil.)

Oil Sands   (*See* Tar Sand.)

Oil Shale   Naturally occurring combination of kerogen and marlstone rock; properly neither an "oil" nor a "shale."

Old Gas   Gas brought into production in or before 1975.

Old Oil   Oil discovered prior to 1973.

Old Scrap (Post-Consumer Scrap)   Waste that is generated from products that have reached the end of their useful lives.

Oleoresin   Natural product containing essential oil and resin.

Onshore   Land area above mean high-tide level.

Open-Cast Mining   (*See* Open-Pit Mining.)

Open-Pit Mining (Open-Cast Mining)   Extraction of shallow ores by removing the overburden and breaking and loading the ore.

Outcropping   Ore bed, seam, or deposit exposed to atmosphere.

Outer Continental Shelf (OCS)   (*See* Continental, Shelf.)

Overburden (Capping; Spoil)   Earth and rock (clay, limestone, shale, slate, etc.) overlying a mineral deposit or coal seam suitable for strip mining.

Overburden Ratio   Ratio of overburden thickness to ore deposit thickness.

Overthrust Belt   Petroleum-rich geological formation extending in a 70-mi-wide band between the Mexican and the Canadian borders (and beyond) of the Rocky Mountains, at depths down to 17,000 ft.

Oxirane   (*See* Epoxy.)

Oxoalcohol   An alcohol containing a keto $C=O$ group.

Oxygenated Hydrocarbons   Oxygen-containing hydrocarbons.

Oxygen Corrosion   Corrosion of well hardware exposed to air, most commonly in offshore wells, shallow producing wells, and brine-handling and injection equipment.

Paint Vehicle   The resin binder and film former of a paint formulation.

Panel   A large rectangular section (block, pillar, face) of coal that is worked independently of other such sections in an underground mine.

Paraho Kiln Processes   Three variations of specific surface retorting methods for extracting shale oil; variations are based on methods of transferring heat in the retort.

Paramagnetic   Magnetized parallel to an applied magnetic field.

Partial Pyrolysis   (*See* Starved Air Combustion.)

Partings   Thin layers of clay, shale, pyrite, and so forth, separating bands of coal.

Pay Thickness   Ore deposit thickness.

Peat   Dark brown or black coal precursor; a highly organic soil combined with a complex mixture of plants, litter, water, burrowing animals, and micro-organisms.

Percolation Technique (Filtration Technique)   A shaftless UCG pregasification method involving driving multiple boreholes and carrying out gasification between pairs; most successful technique for horizontal seams.

Perennial Plant   A plant that has an indefinite life cycle, dying back seasonally and developing new growth.

Permafrost   Perennially frozen subsoil underlying arctic tundra.

Petrobrás   The Brazilian National Oil Co.

Petrochemical   Chemical made from feedstock derived from petroleum or natural gas.

Petroleaching   Extraction of bitumen from tar sands by recycled light hydro-carbons.

Petroleum   Literally, "rock oil." All mobile hydrocarbons recoverable from the subsurface by drilling wells. Includes crude oil ("oil"), oil-associated gas ("casinghead gas" or gas dissolved in oil), natural gas ("free gas") not associated with oil, and liquid condensates (distillates; natural gas liquids or NGL). A flammable, complex mixture of hydrocarbons with small amounts of other substances (e.g., sulfur, metals).

Photosynthesis   Synthesis of carbohydrates with the aid of photosynthetically active radiation (400 to 700 nm) acting on CO_2, and a source of hydrogen (usually water), with the release of oxygen.

Photosynthesis, Net   (*See* Net Photosynthesis.)

Photosynthetically Active Radiation   Portion of spectrum (400 to 700 nm) participating in photosynthesis.

Phreatophytes   Woody plants, perennial grasses, and broad-leaved plants growing at the water's edge or with roots extending into the capillary zone overlying the water table.

Phytohormone   (*See* Plant Growth Regulator.)

Phytoplankton   Weakly motile aquatic plant and animal life.

Piceance Creek Basin   World's richest known deposit of oil shale, in northwest section of Colorado; part of Green River Formation.

Pillar (Rib)   Block of underground mine left undisturbed, to support mine roof.

Pillar-and-Stall (U.K.)   (*See* Room-and-Pillar.)

Pipeline Quality Gas   Gas with a heat content of about 1000 to 1100 Btu per scf.

Plant, C-4   Plant of tropical origin that has an extra preliminary CO_2-fixing pathway in addition to reductive pentose phosphate cycle.

Plant Efficiency   Heating value of product as fraction of total energy input. (*See also* Process Efficiency.)

Plant Growth Regulator (Plant Hormone; Phytohormone)   Organic chemical that affects physiological processes in plants.

Plant Hormone (Phytohormone)   (*See* Plant Growth Regulator.)

Plant Respiration   (*See* Respiration.)

Plastic   A material containing a high-molecular-weight organic substance as an essential ingredient, and which can be shaped by heat and/or pressure into solid finished objects.

Platform (**v.**)   To reform, using platinum catalysts.

Pneumatic Fracture   (*See* Pneumatic Linkage.)

Pneumatic Linkage (Pneumatic Fracture)   Fracture of coal seam by injecting high-pressure air.

Pneumoconiosis   Black lung disease, a chronic inflammation of the lungs caused by accumulation of inhaled coal particulates.

Pollard (**v.**)   To cut back tree to stem to promote growth of dense crown or foliage.

Pollution   Uncontrolled release of waste.

Polybutadiene (PBD)   A synthetic rubber formed by polymerizing 1,3-butadiene, $CH_2:CH\cdot CH:CH_2$.

Polyester Resin   Polymeric product of reaction between organic acid and alcohol.

Saturated: Polyester resin containing no double or triple bonds, for example polyethylene terephthalate (fibers and films) or polybutylene terephthalate (thermoplastic molding resin).

Unsaturated: Polymer of a linear unsaturated resin formed from dibasic acids and dihydric alcohols, cross-linked with an unsaturated vinyl-type monomer (e.g., styrene).

Polyethylene Resin   A thermoplastic resin formed from ethylene monomer.

Polyethylene Terephthalate Resin (PET)   A polymer of ethylene glycol and dimethyl terephthalate or purified terephthalic acid.

Polyisoprene Synthetic "natural" rubber.

Polymer A natural or synthetic, high-molecular-weight organic compound. (From "poly"—many and "mer"—unit or part.)

Poly(Vinyl Chloride) Resin Thermoplastic resin formed by polymerizing vinyl chloride monomer.

Pool (Reservoir) Underground accumulation of petroleum, characterized by a single natural reservoir and natural pressure system. Geologically, it is a self-contained, permeable mineral reservoir totally surrounded by impermeable barriers.

Porphyrins The active nucleus of chlorophylls and hemoglobin.

Possible Reserves (*See* Inferred Reserves.)

Post-Consumer Scrap (*See* Old Scrap.)

Potential Reserves Deposits currently nonexploitable, yet which may become exploitable if technology or economic conditions improve.

Precursor A chemical from which other chemical products are ultimately produced.

Pregasification Process of providing access to coal seam for UCG and linking the blast inlet and gas offtake through the seam.

Preliminary Mining (*See* Longwall Generation.)

Preparation (Coal) (*See* Beneficiation.)

Presumptive Carcinogen (*See* Carcinogen, Presumptive.)

Primary Material Virgin or unused material.

Primary Process Original process of converting raw material to product. (*See also* Secondary Process.)

Primary Production, Gross (*See* Gross Primary Production.)

Primary Production, Net (*See* Net Primary Production.)

Primary Renewable Resource The output of a biological production system designed for the deliberate purpose of supplying plant- or animal-derived materials to industry (e.g., wood products).

Probable Reserves (*See* Indicated Reserves.)

Process Efficiency Heating value of product as fraction of heating value of feedstock.

Producer Gas Low-heating-value raw gas (high in CO and H_2) formed in a gasifier by burning coal with a deficiency of air.

Production, Gross Primary (*See* Gross Primary Production.)

Production, Net Energy (*See* Net Energy Production.)

Production, Net Primary (*See* Net Primary Production.)

Production Well   Shaft from which bitumen is extracted by an in situ process.

Productivity   Gross or net rate of production over a given time period.

Prompt Industrial Scrap (See New Scrap.)

Propylene (Methyl Ethylene; Propene)   Organic compound (monomer) having the composition $CH_3CH=CH_2$; used to manufacture plastics; also a chemical intermediate.

Propylene Alkylates   Additives used with tetraethyllead to improve antiknock properties of gasoline.

Proved Reserves   (*See* Measured Reserves.)

Province   Arbitrary geographical region established for statistical and administrative purposes for dealing with a particular resource.

Puff   (*See* Huff-and-Puff.)

Puffing   Expansion of oil shale during processing. (*See also* Fluffing.)

Pulp   Mixture of tar sand feed and water or other fluid carrier.

Pulpwood   Wood used in the manufacture of paper and paper products. (*See also* Roundwood.)

Pumpherston Process   Original commercial method of retorting oil shale, dating back to about 1860.

Purification (Cleanup)   Removal of contaminants from synthesis gas or product gas stream.

PUROX   Hybrid pyrolytic-incineration process for producing medium-heating-value gas from refuse.

Pyrite   Iron disulfide (FeS_2).

Pyritic Shale   Fine-grained sedimentary rock rich in pyrite.

Pyrolysis (Carbonization)   Thermal cracking or destructive distillation of hydrocarbons in inert atmosphere.

Pyrolysis Gas   Raw gas from pyrolyzer (about 35 percent methane) which is combined with synthesis gas for upgrading to pipeline gas.

Pyrolysis, Partial   (*See* Starved Air Combustion.)

Pyrolytic Oil (Pyrolysis Oil)   A heavy, fluid pyrolysis product derived from MSW; empirically $C_5H_8O_2$, having a low carbon content.

Quad   A large unit of energy—a quadrillion (10^{15}) Btu. (*Note:* Some publications use Q as an abbreviation for quintillion or 10^{18} Btu).

Quarter Carry   Twenty-five percent of working interest in producing well.

Radiation, Photosynthetically Active   (*See* Photosynthetically Active Radiation.)

Radon An elemental, highly toxic radioactive gas associated with coal; a decay product of radium.

Rank Coal classification based on degree of metamorphism.

Raw Material, Primary (*See* Primary Material.)

Reclaim (v.) To recover a used material or product from a waste product or by-product.

Reclamation
1 Return of land to full productive use following mining operations; basic reclamation entails grading, revegetation, and drainage control.
2 Recovery of primary or secondary products for reuse of the material(s).

Recoverable Reserves Deposits expected to be extractable with present techniques and under current economic conditions.

Recoverable Resources Deposits extractable if techniques and economic conditions were to be enhanced slightly.

Recovery Factor Ultimate production as a fraction of resource (oil or gas) initially in place.

Recovery Rate Percent of in-place hydrocarbon recovered from ore by extraction process.

Recyclable Material Material that *can* be reclaimed from waste. (*See also* Recycled Material.)

Recycle To collect and process a waste product to recover the materials for reuse in a manner comparable to its original use, or to recover chemical intermediates or components used in making the product. (Does not include heat recovery from combustion, or reuse of a product for a different purpose.)

Recycled Material Material that *has been* reclaimed from waste.

Recycling Index Measure of recoverability of a material through collection and reprocessing.

Reduced Organic Material Portion of natural rubber latex extractable by acetone and benzene.

Refine To purify or process a material to a useful state.

Reformate Product of petroleum refinery thermal or catalytic refining process.

Reforming The conversion of petroleum feeds to volatile, higher-quality products by use of catalysts and/or high temperatures. Combines cracking, polymerization, dehydrogenation, and isomerization.

Refractory Heat Carrier Process USSR oil shale extraction process employing hot spent shale ash as heat transfer medium for crushed ore.

Refuse Solid waste.

Refuse-Derived Fuel (RDF)   Solid fuel processed from combustibles separated from solid waste by essentially mechanical means.

Regeneration   Photosynthesis.

Regenerator   Fluid-bed coke burner.

Rehabilitation   Restoration of land to predetermined condition and productivity, without substantial environmental deterioration and in conformance with surrounding aesthetic values.

Renewability Ratio   The ratio of stock (wood) renewal rate to stock-depletion rate.

Renewable Resource   A resource whose supply can be replenished or restored when the original stock has been depleted; generally implies biological origin.

Repressure (v.)   To apply external gas pressure drive to enhance oil recovery.

Reserve Base   Quantity in place, as estimated on basis of arbitrary depth and thickness criteria, ignoring recovery factor; intermediate between conventional reserves and resources.

Reserves   Identified deposits known to be recoverable with current technology under present economic conditions.

Reservoir   (*See* Pool.)

Residual Fuel Oil   The petroleum fraction remaining after all lower-boiling fractions have been distilled off. Requires heating to be pumped and handled conveniently. Used in commercial and industrial power generation.

Residue Gas   Gas remaining after removal of higher hydrocarbons.

Residues, Nonused   (*See* Nonused Residues.)

Residues, Returned   (*See* Returned Residues.)

Residuum   Solids remaining from coker. (*See* Coked Solids.)

Resin   A solid, semisolid, or pseudosolid organic material of indefinite (generally high) molecular weight, which flows under stress and usually exhibits conchoidal fracture.

Resource, In-Place   Quantity of hydrocarbon present in ground disregarding recoverability.

Resource, Nonrenewable   (*See* Nonrenewable Resource.)

Resource, Renewable   (*See* Renewable Resource.)

Resources   Includes reserves as well as deposits that have been identified, but cannot be extracted now because of economic or technological constraints; also includes economic or subeconomic materials that have not yet been discovered.

Respiration   Slow, regulated combustion process; oxidation of organic materials.

Restoration   Duplication of the premining conditions of a site.

Retreat Mining   Working of coal seam from boundary toward mine shaft, using shearing and breaking machine. (*See also* Longwall Mining.)

Returned Residues   Portion of residues from crops and manures returned to soil without sale.

Reverse Combustion (Countercurrent Combustion)   Process in which bitumen is ignited at production well and flame front is propagated toward air injection well counter to air flow. (*See also* Thermal In Situ Extraction.)

Reverse Ignition   In situ gasification accomplished by combustion with injected air between two parallel wells 20 to 50 m apart driven to the bottom of a coal seam; flame front advances countercurrent to flow of injected air.

"Reverse Mine"   (*See* Landfill.)

Rib   Narrow wall of coal bed left undisturbed to support roof of underground mine. (*See also* Pillar.)

Rig   Oil or gas drilling platform for OCS exploration, development, or production.

Rolling Stock   Locomotives, freight cars, and so forth, used for transport.

Rollover   Very rapid, spontaneous mixing of stratified layers of LNG having different densities (compositions).

Roof Rock   Impermeable rock layer (generally shale or evaporites—gypsum, anhydrite or halite) that forms a cap on petroleum traps.

Room-and-Pillar (Pillar-and-Block; Pillar-and-Stall)   A deep mining technique in which coal or ore is mined in rooms interspersed with pillars or ribs left for roof support.

Rotary Kiln   Cylindrical kiln rotating on a slightly inclined axis.

Rotation, Mini-   (*See* Minirotation.)

Rotation, Short   (*See* Short Rotation.)

Roundwood   Bolewood for uses other than sawing into lumber.

Rubblize (v.)   To fracture ore formation into mined-out chamber by use of explosives.

Rule, Log   (*See* Log Scale.)

Rule of 70   Rule of thumb stating that 70 divided by the annual rate of increase of a given parameter equals the doubling time or the time in years required for that parameter to double; applies up to annual rate of increase of about 10 percent.

Rule 66   Los Angeles County rule governing use of smog-producing solvents.

Run-of-Mine Coal   Grade of coal delivered without washing or other preparation.

Run-of-Mine Incineration Combustion of solid waste without initial separation of noncombustibles.

Sand Reduction Process Variation of cold-water separation process for tar sands, aimed primarily at providing feed for fluid coking.

Sanitary Landfill Land disposal site engineered to minimize environmental hazards, in which thin layers of refuse are compacted, and covered with earth at the end of each day.

Saturated Polyester (*See* Polyester Resin, Saturated.)

Sawlog Log suitable for cutting into lumber.

Sawtimber Tree(s) sufficiently large to contain at least one sawlog.

Scale, Log (*See* Log Scale.)

Scavenge (v.) To use froth flotation to separate bitumen from middlings stream.

Scavenger Cell Froth flotation cell.

Scrap (*See* particular type—home, new, old, prompt.)

Scrubber A wet collector for the removal of entrained liquid droplets or particulates from a gas stream.

Seam Natural bed or stratum of coal or other mineral.

Secondary Process Process of converting scrap to new product.

Secondary Recovery Recovery of crude oil by water injection (waterflooding) following exhaustion of primary drive.

Secondary Renewable Resource By-products or residues of biological production systems whose essential purpose is to produce food, feed or other nonmaterial goods.

Secured Landfill Land disposal site for hazardous wastes that allows no hydraulic change, segregates the refuse from the environment by containerization, has restricted access, and is continually monitored.

Separation Process of floating bitumen off of pulp mixture, and settling the sand.

Seven Majors (*See* Seven Sisters.)

Seven Sisters Seven major multinational oil companies: Exxon, Mobil, Standard of California, Texaco, Gulf, Royal Dutch Shell, and British Petroleum.

Sewage Sludge A semiliquid waste with a solids content in excess of 2500 ppm, formed in a wastewater treatment system.

Shaftless Methods UCG pregasification technologies involving driving multiple boreholes from the surface; do not require underground labor.

Shaft Methods UCG pregasification technologies involving the driving of large diameter openings; require much underground labor.

Shale Gas   Natural gas trapped in Devonian Basin shale deposits.

Shelf Goods   (*See* Trade Sales.)

Short Rotation (of Forests)   Ten- to 20-year harvest cycle.

Shortwall Mining   Modification of longwall technique, in which relatively short mine face (typically 5 to 30 yd) is worked with a continuous-mining machine.

Silage   Green or mature fodder fermented to inhibit spoilage, and used as winter feed.

Silviculture   Phase of forestry dealing with establishment, development, reproduction, and care of trees.

Sink   A mass that acts as an absorber, for example, of CO_2.

Slag (Spoil; Gob) **(n.)**

  1 Mining debris.

  2 Cinderlike inorganic residue from thermal process.

Slag. **(v.)**   To form a molten mass of mineral waste which solidifies to a clinker (on cooling).

Slagging   That which forms a fused mineral residue.

Slagging Pyrolizer   Reactor producing slag in pyrolysis process.

Slow Rules   Regulations reducing rolling stock speeds, imposed as a result of poor roadbed conditions.

Sludge

  1 Wet product of gas-stack scrubber; largely calcium sulfate, in case of common lime or limestone scrubber.

  2 Semiliquid waste from sewage treatment, with >0.25 percent solids content.

  3 Any semisolid waste from chemical process.

Slug Feeding   Periodic feeding of coarse, high-solids sludge input to digester.

Sod Cutting   Harvesting peat by cutting it into bricks or cylinders for drying.

Softwood   Wood from coniferous trees.

Sol-Frac   Explosive fracturing and solvent extraction process for extracting very heavy oil from midwestern fields.

Solids, Dry Organic   (*See* Dry Organic Solids.)

Solids, Volatile   (*See* Dry Organic Solids.)

Solution Gas Drive   Well drive pressure created by gas dissolved in the crude oil under pressure.

Solvent Refined Coal   Low-melting (about 375°F), solid coal product formed by solvent and hydrogen treatment of coal to reduce ash, sulfur, and other impurities.

Solvent Refining   Elimination of sulfur and mineral impurities by solvent extraction of organic components in coal.

Sour Corrosion   Corrosion by gas or oil containing hydrogen sulfide.

Sour Crude   High-sulfur crude (1 to 4 percent).

Sour Gas   Natural gas high in hydrogen sulfide.

Southern Oscillation   Episodic (3 to 4°C) variation in temperature of Pacific waters below the equator occurring on a time scale of six to seven years.

Speculative Resources   Includes conventional deposits in areas where there have not yet been discoveries, and unconventional resources that have only recently been identified or that have yet to be identified.

Sperm Oil   Oil extracted from sperm whales; component of fine lubricating oils. Sale in United States was banned in early 1970s.

Spherical Agglomeration Process   Extraction method similar to sand reduction; dense agglomerates are separated for upgrading.

Spoil Bank   Accumulation of overburden from surface mining, or combined mineral waste and unrecovered coal from mining, in general.

Stack Gas   Gaseous combustion products passed through a chimney.

Stack-Gas Scrubber   Equipment used to remove entrained liquid droplets or dust from flue gas.

Standardized Reserves   Estimates (coal) normalized for equivalent sulfur and heat contents.

Standing Volume   (*See* Growing Stock.)

Starved Air Combustion (Partial Pyrolysis)   Gasification or liquefaction of combustible materials by heating in the presence of a controlled amount of oxygen.

Steam Cracker   A unit for using high-pressure steam to break an organic compound (e.g., crude oil) into simpler molecules (e.g., naphthas).

Steam Stimulation   Process for in situ extraction of bitumen from tar sands by injecting high-temperature (350°C)/high-pressure (about 2000 psi) steam into wells; injection may be continuous or intermittent (huff-and-puff).

Stem   Trunk or bole of tree.

Stock, Growing   (*See* Growing Stock.)

Stoke Mining   (*See* Modified In Situ Retorting.)

Stomal Aperture (Stoma)   Minute opening in epidermis of higher plants, through which gases and water vapor are passed.

Stope   An essentially horizontal, subsurface working for removal of coal.

Stover   Fodder or feed, such as cornstalks.

Strategic Petroleum Reserve   Stockpile of crude oil held against contingency of strikes, embargoes, and so forth.

Stream Factor   Actual output as a fraction of nameplate capacity, assuming continuous operation.

Stream Technique   A shaft method of UCG pregasification used on steeply pitched seams; requires much underground work to prepare parallel inclined galleries and horizontal fire drifts.

Strike   Discovery of economically recoverable crude oil, natural gas, or other natural resource.

Strip Cropping   Alternating strips of wheat (or other grain crops offering little soil protection) with strips of grass or legumes.

Strip Mining   Surface mining in which overburden is stripped away to expose the coal bed.

Stripper Well   Oil well producing less than 10 bpd.

Styrenated Alkyd   An unsaturated polyester resin cross-linked with styrene monomer.

Styrene-Butadiene Rubber (SBR)   An elastomeric copolymer comprising a ratio of about 1 part styrene and 3 parts butadiene monomers; our major synthetic rubber.

Subbituminous Coal   Black coal intermediate in rank between lignite and bituminous coal; higher in carbon and lower in moisture than lignite.

Subsidence   A sinking of part of the earth's crust as a result of the collapse of subterranean cavities, such as mine workings.

Substitute Natural Gas (SNG)   Methane, 85+ percent; minimum heating value, 900 Btu per scf; delivery pressure, 1000 psig; maximum CO content, 0.1 percent; maximum total S, 10 g per 100 scf; maximum H_2O, 7 lb per 10^6 ft^3; maximum total inerts, 5 percent; specific gravity (air standard), 0.59 to 0.62; HC dew point, 40° at 1000 psig, in accordance with DOE.

Substitute Pipeline Gas (SPG)   High-methane gas (950+ Btu per scf) formed from coal, peat, oil shale, tar sands, refuse, and so forth.

Sucker   Oil well pump.

Sucrochemistry   Derivation of chemicals from sugar.

Sudds   Islands or accumulations of free-floating dead aquatic vegetation.

Suess Effect   Dilution of ^{14}C in the atmosphere, due to combustion of fossil fuel (which is low in ^{14}C).

Sulf-X   Proprietary desulfurization process based on gas–solid reaction between SO_2 and a proprietary sulfide of iron.

Supergiant Field   A continuous producing area which contains at least 10 $\times$ 10^9 bbl (1.4 $\times$ 10^9 tonnes) recoverable crude oil or at least 60 tcf (1.7 $\times$ 10^{12} m^3) recoverable natural gas.

Surface Mining   Mining from coal or mineral bed exposed to surface by removal of overburden.

Surfactant Flooding   (*See* Micellar Flooding.)

Surge Pile (Surge Bunker)   A large material reserve at strategic points or at the head of a process line to ensure a uniform flow of feed.

Suspension Bed   (*See* Entrained Bed.)

Swamp Gas   (*See* Marsh Gas.)

Sweep Efficiency   Fraction of in-place bitumen extracted from portion of tar sands deposit wet by emulsifying fluid or diluent.

Sweet Corrosion   Corrosion in the absence of oxygen and hydrogen sulfide, caused by CO_2 and fatty acids in sweet oil, gas, and condensate products.

Sweet Crude   Low-sulfur crude (<1 percent).

Sweet Gas   Natural gas low in hydrogen sulfide.

Swell Factor   Volume ratio of disturbed to undisturbed overburden and ore. (*See also* Fluffing.)

Swing Fuel   Alternative fuel resorted to during shortage of normal fuel(s).

Syncrude

    1   Synthetic crude oil derived from coal, tar sand, oil shale, refuse, etc.

    2   Syncrude Canada Ltd., conglomerate of Imperial Oil Ltd. (31 percent), Canada Cities Service Ltd. (22 percent), Gulf Oil Canada Ltd. (17 percent), Petro-Canada (Canadian government, 15 percent), Alberta Syncrude Equity (Alberta government, 10 percent), and Ontario Energy Co. (Ontario government, 5 percent).

Synfuel   Fluid fuel derived from coal, tar sands, or oil shale.

Syn Gas   Synthesis gas (CO plus H_2).

Syngas   Gas derived from coal, oil shale, tar sands, biomass, refuse, and so forth.

Synoil   Oil derived from coal, oil shale, tar sands, biomass, refuse, and so forth.

Synthesis Gas   A $CO-H_2$ mixture, in various proportions, used as feedstock for chemical reactions.

Tables, Volume   (*See* Volume Tables.)

Tackifier   Material added to rubber, adhesive, sealant, coating, and so forth, to enhance adhesion or building tack.

Tailings   Waste stream of suspended minerals, bitumen, and toxic chemicals from tar sands extraction plant.

Tall Crude   Standing timber raised specifically for fuel or feedstocks.

Tar Residue from thermal refining of petroleum; a misnomer for the hydro-carbon component of tar sands. (*See* Bitumen.)

Tar Acids Phenols.

Tar Bases Organic nitrogen compounds (e.g., amines).

Tar Sand (Oil Sand, Heavy Oil) Sand or rock formation saturated with crude hydrocarbon substance too viscous to be recovered in its natural state by pump-ing from a well.

Tasmanite An impure coal; regional name for local oil shale variant.

Terpolymer A copolymer resulting from combining three different monomers.

Tertiary Recovery (Enhanced Recovery) (EOR) Recovery of oil by thermal, gas injection, or chemical flooding techniques, following exhaustion of primary and secondary recovery.

Thermal Conversion Conversion by various forms of combustion or pyrolysis.

Thermal In Situ Extraction (Fireflooding) Process entailing ignition of bitu-men at base of well driven into deep tar sands formation and sustaining com-bustion by injection of air; heat increases mobility of bitumen and cracks it to upgraded crude, which is recovered from production or collection well(s).

Thermal NO$_x$ Oxides of nitrogen formed by oxidation of atmospheric nitrogen during combustion.

Thermophilic Digestion Elevated temperature anaerobic digestion carried out at 130 to 140°F (55 to 60°C).

Thermoplastic A plastic capable of being repeatedly softened by increasing the temperature and hardened by decreasing the temperature.

Thermoset Plastic A plastic which, when cured thermally and/or chemically, becomes essentially infusible and insoluble.

Third World (*See* World, Third.)

Tight Sands Relatively impermeable, gas-containing clay, chalk, or sandstone beds interspersed with shale and existing below conventional gas fields; located primarily near the New Mexico–Colorado border.

Time-Surge Coupling Bins and surge piles added to in-line extraction facility to take up slack, and offset mismatching and breakdowns in individual steps.

Toluene An aromatic chemical solvent with the formula $C_6H_5CH_3$; derived from coal tar or by catalytic reforming of petroleum naphthas.

Ton (t) Unit of mass.
   Long: 2240 lb.
   Short: 2000 lb.

Tonne (te) Metric ton; 1000 kg.

Torbanite Kerosene shale; a regional variant of oil shale.

TOSCO Process   Oil shale extraction process employing concurrent flow of hot ceramic balls to heat cold ore in retort.

Town Gas   Gas formed by carbonizing coal in retort.

Trade Sales (Shelf Goods)   Paints sold over the counter for consumer use.

Trap   Pocket or natural reservoir of petroleum capped with impermeable seal of roof rock.

Trash, Cotton Gin   (*See* Cotton Gin Trash.)

Tripolite   A diatomaceous earth; regional variant of oil shale.

Trommel   Cylindrical graded screen which rotates on slightly inclined axis.

Trona   Mineral $Na_2CO_3 \cdot NaHCO_3 \cdot 2H_2O$, interbedded with oil shale in Green River Basin.

Trunk Pipeline   Main transmission pipeline.

Tundra   Fragile combination of mosses, lichens, and grasses that overlays and protects the permafrost from warm summer air.

Turgidity (Turgor)   Distension of plant cell walls and membrane caused by fluid contents.

Uinta Basin   Oil shale deposit site in northeastern Utah; part of Green River Formation.

Underclay   Fine sedimentary mineral material deposit formed by subsidence of coastal areas.

Underground Coal Gasification (UCG)   Gasification of coal in situ, by partial combustion with air or oxygen.

Underground Mining   (*See* Deep Mining.)

Undiscovered Resources   Hypothetical plus speculative resources.

Unipol   Proprietary process for polymerizing ethylene to LDPE.

Unit Train   Coal train comprising about 100 hopper cars pulled by dedicated locomotives, and serving a single round-trip route between mine and large consumer.

Unnatural Gas   (*See* Landfill Gas.)

Up-Dip Mining   Working of a mine face sloping up from shaft, allowing water to flow out and loaded coal cars to roll out by gravity. (*See also* Down-Dip Mining.)

Value-Added   Increase in value of a material brought about by processing it to a higher degree (e.g., natural gas to plastic).

Vegetative Reproduction   Propagation of plant species from clones (q.v.).

Vein   Seam or bed of coal or other mineral.

Vent (v.)   To dissipate waste gas to atmosphere without burning.

Vertical Divestiture (*See* Divestiture, Vertical.)

Vertical Drilling Basically reverse ignition, but using grid of boreholes.

Vintage (v.) To produce gas before the controlled price rises, thus committing it permanently for sale at the price of older gas.

Volatile Solids (*See* Dry Organic Solids.)

Volume, Commericial (*See* Growing Stock.)

Volume, Standing (*See* Growing Stock.)

Volume Tables Tables used to convert log scale data to wood volume.

Washability Measure of flotation and wettability of coal.

Wash-Out Removal of digested carbon solids in supernatant or sludge streams of digester.

Waste Something that is cheaper to discard than to use further; material that is not a prime product.

Water Drive Well drive pressure created by natural influx of water from conterminous aquifer(s).

Water Fern *Salvinia auriculata* Aublet. Free-floating weed which forms floating mats.

Waterflooding Secondary extraction of crude oil by injection of large volumes of water into a natural oil reservoir.

Water Gas An intermediate gas mixture (predominantly CO plus H_2O) formed by passing steam over hot coke or coal at 600 to 1000°C.

Water-Gas Shift Reaction of CO and water to form CO_2 and H_2, and provide proper CO-to-H_2 ratio to produce methane.

Water Hyacinth *Eichhornia crassipes*. A free-floating freshwater weed; native of South America.

Water Lettuce *Pistia stratiotes*. A free-floating freshwater weed.

Wellhead Price Price of gas or oil raised to surface, prior to transporting, refining, distributing, and applying a profit.

Wellhead Technical Cost Cost (at wellhead) of raising oil, exclusive of transportation, government revenues, and producer's profits.

Wet Combustion Incineration of wet organic feed by injection into pressurized reactor at moderately elevated temperature (about 300°F).

Wet Gas Natural gas high in other hydrocarbons. Contains more than 0.1 gal (condensed) natural gasoline vapor per 1000 scf.

Wildcat Well (*See* Exploratory Well.)

Wood Oil Oil derived from wood by pyrolysis; has higher oxygen content than cellulose-derived pyrolysis oil, because of the presence of lignin in the feed.

Working Open area used for underground mining operation; strata excavated in extraction of a seam.

World*

First World: OECD nations plus France. (Basically the Western oil-importing nations.)

Second World: Communist nations.

Third World: OPEC nations.

Fourth World: Less-developed countries (LDCs).

Xanthan Gum A fermentation product of corn dextrose, used in tertiary recovery and as a thickener for drilling muds.

Xylene Any one of a family of three organic isomers having the formula $C_6H_4(CH_3)_2$.

Yellow Boy Ferric hydroxide from acid mine drainage.

Yield

1 Total weight harvested per unit area and time.

2 Output of manufacturing process.

Zeolite Crystalline aluminosilicate.

Zymosis (*See* Fermentation.)

*As defined by H. B. Thorelli, *Bus. Horiz.* **XVIII**(1), 53–56 (Feb. 1975).

INDEX